ADVANCED ELECTROCHEMICAL MATERIALS IN ENERGY CONVERSION AND STORAGE

Emerging Materials and Technologies

Series Editor

Boris I. Kharissov

2D Materials for Surface Plasmon Resonance-Based Sensors
Sanjeev Kumar Raghuwanshi, Santosh Kumar, and Yadvendra Singh

Functional Nanomaterials for Regenerative Tissue Medicines
Mariappan Rajan

Uncertainty Quantification of Stochastic Defects in Materials
Liu Chu

Recycling of Plastics, Metals, and Their Composites
R.A. Ilyas, S.M. Sapuan, and Emin Bayraktar

Viral and Antiviral Nanomaterials
Synthesis, Properties, Characterization, and Application
Devarajan Thangadurai, Saher Islam, Charles Oluwaseun Adetunji

Drug Delivery using Nanomaterials
Yasser Shahzad, Syed A.A. Rizvi, Abid Mehmood Yousaf and Talib Hussain

Nanomaterials for Environmental Applications
Mohamed Abou El-Fetouh Barakat and Rajeev Kumar

Nanotechnology for Smart Concrete
Ghasan Fahim Huseien, Nur Hafizah A. Khalid, and Jahangir Mirza

Nanomaterials in the Battle Against Pathogens and Disease Vectors
Kaushik Pal and Tean Zaheer

MXene-Based Photocatalysts: Fabrication and Applications
Zuzeng Qin, Tongming Su, and Hongbing Ji

Advanced Electrochemical Materials in Energy Conversion and Storage
Junbo Hou

Emerging Technologies for Textile Coloration
Mohd Yusuf and Shahid Mohammad

For more information about this series, please visit: https://www.routledge.com/Emerging-Materials-and-Technologies/book-series/CRCEMT

ADVANCED ELECTROCHEMICAL MATERIALS IN ENERGY CONVERSION AND STORAGE

Edited by
Junbo Hou

CRC Press is an imprint of the
Taylor & Francis Group, an **informa** business

First edition published 2022
by CRC Press
6000 Broken Sound Parkway NW, Suite 300, Boca Raton, FL 33487-2742

and by CRC Press
4 Park Square, Milton Park, Abingdon, Oxon, OX14 4RN

ISBN: 978-0-367-68048-0 (hbk)
ISBN: 978-0-367-68049-7 (pbk)
ISBN: 978-1-003-13397-1 (ebk)

DOI: 10.1201/9781003133971

Typset in Times
by MPS Limited, Dehradun

Contents

Preface

Technologies urge the new materials design and development, and new materials in turn thrust emerging technologies, in such a way that they intercross and amalgamate each other. Especially, today's interdisciplinary research and development from scientists and engineers with various backgrounds push the boundaries of people's knowledge and cognition on the existing technologies and materials, and produce revolutionary or even disruptive technologies. Such activities have become more and more intense than before, with exponential growth. In this book, we will focus on the areas of electrochemical energy conversion and storage, introduce novel materials particularly designed for the specific energy application, present the relationship of materials properties of state-of-the-art processing in device performance, and shed light on the research, development, and deployment (RD&D) trend of emerging materials and technologies in this field.

The technology of alternating current, or a well-known facility term, "grids", is the foundation of human invention in the modern history and the foundation of modern human life. The growing energy demands, the concern of the fossil energy depletion, the anxiety of the global warming and the requirement of national energy safety advance the reform of the supply side of the grids by introducing renewable energy or increasing the portion of renewable energy like solar, wind and biomass energy. As a result, the energy storage at the sides of the grid becomes an efficient way to utilize such a fluctuated energy supply. Furthermore, at the customer side of the grids, the power to power of batteries and supercapacitors, the power to gas like H_2 evolution by water splitting for fuel cells, and the power to X, i.e., CO_2 reduction, are all revolutionary towards the extension of the grids. As manifested by the word itself, "electrochemistry" links electricity and chemistry, and electrochemical energy conversion and storage attract more and more interest and attention due to its important role in either balancing the grid or extending the grid to mobility in human society. Without a narrative on social, economic, and environmental impacts of electrochemical energy conversion and storage, advanced electrochemical materials in energy conversion and storage including fuel cells, water electrolysis, lithium ion batteries, lithium batteries, supercapacitors, and CO_2 reduction reaction will be comprehensively discussed in this book.

Fuel cells, due to their high efficiency and environmental benign, attract worldwide interest and a lot of efforts have been devoted to their RD&D. Fuel cells have demonstrated a successful story "from NASA Apollo to Toyota Mirai". In the 1960s, fuel cells were first used on the NASA Apollo lunar spacecraft as an auxiliary power source, which also marked the real application of the fuel cell from the laboratory stage. Before 2000, it can be called the concept design and principle certification stage of fuel cell vehicles, and many OEMs launched fuel cell vehicles programs. From 2000 to 2010 is the stage of fuel cell vehicle demonstration and technical verification. From 2010 to 2015 is the stage of fuel cell vehicle performance improvement. At this stage, the fuel cell vehicle's power density and life have been improved. After 2015, fuel cell vehicles entered the stage of commercial promotion. The listing of Mirai by Toyota and Clarity by Honda represented the start of sales in the field of private passenger vehicles and officially started the stage of commercialization. Water splitting is a way to obtain high purity H_2 and a successful demonstration of power to gas, in which the electric energy is converted into the chemical energy. H_2 is a medium to "transport" or "extend" the function of the grids. In Chapters 1–3 and 13, we will introduce these two revolutionary technologies, and emphasize electrochemical materials applied in these two areas, as well as the fundamentals and basics beneath shedding light on the materials and device design and development.

Batteries are a typical type of power to power, which transform electricity energy into chemical potential energy (charge) and then dissipate the stored chemical energy into electricity energy (discharge). Assisted by them, the customer side of the grids is no longer fixed by house unit and transforms into a single person or single piece of portable machine. The grid terminal becomes more and more specified. Particularly, lithium ion batteries have manifested themselves as the leading positions either in 3C or in electric vehicles, which can be called "from Sony to Tesla". The first stage (1980s) of lithium ion batteries was mainly led by European and American academic research institutions and was still in the early stage of research and development during this period. The second stage (1990–2010) of 3C commercialization quickly landed. China, Japan, and South Korea staged the "Romance of the Three Kingdoms". In 1991, Sony Corporation took the lead in using lithium cobalt oxide as the cathode and quickly took the majority of the market share. The situation of global lithium ion battery competition among the three countries was formed in the late 2000s. In the third stage (from 2010 to the present), electric vehicles have sprung up, and lithium ion batteries with power performance have become a new "blue ocean", especially after Tesla Model S landed on the market in 2012. With the automotive revolution of the round, many battery manufacturers have invested heavily in the field of power batteries, which has also driven the explosive growth of lithium ion batteries in the past ten years, especially, with the chemistry award of the Nobel Prize in 2019 given to Professors John B. Goodenough and Stanley Whittlingham and Mr. Akira Yoshino for their significant contributions in the field of lithium ion batteries. In Chapter 4, we will summarize anode, cathode, electrolyte, and additive materials developed for lithium ion batteries, specifically, which can further boost the battery capacity, rate performance, and cycling ability. We will also discuss the interface/interphase like solid electrolyte interphase (SEI) and cathode electrolyte interphase (CEI) to show the chemical, physical, electrochemical, and mechanical interactions of these materials. Due to the requirement of further boosting the battery energy density, the possible solution of the safety issues, and the concern of the lithium resources, we will also review beyond lithium ion batteries, including lithium sulfur and solid state batteries in Chapters 5–9. Another electrochemical device, a supercapacitor, can supply even higher power density but less energy density compared to lithium batteries. In Chapter 10, we will talk about advanced materials for supercapacitors, along with some novel emerging materials in this field.

CO_2 reduction reactions (CO_2RR) are examples of power to X, where X can refer to one of the following: power to chemicals, power to fuel, power to gas, or power to ammonia. CO_2RR is the conversion of carbon dioxide to more reduced chemical species by using electrical energy. It is also a method of carbon capture and utilization. The reduced chemical species could be just chemicals, fuels, and gas such as formic acid (HCOOH), carbon monoxide (CO), ethylene (C_2H_4), ethanol (C_2H_5OH), and methane (CH_4) due to different catalysts used. In Chapters 11 and 12, we will highlight catalyst design and development for CO_2RR and some fundamentals about proton-facilitated reduction reactions.

Electrochemical energy conversion and storage have become more hot topics, especially coupled with nanotechnology, which directly link materials and catalysts at a multi-size scale (down to nano or even single atom) to multi-phase (gas, liquid, solid), multi-time scale (different charge transfer and mass transport processes), and multi-components (cathode, anode, and electrolyte) involved in device performance. This suggests electrochemical energy conversion and storage is the cross-discipline of chemistry, physics, chemical engineering, electrochemistry, materials science and engineering, mechanical engineering, catalysis, etc. It is believed that this book covers most important topics in this area and might be very helpful for beginners and experts, scientists and engineers, theorists and experimenters either from academia or from industry. We acknowledge the efforts of Professor Boris I. Kharissov at the Autonomous University of Nuevo León (Monterrey City, Mexico) and Series Editor for the Emerging Materials and Technologies Book Series, and Editor Allison Shatkin at CRC Press/Taylor & Francis, for initiating this project.

Editor

Junbo Hou received his B.S. and M.S. degrees from Harbin Institute of Technology, Harbin, China, in 2003 and 2005, respectively, and his Ph.D. degree from Dalian Institute of Chemical Physics, Chinese Academy of Science, Dalian, China, in 2008, all in chemical engineering, particularly electrochemical engineering.

From 2008 to 2010, he was with Montanuniversität Leoben and Erich-Schmid-Institut für Materialwissenschaft (ESI), ÖAW, as a researcher, working on electroceramic materials and high resolution transmission electron microscopy (HRTEM). From 2010 to 2012, he was with the Institute of Critical Technology and Applied Science (ICTAS), Virginia Tech, as a research scientist, while simultaneously engaged in research work at the Department of Mechanical Engineering and Department of Chemistry, Virginia Tech. In 2013, he started working in a start-up company, SAFCell, Inc., of the California Institute of Technology (Caltech). He made significant contributions to the company: developed new technology to realize the mass production which the company had not been able to realize since 2004, dramatically reduced the cost by 75% while even improving the cell performance, and helped the company obtain funding from ventures, the U.S. government, and the Army. In 2016, he was invited to join CEMT as a chief scientist responsible for the development and production of proton exchange membrane (PEM) fuel cell engines for vehicles. He established a membrane electrode assembly (MEA) production line and metal bipolar plate (BIP) production line. The developed fuel cell engines, with outputs of 30–128 kW, all passed the mandatory testing of the National Automobile Testing Center, China. Working with OEMs, a total of eight fuel cell vehicles obtained production certification from the Ministry of Industry and Information Technology, China. Since May 2008, he was appointed as an associate professor at Shanghai Jiao Tong University, and continued to conduct research on fuel cells and batteries.

He has published 55 peer-reviewed journal papers and contributed two invited book chapters in the area of electrochemical energy conversion and storage. He also holds 19 Chinese patents. He is currently an Outstanding Scientist of Zhejiang, China, a member of the National Fuel Cell Standardization Technical Committee, and a member of the APEC Sustainable Energy Center's domestic expert team.

Contributors

Hanwen An
MIIT Key Laboratory of Critical Materials Technology for New Energy Conversion and Storage
School of Chemistry and Chemical Engineering
Harbin Institute of Technology
Harbin, People's Republic of China

Kedi Cai
College of Chemistry and Materials Engineering
Bohai University
Jinzhou, People's Republic of China

Zhenlian Chen
Jianghan University
Wuhan, Hubei, People's Republic of China

Yongzhu Fu
College of Chemistry
Zhengzhou University
Zhengzhou, People's Republic of China

Junbo Hou
Institute of Fuel Cells
School of Mechanical Engineering
Shanghai Jiao Tong University
Shanghai, People's Republic of China

Jing Hu
School of Chemistry and Chemical Engineering
Harbin Institute of Technology
Harbin, People's Republic of China

Changchun Ke
Institute of Fuel Cells
School of Mechanical Engineering
Shanghai Jiao Tong University
Shanghai, People's Republic of China

Xiaoshi Lang
College of Chemistry and Materials Engineering
Bohai University, Jinzhou, People's Republic of China

Zheng Li
Department of Mechanical and Energy Engineering
Southern University of Science and Technology
Shenzhen, Guangdong, People's Republic of China

Bachirou Guene Lougou
School of Energy Science and Engineering
Harbin Institute of Technology
Harbin, People's Republic of China
MIIT Key Laboratory of Critical Materials Technology for New Energy Conversion and Storage
Harbin Institute of Technology
Harbin, People's Republic of China

Huiyang Ma
College of Chemistry
Zhengzhou University
Zhengzhou, People's Republic of China

Azeem Mustapha
School of Energy Science and Engineering, Harbin Institute of Technology
Harbin, People's Republic of China

Enkhbayar Shagdar
School of Energy Science and Engineering
Harbin Institute of Technology
Harbin, People's Republic of China

Yong Shuai
School of Energy Science and Engineering
Harbin Institute of Technology
Harbin, People's Republic of China

Nan Sun
MIIT Key Laboratory of Critical Materials Technology for New Energy Conversion and Storage
School of Chemistry and Chemical Engineering
Harbin Institute of Technology
Harbin, People's Republic of China

Qiqi Wan
Institute of Fuel Cells
School of Mechanical Engineering
Shanghai Jiao Tong University
Shanghai, People's Republic of China

Yameng Wang
Department of Mechanical and Energy Engineering, Southern University of Science and Technology
Shenzhen, Guangdong, People's Republic of China

Deyu Wang
Jianghan University
Wuhan, Hubei, People's Republic of China

Xufeng Wang
MIIT Key Laboratory of Critical Materials Technology for New Energy Conversion and Storage
School of Chemistry and Chemical Engineering
Harbin Institute of Technology
Harbin, People's Republic of China

Han Wang
MIIT Key Laboratory of Critical Materials Technology for New Energy Conversion and Storage
School of Chemistry and Chemical Engineering
Harbin Institute of Technology
Harbin, People's Republic of China

Jiajun Wang
MIIT Key Laboratory of Critical Materials Technology for New Energy Conversion and Storage
School of Chemistry and Chemical Engineering
Harbin Institute of Technology
Harbin, People's Republic of China

Zhenhua Wang
School of Chemistry and Chemical Engineering
Beijing Institute of Technology
Beijing, People's Republic of China

Zhijiang Wang
MIIT Key Laboratory of Critical Materials Technology for New Energy Conversion and Storage
Harbin Institute of Technology
Harbin, People's Republic of China

Yi Wei
College of Materials Science and Engineering
Zhengzhou University
Zhengzhou, People's Republic of China

Ping Xu
School of Chemistry and Chemical Engineering
Harbin Institute of Technology
Harbin, People's Republic of China

Min Yang
Central Research Institute
Shanghai Electric Group
Zhabei District
Shanghai, People's Republic of China

Shuang Yan
College of Chemistry and Materials Engineering
Bohai University, Jinzhou, People's Republic of China

Lin Zeng
Department of Mechanical and Energy Engineering
Southern University of Science and Technology,
Shenzhen, Guangdong, People's Republic of China

Fang Zhang
MIIT Key Laboratory of Critical Materials Technology for New Energy Conversion and Storage
School of Chemistry and Chemical Engineering
Harbin Institute of Technology
Harbin, People's Republic of China

Jianshuo Zhang
Department of Materials Science and Engineering
Southern University of Science and Technology
Shenzhen, Guangdong, People's Republic of China

Yuanyuan Zhang
School of Chemistry and Chemical Engineering
Harbin Institute of Technology
Harbin, People's Republic of China

Wei Zhao
MIIT Key Laboratory of Critical Materials Technology for New Energy Conversion and Storage
School of Chemistry and Chemical Engineering
Harbin Institute of Technology
Harbin, People's Republic of China

1 Catalyst Support Materials for Proton Exchange Membrane Fuel Cells

Yameng Wang, Zheng Li, and Lin Zeng
Department of Mechanical and Energy Engineering, Southern University of Science and Technology, Shenzhen, Guangdong, People's Republic of China

Jianshuo Zhang
Department of Materials Science and Engineering, Southern University of Science and Technology, Shenzhen, Guangdong, People's Republic of China

CONTENTS

DOI: 10.1201/9781003133971-1

1.1 INTRODUCTION

Proton exchange membrane fuel cell (PEMFC) has been widely employed as electrochemical energy conversion systems in many fields due to its advantages, such as high energy conversion efficiency, high power density, and low operation temperature. Particularly, due to its high power density, PEMFC is suitable for transportation applications, such as sedans and buses. However, its wide commercialization is still hampered by three aspects: high cost, low catalyst activity, and low durability. The high cost is mainly caused by the utilization of platinum catalysts. In the early stage of PEMFC development, platinum black was employed as the catalyst with catalysts loading as high as 28 mg/cm^2. In the late 1990s, researchers began to use the high-surface-area carbon to disperse the platinum with a nanosize structure. Hence, the platinum catalyst loading in the catalyst layer was significantly reduced to 0.3–0.4 mg/cm^2. In this regard, the catalyst supporting material is critical to improving noble catalyst utilization. Besides the dispersion of nanosized catalysts, the catalyst supporting material is also related to the mass transport and durability of PEMFC. On the one hand, with the catalyst loading further reduced, for example, <0.1 mg/cm^2, the mass transport, particularly for the oxygen diffusion in the cathode, arises as a critical issue. Hence, the optimization architecture of catalyst supporting material, including the pore size, pore distribution, and porosity, to meet the mass transport requirement becomes a new research highlight in the research community of PEMFC. On the other hand, the durability of PEMFC highly depends on the catalyst supporting materials. Without a suitable supporting material, the platinum nanoparticles will be agglomerated, dissoluted, and detached, which will deteriorate the performance of PEMFC. Therefore, a suitable catalyst supporting material should meet the following requirements:

1. High surface area;
2. High electrical conductivity;
3. High durability under the harsh chemical environment (high potential and low pH);
4. Suitable nanostructure for platinum dispersion and mass transport.

In this chapter, the cutting-edge development of catalyst supporting materials for platinum in the recent decade will be first introduced. These catalyst supporting materials include functionalized carbon, carbide, nitride, and transition metal oxides. The perspective and future development will then be given at the end of the chapter.

1.2 FUNCTIONALIZED CARBON SUPPORT MATERIALS

The carbon supports play a critical role in determining catalyst performance and stability [1–3]. Porous carbons have internal pores within their primary particles and often have very high surface areas (500–2000 m^2/g), suitable for Pt deposition. Due to the internal porosity of porous carbons, many Pt nanoparticles can be deposited in the interior of the micropores [4]. The carbon micropores normally have very narrow openings (1–4 nm), and the pores themselves can be tortuous [5,6]. Electrochemical measurements showed that when the catalyst was fabricated to form the catalyst layer, the ionomer did not penetrate into the small pores to contact these interior Pt nanoparticles [7–11]. On the basis of this physical structure, Figure 1.1 illustrates the kinetic and transport properties of the catalysts using different carbon supports [5]. Solid carbon-supported Pt catalysts have relatively poor activity due to the poisoning of the active sites by the ionomer but have good transport properties. In contrast, on porous carbon catalysts, the interior Pt nanoparticles uncoated by ionomers have good activity but restricted access to protons and reactant gases. The performance loss due to local transport depends on the required local flux of reactants. Therefore, PEMFC with low Pt loading will become more susceptible to this loss at a high power load. Based on the reasons mentioned, excellent carbon support needs to have some important pore characteristics. It should have accessible internal-connected porous structures that can host catalyst nanoparticles and prevent them from direct contact with ionomers but still allow protons and hydrogen/oxygen to have reasonable access to the catalytic species. Furthermore, the pore depth and tortuosity should not be too large. The accessible carbon pores should have a reasonably small opening to restrict ionomer penetration but not too small to restrict oxygen transport to the catalytic sites. The concept is straightforward, but it will not be easy to use a reasonable method to manufacture carbon support that meets all of the requirements.

1.2.1 Carbon Nanospheres

Carbon spheres with a nanometer diameter are widely employed as the catalyst support for PEMFCs. As one of the products manufactured by pyrolyzing petroleum hydrocarbon, carbon nanospheres (or carbon blacks) are the most commonly used supports for Pt-based catalysts for PEMFCs in many studies and commercial applications. Carbon nanospheres consist of near-spherical particles, <50 nm in diameter.

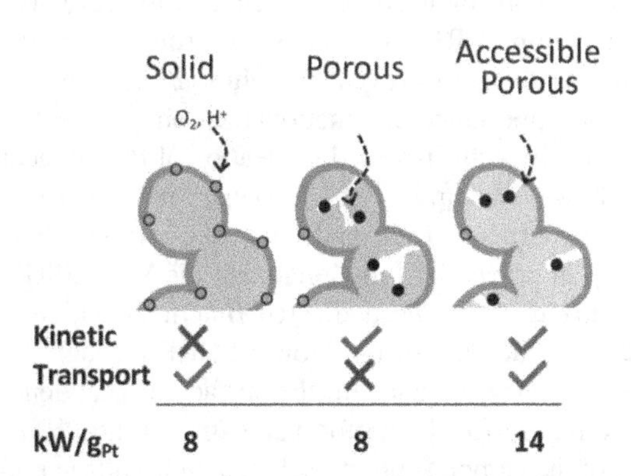

FIGURE 1.1 ORR kinetic and transport (O_2 and proton) characteristics of catalyst layer structures made from three types of carbon (gray). Small black and gray circles represent relatively high and low activity Pt particles, respectively, due to ionomer (blue) adsorption [5].

These coalesce into particle aggregates and agglomerates of around 250 nm in diameter. Typical carbon nanospheres, such as Vulcan XC-72, acetylene black, and Ketjenblack, possess different physicochemical characteristics (specific surface area, electronic conductivity, the surface-to-volume ratio, stability, and surface functionality). Vulcan XC-72 nanoparticles have attracted special attention due to their excellent compromise between adequate surface area (~250 m^2/g) and high electric conductivity (~2.77 S/cm). Acetylene black is a material produced by the incomplete combustion of ethylene cracking tar. Researchers have applied acetylene black as carbon support for a range of applications for many years. Uchida and co-workers reported the high specific activity of Pt/C with acetylene black as carbon support for PEMFCs and direct methanol fuel cells [12,13]. Acetylene black mainly has the following characteristics: (i) low resistivity, (ii) ability to absorb and reserve an effective volume of electrolyte without cutting down its ability to mix with the active material, (iii) high surface area, and (iv) low cost compared to its counterparts. Ketjenblack is a unique electroconductive carbon black that has received favorable evaluations for its superior performance and stability of quality. Ketjenblack has garnered high marks for providing the same level of electroconductivity with a lower loading quantity as conventional carbon black.

However, Pt nanoparticles have poor stability on traditional carbon black supports. One of the most crucial issues in state-of-the-art PEMFCs is the agglomeration of Pt nanoparticles at the cathode, which leads to a significant loss in the electrochemical surface area (ECSA) during operation. It is believed that there are two primary mechanisms that drive the Pt surface area loss: crystal migration of Pt (Pt atoms and nanoparticles) on the support surface and Ostwald ripening (the dissolution of smaller Pt clusters and Pt ion redeposition on larger Pt clusters). However, the migration and agglomeration of Pt nanoparticles on the carbon surface cannot be effectively suppressed by the weak interaction between the carbon support and Pt nanoparticles [14]. Therefore, enormous studies have attempted to solve these problems. Among them, functionalization by heteroatoms is an effective way. The doped-heteroatom of nitrogen contributes to the active activity for the oxygen reduction reaction (ORR) and shows a strong interaction with Pt nanoparticles for entrapping them to prevent dissolution/migration [15]. Badam et al. [16] demonstrated the importance of functionalization or defects on the surface of acetylene black to achieve well-dispersed nucleation of Pt over acetylene black. The resulting material showed enormously high oxygen reduction reactivity compared to its commercial counterparts. Rich and co-workers found that catalytic performance can be greatly improved by functionalizing the Vulcan XR-72 with nitrogen covalently functionalized [17]. The dedicated functionalization was found to enhance the intrinsic electroactivity of Pt/C toward ORR through the even nucleation and growth of platinum nanoparticles on the surface of acetylene black.

Another effective reason for the performance loss in the PEMFC cathode is the oxidation/corrosion of the carbon support itself at a high-voltage range (above 1.0 V). From a thermodynamic point of view, surface carbon atoms are electrochemically activated at potentials greater than 0.207 V vs. NHE. Hereafter, the activated carbon species reacts with nearby water molecules, finally yielding CO_2. The rate of carbon corrosion is continuous under typical operating conditions of PEMFCs. Even worse, researchers have observed that Pt nanoparticles enabled to catalyze carbon corrosion reaction significantly [18,19]. The areas in direct contact with Pt nanoparticles may make a majority of carbon loss. The local carbon degradation isolates Pt nanoparticles and accelerates Pt dissolution. Researchers have recently tried several approaches to alter the nature of the carbon support to improve stability. The most important solution is high-temperature pretreatment of the carbon support. The purpose of high-temperature pretreatment is to increase the graphitization degree of these carbon supports. The higher level of graphitizatio n is thought to yield higher stability due to a lower density of edges and defect sites.

In spite of wide application as catalyst-support, carbon blacks still suffer from some critical problems. The pore size and pore distribution play an important role in the interaction between ionomer and catalyst nanoparticles. Since the size of ionomer micelles (>40 nm) is larger than the pores in the carbon black, catalyst nanoparticles in pores are not accessible to ionomers and have no contribution to the catalytic performance. Deep micropores or recesses trap the catalyst nanoparticles making them inaccessible to reactants, thus reducing catalytic activity. Furthermore, carbon black is thermochemically unstable. Thermochemical stability is required under high temperatures and high acidic conditions of a typical PEMFC. Without the thermochemical stability, the carbon black will be corroded, thus leading to the catalyst nanoparticle detachment from the carbon black and the disintegration of the catalyst layer.

1.2.2 High Surface Area Carbon

High surface area carbon is a type of porous carbon with a high surface area and a high pore volume. The high specific surface area helps to improve the dispersion of Pt nanoparticles in the carbon support. The Johnson Matthey company and Tanaka Kikinzoku company have developed a series of catalysts using high surface area carbon as supports. Ketjenblack EC-300J and EC-600JD are very pure carbon blacks extremely suitable for antistatic and electroconductive applications. Due to their unique morphologies and high surface areas (approximately 800 m^2/g for EC-300J and approximately 1,400 m^2/g for EC-600JD), only a very small amount of Ketjenblack EC-300J or EC-600JD is needed compared to conventional electroconductive blacks in order to achieve the same conductivity. Kim et al. [20] synthesized various 60 wt% Pt/C catalysts using the precipitation method with various carbon supports (Vulcan XC-72, Ketjenblack EC-300J, and Ketjenblack EC 600JD). The average Pt particle size decreases with increasing the surface area of carbon supports. The 60 wt% Pt/C with 1.6 nm of Pt particle size and good dispersion was prepared using Ketjenblack EC 600JD. In a single cell test, the activity of electrode catalysts was enhanced with an increase in Pt surface area when the cell voltage was above 0.6 V. Pt catalyst supported on Ketjenblack EC-600JD showed a good ORR activity in all voltage regions and showed stable cell performance (900 mA/cm^2@0.6 V) without degradation during the durability test (over 180 hours).

In addition, many new types of carbon support with a high specific surface area have been developed as catalyst supports. Bo et al. prepared hollow carbon spheres using poly (ionic liquids) as a carbon precursor and monodisperse silica particles as a template.

The hollow carbon spheres were used as the support of Pt nanoparticles [21]. Pt nanoparticles with an average size of 2.8 nm are uniformly distributed onto the hollow carbon spheres. The high surface area and unique structure facilitated the fine dispersion of nanoparticles. The obtained electrocatalyst exhibited a significant catalytic activity for methanol oxidation. Peter Strasser and co-workers reported the synthesis of a mesoporous nitrogen-doped carbon-supported platinum catalyst (Pt/meso-BMP) based on an ionic liquid as nitrogen/carbon precursor and the evaluation of the catalytic system for ORR [22]. The long-term performance of this new catalyst was compared with a commercial high surface area, carbon-supported platinum catalyst. The new catalyst showed comparable activities with the benchmark (20 wt% Pt/C) while a significantly increased ECSA was established compared to the benchmark catalyst. Yang et al. [23] used a 3D nanoporous carbon (NanoPC) with a high specific surface area of 1,037 m^2/g as carbon support for Pt nanoparticles. The electrocatalyst exhibited high performance in high-temperature PEMFCs. Even after 10,000 start-up/shut-down cycles in a voltage range of 1.0 V to 1.5 V, the catalyst presented almost no loss in ECSA. The power density of this as-prepared catalyst was 342 mW/cm^2, which was much higher than those of the commercial catalysts. Yin et al. [24] reported a kind of mesoporous hollow polypyrrole spheres supported Pt nanoparticles to form Pt/MHPS catalyst via a simple chemical reduction method using formic acid as a reducing agent for the methanol oxidation reaction. Zhao et al. [25] investigated the impact of pore size of ordered mesoporous carbon FDU-15-supported Pt catalysts for ORR. Characterizations showed that the increase in pore size enlarged the specific surface area and the pore volume of FDU-15, but decreased the electrical conductivity. The particle size of the FDU-15-supported Pt catalyst was also influenced by a decreasing trend with the increase of pore size. Electrochemical measurement demonstrated that the FDU-15-supported Pt catalyst with a pore size of 6.5 nm yielded the highest electrocatalytic activity for ORR, which was further confirmed by a single-cell test on DMFC with the as-prepared catalyst as the cathode. Researchers also evaluated the advantages of high-surface-area carbon support through simulation work. The model indicated that the nanoscale pores contributions to oxygen transport resistance were approximately five times higher in high surface area carbon (HSAC) than that of low surface area carbon (LSAC) [26].

Cell voltage losses at high current densities are due to the local-O_2 and bulk-H^+ transport resistances in the catalyst layer when HSACs are employed as the catalyst support. Specifically, both the platinum catalyst and ionomer are dispersed upon the microstructure of HSAC. Hence, the microstructures play a pivotal role in controlling the reactant transport to the active site in the catalyst layer. Ramaswamy et al. [27] performed a systematic analysis of the underlying microstructure of PtCo catalyst dispersed on various high-surface-area carbon support in terms of their surface area and pore-size distribution. They found that carbon microstructures strongly influenced the PtCo nanoparticle dispersion, catalyst layer ionomer distribution, and transport losses governing the performance at high current densities. Catalyst layer electrochemical diagnostics were performed to quantify local-O_2 transport resistance and bulk-H^+ transport resistance in the cathode. It was found that the two resistances were directly correlated to the micropore and macropore surface areas of the carbon support, respectively.

However, HSAC as the catalyst support, also suffers some critical issues. The big concern is inferior stability. The lower degree of graphitization causes it to be easily oxidized. Oxygen-containing functional groups (OCFGs), such as hydroxyl, carboxyl, and carbonyl groups, are easily grafted on the surface of HSAC during the synthesis process or in the operation condition of PEMFCs [28]. In an early study, Giordano et al. [29] showed that the presence of OCFGs on the carbon support accelerated the carbon corrosion rate. The latter was rationalized by transferring oxygenated species from the Pt nanoparticles to the OCFGs to form gaseous products, CO and CO_2. The most common solution is high-temperature pretreatment. Kamitaka et al. [30] modified the dimensional, crystalline, surface, and porous structures of a MgO-templated mesoporous carbon. The electrochemical activities were examined by the RDE method and as well as the MEA method. They studied the effects of different temperature treatments on the pore structure of this kind of mesoporous carbon support, and found that after the first treatment at 2100°C in an inert atmosphere and a second low-temperature treatment in the air to remove unstable components, the optimized carbon support was obtained. The optimized carbon support was used to support Pt nanoparticles. The catalyst exhibited an excellent single cell performance. The first high-temperature treatment was to increase the degree of graphitization of the carrier. The second step of low-temperature air treatment was to remove unstable components and open closed pores. This method has the potential to be extended to other porous carbon support treatment methods.

1.2.3 Graphitized Carbon Materials

Normally, the carbon materials can improve their degree of graphitization with a thermal treatment with inert gas protection. The graphitized carbon materials possess high electron conductivity and high electrochemical corrosion resistance, which enables them to be utilized as catalyst support materials to enhance catalyst stability. Graphitized carbon materials can be divided into graphitized carbon nanospheres (blacks), carbon nanotubes, graphene, and carbon nanofibers. Among them, graphitized carbon nanospheres (blacks) are the most commonly used catalyst support due to the traditional sphere shape and easy mass production.

Mohanta et al. prepared platinum nanoparticles loaded on graphitized carbon (GC) Timcal-167 [31]. Although the as-prepared Pt electrocatalysts with GC supported showed decreased mass activity for ORR in 0.1 M $HClO_4$, when compared with traditional carbon black supported Pt catalysts and commercial Tanaka catalysts, they retained the highest ratio of initial ECSA after the accelerated stress test (AST) for both carbon support and Pt nanoparticles. More importantly, the MEA prepared with Pt/GC showed higher voltage at a high current density range because the lower electrode thickness granted it a superior mass transport property. Selvaganesh et al. compared the MEA performance with various types of electrocatalysts (Pt/Non-GC, Pt/GC, Pt_3Co/Non-GC and Pt_3Co/GC) obtained from Sainergy Tech Inc., USA [32]. They demonstrated that nanoparticles supported on graphitic carbon exhibited superior performance and enhanced stability for Pt and Pt_3Co alloy during the ASTs for catalyst supports. The higher performance and stability were ascribed to the better electrical conductivity and resistance toward carbon corrosion.

Although the conventional graphitized carbon spheres exhibited a high corrosion resistance during the operation of PEMFC, mesoporous graphitized carbon spheres are preferred in recent years due to their inner connected pores, which are beneficial to the transportation of reactants and water.

Kamitaka et al. employed MgO-Templated mesoporous carbon CNovel® as the cathode catalyst support [30]. After being treated at 2100°C, a sharp X-ray diffraction peak around 27° confirmed the graphitization of the mesoporous carbon. The graphited mesoporous carbon support exhibited greatly enhanced stability during the AST, and retained most of its original oxygen reduction activity. Although the Pt/CNovel showed a comparable activity during the RDE test, it exhibited a considerable advantage during the MEA evaluation. The performance enhancement was attributed to the Pt nanoparticles anchored on inner pores of CNovel, which were not covered and poisoned by the ionomer, e.g., Nafion resin.

Graphitized carbon nanofibers (G-CNFs) exhibit higher corrosion resistance in the harsh chemical environment (high temperature, humidity, and acidic conditions) when compared with traditional carbon blacks. According to the stacking pattern of graphite layers, CNFs are divided into three types: plate carbon nanofibers, tube carbon nanofibers, and fishbone carbon nanofibers [33]. The CNF diameters are also much larger than CNTs and may go up to 500 nm, while the length can be up to a few millimeters [34]. The basic difference between CNTs and CNFs is that CNFs either have a very thin cavity or no hollow cavity. Meanwhile, the CNFs possess the exposure of active edge planes. Unlike the CNTs where a predominant basal plane is exposed, CNFs (platelets and herringbone structure) expose more atoms at the edge of graphite planes, which present potential anchoring sites.

Li and co-workers prepared Pt nanoparticles (5–30 wt%, diameter: 2–4 nm) supported on stacked-cup CNFs (SC-CNFs) using a polyol process [35]. The MEAs with Pt/SC-CNFs as the catalysts were proved to have a higher power density than the MEAs with the conventional catalyst (Pt/C with carbon black as the support material, E-tek). The improved performance was attributed to the high aspect ratio (length to diameter ratio) of CNFs, which allowed the formation of continuously conducting networks with the existence of proton-conducting ionomer.

Yli-Rantala et al. prepared G-CNFs supported Pt catalysts using tubular CNFs [36]. The diameter and surface area of these tubular fibers were 150 nm and 13 m^2/g, respectively. Due to the tubular structure, these fibers were assumed to have less exposed graphene sheet edges than other forms of carbon nanofibers. It was observed that the acid treatment to graft OCFGs on the fiber surface was too mild and did not cause a significant increase in the extent of surface oxidation for the tubular structure CNFs. The MEA with the as-prepared Pt/G-CNFs as the cathode catalyst exhibited an enhanced durability compared with conventional catalysts (Pt/C, carbon blacks). However, the initial power performance of the MEA with Pt/CNFs cathode was lower, perhaps because of poor interaction between Pt nanoparticles and CNFs in the catalyst layer.

Although the high electrical conductivity and durability, it is still challenging to support platinum nanoparticles on the G-CNFs. To address this issue, Jung and

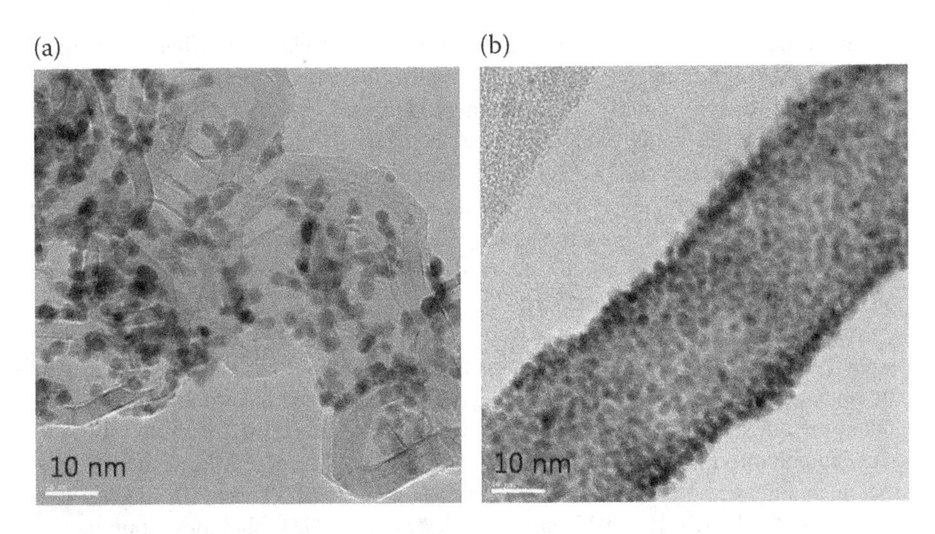

FIGURE 1.2 TEM images of the (a) commercial Pt/C and (b) Pt/CNF catalysts.

co-workers used the herringbone-type CNFs with a diameter of about 20–30 nm to support Pt nanoparticles [37]. Before the preparation of Pt/G-CNFs, the G-CNFs were treated with H_2SO_4/HNO_3 (v/v = 4:1) to increase the density of anchoring sites for Pt precursors. After impregnated with ethylene glycol solution of chloroplatinic acid, the mixture was thermally treated in an argon atmosphere at 350°C. The XRD spectrum of Pt/CNFs exhibited a mean Pt particle size of about 2.8 nm. Due to the acid treatment, more anchoring sites were created that facilitated the formation of small and even platinum nanoparticles. Thus, small Pt nanoparticles can grant the catalysts more catalytic sites and enhance the corresponding catalytic activity. From the TEM images in Figure 1.2, uniformly distributed Pt nanoparticles with mean particles of about 2.5 nm can be observed. The MEA with Pt/CNFs as the anode catalyst exhibited a superior corrosion resistance during fuel starvation tests compared to Pt/CBs as the anode catalyst.

1.2.4 Other Carbon Materials (Carbon Nanotubes and Graphene)

Carbon nanotubes (CNTs) are 2-D nanostructures, typically tubes formed by rolled-up single sheets of hexagonally arranged carbon atoms. They can be divided into single-walled (SWCNTs) and multi-walled (MWCNTs). Depending on the structure, SWCNTs can be conducting, i.e., metallic as well as semi-conducting in nature. SWCNT structure is characterized by an achiral vector (m, n), which defines its metallic or semi-conducting properties. All armchair SWCNTs (n = m) are known to be metallic. SWCNTs with n-m = 3k, where k is a nonzero integer, are semiconductors with a small bandgap. All other SWNTs are semiconductors with a bandgap inversely proportional to the nanotube diameter. MWCNTs can have diameters of a few tens of nanometers with a spacing of 0.34 nm between cylindrical walls [34].

CNTs are more corrosion resistant than carbon blacks in the harsh chemical environment (high temperature, humidity, and acidic conditions) due to the merits of graphitic carbon bonds. Various methods were adapted to prepare CNT-supported Pt-based catalysts to enhance the activity and stability for PEMFCs [38–48]. However, the catalyst layer with disordered CNT-supported catalysts was not exhibited much enhanced initial performance due to the poor pore structure and the established complex mass transport path for reactant gas and produced water. Therefore, more attention should be paid to the ordered catalyst layer based on vertically aligned carbon nanotubes.

Murata and co-workers prepared vertically aligned carbon nanotubes (VACNTs) as the cathode for PEMFCs [49]. Pt deposition was conducted using a conventional impregnation-dry-H_2 reduction method. The well-organized electrode structure promoted the mass transport and achieved a high current density of 2.6 A/cm^2 at 0.6 V. More importantly, the Pt loading was as low as 0.1 mg/cm^2 on the cathode. As is shown in Figure 1.3, the output power per unit mass of Pt was 10.4 kW/g, outperforming the US-DOE's 2017 target of 8 kW/g_{PGM} (PGM: platinum group metal).

Meng et al. prepared PtCo nanoparticles supported on VACNTs with a hydrophilic catalyst-loaded side and a hydrophobic gas diffusion side [50]. When the PtCo/VACNTs were employed as the cathode catalyst in PEMFCs, the hydrophobic gas diffusion side faced the bare carbon paper layer without a microporous layer. Strikingly, the cathode catalyst layer exhibited an enhanced mass transport capability and enabled the MEA to achieve a power density of higher than 1.4 W/cm^2. However, when the corresponding Pt/VACNTs was tested in the MEA, the power performance was decreased dramatically. The authors ascribed this superior performance to the enhanced hydrophilicity and oxygen reduction activity when Co was involved. On the other hand, when Pt/VACNTs instead of PtCo/VACNTs was tested in the MEA, the power performance was decreased dramatically because of degraded hydrophilicity. Due to the ordered electrode structure with super gas-water diffusion capability, the PEMFC delivers a high power density with a Pt loading of 65 $\mu g_{cathode}$/cm^2, and 89% of initial power was retained after 6,000 AST cycles.

Tian et al. prepared ultra-low Pt loading MEA using VACNTs as highly ordered catalyst support for PEMFCs [51]. The MEA with Pt loading of 35 μg/cm^2 was comparable to that of the commercial Pt catalyst on carbon powder with 400 μg/cm^2. The length of VACNTs was about 1.3 μm, which facilitated the transmission of reactants and product water. However, the MEA with Pt/VACNTs outperformed the MEA with commercial Pt/C catalysts when both sides of VACNTs were deposited with Pt nanoparticles. The boosted performance was ascribed to the enhanced hydrophilicity of the catalyst layer, because Pt decorated VACNTs showed higher hydrophilicity when compared with naked VACNTs. Pt/VACNTs catalyst with better hydrophilicity facilitated the transportation of reactant gas to the catalytic sites and produced water leaving the catalytic sites.

Zhu et al. prepared polyelectrolyte poly (diallyl dimethylammonium chloride) (PDDA) functionalized CNTs and used them as the support materials to load Pt nanoparticles [52]. The as-prepared Pt/PDDA-CNTs with an average particle size of about 3.1 nm exhibited enhanced electrocatalytic activity and electrochemical stability. The Pt/PDDA-CNTs catalyst was used to fabricate a double-layered

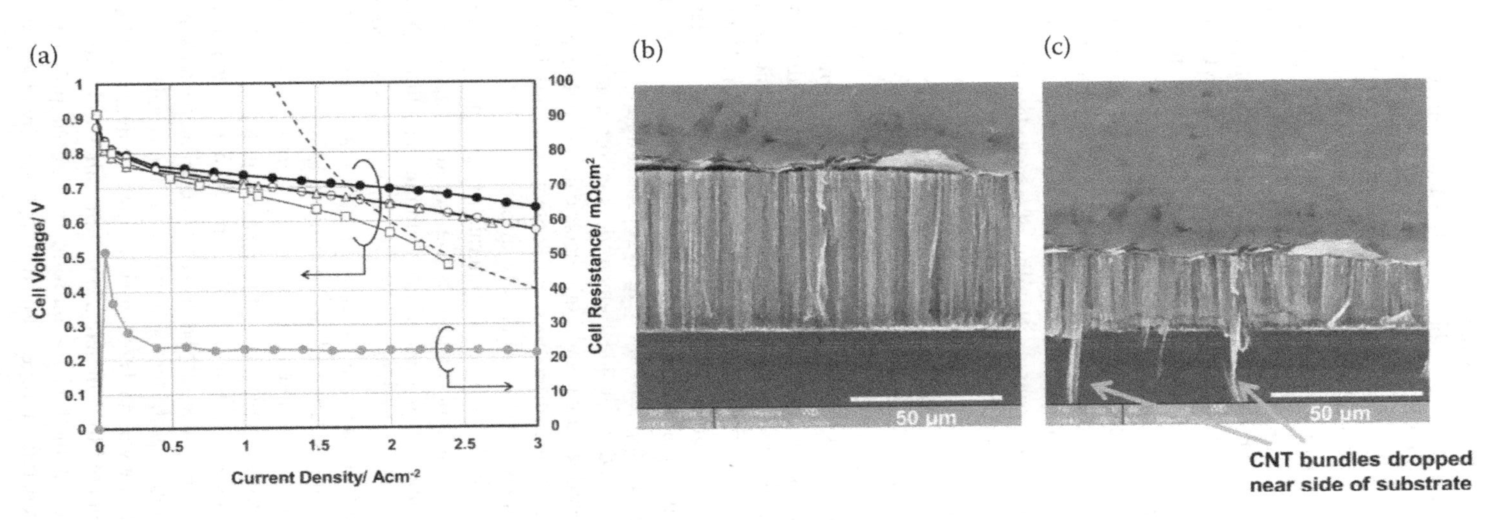

FIGURE 1.3 (a) IV performances of VACNT electrodes comparison to conventional electrode. ○: VACNT electrodes (20 cm^2 cell), ●: VACNT electrodes IR corrected (20 cm^2 cell), Δ: VACNT electrodes (236 cm^2 cell), □: conventional electrodes (236 cm^2 cell), Cell resistance of VACNT electrodes. Dashed line: DOE target 0.125 g_{PGM}/kW; SEM image of VACNT forest under compression. (b) early stage in compression, (c) final stage in compression.

buckypaper catalyst layer (BPCL) with the addition of CNFs through a vacuum filtration method. The authors regulated the porosity of the catalyst layer by adjusting the ratios of CNFs and CNTs, thus forming a gradient structure along the thickness direction. BPCLs with a gradient structure enabled to ensure a high Pt utilization (up to 90%) and improved catalyst durability in the AST cycles.

Mardle et al. prepared Pt nanorods grown on MWCNTs and deposited on carbon paper via a plasma-enhanced chemical vapor deposition (PECVD) and nitrided using active screen plasma (ASP) treatment, which was directly employed as the cathode for PEMFCs [43]. The as-prepared nitrogen-doped MWCNTs had an average diameter of about 10–20 nm and lengths of hundreds of nanometers. The Pt nanorods possessed a diameter of about 3–4 nm and lengths up to 10 nm. The thin cathode catalyst layer effectively enhanced the mass transfer performance and, with less than half of the Pt loading, 1.23-fold power density was achieved compared with that of commercial Pt/C catalysts. A higher durability was also confirmed, which was attributed to the good structure stability of nanorods and the enhancement effects from the nitrogen-doped MWCNT support.

However, despite all the advantages offered by CNTs, their application in PEMFCs is still faced with many challenges. On the one hand, the dispersion of CNTs in the solvents is difficult due to their hydrophobic property. The suitable anionic surfactants, such as sodium dodecyl sulfate, are necessary to disperse the CNTs. These types of surfactants, which are challenging to be removed after the synthesis process, will adsorb on the catalyst surface, thereby leading to inferior activities. On the other hand, the CNTs for the catalyst support should be high-quality with ultra-low metal impurities that lead to poison the catalyst. However, the current synthesis technique for high-quality CNTs is not suitable for large-scale production in a cost-effective approach. Hence, more effort is needed for the mass production of CNTs for broad application in catalyst production.

Graphene has been proposed as an ideal electrode support material due to its corrosion resistance, high surface area, and high electrical conductivity. However, to date, graphene-based electrodes suffer from high defect concentrations and non-uniform nanoparticle coverage, which negatively affects the Pt loading and the mass-transport performance in the MEA. In general, for producing a graphene-supported catalyst, the most common method is via liquid-processable graphene oxide (GO) created from graphite using a modified Hummer's method. Then GO is partially converted to conductive graphene oxide (rGO), which is supported with catalyst nanoparticles. However, the rGO contains a high number of defects, which negatively impacts the as-synthesized catalysts' durability [53].

Angel et al. prepared Pt nanoparticle-coated graphene whereby $PtCl_2$ was reduced directly by negatively charged single-layer graphene sheets in tetrahydrofuran solution [54]. The Pt nanoparticles with a mean size of 2.5 nm were evenly distributed on graphene nanosheets. The Pt-decorated graphene catalysts exhibited excellent activity toward the ORR, comparable with the activity of commercial Pt/C catalyst. Meanwhile, the catalyst stability was analyzed, which showed a decreased ECSA and cell performance when compared to commercial Pt/C after the AST cycles.

Vinayan et al. developed a method for uniform coating of polypyrrole over graphene surface and prepared nitrogen-doped graphene by the subsequent pyrolysis of PPy-modified graphene nanocomposite [55]. Pt and PtCo alloy nanoparticles were uniformly dispersed over the nitrogen-doped graphene by the modified-polyol reduction method, and their electrocatalytic activity towards the ORR was studied by PEMFCs. The Pt_3Co/graphene catalysts exhibited superior MEA performance, which was about four times higher than commercial Pt/C cathode catalysts. This high performance was attributed to the combined effects of the modified oxygen adsorption capability on Pt surface due to the alloying and nitrogen doping, uniform dispersion of catalyst nanoparticles, and high intrinsic catalytic activity of nitrogen-doped support.

Although graphene possesses many advantages in supporting Pt-based catalysts, several aspects need to be improved for real application in PEMFC. Firstly, the two-dimensional structure of graphene may inhibit the transportation of reactants and water in the catalyst layer, deteriorating the performance of MEA. Secondly, due to the strong interactions among graphene layers, it is difficult to disperse the catalyst ink and form an uniform catalyst layer. Thirdly, it is still a challenge in the mass production of graphene environmentally and economically. Therefore, more work is needed to optimize the structure and dispersion of graphene-based catalysts before the real application of these carbon supporting materials.

1.3 CARBIDE SUPPORT MATERIALS

Carbides attract significant interests for their high melting points, relatively low cost as well as excellent physical and chemical stability. Carbides consist of carbon and a less electronegative element with different levels of covalent function. Carbides can be divided into four categories: covalent compounds, salt-like, interstitial compounds, and "intermediate" transition metal carbides (TMCs) [56]. Several typical carbides are particularly discussed as catalyst supporting materials for fuel cells in the following section.

1.3.1 Boron Carbide

Boron carbide (B_4C) is a typical covalent carbide with decent conductivity (0.3 S/cm at room temperature), corrosion resistance and oxidative stability, as well as extremely high hardness (behind cubic boron nitride and diamond) [57]. Even though the typical surface area of B_4C materials is approximately 3–20 m^2/g, Borchardt et al. [58] successfully prepared mesoporous boron carbide with 300–770 m^2/g surface area according to a precursor nanocasting method. It was the first time boron carbide was used as catalyst support to improve the platinum catalytic activity in phosphoric acid fuel cells in the 1960s [59]. After that, more and more researchers have been paying attention to the effective usage of boron carbide as catalyst supports in fuel cells. Jackson et al. [60] synthesized Pt/B_4C with different contents to investigate the electronic metal-support interaction and found that platinum nanoparticles supported on boron carbide exhibited almost doubled activity and improved long-term stability when compared to conventional Pt/C due to the electronic structure development. Furthermore, it was reported that the Pt/B_4C showed an improved CO tolerance than the benchmark (Pt/C) because of the high OH_{ad} coverage and rapid hydroxyl adsorption during the hydrogen desorption process [61]. Especially, the nanosized B_4C was intercalated into graphene interlayer as composite supports (Pt-rGO/B_4C) to achieve higher electrochemical surface area, mass activity, and excellent stability compared with Pt supported on graphene oxide (rGO) (Pt/rGO) and the commercial Pt/C. The improved activity is due to the strong restriction effect between boron carbide and graphene interlayers. It should be pointed out that the combination of boron carbide and other materials is a promising approach to further improve the activity and stability of catalysts in fuel cells [62].

1.3.2 Silicon Carbide

Silicon carbide (SiC), especially β-SiC, is also a promising catalyst support material for its ideal thermal conductivity as well as acid and oxidation-resistant properties [63]. Honji et al. [64] first put forward the application of SiC (surface area: 10 m^2/g) as catalyst support for fuel cells. However, the poor electrical conductivity (approximately 10^{-6} S/cm) of nano-SiC limited its application in catalyst support. The reason is that the bandgap, which is the energy required for the material to start conducting electrons, reaches 2.3 eV for 3C-SiC and 3.2 eV for 4H-SiC [65]. Besides, the Pt nanoparticle agglomeration on SiC and the small specific surface area caused a low electrochemical surface area and inferior mass activity, which hindered the SiC application to a greater extent. Many efforts have been made to improve catalyst performance with SiC as the support. Ma et al. [66] achieved the SiC application as a Pt catalyst support in electrode and MEA level, demonstrating better affinity between the Nafion ionomer and SiC-supported catalyst and less particle size growth as well as more stable performance in comparison to traditional carbon-supported catalyst (Pt/C). Lv et al. [67] attempted to deposit Pt nanoparticles on β-SiC with a chemical reduction method. Meanwhile, carbon powers were introduced to form Pt/SiC/C catalysts to increase electrical conductivity and improve the catalytic performance. The catalyst was stable at 1.20 V for 48 hours, as proved by electron microscopic analysis. Meanwhile, Zang et al. [68] synthesized a core-shell structure support (SiC@C) by vacuum annealing treatment and microwave-assisted reduction method. The as-prepared catalyst (Pt/SiC@C) possessed a higher ORR activity and stability than the commercial counterpart. Morgen et al. [69] successfully developed nano-SiC supports with particle sizes in the range of 50–150 nm (SiC-SPR) and 25–35 nm (SiC-NS) using a commercial carbon black as a template. In addition, an acid treatment is utilized as surface group sources to anchor Pt nanoparticles. The Pt/SiC-NS exhibited higher electrochemical activity than commercial catalysts using the same carbon blacks.

1.3.3 Titanium Carbide

Titanium carbide (TiC) is a super hard material with a typical face-centered cubic crystal structure. As a transition metal carbide (TMC), titanium carbide is reported to exhibit high electrical conductivity, strong thermal stability (no diffraction peak at 200°C in XRD patterns), and low cost. Therefore, TiC is always used as the coating material in the coating metal bipolar plates in PEMFCs. However, it was also reported that a large oxidation peak around 0.95 V was observed in the electrochemical test, resulting in the formation of a thin TiO_2 film around the TiC particles [70]. The TiO_2 film severely decreased the conductivity and led to a performance loss. Chiwata et al. [71] successfully prepared Pt nanoparticles supported on TiC as an alternative catalyst for ORR after a thermal treatment at 600°C in 1% H_2/N_2. They compared the Pt/TiC-600°C with commercial Pt/carbon black and found a comparable mass activity at 0.85 V (507 A/g vs. 527 A/g) and overwhelmingly higher durability than that of Pt/C. Wang et al. [72] used *ab initio* density functional theory calculations to investigate the adsorption and dissociation of oxygen on the Pt modified TiC (001) surfaces and found that the monolayer Pt on TiC (001) promoted the scission of O-O bond and weakened the adsorption of O*, thereby boosting the activity of ORR. Besides, TiC with several unique structures was also employed as the supporting materials. Lee et al. [73] synthesized an ORR catalyst with two-dimensional titanium carbide (Ti_3C_2) as the supporting material (Pt/Ti_3C_2) and proved that the enhancement ORR electroactivity was derived from the expected strong metal-support interaction (SMSI) between Pt and Ti. Particularly, monolayer, few-layer, and multilayer Ti_3C_2 were fabricated as supports for Pt nanoparticles with an average diameter of 4 nm. It was found that edge-dominant multilayer Ti_3C_2 donated more electrons to Pt than others, thereby exhibiting a higher activity toward the ORR. In addition, the electron transfer from the support to Pt nanoparticles with the strong interaction between Pt and Ti greatly enhanced the catalyst stability during the electrochemical cycling test.

1.3.4 Tungsten Carbides

General speaking, tungsten carbides mainly include two types of phases: tungsten monocarbide (WC) and tungsten subcarbide (W_2C). However, W_2C is thermodynamically unstable below 900°C and has relatively worse acidic resistance. Hence, WC is mainly used as catalyst supports in PEMFCs. WC has a super high melting point at 2870°C at 1 bar as well as good thermal conductivity. Pt/WC effectively enhanced ORR's catalytic activity because of the synergetic effect between the Pt nanoparticles and the WC. The similar electronic structure of platinum and WC was the main reason for Pt/WC with improved ORR activity [74]. However, the high voltage (>0.8 V) will lead to the oxidation of WC and the oxidation products, namely tungsten oxides, will lead to the detachment of Pt nanoparticles, thereby losing the ECSA. The oxidation of WC at high potential follows the reactions below.

$$WC_{(s)} + 4H_2O \rightarrow WO_{2(s)} + CO_{2(g)} + 8H^+ + 8e^- \tag{1.1}$$

$$WO_{2(s)} + H_2O_{(l)} \rightarrow WO_{3(s)} + 2H^+ + 2e^- \tag{1.2}$$

It has been proved that the bonding strength between tungsten carbide and CO is much weaker than the bonding strength between Pt and CO. Hence, the CO is less likely to absorb on the tungsten carbide surface. Besides, even though absolutely avoiding the CO poisoning of Pt is impossible, the CO adsorbed on the Pt surface with the existence of tungsten carbide would be oxidated through the following reaction and mitigate the CO poisoning to a great extent.

$$Pt\text{–}CO + WC\text{–}OH \rightarrow Pt + CO_2 + WC + H^+ + e^- \quad (1.3)$$

Meanwhile, the Pd/WC catalyst was employed as the HOR catalyst and showed almost the same efficiency and similar behavior as the commercial Pt/C catalyst [75]. Hassan and co-workers [76] synthesized Pt supported on tungsten carbide-impregnated carbon (Pt/WC/C) as a hydrogen oxidation catalyst in the anode. It was found that the overpotential at a current density of 1 A/cm^2 was reduced from 398 mV for Pt/WC to 364 mV for Pt/WC/C in the presence of 100 ppm CO. In addition, the following AST proved that the Pt/WC/C catalyst possessed an improved electrochemical stability.

1.3.5 Molybdenum Carbides

Molybdenum carbides (MoC/Mo_2C) consist of various kinds of crystalline structures, such as α-MoC_{1-x}, β-Mo_2C, γ-MoC, and η-MoC [77]. The α-MoC_{1-x} is a typical face-centered cubic (fcc) structure and the others are hexagonal crystal structures with different stacking sequences [78]. The main approach to synthesize molybdenum carbide nanostructures is reductive carburization of molybdenum oxide in the gas mixture (CH_4/H_2), such as a temperature-programmed reduction method. This method could synthesize α-MoC_{1-x} nanoparticles with surface areas of higher than 150 m^2/g even though the contamination chars were stubborn to remove [79]. In terms of the catalyst support for PEMFC, the molybdenum carbides were not as popular as other carbides. Elbaz and co-workers synthesized molybdenum carbide supported platinum nanorafts with a high electrocatalytic activity by a novel reduction-expansion-synthesis of catalysts method [80]. In this process, the platinum nanoparticles and the α-MoC_{1-x} nanoparticles were produced simultaneously. The platinum nanorafts possessed a size of fewer than six atoms on the molybdenum carbide and were verified to have a stronger bonding to the support, which resulted in higher durability of the catalyst (only 10% ECSA loss after 5000 accelerated stress tests cycles). Meanwhile, the activity of the platinum nanorafts was compared to commercial Pt/C and presented an improved electroactive activity towards ORR, which was ascribed to the electronic interaction between the support and Pt nanoparticles. It was also reported that Pt nanoparticles supported on Mo_2C hollow nanotubes via atomic layer deposition (ALD) with ultralow platinum loading (0.02 mg/cm^2) was employed as both anode catalyst and cathode catalyst for a realistic MEA, which generated a peak power density of 414 mW/cm^2. It corresponded to 10.35 kW/g_{Pt} (DOE 2020 target: 8 kW/g_{Pt}). Besides, the AST of MEA with the as-prepared Pt/Mo_2C catalysts showed 1.11-fold power density than that of MEA with commercial 20 wt% Pt/C [81,82] (Table 1.1).

TABLE 1.1
Summary of TMC-supported catalysts materials reported for ORR in the acidic environment

Catalyst	Electrolyte	E_{onset} (V vs RHE)	Tafel Slope (mV/decade)	MA (A/mg_{Pt}@ 0.9 V vs RHE)	SA (mA/cm^2@ 0.9 V vs RHE)	Decay	Ref.
Pt/BC	0.1 M $HClO_4$	–	–	0.168	0.3	12%, 6,000 cycles	[60]
Pt/RGO/B_4C	0.1 M $HClO_4$	1.03	–	0.185	0.15	54.8% in ECSA, 10,000 cycles	[62]
Pt/TiC	0.1 M $HClO_4$	0.98	60, E>0.9 V120, E ≈ 0.85 V	0.507@0.85 V	0.70@0.85 V	14% in ECSA, 1,000 cycles	[71]
Pt/WC	0.1 M $HClO_4$	0.95	77	0.0033	0.0073	–	[83]
Pt/Mo_2C	0.1 M $HClO_4$	1.11	Two sections with 60 and 126	0.29	–	10% in ECSA, 5,000 cycles	[80]
Pt/WC/CMS	0.5 M H_2SO_4	0.7 V vs Ag/AgCl	–	–	5.2	78.5% at 0.65 V for 30 minutes	[84]
Pt-β-WC_{40}/C	0.5 M H_2SO_4	1.0	68 at low current density and 117 at high current density	0.01	0.204	–	[85]
Pt-α-WC_{20}/C	0.5 M H_2SO_4	1.0	76 at low current density and 122 at high	0.01	0.068	–	[85]
Pt/OMC/TiC	0.5 M $HClO_4$	1.05	–	0.26	0.26	28% in ECSA, 6,000 cycles	[86]
Pt/TiC	0.1 M $HClO_4$	0.80	106	0.16	0.25	10.1% increase in ECSA, 5,000 cycles	[87]

1.4 NITRIDE SUPPORT MATERIALS

The combination of nitrogen with less electronegative elements can form the nitride compounds. The ionic nitrides, including alkali and alkaline metals nitrides, have poor thermal stability, which will cause hydrolysis to produce ammonia and hydroxide. Therefore, they are of little practical significance to use as catalyst support. The most widely used nitrides as catalyst support are transition metal nitrides (TMNs), thanks to the three different interactions between atoms: metallic bond, ionic bond, and covalent bond. The metallic bond is related to the high electrical conductivity, while the covalent bond gives electrochemical durability, corrosion resistance, and great hardness. Besides, the ionic bond endows the TMNs with high stability and high melting point [88,89]. Traditionally, the TMNs are synthesized by the heat-treatment of metal precursors in the nitrogen sources environment, such as ammonia (NH_3), nitrogen (N_2), and urea. Recently, the reduction of transition metal oxides (TMOs) is intensively reported as an alternative method to synthesize transition metal nitrides [90–92].

Among the TMNs, the TiN is the most popular one as a catalyst support material in PEMFCs, even though CrN [93] and VN [94] are also occasionally selected as the Pt or Pd support. The TiN has a high electrical conductivity of approximately 400 S/m with good corrosion resistance resulting from the oxy-nitride layer. It was reported that the TiN-supported catalysts would exhibit higher CO tolerance because the TiN would form Ti-OH by decomposing water, which could oxidize the absorbed CO on platinum into CO_2, as shown in Figure 1.4 [95]. It was reported by Jeong and co-workers [96] that the adsorption strength of Pt/TiN was about 2.4–2.6 times higher than Pt/C using functional theory calculations, which was ascribed to a larger degree of charge transfer between Pt and TiN. They also demonstrated that the d-band centers of Pt absorbed on TiN and graphene were –1.91 eV and –0.97 eV,

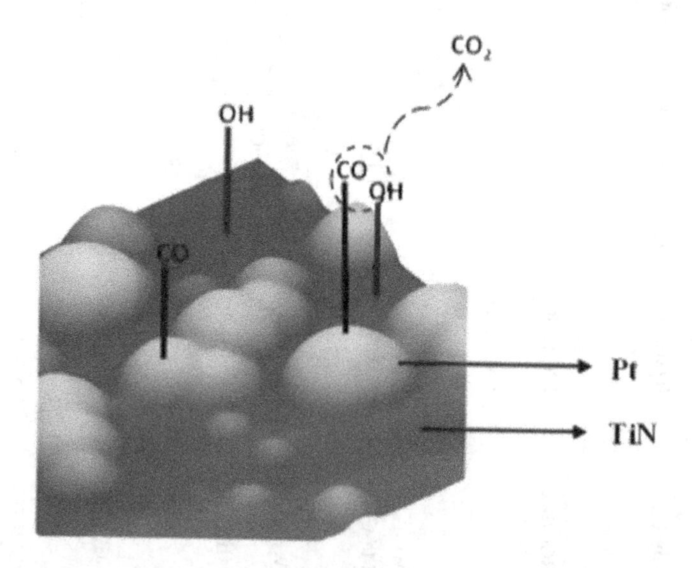

FIGURE 1.4 Scheme of carbon monoxide oxidation on Pt-TiN electrocatalysts.

which is indicative of not only improved stability of Pt/TiN, but also a potential increase in the ORR electroactivity by moderately weakening the absorbed oxygen-binding strength. Chi and co-workers fabricated a layer of TiN with microporous/mesoporous structures on the carbon papers (CPs) via a sol-gel method and deposited platinum nanoparticles by atomic layer deposition [97]. The Pt@TiN@CPs with 50 wt% of PVP and 50 cycles ALD, which presented 9.2 ug/cm^2 platinum loading, was directly employed as the anode catalyst for PEMFCs and showed better performance than the commercial Pt/C E-Tek catalyst. The mass and specific power density of MEA with Pt@TiN@CPs as catalyst were 62 and 3 times higher than that of the MEAs with commercial Pt/C. There are several other researches that reported the electrochemical performance in the rotating-disk-electrode experiments, as summarized in Table 1.2.

However, the TiN would still be corroded into oxide under the PEMFC operation conditions (high temperature, high humidity, and high potential), causing the apparently lower conductivity [98]. Avasarala and co-workers found that there were three main degradation mechanisms for the Pt/TiN catalyst, i.e., support oxidation, platinum agglomeration, and platinum dissolution. Among these three mechanisms, the platinum nanoparticle agglomeration was considered as the dominant factor [99].

1.5 TRANSITION METAL OXIDES

The transition metal oxides show great potential as platinum nanoparticles support due to the high chemical stability in acidic and alkaline environments and remarkable mechanical and thermal stability, although the majority of them are poor electrical conductors and possess relatively low surface area.

1.5.1 Titanium Oxides

Compared with semi- or non-conducting titanium oxides (such as TiO_2), the Ti_nO_{2n-1} ($4<n<10$) group attracts more attention due to the high electrical conductivity, exceeding 10^3 S/cm at 25°C, and the inert property to form hydrides in contact with hydrogen. Ioroi and co-workers initially studied the Ti_4O_7 as catalyst support for platinum nanoparticles [110]. They found that the Ti_4O_7 showed improved resistance to electrochemical oxidation than traditional carbon black under the positive direction sweep of CV. The cell performance of MEA with the Pt/Ti_4O_7 as cathode catalyst presented a similar activity for the ORR as the benchmark catalyst (Pt/C) at identical operating conditions. The same groups subsequently synthesized sub-stoichiometric titanium oxide (TiOx) with a nanoscale diameter size by pulsed UV laser irradiation of TiO_2 as the cathode catalyst support [111]. The experimental results exhibited that the platinum nanoparticles deposited on the TiO_x support were alloyed with titanium to form an ordered Pt_3Ti phase, thereby resulting in a twofold higher specific activity than the conventional catalyst (Pt/XC72). The catalyst stability was analyzed by potential cycling from 1.0 V to 1.5 V vs. RHE. Almost no ECSA loss for Pt/TiOx after 10,000 cycles but the Pt/XC72 lost 30–50% of its initial ECSA. The high performance of Pt/Ti_4O_7 was ascribed to the increased ECSA by the reduced platinum nanoparticle size and the formation of the ordered Pt_3Ti phase. Meanwhile, the

TABLE 1.2
Summary of TMN-supported catalysts materials reported for ORR in the acidic environment [100]

Catalyst	Electrolyte	E_{onset} (V vs RHE)	$E_{1/2}$ (V vs RHE)	MA (A/ mg_{Pt}) (0.9 V vs RHE)	SA (mA/ cm^2) (0.9 V vs RHE)	Decay	Ref.
Pt/TMNs	0.1 M $HClO_4$	0.95	0.85	–	–	–	[101]
Pt/TiN NTs	0.1 M $HClO_4$	0.96	0.84	0.056	0.111	8%, 5,000 cycles	[102]
Pt/3D-TiN	0.5 M $HClO_4$	1.05	0.93	0.65	1.06	12% in ECSA, 15,000 cycles	[103]
Pt/$Ti_{0.5}Nb_{0.5}N$	0.1 M $HClO_4$	1.04	0.94	0.256	0.530	19.2%, 5,000 cylcles	[104]
Pt/$Ti_{0.9}Co_{0.1}N$	0.1 M $HClO_4$	1.05	0.90	0.460	0.540	35%, 10,000 cycles	[105]
TiNiN@Pt	0.1 M $HClO_4$	1.05	0.89	0.830	0.490	10 mV in $E_{1/2}$, 10,000 cycles	[106]
Pt/TiN	0.1 M $HClO_4$	0.80	0.62 V	0.055	0.776	10.9% in ECSA, 5,000 cycles	[87]
Pt/$Ti_{0.9}Ni_{0.1}N$ NTs	0.1 M $HClO_4$	1.05	0.93	0.780	1.300	9 mV in $E_{1/2}$, 15,000 cycles	[107]
Pt_3Cu/TiN	0.1 M $HClO_4$	1.02	0.93	2.43	5.320	16.1%, 10,000 cycles	[108]
Fe_3Pt/$Ti_{0.5}Cr_{0.5}N$	0.1 M $HClO_4$	1.03	0.92	0.6730	1.280	5.8%, 12 hours	[109]

modified electronic structure due to the electronic interaction between platinum and Ti_4O_7 was also of great importance to change the oxygen adsorption conditions. The interaction increased the 5d vacancy of Pt and decreased the Pt-Pt bond distance, thereby inhibiting the chemisorption of OH^-. The Pt-OH formation was also shifted to more positive potentials to improve the interaction between oxygen molecules and platinum [112].

In spite of the low conductivity, the TiO_2 was still widely investigated as catalyst support with different modified methods and different morphologies. Dhanasekaran and co-workers synthesized the nitrogen and carbon co-doped TiO_2 (TiON) as the platinum support to form Pt/TiON catalyst by a colloidal method [113]. The heteroatom dopants modified both the electronic and structural properties of TiO_2, which boosted the ORR activity. Similarly, the nitrogen and cobalt [114], iron and nitrogen [115] co-doped titania frameworks were also studied and exhibited better corrosion stability when compared with the benchmark (Pt/C catalyst). On the other hand, the TiO_2 with a mesoporous structure was also synthesized via a template-assisted route [116]. Compared with Pt/C (0.13 mA/cm^2), the Pt/TiO_2 exhibited a nearly tenfold higher ORR activity (1.20 mA/cm^2 vs. 0.13 mA/cm^2) after the accelerated durability test.

The strong metal-support interaction (SMSI) between the Pt nanoparticles and the titanium oxides was generally considered as one of the factors contributing to the improved stability of Pt/TiO_x. However, the function of SMSI in the role of catalyst stability is still debatable. Nail and co-workers prepared carbon-free multifunctional oxygen-deficient titanium oxide nanosheets and applied them as the cathode support in PEMFCs [117]. They believed that the excellent activity was ascribed to the SMSI. However, it was also reported that the ruthenium-titanium mixed-oxide supports might cause the gradual growth of a thin oxide layer on the Pt nanoparticles due to the SMSI, which suppressed the catalytic activity apparently (Figure 1.5) [118].

1.5.2 Tungsten Oxides

There are several oxidation states (usually 2 to 6) for tungsten. When it comes to the catalyst support of PEMFCs, it always refers to the tungsten oxide (WO_x) with

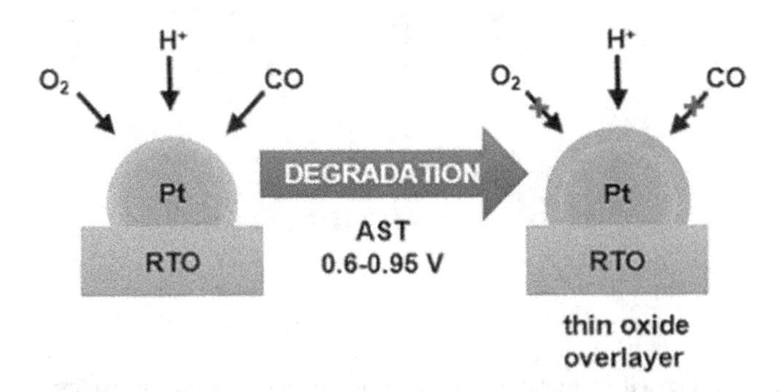

FIGURE 1.5 Illustration of the hypothetical degradation process: formation of a thin oxide overlayer on top of the RTO-supported Pt nanoparticle. The overlayer can be penetrated by a proton, but is impermeable for oxygen and carbon monoxide [118].

relatively high conductivity coming from the non-stoichiometric composition with oxygen-vacancy defects in the lattice [112]. It was proven that WO_x supported catalysts (Pt/WO_x) showed much lower degradation rates compared to carbon materials by potential cycling experiments [119]. Wickman and co-workers fabricated gas diffusion electrodes with thin films (≤40 nm) of WO_x and Pt on WO_x as the anode catalyst layers [120]. The results showed that an increased amount of hydrogen tungsten bronzes (H_xWO_3) formed for increasing WO_x thicknesses, and the oxidation of pre-adsorbed CO was shifted to lower potentials for WO_x containing electrodes, which verified that the Pt-WO_x possessed a higher CO tolerance than Pt. In addition, Pt on thicker films of WO_x exhibited an increased limiting current as the HOR catalyst owing to the larger electrochemically active surface area, which was ascribed to well conductivity and hydrogen permeability in the WO_x film. However, the biggest issue of using WO_x as catalyst support is that the Pt/WO_X showed noticeable electroactivity degradation, mainly due to the Pt detachment, which was caused by the formation of a water-soluble compound (H_xWO_3) on the support surface [121].

Meanwhile, the carbon-tungsten oxide composite support was also intensively investigated to increase electrical conductivity. Saha et al. synthesized a Pt/C-$W_{18}O_{49}$ NWs/carbon-paper composite electrode by the chemical vapor deposition method [122]. Experimental results proved that a 100 mV shift of onset potential for ORR was achieved when compared with Pt/C. The mass activity and specific activities were 0.83 mA/mg_{Pt} and 1.3 uA/cm^2, which were 75% and 160% higher than the commercial counterpart. Yang and co-workers physically mixed a commercial Pt/C catalyst with 20 wt% of two-dimensional WO_3 nanoplates and found that the stability of Pt/C-WO_3 was enhanced [123]. The ECSA and electrochemical activity were twice higher than the pristine Pt/C after a long-time accelerated degradation test. Moreover, the platinum nanoparticles were preferred to migrate or re-nucleate on the WO_3 nanoplate surfaces from the carbon surface during degradation, which circumstantially triggered the "hydrogen spillover" effect between Pt and WO_3, as shown in Figure 1.6.

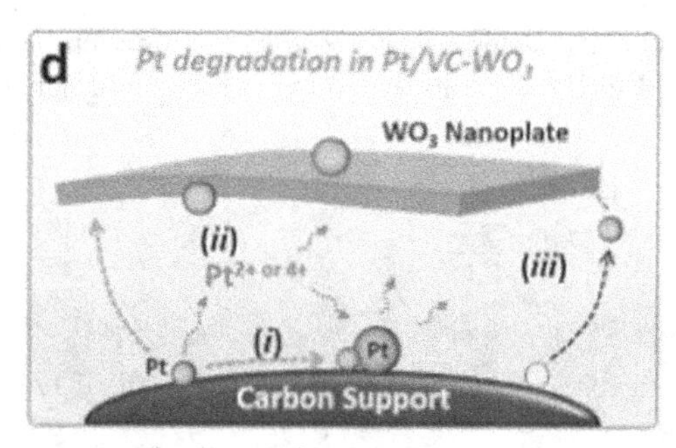

FIGURE 1.6 Proposed degradation mechanisms of platinum nanoparticles in the heterogeneous Pt/C-WO_3 catalyst system. Three major degradation mechanisms are considered: (I) Pt agglomeration via coalescence or particle migration; (II) Pt dissolution and re-nucleation via Oswald Ripening effect; (III) Pt particle detachment from carbon support [123].

1.6 PERSPECTIVES AND OUTLOOK

Herein, the catalyst supporting material is critical for boosting the electrochemical activity and platinum utilization, particularly under the circumstances of lowering the platinum loading in the catalyst layer for the commercialization of PEMFCs. The commercialized catalyst for PEMFCs is Pt supported on different categories of carbon supporting materials. Due to the limitation of electrochemical stability at high voltage, non-carbon supporting materials, such as carbides and nitrides, have been comprehensively investigated in recent years. However, the dispersion of platinum catalysts on the non-carbon supporting materials is usually unsatisfactory. Meanwhile, the long-term durability of these catalysts in the MEA is not thoroughly investigated. The following aspects will be the future development direction:

1. **Carbon-based supporting materials**: The main research direction of carbon-based supporting materials is to balance the high surface area and the graphitization degree. The graphitization degree of carbon materials should be improved without sacrificing the dispersion capability of platinum nanoparticles. Therefore, the carbon corrosion rate at the high voltage will be alleviated. On the other hand, the pore structure, including pore size, pore distribution and porosity, of carbon-based materials should be carefully tailored to meet the requirement of mass transport and to avoid the poisoning effect of sulphonic groups on the PFSA resin. One promising solution is to enlarge the percentage of mesopore in carbon-based materials.
2. **Non-carbon supporting materials**: Due to the enhanced interaction between platinum nanoparticles and non-carbon supporting materials, the non-carbon supporting materials are investigated as the alternative to carbon-based supporting materials. However, their specific surface areas usually are very low, which restricts the loading of platinum nanoparticles. A suitable solution is the fabrication of composite supporting materials with high surface area as well as superior corrosion resistance. On the other hand, most of the performance of platinum supported on non-carbon supporting materials was evaluated in the conventional three-electrode cell, while fewer performance was verified in the MEA. In this regard, the so-called high stability of these catalysts is questionable. All in all, the non-carbon-based supporting materials are still in the R&D stage.

REFERENCES

[1] Park YC, Tokiwa H, Kakinuma K, Watanabe M, Uchida M. Effects of carbon supports on Pt distribution, ionomer coverage and cathode performance for polymer electrolyte fuel cells. *Journal of Power Sources*. 2016;315:179–191.

[2] Tuaev X, Rudi S, Strasser P. The impact of the morphology of the carbon support on the activity and stability of nanoparticle fuel cell catalysts. *Catalysis Science Technology*. 2016;6:8276–8288.

[3] Yu PT, Gu W, Makharia R, Wagner FT, Gasteiger HA. The impact of carbon stability on PEM fuel cell startup and shutdown voltage degradation. *ECS Transactions*. 2006;3:797–809.

[4] Jinnai H, Spontak RJ, Nishi T. Transmission electron microtomography and polymer nanostructures. *Macromolecules.* 2010;43:1675–1688.
[5] Yarlagadda V, Carpenter MK, Moylan TE, Kukreja RS, Koestner R, Gu W, et al. Boosting fuel cell performance with accessible carbon mesopores. *ACS Energy Letters.* 2018;3:618–621.
[6] Ito T, Matsuwaki U, Otsuka Y, Hatta M, Hayakawa K, Matsutani K, et al. Three-dimensional spatial distributions of Pt catalyst nanoparticles on carbon substrates in polymer electrolyte fuel cells. *Electrochemistry.* 2011;79:374–376.
[7] Kongkanand A, Yarlagadda V, Garrick TR, Moylan TE, Gu W. (Plenary) electrochemical diagnostics and modeling in developing the PEMFC cathode. *ECS Transactions.* 2016;75:25–34.
[8] Ohma A, Mashio T, Sato K, Iden H, Ono Y, Sakai K, et al. Analysis of proton exchange membrane fuel cell catalyst layers for reduction of platinum loading at Nissan. *Electrochimica Acta.* 2011;56:10832–10841.
[9] Garrick TR, Moylan TE, Yarlagadda V, Kongkanand A. Characterizing electrolyte and platinum interface in PEM fuel cells using CO displacement. *Journal of the Electrochemical Society.* 2016;164:F60–F64.
[10] Iden H, Sato K, Ohma A, Shinohara K. Relationship among microstructure, ionomer property and proton transport in pseudo catalyst layers. *Journal of the Electrochemical Society.* 2011;158:B987.
[11] Shinozaki, Kazuma, Yamada, Haruhiko, Morimoto, Yu. Relative humidity dependence of Pt utilization in polymer electrolyte fuel cell electrodes: Effects of electrode thickness, ionomer-to-carbon ratio, ionomer equivalent weight, and carbon support. *Journal of the Electrochemical Society.* 2011;158:B467.
[12] Uchida M, Aoyama Y, Tanabe M, Yanagihara N, Eda N, Ohta A. Influences of both carbon supports and heat-treatment of supported catalyst on electrochemical oxidation of methanol. *Journal of The Electrochemical Society.* 1995;142: 2572–2576.
[13] Uchida M, Fukuoka Y, Sugawara Y, Ohara H, Ohta A. Improved preparation process of very-low-platinum-loading electrodes for polymer electrolyte fuel cells. *Journal of The Electrochemical Society.* 1998;145:3708–3713.
[14] Meier JC, Galeano C, Katsounaros I, Topalov AA, Kostka A, Schüth F, et al. Degradation mechanisms of Pt/C fuel cell catalysts under simulated start–stop conditions. *ACS Catalysis.* 2012;2:832–843.
[15] Shi Q, Zhu C, Engelhard MH, Du D, Lin Y. Highly uniform distribution of Pt nanoparticles on N-doped hollow carbon spheres with enhanced durability for oxygen reduction reaction. *RSC Advances.* 2017;7:6303–6308.
[16] Badam R, Vedarajan R, Matsumi N. Platinum decorated functionalized defective acetylene black; a promising cathode material for the oxygen reduction reaction. *Chemical Communications.* 2015;51:9841–9844.
[17] Rich SS, Burk JJ, Kong CS, Cooper CD, Morse DE, Buratto SKJC. Nitrogen functionalized carbon black: A support for Pt nanoparticle catalysts with narrow size dispersion and high surface area. *Carbon.* 2015;81:115–123.
[18] Roen LM, Paik CH, Jarvi TD. Electrocatalytic corrosion of carbon support in PEMFC cathodes. *Electrochemical and Solid-State Letters.* 2004;7:A19.
[19] Kangasniemi KH, Condit DA, Jarvi TD. Characterization of vulcan electrochemically oxidized under simulated PEM fuel cell conditions. *Journal of The Electrochemical Society.* 2004;151:E125.
[20] Kim M, Park J-N, Kim H, Song S, Lee W-H. The preparation of Pt/C catalysts using various carbon materials for the cathode of PEMFC. *Journal of Power Sources.* 2006;163:93–97.

[21] Bo X, Bai J, Ju J, Guo L. Highly dispersed Pt nanoparticles supported on poly (ionic liquids) derived hollow carbon spheres for methanol oxidation. *Journal of Power Sources*. 2011;196:8360–8365.

[22] Hasché F, Fellinger TP, Oezaslan M, Paraknowitsch JP, Antonietti M, Strasser P. Mesoporous nitrogen doped carbon supported platinum PEM fuel cell electrocatalyst made from ionic liquids. *ChemCatChem*. 2012;4:479–483.

[23] Yang Z, Moriguchi I, Nakashima N. Durable Pt electrocatalyst supported on a 3D nanoporous carbon shows high performance in a high-temperature polymer electrolyte fuel cell. *ACS Applied Materials & Interfaces*. 2015;7:9800–9806.

[24] Yin F, Wang D, Zhang Z, Zhang C, Zhang Y. Synthesis of mesoporous hollow polypyrrole spheres and the utilization as supports of high loading of Pt nanoparticles. *Materials Letters*. 2017;207:225–229.

[25] Zhao G, Zhao TS, Xu J, Lin Z, Yan X. Impact of pore size of ordered mesoporous carbon FDU-15-supported platinum catalysts on oxygen reduction reaction. *International Journal of Hydrogen Energy*. 2017;42:3325–3334.

[26] Darling RM, Burlatsky S. Modeling oxygen transport in high surface area carbon supports for polymer-electrolyte fuel cells. *Journal of the Electrochemical Society*. 2020;167.

[27] Ramaswamy N, Gu W, Ziegelbauer JM, Kumaraguru S. Carbon support microstructure impact on high current density transport resistances in PEMFC cathode. *Journal of The Electrochemical Society*. 2020;167:064515.

[28] Maillard F, Bonnefont A, Micoud F. An EC-FTIR study on the catalytic role of Pt in carbon corrosion. *Electrochemistry Communications*. 2011;13:1109–1111.

[29] Giordano N, Antonucci PL, Passalacqua E, Pino L, Aricò AS, Kinoshita K. Relationship between physicochemical properties and electrooxidation behaviour of carbon materials. *Electrochimica Acta*. 1991;36:1931–1935.

[30] Kamitaka Y, Takeshita T, Morimoto Y. MgO-templated mesoporous carbon as a catalyst support for polymer electrolyte fuel cells. *Catalysts*. 2018;8:230.

[31] Mohanta PK, Regnet F, Jorissen L. Graphitized carbon: A promising stable cathode catalyst support material for long term PEMFC applications. *Materials*. 2018;11.

[32] Selvaganesh SV, Sridhar P, Pitchumani S, Shukla AK. A durable graphitic-carbon support for Pt and Pt3Co cathode catalysts in polymer electrolyte fuel cells. *Journal of The Electrochemical Society*. 2012;160:F49–F59.

[33] Wang Zijun, Zhu Yi'an, Cheng Hongye, Yang Qinmin, Sui Zhijun, Zhou Xinggui. Microstructure of fishbone carbon nanofibers. *Petrochemical Industry*. 2016; 45:1037–1042.

[34] Sharma S, Pollet BG. Support materials for PEMFC and DMFC electrocatalysts—A review. *Journal of Power Sources*. 2012;208:96–119.

[35] Li W, Waje M, Chen Z, Larsen P, Yan Y. Platinum nanoparticles supported on stacked-cup carbon nanofibers as electrocatalysts for proton exchange membrane fuel cell. *Carbon*. 2010;48:995–1003.

[36] Yli-Rantala E, Pasanen A, Kauranen P, Ruiz V, Borghei M, Kauppinen E, et al. Graphitised carbon nanofibres as catalyst support for PEMFC. *Fuel Cells*. 2011;11:715–725.

[37] Jung J, Park B, Kim J. Durability test with fuel starvation using a Pt/CNF catalyst in PEMFC. *Nanoscale Research Letters*. 2012;7:34.

[38] Huang S-Y, Li Q, Zhu Y, Fedkiw PS. Investigation and modification of carbon buckypaper as an electrocatalyst support for oxygen reduction. *Journal of Applied Electrochemistry*. 2017;47:105–115.

[39] Guerrero Moreno N, Gervasio D, Godinez Garcia A, Perez Robles JF. Polybenzimidazole-multiwall carbon nanotubes composite membranes for

polymer electrolyte membrane fuel cells. *Journal of Power Sources.* 2015;300:229–237.
[40] Wang J, Li B, Yang D, Lv H, Zhang C. Preparation of an octahedral PtNi/CNT catalyst and its application in high durability PEMFC cathodes. *Rsc Advances.* 2018;8:18381–18387.
[41] Du H-Y, Yang C-S, Hsu H-C, Huang H-C, Chang S-T, Wang C-H, et al. Pulsed electrochemical deposition of Pt NPs on polybenzimidazole-CNT hybrid electrode for high-temperature proton exchange membrane fuel cells. *International Journal of Hydrogen Energy.* 2015;40:14398–14404.
[42] Bharti A, Cheruvally G. Surfactant assisted synthesis of Pt-Pd/MWCNT and evaluation as cathode catalyst for proton exchange membrane fuel cell. *International Journal of Hydrogen Energy.* 2018;43:14729–14741.
[43] Mardle P, Ji X, Wu J, Guan S, Dong H, Du S. Thin film electrodes from Pt nanorods supported on aligned N-CNTs for proton exchange membrane fuel cells. *Applied Catalysis B: Environmental.* 2020;260.
[44] Chandran P, Puthusseri D, Ramaprabhu S. 1D-2D integrated hybrid carbon nanostructure supported bimetallic alloy catalyst for ethanol oxidation and oxygen reduction reactions. *International Journal of Hydrogen Energy.* 2019; 44:4951–4961.
[45] Louisia S, Thomas YRJ, Lecante P, Heitzmann M, Axet MR, Jacques P-A, et al. Alloyed Pt3M (M = Co, Ni) nanoparticles supported on S- and N-doped carbon nanotubes for the oxygen reduction reaction. *Beilstein Journal of Nanotechnology.* 2019;10:1251–1269.
[46] Kanninen P, Eriksson B, Davodi F, Buan MEM, Sorsa O, Kallio T, et al. Carbon corrosion properties and performance of multi-walled carbon nanotube support with and without nitrogen-functionalization in fuel cell electrodes. *Electrochimica Acta.* 2020;332.
[47] Rivera-Lugo YY, Salazar-Gastelum MI, Lopez-Rosas DM, Reynoso-Soto EA, Perez-Sicairos S, Velraj S, et al. Effect of template, reaction time and platinum concentration in the synthesis of PtCu/CNT catalyst for PEMFC applications. *Energy.* 2018;148:561–570.
[48] Kaewsai D, Lin HL, Yu TL. Influence of pyridine-polybenzimidazole film thickness of carbon nanotube supported platinum on fuel cell applications. *Fuel Cells.* 2015;15:361–374.
[49] Murata S, Imanishi M, Hasegawa S, Namba R. Vertically aligned carbon nanotube electrodes for high current density operating proton exchange membrane fuel cells. *Journal of Power Sources.* 2014;253:104–113.
[50] Meng X, Deng X, Zhou L, Hu B, Tan W, Zhou W, et al. A highly ordered hydrophilic–hydrophobic janus bi-functional layer with ultralow Pt Loading and Fast Gas/Water Transport for Fuel Cells. *Energy & Environmental Materials.* 2020;4:12–133.
[51] Tian ZQ, Lim SH, Poh CK, Tang Z, Xia Z, Luo Z, et al. A highly order-structured membrane electrode assembly with vertically aligned carbon nanotubes for ultra-low Pt loading PEM fuel cells. *Advanced Energy Materials.* 2011;1:1205–1214.
[52] Zhu S, Zheng J, Huang J, Dai N, Li P, Zheng JP. Fabrication of three-dimensional buckypaper catalyst layer with Pt nanoparticles supported on polyelectrolyte functionalized carbon nanotubes for proton exchange membrane fuel cells. *Journal of Power Sources.* 2018;393:19–31.
[53] Speder J, Zana A, Spanos I, Kirkensgaard JJK, Mortensen K, Hanzlik M, et al. Comparative degradation study of carbon supported proton exchange membrane

fuel cell electrocatalysts – The influence of the platinum to carbon ratio on the degradation rate. *Journal of Power Sources*. 2014;261:14–22.

[54] Angel GMA, Mansor N, Jervis R, Rana Z, Gibbs C, Seel A, et al. Realising the electrochemical stability of graphene: scalable synthesis of an ultra-durable platinum catalyst for the oxygen reduction reaction. *Nanoscale*. 2020; 12:16113–16122.

[55] Vinayan BP, Nagar R, Rajalakshmi N, Ramaprabhu S. Novel platinum-cobalt alloy nanoparticles dispersed on nitrogen-doped graphene as a cathode electrocatalyst for PEMFC applications. *Advanced Functional Materials*. 2012; 22:3519–3526.

[56] Antolini E, Gonzalez ER. Ceramic materials as supports for low-temperature fuel cell catalysts. *Solid State Ionics*. 2009;180:746–763.

[57] Mazza F, Trassatti S. Tungsten, titanium, and tantalum carbides and titanium nitrides as electrodes in redox systems. *Journal of The Electrochemical Society*. 1963;110:847.

[58] Borchardt L, Kockrick E, Wollmann P, Kaskel S, Guron MM, Sneddon LG, et al. Ordered mesoporous boron carbide based materials via precursor nanocasting. *Chemistry of Materials*. 2010;22:4660–4668.

[59] Grubb WT, McKee DW. Boron carbide, a new substrate for fuel cell electrocatalysts. *Nature*. 1966;210:192–194.

[60] Jackson C, Smith GT, Inwood DW, Leach AS, Whalley PS, Callisti M, et al. Electronic metal-support interaction enhanced oxygen reduction activity and stability of boron carbide supported platinum. *Nature Communications*. 2017;8:15802.

[61] Lv H, Peng T, Wu P, Pan M, Mu S. Nano-boron carbide supported platinum catalysts with much enhanced methanol oxidation activity and CO tolerance. *Journal of Materials Chemistry*. 2012;22:9155–9160.

[62] Mu S, Chen X, Sun R, Liu X, Wu H, He D, et al. Nano-size boron carbide intercalated graphene as high performance catalyst supports and electrodes for PEM fuel cells. *Carbon*. 2016;103:449–456.

[63] Heinrich B, Elina Harlin M, Pham-Huu C, Outi A, Krause I, Ledoux MJ. Characterization of the deactivation of MoO3-carbon-modified supported on SiC for n-butane dehydrogenation reaction. In: Delmon B, Froment GF, editors. *Studies in surface science and catalysis*. Elsevier; 1999. p. 163–170.

[64] Honji A, Mori T, Hishinuma Y, Kurita K. Platinum supported on silicon carbide as fuel cell electrocatalyst. 1988;135:917.

[65] Wijesundara MB, Azevedo RG. SiC materials and processing technology. In: *Silicon carbide microsystems for harsh environments*. Springer; 2011. p. 33–95.

[66] Andersen SM, Larsen MJ. Performance of the electrode based on silicon carbide supported platinum catalyst for proton exchange membrane fuel cells. *Journal of Electroanalytical Chemistry*. 2017;791:175–184.

[67] Lv H, Mu S, Cheng N, Pan MJACBE. Nano-silicon carbide supported catalysts for PEM fuel cells with high electrochemical stability and improved performance by addition of carbon. *Applied Catalysis B: Environmental*. 2010;100:190–196.

[68] Zang J, Dong L, Jia Y, Pan H, Gao Z, Wang Y. Core–shell structured SiC@C supported platinum electrocatalysts for direct methanol fuel cells. *Applied Catalysis B: Environmental*. 2014;144:166–173.

[69] Dhiman R, Johnson E, Skou EM, Morgen P, Andersen SM. SiC nanocrystals as Pt catalyst supports for fuel cell applications. 2013;1:6030–6036.

[70] Ignaszak A, Song C, Zhu W, Zhang J, Bauer A, Baker R, et al. Titanium carbide and its core-shelled derivative TiC@ TiO2 as catalyst supports for proton exchange membrane fuel cells. *Electrochimica Acta*. 2012;69:397–405.

[71] Chiwata M, Kakinuma K, Wakisaka M, Uchida M, Deki S, Watanabe M, et al. Oxygen reduction reaction activity and durability of Pt catalysts supported on titanium carbide. 2015;5:966–980.

[72] Wang S, Chu X, Zhang X, Zhang Y, Mao J, Yang Z. A first-principles study of O2 dissociation on platinum modified titanium carbide: A possible efficient catalyst for the oxygen reduction reaction. *Journal of Physical Chemistry C*. 2017;121:21333–21342.

[73] Lee Y, Ahn JH, Park H-Y, Jung J, Jeon Y, Lee D-G, et al. Support structure-catalyst electroactivity relation for oxygen reduction reaction on platinum supported by two-dimensional titanium carbide. *Nano Energy*. 2021;79:105363.

[74] Liu Y, Mustain WE. Structural and electrochemical studies of Pt clusters supported on high-surface-area tungsten carbide for oxygen reduction. *ACS Catalysis*. 2011;1:212–220.

[75] Nikolic VM, Zugic DL, Perovic IM, Saponjic AB, Babic BM, Pasti IA, et al. Investigation of tungsten carbide supported Pd or Pt as anode catalysts for PEM fuel cells. *International Journal of Hydrogen Energy*. 2013;38:11340–11345.

[76] Hassan A, Paganin VA, Ticianelli EA. Pt modified tungsten carbide as anode electrocatalyst for hydrogen oxidation in proton exchange membrane fuel cell: CO tolerance and stability. *Applied Catalysis B: Environmental*. 2015;165: 611–619.

[77] Oyama ST. Introduction to the chemistry of transition metal carbides and nitrides. In: *The chemistry of transition metal carbides and nitrides*. Springer; 1996. p. 1–27.

[78] Hugosson HW, Eriksson O, Nordström L, Jansson U, Fast L, Delin A, et al. Theory of phase stabilities and bonding mechanisms in stoichiometric and substoichiometric molybdenum carbide. *Journal of Applied Physics*. 1999;86: 3758–3767.

[79] Claridge JB, York APE, Brungs AJ, Green MLH. Study of the temperature-programmed reaction synthesis of early transition metal carbide and nitride catalyst materials from oxide precursors. *Chemistry of Materials*. 2000;12:132–142.

[80] Elbaz L, Phillips J, Artyushkova K, More K, Brosha EL. Evidence of high electrocatalytic activity of molybdenum carbide supported platinum nanorafts. *Journal of The Electrochemical Society*. 2015;162:H681.

[81] Saha S, Martin B, Leonard B, Li D. Probing synergetic effects between platinum nanoparticles deposited via atomic layer deposition and a molybdenum carbide nanotube support through surface characterization and device performance. *Journal of Materials Chemistry A*. 2016;4:9253–9265.

[82] Saha S, Cabrera Rodas JA, Tan S, Li D. Performance evaluation of platinum-molybdenum carbide nanocatalysts with ultralow platinum loading on anode and cathode catalyst layers of proton exchange membrane fuel cells. *Journal of Power Sources*. 2018;378:742–749.

[83] Zhu W, Ignaszak A, Song C, Baker R, Hui R, Zhang J, et al. Nanocrystalline tungsten carbide (WC) synthesis/characterization and its possible application as a PEM fuel cell catalyst support. *Electrochimica Acta*. 2012;61:198–206.

[84] Xiong L, Zheng L, Liu C, Jin L, Liu Q, Xu J. Tungsten carbide microspheres with high surface area as platinum catalyst supports for enhanced electrocatalytic activity. *Journal of The Electrochemical Society*. 2015;162:F468–F473.

[85] Bott-Neto JL, Beck W, Varanda LC, Ticianelli EA. Electrocatalytic activity of platinum nanoparticles supported on different phases of tungsten carbides for the oxygen reduction reaction. *International Journal of Hydrogen Energy*. 2017;42: 20677–20688.

[86] Zhao G, Zhao T, Yan X, Zeng L, Xu JJET. Ordered mesoporous carbon/titanium carbide composites as support materials for platinum catalysts. *Energy Technology.* 2016;4:1064–1070.

[87] Mirshekari GR, Shirvanian AP. A comparative study on catalytic activity and stability of TiO2, TiN, and TiC supported Pt electrocatalysts for oxygen reduction reaction in proton exchange membrane fuel cells environment. *Journal of Electroanalytical Chemistry.* 2019;840:391–399.

[88] Yang M, DiSalvo FJ. Template-free synthesis of mesoporous transition metal nitride materials from ternary cadmium transition metal oxides. *Chemistry of Materials.* 2012;24:4406–4409.

[89] Yang M, Van Wassen AR, Guarecuco R, Abruña HD, DiSalvo FJ. Nanostructured ternary niobium titanium nitrides as durable non-carbon supports for oxygen reduction reaction. *Chemical Communications.* 2013;49:10853–10855.

[90] Kreider ME, Gallo A, Back S, Liu Y, Siahrostami S, Nordlund D, et al. Precious metal-free nickel nitride catalyst for the oxygen reduction reaction. *ACS Applied Materials & Interfaces.* 2019;11:26863–26871.

[91] Li Z, Wang X, Liu J, Gao C, Jiang L, Lin Y, et al. 3D honeycomb nanostructure comprised of mesoporous N-doped carbon nanosheets encapsulating isolated cobalt and vanadium nitride nanoparticles as a highly efficient electrocatalyst for the oxygen reduction reaction. *ACS Sustainable Chemistry & Engineering.* 2020; 8:3291–3301.

[92] Tang H, Tian X, Luo J, Zeng J, Li Y, Song H, et al. A Co-doped porous niobium nitride nanogrid as an effective oxygen reduction catalyst. *Journal of Materials Chemistry A.* 2017;5:14278–14285.

[93] Yang M, Guarecuco R, DiSalvo FJ. Mesoporous chromium nitride as high performance catalyst support for methanol electrooxidation. *Chemistry of Materials.* 2013;25:1783–1787.

[94] Yang M, Cui Z, DiSalvo FJ. Mesoporous vanadium nitride as a high performance catalyst support for formic acid electrooxidation. *Chemical Communications.* 2012;48:10502–10504.

[95] Thotiyl MMO, Sampath S. Electrochemical oxidation of ethanol in acid media on titanium nitride supported fuel cell catalysts. *Electrochimica Acta.* 2011;56: 3549–3554.

[96] Kwon JA, Kim M-S, Shin DY, Kim JY, Lim D-H. First-principles understanding of durable titanium nitride (TiN) electrocatalyst supports. *Journal of Industrial and Engineering Chemistry.* 2017;49:69–75.

[97] Chi Y-M, Mishra M, Chin T-K, Liu W-S, Perng T-P. Fabrication of macroporous/mesoporous titanium nitride structure and its application as catalyst support for proton exchange membrane fuel cell. *ACS Applied Energy Materials.* 2019;2:398–405.

[98] Avasarala B, Haldar P. On the stability of TiN-based electrocatalysts for fuel cell applications. *International Journal of Hydrogen Energy.* 2011;36:3965–3974.

[99] Avasarala B, Haldar P. Durability and degradation mechanism of titanium nitride based electrocatalysts for PEM (proton exchange membrane) fuel cell applications. *Energy.* 2013;57:545–553.

[100] Zheng J, Zhang W, Zhang J, Lv M, Li S, Song H, et al. Recent advances in nanostructured transition metal nitrides for fuel cells. *Journal of Materials Chemistry A.* 2020;8:20803–20818.

[101] Avasarala B, Murray T, Li W, Haldar P. Titanium nitride nanoparticles based electrocatalysts for proton exchange membrane fuel cells. *Journal of Materials Chemistry.* 2009;19:1803–1805.

[102] Ding Z, Cheng Q, Zou L, Fang J, Zou Z, Yang H. Controllable synthesis of titanium nitride nanotubes by coaxial electrospinning and their application as a durable support for oxygen reduction reaction electrocatalysts. *Chemical Communications*. 2017;53:13233–13236.

[103] Zheng Y, Zhang J, Zhan H, Sun D, Dang D, Tian XL. Porous and three dimensional titanium nitride supported platinum as an electrocatalyst for oxygen reduction reaction. *Electrochemistry Communications*. 2018;91:31–35.

[104] Cui Z, Burns RG, DiSalvo FJ. Mesoporous Ti0.5Nb0.5N ternary nitride as a novel noncarbon support for oxygen reduction reaction in acid and alkaline electrolytes. *Chemistry of Materials*. 2013;25:3782–3784.

[105] Xiao Y, Zhan G, Fu Z, Pan Z, Xiao C, Wu S, et al. Titanium cobalt nitride supported platinum catalyst with high activity and stability for oxygen reduction reaction. *Journal of Power Sources*. 2015;284:296–304.

[106] Tian X, Luo J, Nan H, Zou H, Chen R, Shu T, et al. Transition metal nitride coated with atomic layers of pt as a low-cost, highly stable electrocatalyst for the oxygen reduction reaction. *Journal of the American Chemical Society*. 2016; 138:1575–1583.

[107] Nan H, Dang D, Tian XL. Structural engineering of robust titanium nitride as effective platinum support for the oxygen reduction reaction. *Journal of Materials Chemistry A*. 2018;6:6065–6073.

[108] Wu Z, Dang D, Tian X. Designing robust support for Pt alloy nanoframes with durable oxygen reduction reaction activity. *ACS Applied Materials & Interfaces*. 2019;11:9117–9124.

[109] Liu Q, Du L, Fu G, Cui Z, Li Y, Dang D, et al. Structurally ordered Fe3Pt nanoparticles on robust nitride support as a high performance catalyst for the oxygen reduction reaction. *Advanced Energy Materials*. 2019;9:1803040.

[110] Ioroi T, Senoh H, Yamazaki S-i, Siroma Z, Fujiwara N, Yasuda K. Stability of corrosion-resistant magnéli-phase Ti[sub 4]O[sub 7]-supported PEMFC catalysts at high potentials. *Journal of The Electrochemical Society*. 2008;155:B321.

[111] Ioroi T, Akita T, Yamazaki S-i, Siroma Z, Fujiwara N, Yasuda K. Corrosion-resistant PEMFC cathode catalysts based on a magnéli-phase titanium oxide support synthesized by pulsed UV laser irradiation. *Journal of The Electrochemical Society*. 2011;158:C329.

[112] Lori O, Elbaz LJC. Advances in ceramic supports for polymer electrolyte fuel cells. *Catalysts*. 2015;5:1445–1464.

[113] Dhanasekaran P, Vinod Selvaganesh S, Bhat SD. Nitrogen and carbon doped titanium oxide as an alternative and durable electrocatalyst support in polymer electrolyte fuel cells. *Journal of Power Sources*. 2016;304:360–372.

[114] Dhanasekaran P, Selvaganesh SV, Bhat SD. A nitrogen and cobalt co-doped titanium dioxide framework as a stable catalyst support for polymer electrolyte fuel cells. *RSC Advances*. 2016;6:88736–88750.

[115] Dhanasekaran P, Selvaganesh SV, Giridhar VV, Bhat SD. Iron and nitrogen co-doped titania framework as hybrid catalyst support for improved durability in polymer electrolyte fuel cells. *International Journal of Hydrogen Energy*. 2016;41:18214–18220.

[116] Huang S-Y, Ganesan P, Popov BN. Titania supported platinum catalyst with high electrocatalytic activity and stability for polymer electrolyte membrane fuel cell. *Applied Catalysis B: Environmental*. 2011;102:71–77.

[117] Naik KM, Higuchi E, Inoue H. Two-dimensional oxygen-deficient TiO2 nanosheets-supported Pt nanoparticles as durable catalyst for oxygen reduction reaction in proton exchange membrane fuel cells. *Journal of Power Sources*. 2020;455:227972.

[118] Hornberger E, Bergmann A, Schmies H, Kühl S, Wang G, Drnec J, et al. In situ stability studies of platinum nanoparticles supported on ruthenium–titanium mixed oxide (RTO) for fuel cell cathodes. *ACS Catalysis*. 2018;8:9675–9683.
[119] Perchthaler M, Ossiander T, Juhart V, Mitzel J, Heinzl C, Scheu C, et al. Tungsten materials as durable catalyst supports for fuel cell electrodes. *Journal of Power Sources*. 2013;243:472–480.
[120] Wickman B, Wesselmark M, Lagergren C, Lindbergh G. Tungsten oxide in polymer electrolyte fuel cell electrodes—A thin-film model electrode study. *Electrochimica Acta*. 2011;56:9496–9503.
[121] Liu Y, Shrestha S, Mustain WE. Synthesis of nanosize tungsten oxide and its evaluation as an electrocatalyst support for oxygen reduction in acid media. *ACS Catalysis*. 2012;2:456–463.
[122] Saha MS, Zhang Y, Cai M, Sun X. Carbon-coated tungsten oxide nanowires supported Pt nanoparticles for oxygen reduction. *International Journal of Hydrogen Energy*. 2012;37:4633–4638.
[123] Yang C, Zhou M, Zhang M, Gao L. Mitigating the degradation of carbon-supported Pt electrocatalysts by tungsten oxide nanoplates. *Electrochimica Acta*. 2016;188:529–536.

2 Recent Advances in Low PGM for Fuel Cell Electrocatalysis

Junbo Hou
Institute of Fuel Cells, School of Mechanical Engineering, Shanghai Jiao Tong University, Shanghai, People's Republic of China

Min Yang
Central Research Institute, Shanghai Electric Group, Zhabei District, Shanghai, People's Republic of China

CONTENTS

2.1 INTRODUCTION

Proton exchange membrane (PEM) fuel cells attract continuous attention all over the word for last three decades due to their high power conversion efficiency, environmentally friendly emission and technically challenging issues. Fuel cell vehicles have been offered to the market by Japanese and Korean first lines of OEMs, spurred by a wave of commercialization of fuel cell vehicles that has been widely spread in China. Despite this pre-marketization activity, the large scale of PEM fuel cell deployment is still cumbered by the initial high cost and the durability of key components.

DOI: 10.1201/9781003133971-2

Thanks to Los Alamos national lab's work [1,2] on the thin film catalyst layer fabrication, the carbon supported Pt nanoparticles blended with Nafion ionomer were used in the catalyst layer, from which the fuel cell performance was improved and thus the mass specific activity of Pt was significantly increased. After that, lots of efforts have been devoted to increase cell performance, reduce cost, and increase the durability [3–7]. According to DOE 2020 target, platinum group metal (PGM) loading should not be higher than 0.125 g/kW [8], which gives MEA area loading of 0.125 mg/cm^2, assuming the rated power is 1 W/cm^2. For a 100 kW fuel cell vehicle, the PGM loading is 12.5 g and this will compete with the traditional internal combustion engine (ICE) which usually contains about 5–10 g PGM for the exhaust treatment. At such loading level, the fuel cell stack components might be within the optimum cost range based on Strategic Analysis Inc. [9].

In this review, the fuel cell electrocatalysis with low PGM loading is discussed especially fundamentals and basics behind are comprehensively explored. Starting from the catalysts, many efforts to decrease the Pt loading and improve the mass specific activity, and the reasons why the catalysts are designed including the size effect, morphology and structure effects, ligand and strain effects are summarized. And then, the state-of-the-art carbon supports include the basics on the Pt-support interaction, the reason for the catalyst particle distribution containing nucleation, carbon surface properties and the pore structure, and the agglomeration-aggregates-diffusion are discovered. Finally, the ionomer as the binder and proton provider in the catalyst layer, the catalyst layer structure in the presence of ionomer, ink processing and state-of-the-art deposition techniques, and the drying process are discussed.

2.2 WHY PT-BASED CATALYSTS ARE THE BEST FOR ORR

Based on the simple dissociation mechanism, which is O_2 adsorbed on a catalysts surface, followed by one electron charge transfer step by step forming H_2O, the binding energies of the intermediates at each step can be calculated by Gibbs free energies [10]. Thus, the ORR activity can be plotted as a function of O or OH binding energy, which gives the well-known "volcano"-type plot, as shown in Figure 2.1. For the metals showing smaller binding energy than Pt, like Ni, they bind O or OH strongly and thus the proton transfer step becomes slow. For Au, which has a larger binding energy than Pt, O or OH binds on the surface loosely; thus, almost no transfer of protons and electrons to oxygen can occur. This is the main reason for Pt at the top of the volcano. As for the Pt-based alloys, there exists the same volcano trend if the d-band center is used [11]. Similar to the dissociation energy for the single metal catalysts, two opposing effects counterbalance Pt alloy catalysts: a relatively strong adsorption energy of O_2 and reaction intermediates, and a relatively low coverage by adsorbed anions. Thus, for Pt that binds O_2 too strongly, the rate of the ORR is limited by the availability of OH_{ad}/anion-free Pt sites. When the d-band center is too far from the Fermi level, as in the case of Pt_3V and Pt_3Ti, the surface is covered less by OH_{ad} and anions, but the adsorption energy of O_2 is too low and limits the ORR rate.

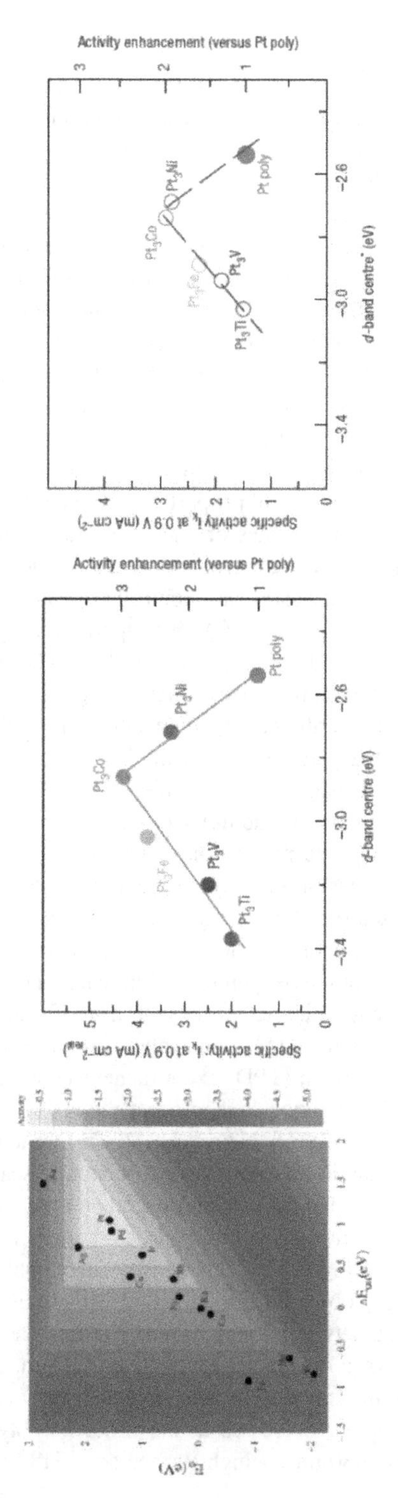

FIGURE 2.1 Trends in oxygen reduction activity plotted as a function of both the O and the OH binding energy [10]; relationships between experimentally measured specific activity for the ORR on Pt3M surfaces versus the d-band center position for the Pt-skin (a) and Pt-skeleton (b) surfaces [11].

2.2.1 Size Effect

Going to nano scale not only makes Pt or Pt alloys particles evenly distribute conductive supports but also provides more available geometrical surface area, and thus possibly reduces PGM loading. In general, the electrochemical active surface area of Pt or Pt alloys in the liquid electrolytes proportionally increases with the total geometrical surface area. Now the question arises: How will the particle size of Pt or Pt alloys affect the ORR? Or, in other words, how the activity of the catalysts correlates to the particle size. Here the activity of the catalysts can be classified as specific activity and mass activity [12]. The former one is the activity per the real surface area of the catalysts, while the latter is the activity per mass, which is important from the cost point of view. Referring back to the nano scale, the particle undergoes the decrease of the composed atoms, and the limited atoms stack together that show specific crystal facets, terraces, and edges. This can be called geometric factors, associated with the topography of atom distribution of the catalyst particles. The second factor is the electrochemical properties of the catalysts' topography, i.e., adsorption-desorption of reactant species on facets, terraces, and edges, which can be called the electronic factors, basically related to the surface electronic structure.

By using the CO displacement charge at a controlled potential method, it is found that the potentials of total zero charge (PTZC) shifts approximately 35 mV negative by decreasing the particle size from 30 nm down to 1 nm [13]. The surface coverage with OH increases by decreasing the particle size, which was demonstrated at the CO bulk oxidation. And thus the ORR regarding the specific activity is hindered after decreasing the catalysts' size. Further, the parallel shift in the Tafel plots found by the authors indicates the same reaction mechanism on the different Pt catalysts. Later, the highest mass activity of ORR at 2.2 nm was found in $HClO_4$ solutions on the Pt particles in the range of 1–5 nm [14]. Because decreasing the edge sites become a majority in the catalyst particles, and these sites show very strong oxygen-binding energies, the Pt nanoparticles exhibit decreased specific activity. The {111} facets contribute to the high activity observed on the 2.2 nm Pt particle due to a proper oxygen-binding energy. Similarly, the specific activity of the oxygen reduction reaction on Pt nanoparticles was found to decrease with decreasing particle size, with a maximum in mass activity for particles with a diameter of 3 nm [15]. The authors implemented the vacuum CO temperature programmed desorption (TPD) experiments to correlate the proportion of the terrace sites with the ORR activity. It seems that the active sites for the ORR are only located on the terrace sites of the nanoparticles. To further understand particle size effects on the activity of catalytic nanoparticles in an atomic-scale description of the surface microstructure, a model particle was introduced and the fractional population n of the (100), (111) and step surface sites of the particle model was correlated with the particle size [16], as shown in Figure 2.2. The density functional theory (DFT) model reproduces the experimentally observed trends in both the specific and mass activities for particle sizes in the range between 2 and 30 nm. The mass activity is calculated to be maximized for particles of a diameter between 2 and 4 nm. In [17], the effect of particle size on ORR was investigated in acid and alkaline mediums. The specific activity toward the ORR rapidly decreases in the order of polycrystalline Pt > unsupported Pt black particles (~30 nm) > high surface area (HSA) carbon-supported Pt

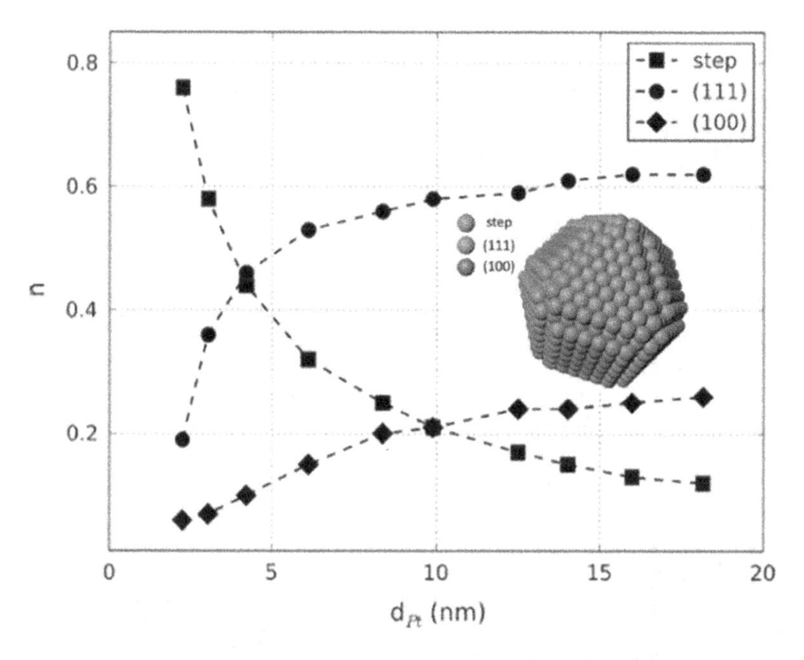

FIGURE 2.2 Fractional population n of the (100), (111) and step surface sites of the particle model; insets: Model particle with truncated octahedral shape with dissolved edges and corners. Different-colored atoms correspond to active sites of different activity for the oxygen reduction reaction [16].

nanoparticle catalysts (of various size between 1 and 5 nm) in all three mediums, and the absolute reaction rates decrease in the order $HClO_4 > KOH > H_2SO_4$ in line with an increasing anionic adsorption strength in the case of the acid solutions. Simulation results show a maximum mass activity at 2.4 nm, as shown in Figure 2.2.

Based on the previous discussion, it can be concluded that the size effect with specific activity increases with increasing catalyst particle diameter, and the maximum mass activity appears within 2–4 nm. However, the particle size effect begins to lose its clarity as catalyst ages, where Pt dissolution initiated by the formation of irreversible surface oxides results in dissolution/deposition and a broadening of the particle size distribution: structural evolution under moderate reaction conditions yielded particles of a round geometry, as shown in Figure 2.3 [18]. The authors suggested mono-dispersed 7 nm cubo-octahedral Pt NPs would be recommended as a trade-off between initial mass activity and durability performance. This result is confirmed by others [19]. It suggests smaller particles (2.2 and 3.5 nm) undergoes dissolution/deposition process during electrochemical cycling giving broadening of the size distribution, while larger particles (5.0, 6.7, and 11.3 nm) appear to be stable even after 10,000 cycles. However, later research found Pt 2 nm/CB maintained the highest mass activity after 30,000 cycles, and implies the carbon support, narrow particle size distribution (10%) and evenly dispersed over the carbon surface are very important to support the stability of the Pt NPs [20].

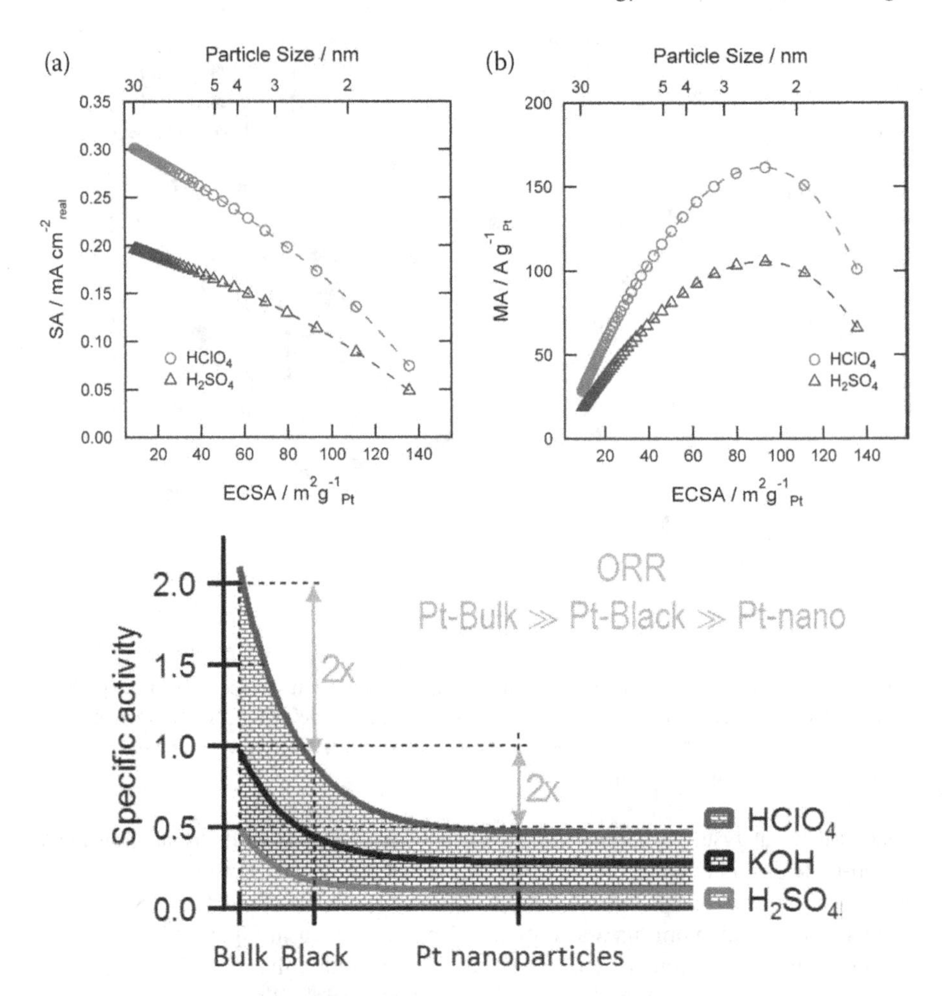

FIGURE 2.3 Simulation of expected SA (a) and MA (b) at 0.9 VRHE as a function of the ECSA for perchloric and sulfuric acid solutions. It is assumed that only (100) sites are active and that 30% of the particle is covered by the support [17].

2.2.2 Morphology and Structure Effects

2.2.2.1 Core-Shell Structures

Driven by the high activity and durability, the material structure at the nano scale has been explored for Pt-bimetallic catalysts; for example, core-shell structures show the superior electrocatalytic performance for the ORR. The core shell is a general structure describing the Pt-enriched surface of the catalyst particle. Actually, it can be classified into three situations: Pt skin, Pt mono layer, and Pt nano porous skeleton, as shown in Figure 2.4 [21]. The synthesis methods have been summarized in [12,22] as dealloying/leaching, thermal and adsorbate segregation, and deposition on seeds for all those three kinds of core-shell structures.

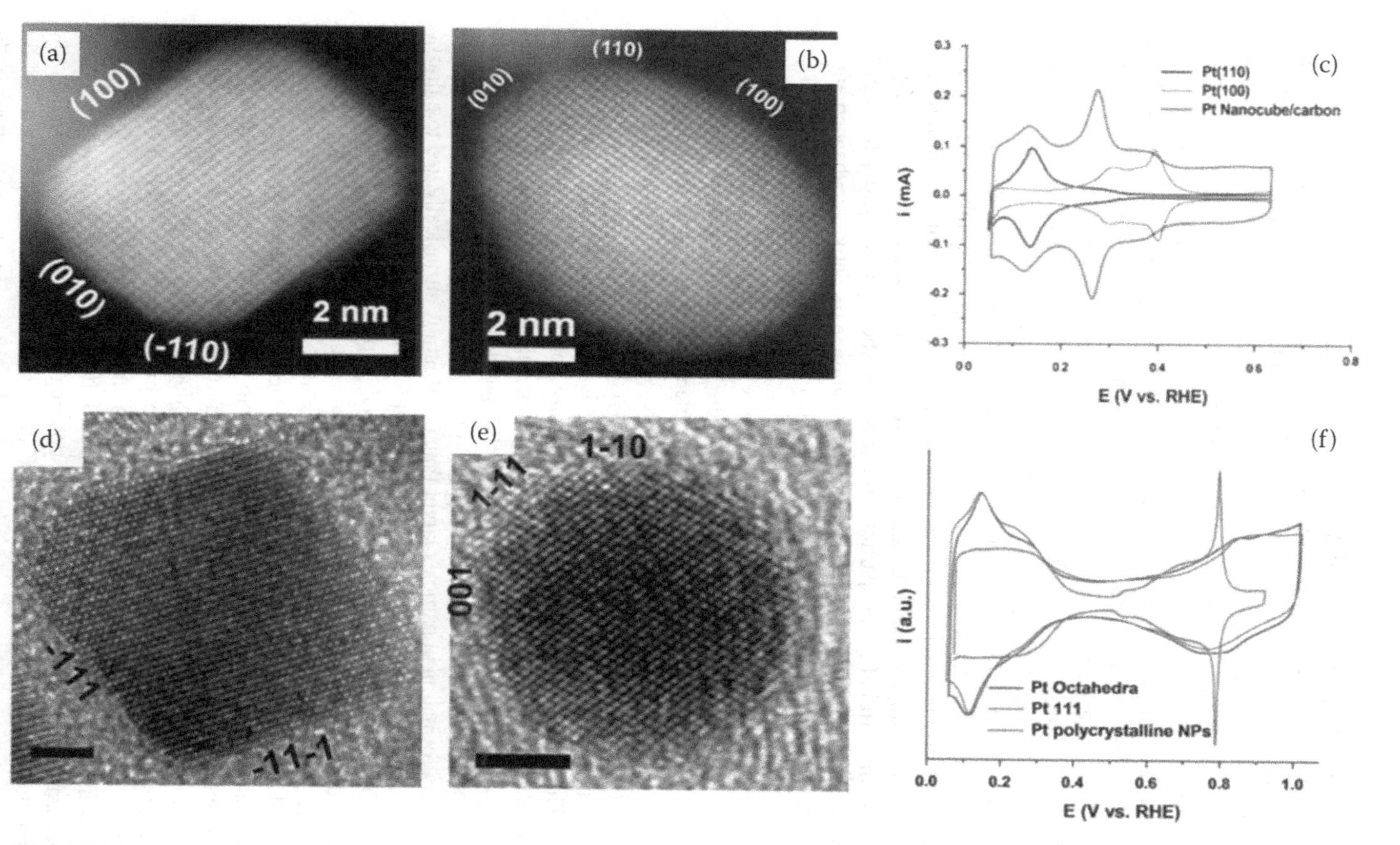

FIGURE 2.4 STEM images of Pt nanocube (a) before and (b) after potential cycling, (c) cyclic voltammograms of Pt nanocubes (red line) and corresponding Pt (110) (blue line), Pt(100) (green line) surfaces from 0.5 M H2SO4; HRTEM images of Pt nano-octahedron (d) before and (e) after potential cycling, (f) cyclic votammograms of Pt nano-octahedrons (blue line) compared with Pt polycrystalline nanoparticles (green line) and Pt(111) (red line) from 0.1 M HClO4 [18].

2.2.2.2 Pt Skin

Two different polycrystalline alloys of Pt_3Ni and Pt_3Co with 75% Pt and 100% Pt on the sufaces were prepared in ultrahigh vacuum (UHV). The latter is a "Pt-skin" structure and is produced by an exchange of Pt and Co in the surface layers [23]. It was found that the activity enhancement for the "Pt-skin" on Pt_3Co in 0.1 M $HClO_4$ being 3–4 times that for pure Pt. The reason why such enriched Pt-skin structure shows superior performance is that the alloying Pt with 3d transition metals tunes the electronic structure of catalyst surfaces. The activity correlates well with the strength of the oxygen–metal bond interaction, which in turn depends on the position of the metal d states relative to the Fermi level [24]. This is now a well-known "volcano"-type activity graph. The principle is searching surfaces that bind oxygen a little weaker than Pt. Specifically for Pt skins, Pt d state must shift downward, thus giving improved ORR performance. To further advance the ORR activity, Pt-ternary alloys ($Pt_3(MN)_1$ with M, N = Fe, Co, or Ni) as electrocatalysts was studied. It indicates that Pt-ternary alloys achieve higher catalytic activities than bimetallic Pt alloys and improvement factors of up to four versus monometallic Pt [25]. Multimetallic Au/$FePt_3$ nanoparticles possess both the high catalytic activity of Pt-bimetallic alloys and the superior durability, with mass-activity enhancement of more than one order of magnitude over Pt catalysts [26].

2.2.2.3 Dealloyed-Nano Porous

Prolonged exposure to reaction conditions, the Pt-bimetallic catalyst with multi-layered Pt-skin surface exhibited an improvement factor of more than one order of magnitude in activity versus conventional Pt catalysts [21]. Strictly, the Pt-skin surface here is actually electrochemical dealloyed nano porous structure. Generally, the enhanced catalytic activity and durability is from the increased active surface area. This roughing or even the nanoporosity formation on the single catalyst particle by leaching or electrochemical dealloying will also possibly cause the activity and durability loss depending on the particle size, the composition of the outlayers of the particle, and the structure after dealloying. It is uncovered that nanoporosity formation in particles larger than ca. 10 nm is intrinsically tied to a drastic dissolution of Ni and, as a result of this, a rapid drop in intrinsic catalytic activity during ORR testing, translating into severe catalyst performance degradation. A O_2-free acid leaching can suppress nanoporosity. The authors suggest that catalytic stability could further improve by controlling the particle size below ca. 10 nm to avoid nanoporosity [27]. Specifically, the particle size, the dealloying protocol and post-acid-treatment annealing was investigated to correlate with the nanoporosity and passivation of the alloy nanoparticles. It was found that smaller size, less-oxidative acid treatment, and annealing significantly reduced Ni leaching and nanoporosity formation resulting in improved stability and higher catalytic ORR activity [28]. Another important research topic in this area is clarifying the dealloying of the facets and its influence on ORR. It was demonstrated that the $Pt_3Ni(111)$ surface is tenfold more active for the ORR than the corresponding Pt(111) surface and 90-fold more active than the current state-of-the-art Pt/C catalysts for PEMFC [29]. Because of the electronic structure (d-band center position) and arrangement of surface atoms in the near-surface region, the nonreactive oxygenated species interact weekly with the Pt surface atoms, and thus increasing the number of active sites for O_2 adsorption.

2.2.2.4 Pt-ML

An effective strategy to reduce the Pt content while retaining the activity of a Pt-based catalyst is to deposit a few atomic layers of Pt atoms on top of nanoscale substrates. Through this strategy, the available active surface area of Pt catalysts is maintained but the total mount can dramatically decrease. Further, if the outmost top layers can be limited as a few atomic layers, the activity of such a layer for ORR may be significantly differ from the bulk materials. Pt monolayer on Au(111), Rh(111), Pd(111), Ru(0001), and Ir(111) surfaces was synthesized and investigated regarding ORR activity [30]. By experimental data and density functional theory (DFT) calculations, ORR shows a volcano-type dependence on the center of their d-bands. Pt-ML/Pd (111) shows improved ORR due to facilitated O-O bond breaking and hydrogenation. Later on, the same group used the same core Pd (111) and varied the Pt monolayer composition by adding a transition metal (Ir, Ru, Rh, Re, or Os), by which a mixed monolayer of Pt with the transitional metals was formed [31]. Some of these catalysts exhibit very high activity, i.e., 20-fold increase in a Pt mass-specific activity. The reason that the catalyst shows superior activity and stability is explained as a low-OH coverage on Pt due to the spacing limitation, in detail, the lateral repulsion between the OH adsorbed on Pt and the OH or O adsorbed on neighboring, other than Pt, late transition metal atoms limit OH coverage on Pt. Based on the Pt-monolayer depositions on Pd and Pd3Co nanoparticles, the effects of the thickness of the Pt shell, lattice mismatch, and particle size on specific and mass activities were studied. It was found that the enhancements in specific activity are largely attributed to the compressive strain effect revealing the effect of nanosize-induced surface contraction on facet-dependent oxygen binding energy [32]. The authors also suggested that moderately compressed (111) facets are most conducive to an oxygen reduction reaction on small nanoparticles. Based on the Pd nanocubes, the number of layers, i.e., 1–6 atomic layer of Pt was controlled and deposited on top of them Pd@PtnL (n = 1–6) [33]. Both theoretical and experimental studies indicate that the ORR specific activity was maximized for the catalysts based on Pd@Pt2–3L nanocubes. Because of the reduction in Pt content used and the enhancement in specific activity, the Pd@Pt1L nanocubes showed a Pt mass activity with almost a threefold enhancement relative to the Pt/C catalyst. The DFT calculations on model (100) surfaces suggest that the enhancement in specific activity can be attributed to the weakening of OH binding through ligand and strain effects, which, in turn, increases the rate of OH hydrogenation. Although Pt-ML nanocatalysts have demonstrated potential as highly active low-Pt fuel cell cathodes for the oxygen reduction reaction (ORR) challenges remain in optimizing their surface and interfacial structures, which often exhibit undesirable structural degradation and poor durability; 3.5–5 nm of Pd and Pd_9Au_1 alloy core/Pt monolayer was developed and no loss of platinum was observed in 200,000 potential cycles [34]. From the interaction of Pt with Pd either coordinated (Pd solid) or not well coordinated (Pd mono layer), and their dissolution potentials ($U_{Pd}{}^0$ = 0.92 V vs $U_{Pt}{}^0$ = 1.19 V) at different particle size and model structure, the stability is prove to derive the core protecting the shell from dissolution, as shown in Figure 2.5. Later, by the same group adding a small amount of gold to

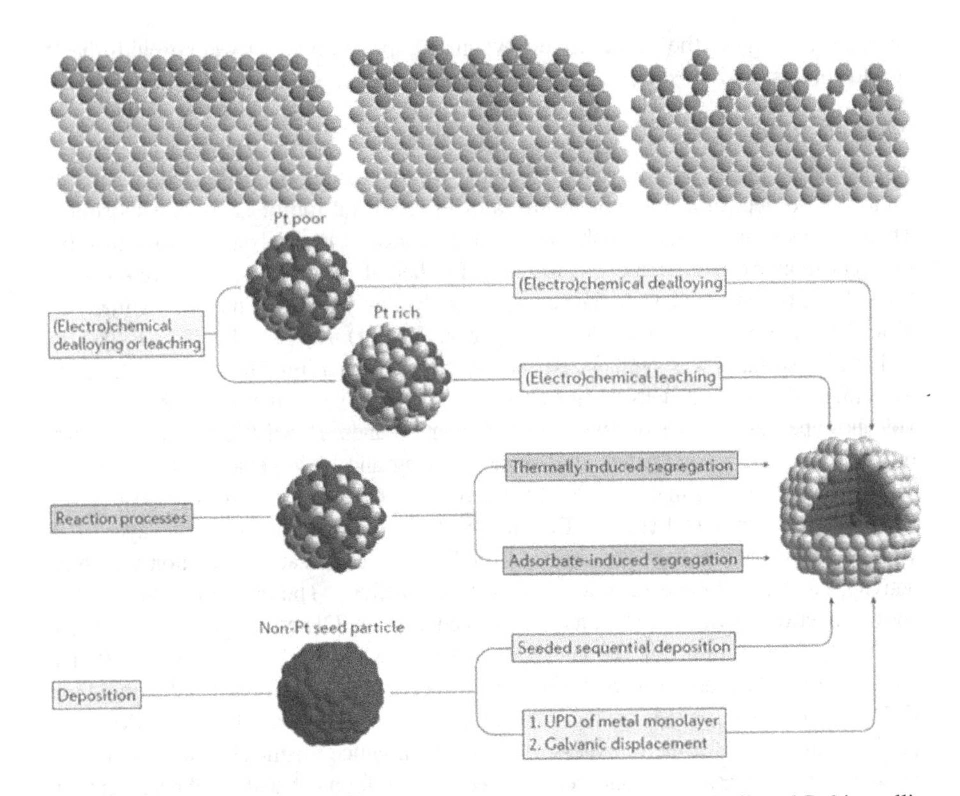

FIGURE 2.5 Core-shell structures: Pt skin, Pt mono layer, and Pt dealloyed Pt-bimetallic nanoparticles, modified from [21]; synthesis of Pt-skin, Pt-monolayer, and dealloyed Pt-bimetallic nanoparticles with a core-shell structure [12,22].

palladium and forming highly uniform nanoparticle cores, no marked losses in platinum and gold despite the dissolution of palladium was observed after 100,000 cycles between potentials 0.6 and 1.0 V or even more severe conditions with a potential range of 0.6–1.4 V [35]. Besides the influence of the core composition and structure on the Pt-ML catalyst durability, the outmost monolayer may also form an interphase with the intermediate layer together with the substrate material. An unsupported nanoporous catalyst with a Pt–Pd shell of sub-nanometer thickness on Au was synthesized and demonstrated as stable over a 100,000 cycles [36]. It is revealed to be an atomic-scale evolution of the shell from an initial Pt–Pd alloy into a bilayer structure with a Pt-rich trimetallic surface, and finally into a uniform and stable Pt–Pd–Au alloy, as shown in Figure 2.5.

2.2.2.5 Polyhedron Facets

It has been verified and widely accepted that the ORR activity highly relies on Pt single-crystal surface facets and it follows an order of structure-dependent ORR activity: {100} <{111} <{110} [37]. After the transitional metals were introduced to form Pt-based bimetal, the ORR activity of the catalysts has been further boosted, i.e., PtNi and PtCo on the top of the "volcano"-based d-band center position of Pt relative to Fermi level. Furthermore, these bimetallic Pt-based catalysts also show surface facet preference or superiority for ORR, i.e., Pt_3Ni {111}) demonstrates an ORR activity in $HClO_4$ in the order of {100} < {110} < {111} [11,24,29]. And therefore, such nanoparticles with shape control, specifically polyhedron facets, as shown in Figure 2.6 [38], have been synthesized and investigated regarding ORR. Three octahedral Pt_xNi_{1-x} alloy nanoparticle electrocatalysts were fabricated featuring a Pt-rich frame along their edges and corners, whereas their Ni atoms preferentially segregated in their {111} facet region, as shown in Figure 2.6 [39]. The octahedra preferentially leach in their facet centers and evolve into "concave octahedra". Later, the mechanism of such anisotropic growth of the octahedral and the evolution of the structure change after the Ni leaching was interpreted, is shown in Figure 2.6. In particularly, a Pt-rich phase evolves into precursor nanohexapods, followed by a slower step-induced deposition of an M-rich (M = Ni, Co, etc.) phase at the concave hexapod surface forming the octahedral facets [40]. Although PtNi octahedra represent an emerging class of electrocatalysts, the durability cannot meet the practical use in PEM fuel cells. By doping Pt_3Ni octahedra with transition metals, including vanadium, chromium, manganese, iron, cobalt, molybdenum (Mo), tungsten, or rhenium, the ORR activity was characterized [41]. The Mo-doped Pt3Ni/C showed the best ORR performance, with a specific activity of 10.3 mA/cm^2 and mass activity of 6.98 A/mg Pt, which are 81- and 73-fold enhancements compared with the commercial Pt/C catalyst (0.127 mA/cm^2 and 0.096 A/mg Pt). Theoretical calculations suggest that Mo prefers subsurface positions near the particle edges in a vacuum and surface vertex/edge sites in oxidizing conditions, where it enhances both the performance and the stability of the Pt_3Ni catalyst, as shown in Figure 2.6.

2.2.2.6 Nanostructure

2.2.6.1 Nanocage

Designing a hollow structure of a Pt catalyst offers a great opportunity to enhance the electrocatalytic performance and maximize the use of precious Pt. Hollow structures can be obtained from four different methods: heteroepitaxial growth on templates which is usually with flat facets, i.e., cubic template, non-epitaxial/ random growth which is generally on a sphere template, galvanic replacement and Kirkendall effect [42]. Based on the Kirkendall effect, compact and smooth Pt hollow nanocrystals were fabricated and exhibited an enhancement in Pt mass activity for ORR [43]. The ORR improvement is ascribed by the authors to the hollow-induced lattice contraction, high surface area per mass, and oxidation resistant surface morphology. By heteroepitaxial growth of Pt layers on Pd templates, Pt nanocages with {100} (cubic nanocages) and {111} (octahedral nanocages) facets, shown in Figure 2.7, which exhibited distinctive catalytic activities toward oxygen reduction [44]. By using the similar method, Pt-based icosahedral nanocages with {111} facets and six atomic layers of Pt atoms were synthesized, as shown in Figure 2.7 and the catalysts show a specific activity of 3.50 mA cm^{-2}, larger than those of the Pt-based octahedral nanocages (1.98 mA cm^{-2}) and a state-of-the-art commercial Pt/C catalyst (0.35 mA cm^{-2}). After 5,000 cycles of

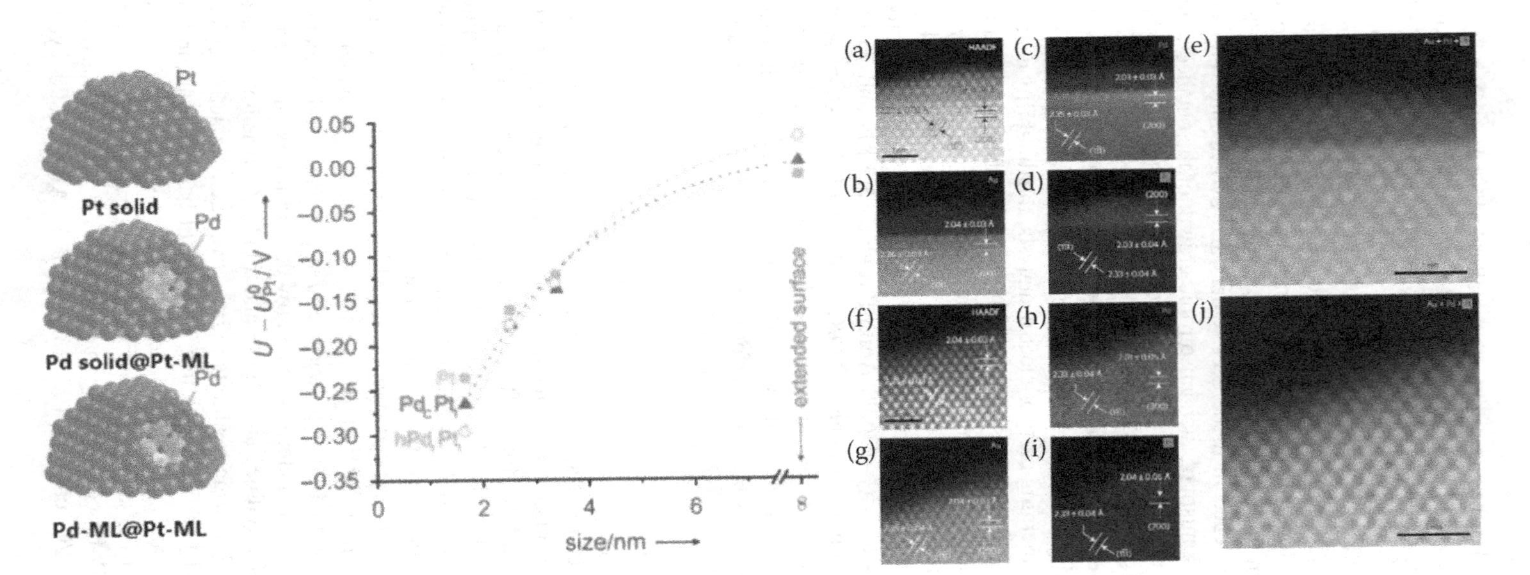

FIGURE 2.6 Surface models for the nanoparticles and predicted dissolution potentials of a 1 ML Pt shell from nanoparticles and extended surfaces of Pt, PdCPt1, and hPd1Pt1 as a function of particle sizes [34]; Atomically resolved elemental mapping of the surfaces of NPG–Pd–Pt10,000 and NPG–Pd–Pt30,000 electrocatalysts. (a) High-resolution HAADF image of a surface region of NPG–Pd–Pt10,000. (b–d) Atomically resolved elemental mapping images of a using the EDS signals of Au, Pd, and Pt, respectively [36].

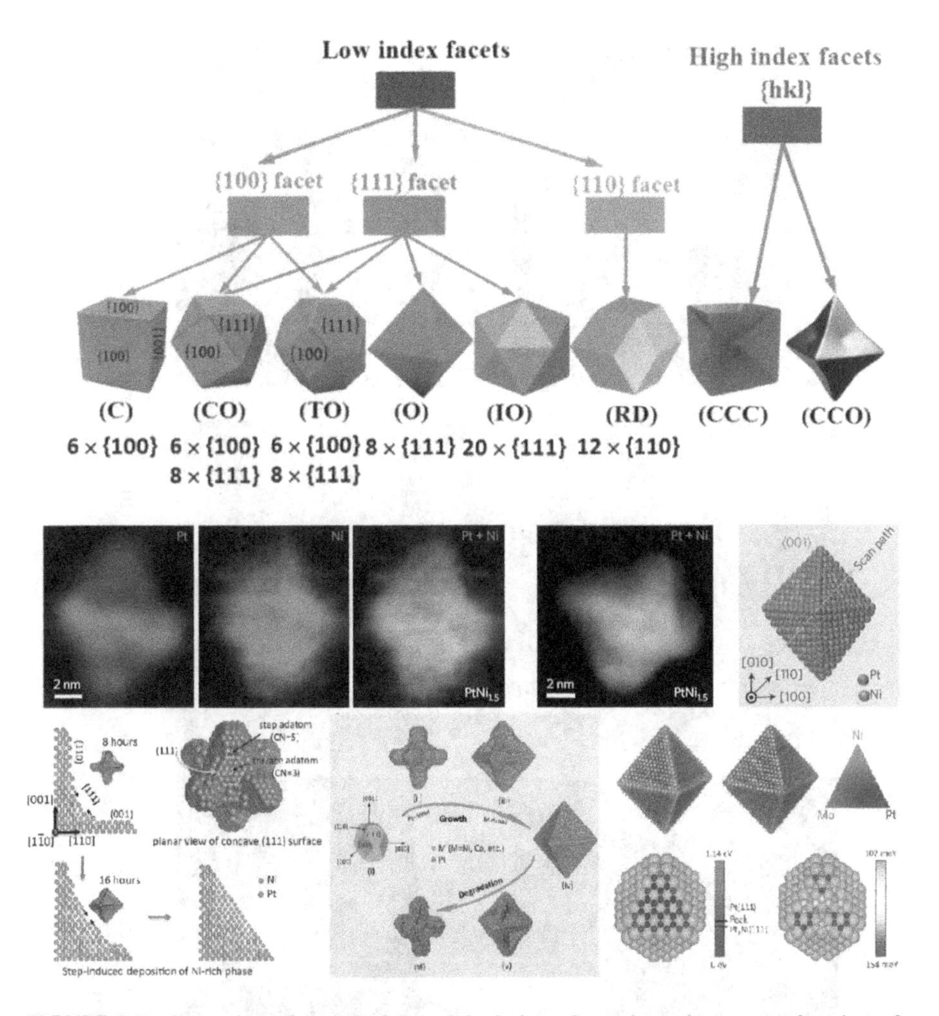

FIGURE 2.7 llustration of various 3-D polyhedral configurations shown as a function of low index {100}, {111}, and/or {110} facets and high index {hkl} facets [notes: C = cube, CO = cuboctahedron, TO = truncated octahedron, O = octahedron, IO = icosahedron, RD = rhombic dodecahedron, CCC = concave cube, CCO = concave octahedron] [38]; atomic-scale Z-contrast STEM images and composition profile analysis of $PtNi_{1.5}$ octahedral nanoparticles [39]; atomic structural models of octahedral Pt bimetallic alloy NCs (Pt-M; M = Ni, Co, etc.) during the solution-phase co-reduction and during the acidic ORR electrocatalysis [40]; Mo doping at the edge and corner of the octahedron of PtNi particles by Monte Carlo simulation, binding energies for a single oxygen atom on all fcc and hcp sites on the (111) facet and the change of binding energies once adding Mo [41].

accelerated durability test, the mass activity of the Pt-based icosahedral nanocages drops from 1.28 to 0.76 A mg^{-1} Pt, which is still about four times greater than that of the original Pt/C catalyst (0.19 A mg^{-1} Pt), as shown in Figure 2.8 [45]. Ultrathin icosahedral Pt-enriched nanocages were fabricated using Pd icosahedral seeds [46]. This catalyst shows extraordinary ORR activity than the commercial Pt/C, with a

FIGURE 2.8 TEM and HAADF-STEM of Pt cubic and octahedral nanocages [44]; (a-g) TEM and (b) low-magnification HAADF-STEM images and EDX mapping of the Pt icosahedral nanocages [45].

10 times higher specific activity and 7 times higher mass activity, outperforming the cubic and octahedral nanocages reported above.

2.2.6.2 Nanoframe

Starting from the crystalline $PtNi_3$ polyhedra, the edges of the Pt-rich $PtNi_3$ polyhedra are maintained in the final Pt_3Ni nanoframes after immersing in corrosion medium [47]. The nanoframe catalysts achieved a factor of 36 enhancement in mass activity and a factor of 22 enhancement in specific activity. It is that both the interior and exterior catalytic surfaces of this open-framework structure are composed of the nanosegregated Pt-skin structure exhibits enhanced oxygen reduction reaction (ORR) activity. Using a facet-controlled Pt@Ni core–shell octahedron nanoparticle, the nanoscale phase segregation can have directionality and be geometrically controlled to produce a Ni octahedron that is penetrated by Pt atoms along three orthogonal Cartesian axes and is coated by Pt atoms along its edges. The selective removal of the Ni-rich phase by etching then results in structurally fortified Pt-rich skeletal PtNi alloy framework nanostructures. Electrochemical evaluation of this hollow nanoframe suggests that the oxygen reduction reaction (ORR) activity is greatly improved compared to conventional Pt catalysts [48]. By using the hydrothermal method, PtCu NFs with nanothorns protruding from their edges were synthesized [49]. Pure Cu nanodecahedra are first formed, and then the galvanic replacement reaction happens between Cu nanodecahedra and Pt precursors. Finally the co-deposition of Pt and Cu atoms are responsible for the formation of highly anisotropic five-fold-twinned PtCu NFs, which shows impressive ORR activity.

2.2.6.3 Nanowires

By three steps, Pt/NiO core/shell nanowires solution-synthesized was first synthesized, and then they were converted into PtNi alloy nanowires through a thermal annealing process. Finally, the jagged Pt nanowires were obtained by electrochemical dealloying [50]. The jagged nanowires exhibit an ECSA of 118 m^2/g_{Pt} and a specific activity of 11.5 mA/cm^2 for ORR, yielding a mass activity of 13.6 A/mg_{Pt}, as shown in Figure 2.9. This is the second-highest ORR activity in the precious report except PtNi {111}. It is suggested that highly stressed, undercoordinated rhombus-rich surface configurations of the jagged nanowires enhanced ORR activity versus more relaxed surfaces.

2.2.3 Ligand and Strain Effects

Introducing heterogeneous atoms into Pt metal forms bimetallic or multimetallic Pt-based catalysts will result in electronic change in Pt metal, thus modifying the chemical properties of catalyst surfaces. Two critical effects contribute to the modification of the electronic properties of a metal in a bimetallic surface. The first one is a strain effect or geometric effect (shown in Figure 2.10), which is usually caused by the atomic arrangement of surface atoms and it generally includes compressed or expanded arrangements of surface atoms. The second one is a ligand effect or electronic effect (Figure 2.10), which is introduced by the atomic vicinity

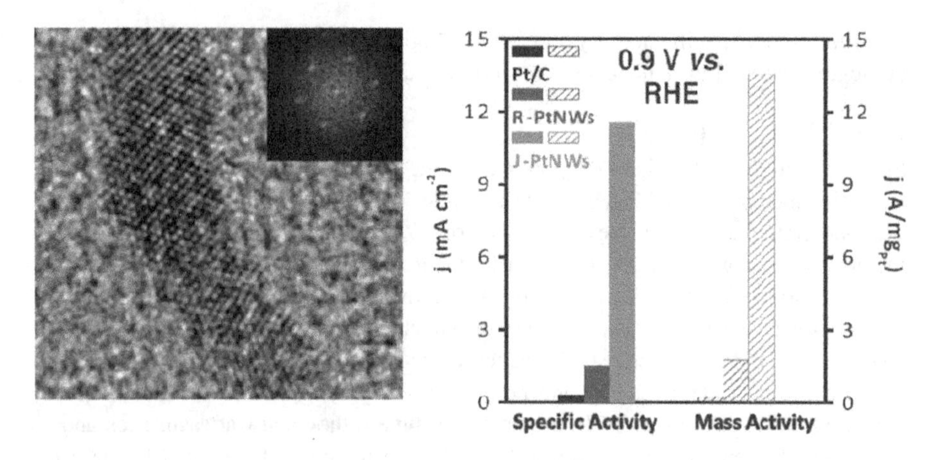

FIGURE 2.9 HRTEM characterization of the J-PtNW structure, and the ORR activity comparison [50].

of two dissimilar surface metal atoms and it generally involves electron transfer between the two metal atoms. Both effects affect the electronic band structure of the Pt-based catalysts. By DFT calculation and not considering the strain effect, it was found that the Pt surface d-band was broadened and lowered in energy by interactions with the subsurface 3d metals, resulting in weaker dissociative adsorption energies of hydrogen and oxygen on these surfaces. The magnitude of the decrease in adsorption energy was largest for the early 3d transition metals and smallest for the late 3d transition metals [51]. After that, the same authors showed how the combination of strain and ligand effects modify the electronic and surface chemical properties of Ni, Pd, and Pt monolayer supported on other transition metals [52]. Strain and the ligand effects are shown to change the width of the surface d band, which subsequently moves up or down in energy to maintain a constant band filling. By tuning the core-shell structure and composition of the catalyst, the strain effect was experimently demonstrated [53]. The platinum-rich shell exhibits compressive strain, which results in a shift of the electronic band structure of platinum and weakening chemisorption of oxygenated species. Pt_5La is firstly synthesized and demonstrated with a 3.5- to 4.5-fold improvement in activity over Pt in the range 0.9 to 0.87 V, the strain and ligand effects were used to understand the activity of Pt_5La [54] (Figure 2.10).

2.3 CONCLUSIONS

The state-of-the-art electrode structure in PEM fuel cells consists of the carbon-supported Pt-based catalysts and ionomer. Although it is already a very thin catalyst layer due to the carbon support, it still requires the ionomer to enforce the 3d structure. For this structuring, the Pt-based catalysts have to go down to nanosize, and the optimum size 2–4 nm gives the highest mass specific activity. To further decrease the catalyst cost, the core-shell, Pt skin, de-alloyed, Pt monolayer, polyhedron facets are the main focus of research. Ligand and strain effects might need further exploration and

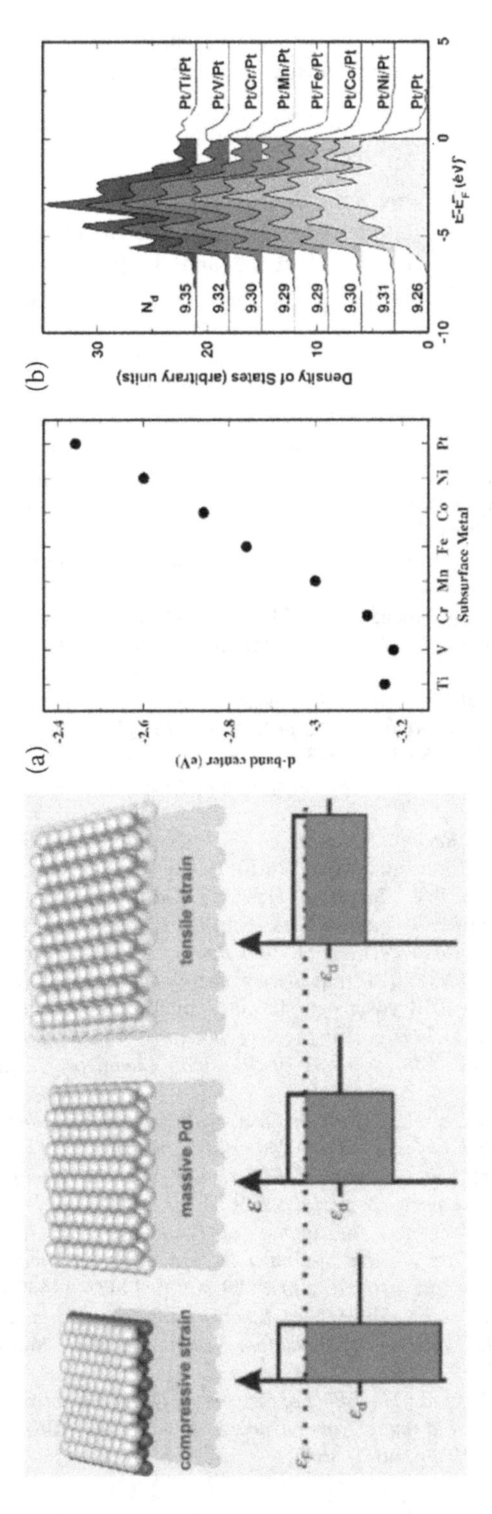

FIGURE 2.10 Strain effect and its influence on d band [55]; ligand effect with early and later transition metals [51].

could be applied to design such binary or triple Pt-based catalysts. Another concern besides the activity is the durability, which will be a critical topic.

ACKNOWLEDGMENT

This work is supported by the National Natural Science Foundation of China (21503134, 21406220), the National Key Research and Development Program of China (2016YFB0101201), and the Science and Technology Commission of Shanghai Municipality (15YF1406500).

REFERENCES

[1] Wilson, M.S. and S. Gottesfeld, *Thin Film Catalyst Layers for Polymer Electrolyte Fuel Cell Electrodes. Journal of Applied Electrochemistry*, 1992. **22**(1): p. 1–7.
[2] Wilson, M.S. and S. Gottesfeld, *High Performance Catalyzed Membranes of Ultra - Low Pt Loadings for Polymer Electrolyte Fuel Cells. Journal of the Electrochemical Society*, 1992. **139**(2): p. L28–L30.
[3] Debe, M.K., *Electrocatalyst Approaches and Challenges for Automotive Fuel Cells. Nature*, 2012. **486**: p. 43.
[4] Wagner, F.T., B. Lakshmanan, and M.F. Mathias, *Electrochemistry and the Future of the Automobile. The Journal of Physical Chemistry Letters*, 2010. **1**(14): p. 2204–2219.
[5] Gasteiger, H.A., et al., *Activity Benchmarks and Requirements for Pt, Pt-Alloy, and Non-Pt Oxygen Reduction Catalysts for PEMFCs. Applied Catalysis B: Environmental*, 2005. **56**(1): p. 9–35.
[6] Nørskov, J.K., et al., *Towards the Computational Design of Solid Catalysts. Nature Chemistry*, 2009. **1**: p. 37.
[7] Shao, M., et al., *Recent Advances in Electrocatalysts for Oxygen Reduction Reaction. Chemical Reviews*, 2016. **116**(6): p. 3594–3657.
[8] Honda Kazuyoshi, S.T., Suetsugu Daisuke, Kashiwagi Katsumi, and Bessho Kunihiko, *Lithium Ion Secondary Battery.* In *US Patent 8486549*, 2013.
[9] James, B., *2018 Cost Projections of PEM Fuel Cell Systems for Automobiles and Medium-Duty Vehicles*. 2018. https://www.energy.gov/sites/prod/files/2018/04/f51/fcto_webinarslides_2018_costs_pem_fc_autos_trucks_042518.pdf.
[10] Nørskov, J.K., et al., *Origin of the Overpotential for Oxygen Reduction at a Fuel-Cell Cathode. The Journal of Physical Chemistry B*, 2004. **108**(46): p. 17886–17892.
[11] Stamenkovic, V.R., et al., *Trends in Electrocatalysis on Extended and Nanoscale Pt-Bimetallic Alloy Surfaces. Nature Materials*, 2007. **6**: p. 241.
[12] Mistry, H., et al., *Nanostructured Electrocatalysts with Tunable Activity and Selectivity. Nature Reviews Materials*, 2016. **1**: p. 16009.
[13] Mayrhofer, K.J.J., et al., *The Impact of Geometric and Surface Electronic Properties of Pt-Catalysts on the Particle Size Effect in Electrocatalysis. The Journal of Physical Chemistry B*, 2005. **109**(30): p. 14433–14440.
[14] Shao, M., A. Peles, and K. Shoemaker, *Electrocatalysis on Platinum Nanoparticles: Particle Size Effect on Oxygen Reduction Reaction Activity. Nano Letters*, 2011. **11**(9): p. 3714–3719.
[15] Perez-Alonso, F.J., et al., *The Effect of Size on the Oxygen Electroreduction Activity of Mass-Selected Platinum Nanoparticles. Angewandte Chemie International Edition*, 2012. **51**(19): p. 4641–4643.

[16] Tritsaris, G.A., et al., *Atomic-Scale Modeling of Particle Size Effects for the Oxygen Reduction Reaction on Pt. Catalysis Letters*, 2011. **141**(7): p. 909–913.
[17] Nesselberger, M., et al., *The Particle Size Effect on the Oxygen Reduction Reaction Activity of Pt Catalysts: Influence of Electrolyte and Relation to Single Crystal Models. Journal of the American Chemical Society*, 2011. **133**(43): p. 17428–17433.
[18] Li, D., et al., *Functional Links Between Pt Single Crystal Morphology and Nanoparticles with Different Size and Shape: The Oxygen Reduction Reaction Case. Energy & Environmental Science*, 2014. **7**(12): p. 4061–4069.
[19] Yu, K., et al., *Degradation Mechanisms of Platinum Nanoparticle Catalysts in Proton Exchange Membrane Fuel Cells: The Role of Particle Size. Chemistry of Materials*, 2014. **26**(19): p. 5540–5548.
[20] Yano, H., et al., *Particle-Size Effect of Pt Cathode Catalysts on Durability in Fuel Cells. Nano Energy*, 2016. **29**: p. 323–333.
[21] Wang, C., et al., *Design and Synthesis of Bimetallic Electrocatalyst with Multilayered Pt-Skin Surfaces. Journal of the American Chemical Society*, 2011. **133**(36): p. 14396–14403.
[22] Oezaslan, M., F. Hasché, and P. Strasser, *Pt-Based Core–Shell Catalyst Architectures for Oxygen Fuel Cell Electrodes. The Journal of Physical Chemistry Letters*, 2013. **4**(19): p. 3273–3291.
[23] Stamenković, V., et al., *Surface Composition Effects in Electrocatalysis: Kinetics of Oxygen Reduction on Well-Defined Pt3Ni and Pt3Co Alloy Surfaces. The Journal of Physical Chemistry B*, 2002. **106**(46): p. 11970–11979.
[24] Stamenkovic, V., et al., *Changing the Activity of Electrocatalysts for Oxygen Reduction by Tuning the Surface Electronic Structure. Angewandte Chemie International Edition*, 2006. **45**(18): p. 2897–2901.
[25] Wang, C., et al., *Rational Development of Ternary Alloy Electrocatalysts. The Journal of Physical Chemistry Letters*, 2012. **3**(12): p. 1668–1673.
[26] Wang, C., et al., *Multimetallic Au/FePt3 Nanoparticles as Highly Durable Electrocatalyst. Nano Letters*, 2011. **11**(3): p. 919–926.
[27] Gan, L., et al., *Understanding and Controlling Nanoporosity Formation for Improving the Stability of Bimetallic Fuel Cell Catalysts. Nano Letters*, 2013. **13**(3): p. 1131–1138.
[28] Han, B., et al., *Record Activity and Stability of Dealloyed Bimetallic Catalysts for Proton Exchange Membrane Fuel Cells. Energy & Environmental Science*, 2015. **8**(1): p. 258–266.
[29] Stamenkovic, V.R., et al., *Improved Oxygen Reduction Activity on Pt3Ni(111) via Increased Surface Site Availability. Science*, 2007. **315**(5811): p. 493–497.
[30] Zhang, J., et al., *Controlling the Catalytic Activity of Platinum-Monolayer Electrocatalysts for Oxygen Reduction with Different Substrates. Angewandte Chemie*, 2005. **117**(14): p. 2170–2173.
[31] Zhang, J., et al., *Mixed-Metal Pt Monolayer Electrocatalysts for Enhanced Oxygen Reduction Kinetics. Journal of the American Chemical Society*, 2005. **127**(36): p. 12480–12481.
[32] Wang, J.X., et al., *Oxygen Reduction on Well-Defined Core−Shell Nanocatalysts: Particle Size, Facet, and Pt Shell Thickness Effects. Journal of the American Chemical Society*, 2009. **131**(47): p. 17298–17302.
[33] Xie, S., et al., *Atomic Layer-by-Layer Deposition of Pt on Pd Nanocubes for Catalysts with Enhanced Activity and Durability toward Oxygen Reduction. Nano Letters*, 2014. **14**(6): p. 3570–3576.
[34] Sasaki, K., et al., *Core-Protected Platinum Monolayer Shell High-Stability Electrocatalysts for Fuel-Cell Cathodes. Angewandte Chemie International Edition*, 2010. **49**(46): p. 8602–8607.

[35] Sasaki, K., et al., *Highly Stable Pt Monolayer on PdAu Nanoparticle Electrocatalysts for the Oxygen Reduction Reaction. Nature Communications*, 2012. **3**: p. 1115.
[36] Li, J., et al., *Surface Evolution of a Pt–Pd–Au Electrocatalyst for Stable Oxygen Reduction. Nature Energy*, 2017. **2**: p. 17111.
[37] Wu, J. and H. Yang, *Platinum-Based Oxygen Reduction Electrocatalysts. Accounts of Chemical Research*, 2013. **46**(8): p. 1848–1857.
[38] Wang, Y.-J., et al., *Unlocking the Door to Highly Active ORR Catalysts for PEMFC Applications: Polyhedron-Engineered Pt-Based Nanocrystals. Energy & Environmental Science*, 2018. **11**(2): p. 258–275.
[39] Cui, C., et al., *Compositional Segregation in Shaped Pt Alloy Nanoparticles and Their Structural Behaviour During Electrocatalysis. Nature Materials*, 2013. **12**: p. 765.
[40] Gan, L., et al., *Element-Specific Anisotropic Growth of Shaped Platinum Alloy Nanocrystals. Science*, 2014. **346**(6216): p. 1502–1506.
[41] Huang, X., et al., *High-Performance Transition Metal–Doped Pt3Ni Octahedra for Oxygen Reduction Reaction. Science*, 2015. **348**(6240): p. 1230–1234.
[42] Park, J., et al., *Hollow Nanoparticles as Emerging Electrocatalysts for Renewable Energy Conversion Reactions. Chemical Society Reviews*, 2018. **47**: p. 8173–8202.
[43] Wang, J.X., et al., *Kirkendall Effect and Lattice Contraction in Nanocatalysts: A New Strategy to Enhance Sustainable Activity. Journal of the American Chemical Society*, 2011. **133**(34): p. 13551–13557.
[44] Zhang, L., et al., *Platinum-Based Nanocages with Subnanometer-Thick Walls and Well-Defined, Controllable Facets. Science*, 2015. **349**(6246): p. 412–416.
[45] Wang, X., et al., *Pt-Based Icosahedral Nanocages: Using a Combination of {111} Facets, Twin Defects, and Ultrathin Walls to Greatly Enhance Their Activity Toward Oxygen Reduction. Nano Letters*, 2016. **16**(2): p. 1467–1471.
[46] He, D.S., et al., *Ultrathin Icosahedral Pt-Enriched Nanocage with Excellent Oxygen Reduction Reaction Activity. Journal of the American Chemical Society*, 2016. **138**(5): p. 1494–1497.
[47] Chen, C., et al., *Highly Crystalline Multimetallic Nanoframes with Three-Dimensional Electrocatalytic Surfaces. Science*, 2014. **343**(6177): p. 1339–1343.
[48] Oh, A., et al., *Skeletal Octahedral Nanoframe with Cartesian Coordinates via Geometrically Precise Nanoscale Phase Segregation in a Pt@Ni Core–Shell Nanocrystal. ACS Nano*, 2015. **9**(3): p. 2856–2867.
[49] Zhang, Z., et al., *One-Pot Synthesis of Highly Anisotropic Five-Fold-Twinned PtCu Nanoframes Used as a Bifunctional Electrocatalyst for Oxygen Reduction and Methanol Oxidation. Advanced Materials*, 2016. **28**(39): p. 8712–8717.
[50] Li, M., et al., *Ultrafine Jagged Platinum Nanowires Enable Ultrahigh Mass Activity for the Oxygen Reduction Reaction. Science*, 2016. **354**(6318): p. 1414–1419.
[51] Kitchin, J.R., J. K. Nørskov, M. A. Barteau, and J. G. Chen, *Modification of the Surface Electronic and Chemical Properties of Pt(111) by Subsurface 3d Transition Metals. The Journal of Chemical Physics*, 2004. **120**(21): p. 10240–10246.
[52] Kitchin, J.R., et al., *Role of Strain and Ligand Effects in the Modification of the Electronic and Chemical Properties of Bimetallic Surfaces. Physical Review Letters*, 2004. **93**(15): p. 156801.
[53] Strasser, P., et al., *Lattice-Strain Control of the Activity in Dealloyed Core–Shell Fuel Cell Catalysts. Nature Chemistry*, 2010. **2**: p. 454.
[54] Stephens, I.E.L., et al., *Understanding the Electrocatalysis of Oxygen Reduction on Platinum and its Alloys. Energy & Environmental Science*, 2012. **5**(5): p. 6744–6762.
[55] Kibler, L.A., et al., *Tuning Reaction Rates by Lateral Strain in a Palladium Monolayer. Angewandte Chemie International Edition*, 2005. **44**(14): p. 2080–2084.

3 Ultralow Pt Loading for a Completely New Design of PEM Fuel Cells

Junbo Hou
Institute of Fuel Cells, School of Mechanical Engineering, Shanghai Jiao Tong University, Shanghai, People's Republic of China

Min Yang
Central Research Institute, Shanghai Electric Group, Zhabei District, Shanghai, People's Republic of China

CONTENTS

3.1 INTRODUCTION

World energy consumption is continuing to increase. To avoid resource depletion and long-term damage to the environment, renewables have come to the stage, and it is believed that their use will be crucial. The development of high-performance, low-cost, and environmentally benign energy systems will inevitably be needed.

DOI: 10.1201/9781003133971-3

Fuel cells for the direct production of electricity from chemical energy represent such a system. The science and technology of fuel cell engines are evolving, and the desire for decreased dependence on petroleum supplies, lower pollution, and potential for high efficiency are driving this trend toward an alternative power generation technology. The proton exchange membrane (PEM) fuel cell has been considered an ideal power source for automotive applications and has attracted much attention worldwide in the past three decades [1–5]. Although some critical technical development is required before PEM fuel cells reach the stage of large-scale commercialization, they are believed to be a promising pure electrification solution for different mass and usage segments of automotive applications, especially in large vehicles that are used regularly for long trips [6–8].

The PEM fuel cell consists of two critical key components: membrane electrode assemblies (MEAs) in which electrochemical oxidation of H_2 at the anode and reduction of O_2 at the cathode occur and plates that support and compress the MEAs to form a stack. During the operation of a polymer-electrolyte fuel cell, many interrelated and complex phenomena occur. These processes include mass and heat transfer, electrochemical reactions, and ionic and electronic transport.

Precious metal catalysts such as Pt are actually needed, and their high cost is the fundamental obstacle to the commercialization of PEMFCs. Vehicles with internal combustion engines (ICEs) usually require approximately 10 g of Pt-Pd-Rh ternary precious catalyst in the exhaust gas converter to clean the exhaust gas [9]. To conserve Pt use and achieve PEM fuel cell vehicle cost acceptance, the U.S. Department of Energy (DOE) set a Pt group metal (PGM) target of 0.125 g_{PGM}/kW_{rated} or approximately 11.3 g_{PGM} for a 90 kW_{gross} vehicle by the year 2020 [10]. This means that PGM loading of the MEA should decrease to 0.125 mg/cm^2, while the rated performance should reach 1 W_{rated}/cm^2. To operate the fuel cell more efficiently and reduce the amount of electrochemically generated heat, an operating voltage of 0.67 V is chosen, thus requiring MEA to produce a current density of at least 1.493 A/cm^2. Although a Mirai stack conveys 1.85 A/cm^2 @ 0.67 V, a value that exceeds the rated power density target, the MEA PGM loading is still approximately 0.33 mg/cm^2 [11]. Reduction of the catalyst loading is likely to undermine the performance of the cathode due to the sluggish oxygen reduction reaction (ORR), which demonstrates a low exchange current density (i_0) that is usually several orders of magnitude lower than that of the hydrogen oxidation reaction (HOR) [12]. Using the terminology of catalysis, turnover frequency (TOF) times the number of active sites indicates the activity of the catalyst [13]. Thus, general improvement of the TOF of the catalyst itself and increasing the number of active sites, or both, can improve the ORR and thus the MEA performance. When the number of active sites is increased, it is necessary to avoid mass transport loss, which usually depends on the structuring effect of the electrode. From the point of view of electrochemistry, the general apparent activity can be measured by the current density at a specific overpotential, which indicates either performance, activity, or current density and can be depicted based on the idea of the general catalysis. The general apparent activity can be roughly depicted by $j = f \cdot S \cdot i$, in which f refers to a structuring function that is usually linked to the concentration of the adsorbed oxygen species, *S* is the electrochemical active surface area (ECA), and *i* indicates the specific activity of the electrocatalyst. Thus, the more current is present within a specific matrix of the electrode, the more cost-effective is the fuel cell. Based on widely cited references on electrocatalysts for fuel cells, this can be expressed in units of A/cm^3 electrode; it corresponds to a development target of approximately 100 A/cm^3 at 0.90 V, 80°C, and 100 kPa$_{abs}$ O_2 [14]. For a 11.3 μm-thick electrode, the mass activity of the MEA requires 0.44 A/mg PGM if the PGM loading of the MEA is set at 0.2 mg/cm^2. This is also the target set by the DOE for 2020 for the mass activity, while the specific activity is set at 0.7 mA/cm^2 PGM [15], giving an ECA of 62.85 m^2/g PGM.

To date, a great deal of effort has been devoted to research on and development of electrocatalysts with high mass activity [16–22]. Four principal platinum-based heterogeneous electrocatalyst approaches are summarized by Debe; they include i) extended surface area catalysts; ii) discrete low-aspect-ratio nanoparticles dispersed on low-aspect-ratio supports; iii) discrete low-aspect-ratio nanoparticles dispersed on high-aspect-ratio supports; and iv) unsupported nanoparticles [23]. Regarding the rotating disk electrode (RDE) activity and MEA activity, some promising data from these four types is extracted from [23] and replotted together with the recent advances made in DOE national laboratories and at 3M, General Motors (GM), and Ford [15]. It can be seen from Figure 3.1 that the activity of RDEs is generally higher than that of MEAs; this is usually due to the structuring and poisoning effects of incorporating the ionomer into the catalyst layer. Some shape- and size-controlled electrocatalysts exhibit high RDE activity but are generally not stable in MEA testing. None of the commercial Pt/C or Pt blacks meet the DOE target, and without any doubt it appears that only binary Pt-based electrocatalysts, either standard Pt alloy, dealloyed, core-shell, or annealed, will be able to meet the DOE 2020 RDE activity standard. Although recent progress involving extended surface area catalysts such as nanostructured thin films (NSTF) developed by 3M, PtNi nanowires (PtNi NW) developed by NREL, and discrete low-aspect-ratio nanoparticles dispersed on low-aspect-ratio supports (core shell structure Pt@NiN/C developed by BNL, core shell Pt@PtNi/C developed by LANL, atom layer deposition (ALD)-prepared Pt/NbOx/C developed by ORNL and Ford, alloy PtCo on high-surface-area carbon (HSC) developed by GM, and PtNi/HSC developed by ANL) all meet the RDE activity target; for practical use, their MEA activity shows that the PtNi NW, Pt@NiN/C, and ALD assist catalysts still require further effort to improve their electrode structure. Figure 3.1b shows that most electrocatalysts succeeding DOE target demonstrate ECA within the range of 20–80 m^2/g. Very recently, ANL developed low-Pt@PGM-free catalysts with ultralow Pt content using Co or Co/Zn zeolitic imidazolate frameworks as precursors; these catalysts manifest extremely high MEA mass activity of 1~1.85 A/mg PGM [24]. The fact that extended surface area catalysts such as NSTF are free-standing suggests that this catalyst could be applied to a proton exchange membrane (PEM) alone to form a catalyst-coated membrane (CCM). As with discrete low-aspect-ratio nanoparticles dispersed on low-aspect-ratio supports, they are usually nanoparticles on carbon black supports, standard and graphitized; this implies that an ionomer is required to bind them together and to provide the proton pathway in the electrode.

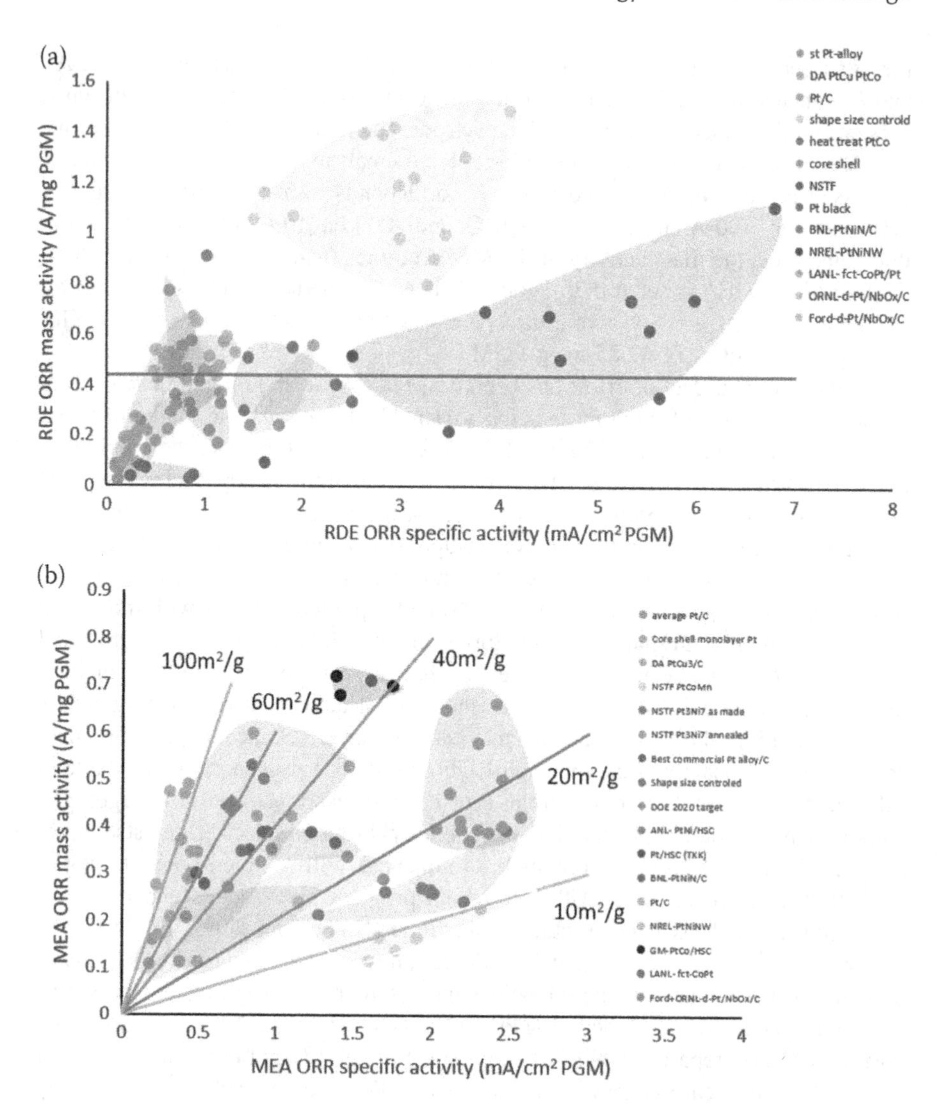

FIGURE 3.1 ORR mass activity and specific activity of the major recently developed Pt-based electrocatalysts, including the electrocatalysts most recently developed at the DOE National Laboratories. a) Activities measured by RDE at 900 mV; b) activities measured in MEAs at 900 mV, 80°C, and 150 kPa saturated O_2. Modified by permission from ref. [23]; Copyright 2012, Macmillan Publishers Limited.

This ionomer-impregnated thin-film electrode structure, especially for electrocatalysts that consist of discrete low-aspect-ratio nanoparticles dispersed on low-aspect-ratio supports, was developed by LANL scientists and represents a breakthrough in fuel cell technology. It reduces Pt loading from 4 mg/cm^2 to 0.4 mg/cm^2 [25,26]. For such a structure, it is noted that with further reduction in PGM loading of the cathode, the voltage loss at high current density (HCD) is significantly larger than would be expected from the known oxygen reduction

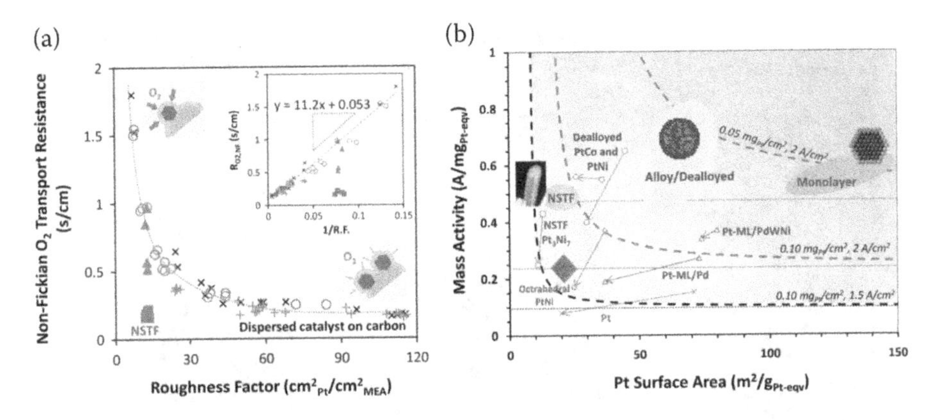

FIGURE 3.2 (a) Non-Fickian O_2 transport resistance as a function of total Pt area on an MEA cathode; (b) ORR mass activity and Pt ECSA targets that enable the cathode to meet vehicle requirements (0.58 V) at the indicated current density and cathode catalyst loadings. Reproduced by permission from ref. [30]; Copyright 2016, American Chemical Society.

kinetics or oxygen transport [27–31]. This impedes advanced downsizing of the fuel cell stack and its cost. Almost every supported electrocatalyst displays an intrinsic Pt-area-specific oxygen transport resistance of 11.2 s/cm [29], as shown in Figure 3.2a. When ECA decreases to 50 m^2/g, the non-Fickian oxygen transport resistance apparently begins to increase. It is obvious that NSTF does not display such local oxygen transport resistance, and even with an ionomer film adhered to it, the intrinsic Pt-area-specific oxygen transport resistance is much lower than 11.2 s/cm. In terms of modeling studies, ORR mass activity and Pt ECA targets can be specified for the cathode to approach 0.58 V under the indicated conditions. For example, a mass activity of 0.24 A/mg PGM is sufficient to provide 2 A/cm^2 at a Pt loading of 0.1 mg Pt/cm^2 with no local O_2 transfer resistance (indicated by the red dashed horizontal line in Figure 3.2b). However, assuming a local oxygen transport resistance of 12 s/cm, only the right-top area enclosed by the red parabolic dashed lines can enable 2 A/cm^2 at 1 mg Pt/cm^2 [30]. Again, one should note that NSTF seems to have the smallest constraint.

3.2 ELECTRODE STRUCTURES

A commercial cation exchange membrane was first demonstrated by Grubb at GE as the acid electrolyte in fuel cells, giving rise to PEM fuel cells [32]. The electrodes used Pt coated on a Ni screen, and the peak power density at room temperature was only 10 mW/cm^2. Later, the GE group prepared a slurry containing PTFE and Pt blacks and coated it onto Pt, Ni, and Ag screen meshes. The "conducting-porous-Teflon electrodes" containing 34 mg Pt blacks and 3.1 mg/cm^2 PTFE yielded a peak power density of 350 mW/cm^2 using O_2 in a base medium [33]. GE built PEM fuel cells by integrating the Kel-F film, P(S-DVB)SA membrane that was used in the Gemini space mission in 1962, but it is not sufficiently chemically stable [34]. Thanks to the invention of the Nafion membrane, ionomer

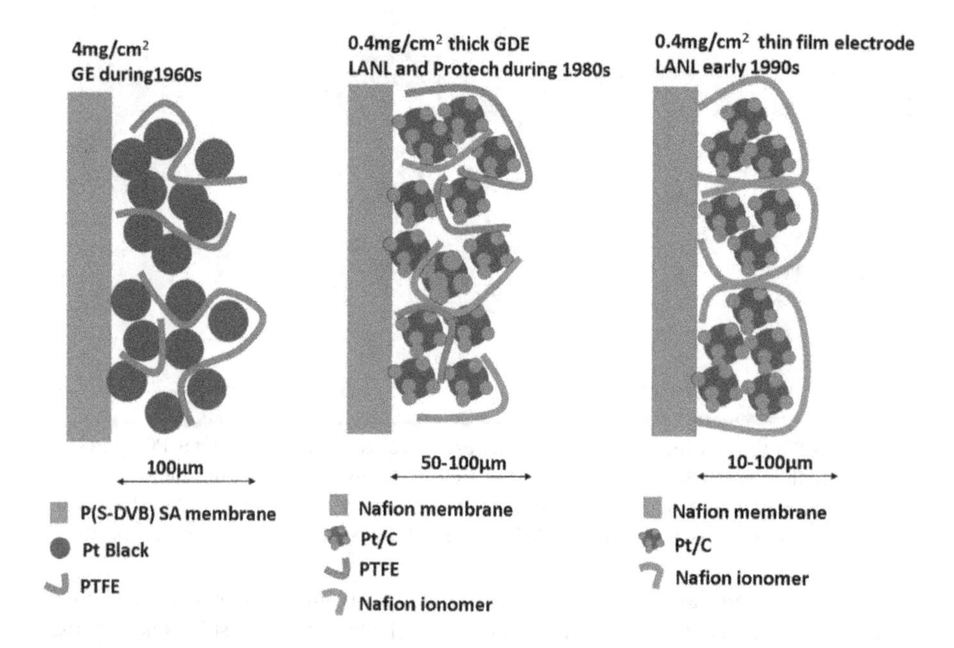

FIGURE 3.3 Structural evolution of PEM fuel cell electrodes.

suspension and carbon-supported Pt, Protech, and LANL started with a PTFE and Pt/C slurry sandwiched between the gas diffusion layer (GDL) and impregnated with a Nafion ionomer suspension, forming thick gas diffusion electrodes (GDEs) [35]. By eliminating PTFE and using the Nafion ionomer as the sole binder, novel electrode structures for the Pt/C catalyst layer of PEM fuel cell electrodes were developed that yield improved utilization efficiency of the catalyst [25,26]. This thin-film electrode can be coated on GDL or deposited on the membrane by spraying or decal transferring [36–38] to form the widely used CCM. Figure 3.3 shows the structural evolution of the PEM fuel cell catalyst layer.

The agglomerate-aggregate-thin-film model accurately describes the micro-meso-macro structure of the state-of-the-art catalyst layer and has been widely studied to determine the relationships between material properties, processing, and electrochemical performance [39–43]. Due to the structure of the catalyst, some Pt sites located deep inside the agglomerates are not proton accessible due to limited ionomer penetration and not mass approachable due to the blocking of oxygen. Thus, an ideal electrode structure is brought in that allows perpendicular electron and proton pathways to the membrane. The oriented electrode layer is composed of long chain-like strings of basic carbon particles that are coated with finely dispersed platinum covered with a thin layer of a proton conductor [44]. Such an ordered catalyst layered structure can facilitate electron, proton, and gas transport. Aimed at cylindrical electrode structure, a steady-state, one-dimensional numerical model was developed and used to analyze the performance of the ordered cathode catalyst layer in PEM fuel cells [45]. The simulation results reveal that ordered catalyst layers exhibit better performance than conventional ones due to the improvements in mass transport and to the uniformity of the electrochemical reaction rate across the entire width of the catalyst layer. Using the modified agglomerate model, ordered-catalyst layers in PEMFC cathodes that yield maximum power density were designed [46]. At higher Pt loading, the cell performance is not very sensitive to the Pt content. At lower Pt loading, electron-conducting tubes should be arranged as densely as possible, and the size of the ordered arrangements should be reduced, thereby avoiding deterioration in performance. The superior performance of the ordered active layer compared to that of the conventional one is reaffirmed by numerical modeling work [47], which shows that the ordered layer has a more uniform oxygen concentration distribution thanks to the more facile oxygen transport in the active layer. Later, the simple cylinder model is made more sophisticated by the use of arrays of Pt-decorated aligned carbon nanotubes (CNTs) as the catalyst layer [48]. Taking into account both multicomponent and Knudsen diffusion, the modeling results implied that shorter lengths give better electrode performance because they present lower diffusion barriers and better catalyst utilization. The compromise between gas transport pore size and the reduction in the array number per unit geometrical area suggests that spacing at approximately 400 nm produces the best limiting current density. A 3-D mathematical model of such an array-ordered catalyst layer is further expatiated on, and oxygen transport through the ionomer film rather than through the spaces along the arrays might be the limiting factor [49]. In contrast, it is found that performance wanes if the spacing between Pt/CNT is increased from 25 nm to 100 nm. A model focused on arrays of ordered Pt electrodes was used to investigate whether the improved oxygen transport can offset the decreased roughness of the ordered electrode array [50]. The authors reported that with a nanorod diameter of 60 nm, the electrode array can provide a limiting current of 11,000 A/cm^3, omitting ohmic loss and mass transport loss outside the electrode. From the agglomerate-aggregate-thin-film model to the ordered nano thin film electrode model, as shown in Figure 3.4 [51]

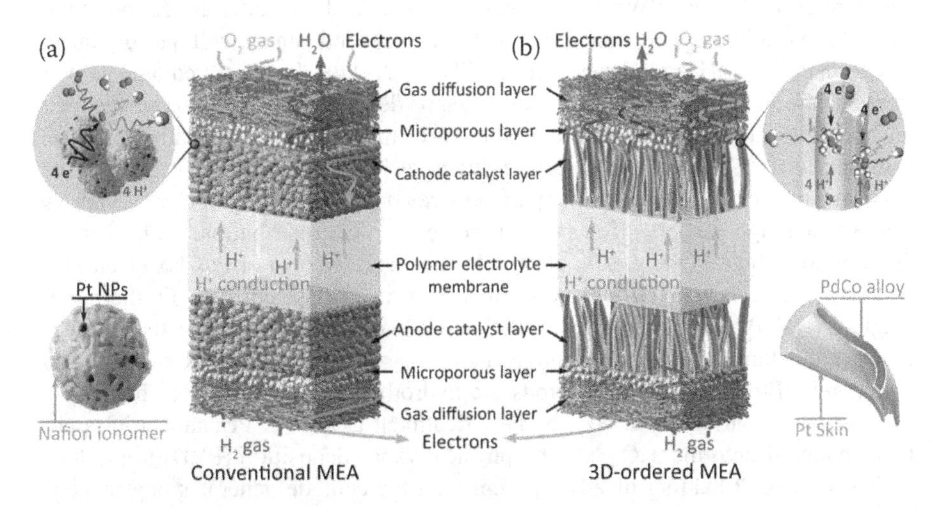

FIGURE 3.4 Conceptual diagrams of (a) conventional MEA and (b) 3-D-ordered MEA. Reproduced by permission from ref. [51]; Copyright 2018, Royal Society of Chemistry.

based on the theoretical studies discussed above, one can expect high Pt utilization due to the highly open structure, enhanced mass transport due to the lower tortuosity, and possibly improved durability due to the absence of a carbon support.

3.3 PREPARATION APPROACHES

3.3.1 Pt on High-Aspect-Ratio Supports

To effectively utilize Pt active sites and boost the support's oxidative resistivity, discrete dispersed Pt nanoparticles are deposited on extended surface areas and high-aspect-ratio supports, particularly on some 3D structures. Here, we will specifically review catalysts with integrated electrode structures, excluding catalysts such as Pt nanoparticles deposited on carbon nanotubes (CNT) due to the fact that such supported catalysts are usually composed of agglomerated electrode parts. The first report of such an electrode design uses a Pt catalyst supported on a nanowire-based electrode [52]. SnO_2 nanowires are grown on the carbon fibers of a carbon paper by thermal evaporation, and Pt nanoparticles are then electrochemically deposited onto the SnO_2 nanowires. Although the authors do not indicate how to avoid Pt deposition on the carbon fiber in the backing layer, the claimed 0.2 mg/cm^2 Pt loading exhibits higher electrocatalytic activity both for the oxygen reduction reaction and the methanol oxidation reaction compared to a standard formula electrode. Unfortunately, the single cell performance is not provided. By chemical vapor deposition (CVD), tungsten oxide nanowires ($W_{18}O_{49}$ NWs) are directly grown on carbon paper, and Pt nanoparticles are then coated on the $W_{18}O_{49}$ NWs. Such a Pt/$W_{18}O_{49}$ NW/carbon paper composite electrode has higher electrocatalytic activity toward ORR and better CO tolerance in a single-cell polymer electrolyte membrane fuel cell [53]. Later, the same group used a similar method to fabricate Pt nanoparticles deposited on Sn nanowires grown on carbon fibers [54]. The 3-D open structure of the Pt/SnC NW/carbon paper composite and the better utilization of the catalytic particles ensure enhanced ORR activity and single cell performance. Rather than Sn, CNT and nitrogen-doped CNT (CNx) can be deposited onto carbon fibers using an aerosol-assisted chemical vapor deposition (AA-CVD) method; Pt nanoparticles are then deposited onto the CNTs through an impregnation process, resulting in a 3D composite electrode structure [55]. At a similar Pt loading of 0.11 mg/cm^2, the Pt shows smaller particle sizes on CNx than on CNT (2.63 nm vs 5.89 nm average particle size) and a narrower particle distribution; both of these features contribute to the greater electrochemical surface area of Pt/CNx electrodes and produce higher single cell performance in an H_2/O_2 fuel cell. Furthermore, single-crystal Pt nanowires grown on Sn-coated carbon fibers using the impregnation procedure demonstrate 1.2-fold higher mass activity for ORR compared to commercial Pt/C [56]. TiO_2 nanorods are hydrothermally grown on carbon fibers and converted into TiO_2–C NRs by heat treatment at 900°C. Pt nanoparticles are then sputtered onto the TiO_2 NRs by physical vapor deposition (PVD) to produce Pt–TiO_2–C. A Pt loading of 28.67 $\mu g\ cm^{-2}$ on the cathode generates power 4.84-fold that of the commercial GDE (11.9 kW g_{Pt}^{-1}) along with enhanced durability [57]. Progress in similar work was achieved using titanium nitride (TiN) nanorods

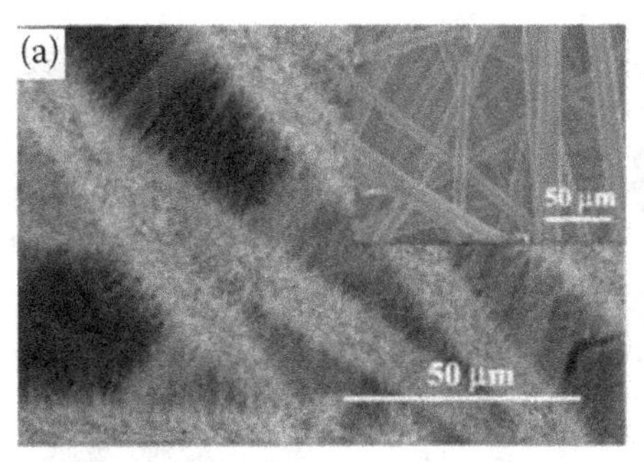

FIGURE 3.5 (a) SnO_2 nanowires grown on carbon fiber; (b) Pt nanoparticles deposited on SnO_2 nanowires by permission from ref. [52]; Copyright 2007, American Electrochemical Society.

on which a thin-film platinum–palladium–cobalt (PtPdCo) catalyst with a Pt loading of 66.9 μm cm^{-2} is deposited [58]. This electrode architecture delivers a maximum power density of 390.5 mW cm^{-2} with a specific power density of 5.84 W ${mg_{Pt}}^{-1}$. Overall, these novel structures have the potential to offer high Pt utilization, high activity, and high durability when employed as fuel cell electrodes. Pt nanoparticles or nanowires have been demonstrated to be capable of depositing on 1-D nanostructures such as CNT, SnO_2, WO_3, TiO_2, etc. which grows on carbon fibers. Such electrode structure is an attempt to modify the traditional agglomerate-aggregate-thin film electrode structure, since at the micro-scale Pt nanoparticles are randomly distributed while at meso-scales the 1-D structure is regulated to distribute evenly. At the macro-scale the carbon fibers randomly overlay together, which builds the root support for such a 3-D electrode structure, as shown in Figure 3.5.

3.3.2 Extended Surface Area of Wires

3.3.2.1 Confined Agglomerates

This strategy originates from the idea that if the traditional agglomerate size can be confined to a 1-D nanowire domain and overlaid, a controlled 3-D electrode structure can thereby be formed. In accordance with this idea, electrospinning is a versatile fabrication method that can be used to prepare 1-D random or oriented continuous nanowires with the possibility of obtaining ordered underlying morphologies such as core-shell, hollow, or porous fibers or even multi-channeled arrangements [59–61]. Using Pt/C catalyst powder with Nafion and PAA binders, a new nanowire-structured electrode was developed and tested as the cathode in PEM H_2/air and H_2/O_2 fuel cells, as shown in Figure 3.6 [62]. The average diameter of the nanowires is 460 nm, and a Pt loading of 0.1 mg Pt/cm^2 provides 524 mW/cm^2 at 0.6 V at 80°C and ambient pressure. The MEA mass activity reaches 0.23 A/ mg_{Pt}. A high ECA of 81.17 m^2/g_{Pt} is achieved with ultra-low Pt loading (4.03 wt%) based on Pt/SnO_2 nanofibers synthesized via electrospinning [63]. Unfortunately, the authors only demonstrated the HER activity of these fibers. An electrode layer structured with a lower noble metal catalyst loading consisting of PtRu nanofibers

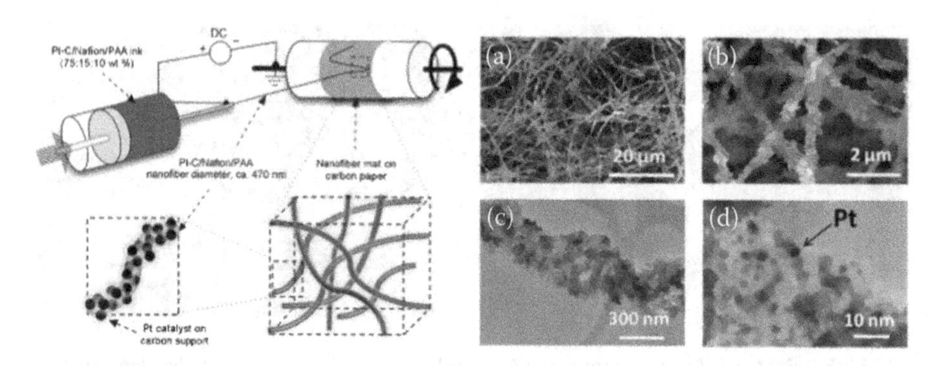

FIGURE 3.6 Scheme for an electrospinning nanowire electrode composed of Pt-C/Nafion/PAA. SEM images (a, b) and TEM images (c, d) of electrospun Pt-C/Nafion/PAA nanofibers. Reproduced by permission from ref. [62]; Copyright 2011, Wiley-VCH.

was designed by electrospinning and used as the anode of a passive direct methanol fuel cell (DMFC) [64]. Subsequently, a systematic investigation of the composition of nanowires comprising three types of electrocatalysts, TKK TEC10E50E (46.1% Pt on Ketjen Black), TKK TEC10E50E-HT (50.5% Pt on Ketjen Black), and Johnson Matthey HiSpec 4000 catalyst (40% Pt on Vulcan carbon), at varied ratios of ionomer to carbon (I/C values) was conducted [65]. With the same Pt loading of 0.10 mg/cm^2, the electrode with electrospun TKK TEC10E50E generated 470 mW/cm^2 at 0.65 V, an improvement of 35% over the 349 mW/cm^2 generated by the traditional electrode. The I/C value seems to have little influence on cell performance. Durability tests using the automotive-specific load cycling (Pt dissolution) and start-stop cycling (carbon corrosion) cathode durability protocols indicate that electrospun nanofiber mat electrodes composed of Pt/C, Nafion, and PAA give high beginning-of-life (BoL) performance and maintain high end-of-life (EoL) performance compared to electrodes with traditional GDEs [66]. The authors attributed the above phenomena to the better oxygen and water transport within the nanofiber electrode and the higher electrochemical surface area of the fiber cathode. An electrospun nanofiber electrode with an ultralow Pt loading of 0.087 mg/cm^2 was constructed and showed a peak power density of 0.692 W/cm^2 [67]. The improved cell performance arises mainly from the uniform distribution of the Pt catalyst and Nafion ionomer within the confined domain and the extraordinarily highly porous structure. The durability study gives hints that PAA might interact with the carbon support and decrease the rate of carbon corrosion. A novel idea seeking to further reduce the Pt loading is the use of a core-shell structured catalyst. In this method, the Pd/C catalyst layer is electrospun-prepared and then galvanically exchanged with Pt to form Pd/C@Pt skin catalysts [68]. The peak power density of such an electrode is 0.62 W cm^{-2} with a Pt loading of 19 μg cm^{-2}, higher than that of the conventional electrode (0.55 W cm^{-2}) with a Pt loading of 100 μg cm^{-2}. Most importantly, the degradation of peak power density of the electrode is only 4.8% after an accelerated stability test (AST) for 30,000 cyclic voltammetry (CV) cycles.

3.3.2.2 Nanowires

Here, the nanowires differ from the common ones discussed in Ref. [69] and specifically refer to those with diameters of several tens of nanometers to several hundreds of nanometers and usually with a length of 100 micrometers or more; these nanowires can form mat-type structures. Following the development of spontaneous galvanic displacement [70], Pt-coated Ni nanowires (PtNiNWs) with diameters of 150–250 nm and lengths of 100–200 μm are synthesized [71]. PtNiNWs show exceptionally high ECA (90 $m^2 g_{Pt}^{-1}$) with a peak mass activity of 0.917 A mg_{Pt}^{-1} at ~7 wt% Pt; this decreases with Pt loading, as expected, and is 3.0 times greater than Pt/HSC and 2.1 times greater than the DOE mass activity target. The morphology and composition of PtNiNWs are shown in Figure 3.7. Later, the substrate NiNWs are replaced by Co nanowires (CoNWs), and Pt is coated by partial galvanic displacement, forming core/shell wires 200–300 nm in diameter and 100–200 μm in length [72]. PtCoNWs exhibit a maximum mass activity of 0.793 A mg_{Pt}^{-1} at 5.5 wt% Pt loading, 2.6 times greater than that of carbon-supported Pt nanoparticles. Although the PtNiNWs electrocatalyst shows high specific activity (>6 mA cm_{Pt}^{-2} and high surface area (>90 m^2 g_{Pt}^{-1}), it appears not to be stable enough. As a result, the authors take advantage of post-synthesis practices such as acid leaching and heat treatments in various atmospheres, and significant improvements in the Pt—Ni nanowires are obtained, including increased activity (threefold increase in specific activity) and durability (mass activity loss reduced from 21% to 3%) and decreased Ni leaching (reduced from 7% to 0.3%) [73]. Compared to traditional Pt/C, these materials show more than a tenfold improvement in mass activity. Further examination showed that leaching of the nickel core during MEA testing contaminates the ionomer in the electrode and the membrane. After pre-leaching to 80 wt% Pt, 9% (vs 16% and 23%) ionomer incorporation yielded the highest performance, 0.238 A/mg_{Pt} [74]. Figure 3.8 shows the morphology of MEA cross-sectional and top views of such mat electrodes. Acid ion exchange can significantly improve the cell performance, indicating the occurrence of severe Ni poisoning. Another method, atom layer deposition (ALD), is implemented instead of galvanic displacement to deposit Pt onto NiNW substrates [75]. A total of 30 cycles of Pt ALD followed by the H_2 annealing process yielded a catalyst with mass activity more than four times greater than the DOE 2020 target.

3.3.3 Ordered Arrays

As discussed in Sections 3.1 and 3.2, either Pt on high aspect ratio supports, or extended surface area of wires that includes confined agglomerates and nanowires, a nonwoven-fabric-like mat-type structure is needed to achieve a free-standing electrode. This structure intrinsically finitely limits the traditional agglomerate-aggregate-thin-film model to one dimension and thus includes highly porous domains created by the packed wires or fibers and the internal dimensions of the fibers. However, the wires or fibers are overlaid randomly parallel to the membrane. In this section, we will discuss approaches to creating a much more ordered electrode structure; namely, causing these wires to be distributed evenly perpendicular

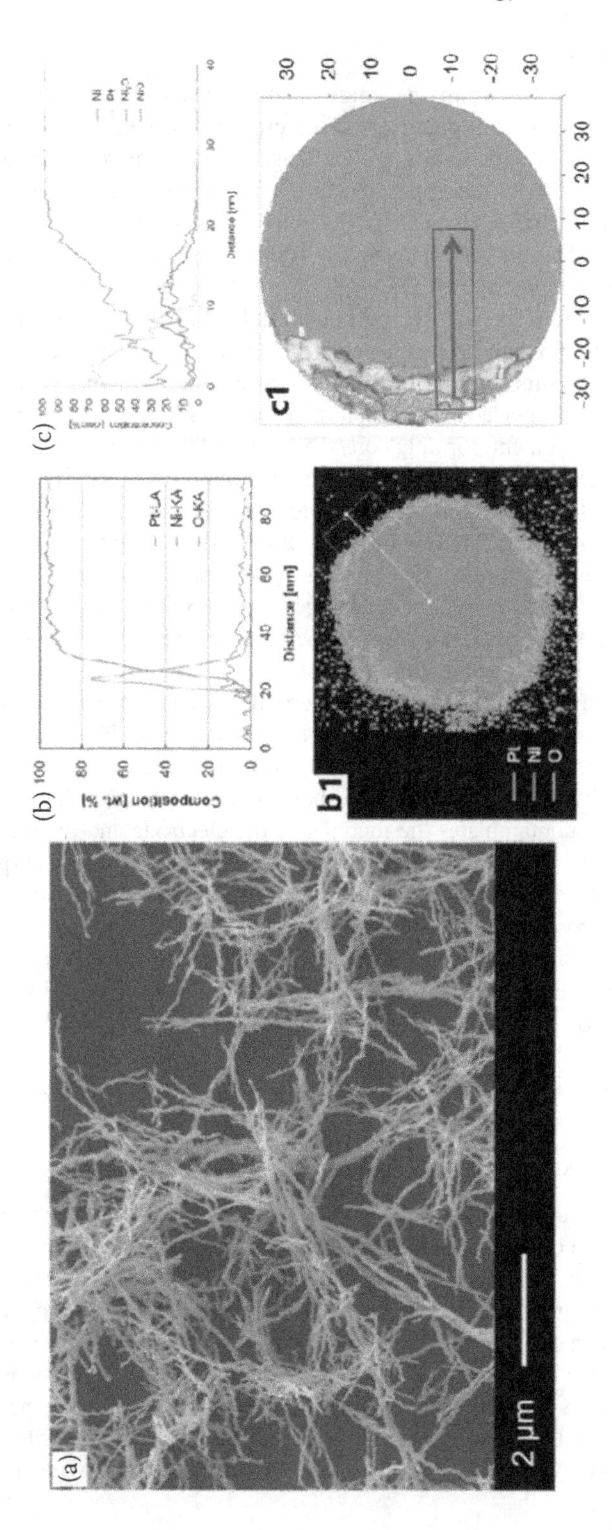

FIGURE 3.7 (a) SEM image of PtNiNWs; (b-b1) EDX line scan and mapping; (c-c1) atom probe line scan and atom probe reconstruction. Reproduced by permission from ref. [71]; Copyright 2014, American Chemical Society.

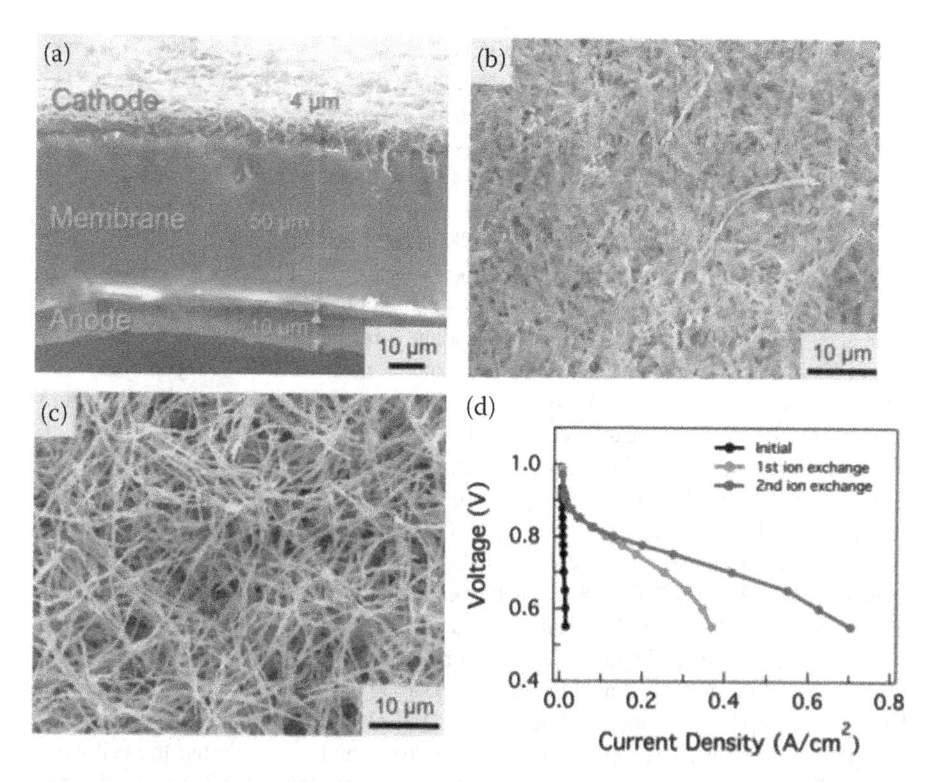

FIGURE 3.8 (a) SEM images of cross-sections of a PtNiNWs MEA at 16% ionomer content; (b) top view after first acid ion exchange; (c) top view after second acid ion exchange; (d) performance curve. Reproduced by permission from ref. [74]; Copyright 2018, American Electrochemical Society.

to the membrane. Physically, there exist an electron pathway and a proton pathway that might provide mechanical support to realize such an ordered structure. We will first discuss the electron arrays and then present the newest findings on proton arrays to intrigue any possible inspiration.

3.3.3.1 Electron Conductor Arrays

3.3.3.1.1 Carbon Tube-Assisted Arrays

By glancing angle deposition (GAD), carbon nanorod arrays (0.1 mg/cm^2) are grown on flat and patterned Si wafers and then coated with Pt catalysts (0.1 mg/cm^2) by magnetron sputtering. Flat and patterned substrates yield different array structures. Generally, the flat one features higher rods and decreased space between nanorods, while the patterned one generates shorter rods and regular spaces with a size of 50 nm [76]. Both give almost the same performance at low current density region, while those that sit on a patterned base yield almost twofold higher limiting current density than those on a flat base. The authors attributed the enhancement of activity to facilitated mass transport, and this was demonstrated by electrochemical impedance spectroscopy (EIS). By pyrolysis of iron(II) phthalocyanine, aligned carbon nanotubes (CNT) arrays with an average height of 4 mm are grown on a quartz plate, and Pt catalysts are then deposited onto the individual CNTs in the aligned arrays at a loading of 0.142 mg/cm^2. A single cell with this electrode and a commercial Pt/C electrode provide 397.23 mW/cm^2 and 210.79 mW/cm^2, respectively, at 0.65 V [77]. Later, the same group modified Pt deposition by functioning aligned CNT arrays with PDDA to achieve the electrostatic adsorption of the Pt precursor and then reducing the Pt precursor with ascorbic acid [78]. The Pt loading is varied from 0.075 mg/cm^2 to 0.30 mg/cm^2; over this range, the morphology of the Pt catalyst, as assessed by SEM, changes from discrete nanoparticles to a continuous film on the single CNT. The single cell performance increases as the Pt loading is increased, reaching a maximum of 680 mW/cm^2. By plasma-enhanced chemical vapor deposition (PECVD) combined with a Fe/Co bimetallic catalyst, vertically aligned CNT arrays are arranged on the aluminum foil, and the Pt catalyst is sputter-coated onto the CNT arrays [79]. This cell shows an excellent performance with ultralow Pt loading (35 μg/cm^2), comparable to that of commercial Pt/C at 400 μg/cm^2. Toyota pioneered the large surface area (236 cm^2) fabrication of Pt-coated vertically aligned CNT arrays (VACNTs) [80]. The VACNTs were grown on a stainless steel substrate coated with iron catalysts and an inactive oxide such as alumina by the CVD method; by the impregnation method, Pt catalysts were then deposited on each CNT. The vertically aligned structure is extremely critical to the MEA fabrication, as shown in Figure 3.9a. Both small single cells (20 cm^2) and large ones (236 cm^2) perform very well at high current density regions (2.6 A cm^{-2} @ 0.6 V) with low Pt loading at the cathode (0.1 mg cm^{-2}). The electrode structure favors the mass transport of reactants, i.e., oxygen, protons, electrons, and water, during the cell's operation. The authors claim that the CNTs displayed anti-agglomerative characteristics during the wet manufacturing process and that they maintained a continuous pore framing the layered catalyst structure. Due to the elastic characteristics of CNTs, they might fill the space between the catalyst layer and the microporous layer, preventing water flooding. It is the point of this reference that the CNT number density, the length of CNT and the diameter of CNT all contribute to the characteristics of the electrode structure and play a very important role in determining the electrode performance.

3.3.1.2 TiO_2-Assisted Arrays

TiO_2 nanotube arrays provide another template that can be used to create aligned and ordered 3-D electrode structures. Due to durability issues associated with the electrocatalysts in PEM fuel cells, the early studies of Pt supported on TiO_2 nanotube arrays (Pt/TNT) mainly addressed the problem of the durability of electrocatalyst supports. TiO_2 nanotube arrays can be formed by electrochemical anodization, and a Pt layer can be deposited by dc sputtering or evaporation [81]. Cyclic voltammogram (CV) tests indicate that the sputtered Pt electrode has better electrochemical catalytic activity, which is derived from the high surface area of the Pt layer. Through acceleration degradation tests (ADT) and long-term chronopotentiometry, the durability of Pt/TNT can be related to the ECA degradation [82]. It is observed that the ESA of Pt/TNT degrades only by 8.8% in sulfuric acid, and the power obtained from a single cell decreases by 11.3%. In essence, the electrocatalytic activity of Pt fades little due to partial oxidation since Pt bind strongly to TNT supports; thus, almost no dissolution or agglomeration of Pt occurs,

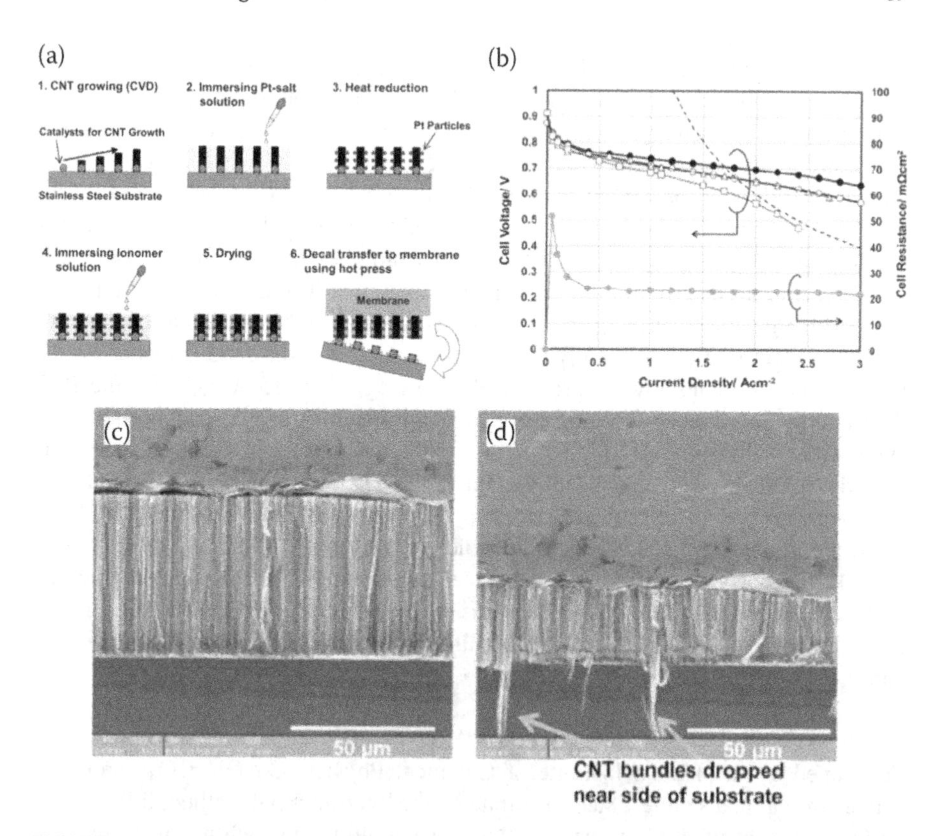

FIGURE 3.9 (a) MEA preparation procedure based on Pt-decorated vertically aligned CNT; (b) cell performance ○-VACNT electrodes (20 cm^2 cell), •-VACNT electrodes, IR corrected (20 cm^2 cell), Δ-VACNT electrodes (236 cm^2 cell), □-conventional electrodes (236 cm^2 cell), dashed line-DOE target 0.125 g PGM kW^{-1}; (c and d) SEM of a cross-section of MEA with VACNT. Reproduced by permission from ref. [80]; Copyright 2014, Elsevier.

leading to greatly improved durability. The high ORR activity of the Pt/TNT catalyst can be accounted for by the synergistic effect between Pt and TNT and the increase in ECA, yielding a charge transfer number of 3.94. Moreover, a minor decrease in the Pt active surface implies that there is a strong and stabilized physical interaction between the Pt catalyst and the TNT support [83]. However, there are two concerns regarding the application of Pt/TNT arrays; these are the relatively low conductivity of TiO_2 tubes and the not-very-uniform distribution of Pt on the TiO_2 nanotubes. After H_2 treatment, the electrical conductivity of H-TNTs increases by approximately one order of magnitude in comparison to air-treated TNTs, and the increase in the number of oxygen vacancies and hydroxyl groups on the H–TNTs helps anchor a greater number of Pt atoms during Pt electrodeposition [84]. The double-annealed Pt/H-TNTs exhibits a minimal decrease in ECA after 1,000 cycles compared with a 68% decrease for the commercial JM 20% Pt/C electrode after 800 cycles. The single fuel cell test indicates that the new electrode generates a specific power density of 2.68 kWg_{Pt}^{-1}. To further reduce Pt loading, Ni nanoparticles approximately 30 nm in diameter are first added to TNTs using a pulse electrodeposition technique followed by coating of the Ni nanoparticles with a thin film of Pt by galvanic exchange with the Pt precursor. The Ni@Pt/TNT array electrode exhibits high activity in the half-cell test, and ECA is reduced by 28% after an accelerated durability test, compared to 57% for commercial Pt/C (JM) [85]. A dense array of Pt nanotubes is produced by coating the surface of a porous anodized Al_2O_3 template using ALD; the Pt nanotubes have a length of approximately 2 μm, an external diameter of approximately 180 nm and are 20 nm thick [86]. Although the surface specific activity is slightly higher than that of the conventional electrode (37 μA cm_{Pt}^{-2} vs 28 μA cm_{Pt}^{-2} for the Pt/C dispersion), the mass activity remains low (approximately 5.5 A g_{Pt}^{-1} vs 15 A g_{Pt}^{-1} for the Pt/C dispersion). This is due to the continuous single metal Pt in a relatively thick film synthesized in this report. To improve the conductivity of the TNT arrays, carbon is coated on TNT by plasma-enhanced chemical vapor deposition (PECVD), and Pt is then deposited by radiofrequency (RF) sputtering to form the oriented C-TNTs-Pt electrode. Based on this 3-D electrode structure, an oriented ultrathin catalyst layer (UTCL) has been demonstrated as a fuel cell electrode [87]. The prepared, oriented UTCL displays maximum power densities of 206 and 305 mW cm^{-2} with ultralow Pt loadings of 24.8 and 50.1 μg cm^{-2}, resulting in relatively high Pt utilization of 8.3 and 6.1 kW $g^{-1}{}_{Pt}$.

3.3.3.1.3 Co–OH–CO_3–Assisted Nanowire Arrays

Developed by the Dalian Institute of Chemical Physics, Co–OH–CO_3 nanowires can be grown on a stainless steel substrate by the hydrothermal method, followed by PVD deposition of Pt-based nanoparticles or nanofilms on each nanowire to form Pt/Co–OH–CO_3 nanowires. The arrays of electrocatalysts can be hot-pressed onto Nafion membranes, producing an ultrathin catalyst layer (UTCL), as shown in Figure 3.10 [88]. Without any ionomers, the UTCL produces a maximum power density of 481 mW cm^{-2} at an ultra-low Pt loading of 43 μg cm^{-2} with a relatively high Pt utilization of 11.2 kW $g^{-1}{}_{Pt}$. SEM images of the UTCL on the Nafion 212 membrane and the UTCL itself are shown in Figure 3.10. On the template of Co–OH–CO_3 nanowire arrays, a continuous Nb_2O_5 film is formed by magnetron sputtering; a Pt nano thin film is then grown on the Nb_2O_5 film by the same technique [89]. Pt/Nb_2O_5 on Co–OH–CO_3 nanowire arrays can be directly aligned onto a Nafion membrane without an ionomer as a binder. Single cell performance reaches 5.80 kW g_{Pt}^{-1} (cathode) and 12.03 kW g_{Pt}^{-1} (anode) at a loading of 66.0 μg Pt cm^{-2}. As expected, the ADT indicates that this UTCL is far more stable than conventional Pt/C-based MEA. In a similar synthesis method, Pt supported on Co–OH–CO_3 nanowire arrays is prepared; by subsequent thermal annealing, the Co in the nanowire support segregates to the surface and alloys with Pt, forming PtCo/Co–OH–CO_3 nanowire arrays [90]. The open-walled PtCo with a diameter of ca. 100 nm can be transferred onto the PEM, forming a UTCL ca. 300 nm thick. A maximum power density of 14.38 kW g_{Pt}^{-1}, 1.7-fold that of the conventional CCM, is achieved at a Pt loading of 52.7 μg cm^{-2}. A further Pt loading reduction can be realized by the use of a Pt shell structure in which the PdCo coated on the Co–OH–CO_3 nanowire arrays is first prepared by the hydrothermal and PVD

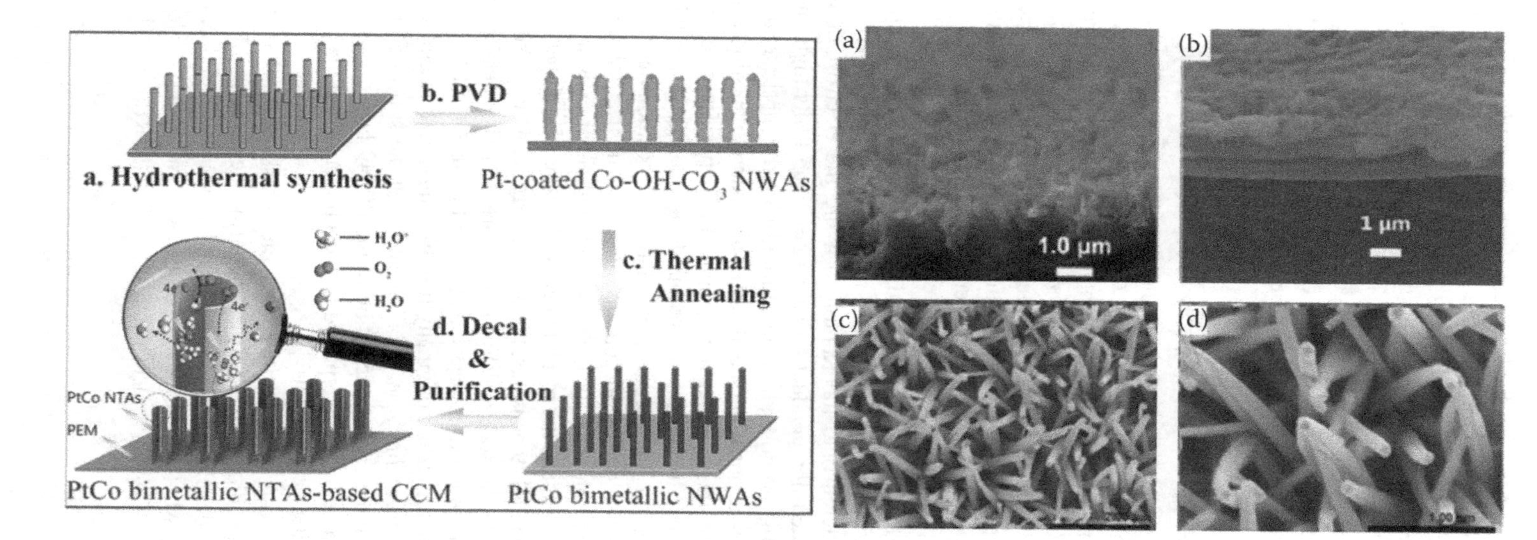

FIGURE 3.10 Scheme of the UTCL fabrication process. Reproduced by permission from ref. [90]; Copyright 2017, Elsevier. SEM images of (a, b) UTCL on a Nafion 212 membrane; (c, d) PtPd coated on Co–OH–CO_3 nanowire arrays. Reproduced by permission from ref. [88]; Copyright 2014, Royal Society of Chemistry.

methods followed by Pt galvanic displacement. Such Pt skin@PdCo/Co–OH–CO_3 nanowire arrays exhibit a maximum power density of 222.5 kW g_{Pt}^{-1} at a cathodic Pt loading of 3.5 μg cm^{-2}, 13.7-fold higher than that of a conventional MEA [51]. This UTCL demonstrates impressive stability during durability tests, and a self-healing mechanism is introduced: during the potential sweeping, Co and Pd leaching out from the core prevent the local potential from going to high values and hence prevent dissolution of the Pt skin. As the dissolution of Pd and Co proceeds from the core, the excess of Pt atoms from the shell can form a partial bi/multilayer due to the Kirkendall effect [91].

3.3.3.1.4 Pigment Red 149-Assisted Arrays

Initiated by 3M, NSTF electrocatalysts grounded on perylene red support whiskers were developed in the late 1990s [92–95]. Concerning the NSTF electrocatalyst technique platform, including support preparation, Pt-based sputtering, catalyst characterization, UTCL processing, and issues related to UTCLs, Debe and Steinbach have already given very detailed tutorial reviews [96–99]. Here, we will briefly discuss NSTF electrocatalyst preparation methods, UTCL structures, and the most attractive composition giving high performance. A Kapton is used as the substrate; for the purposes of enlarging the effective geometrical surface area and achieving a uniform distribution of PR149, the Kapton can be folded to give a $\sqrt{2}$ larger patterned surface. Usually, a very thin layer of metal such as Al, Co, Mo, Ni, Pt, or Si is first sputtered onto the Kapton to avoid the electrostatic effects that generally occur on such non-conducting substrates. Due to its high chemical and thermal stability, PR149 is chosen to form a thin film on the substrate; this is usually implemented by vapor deposition, specifically sublimation deposition. During this sublimation deposition, the substrate temperature, the deposition rate, and the angle of incidence are the keys to forming the required morphology of the thin film. The substrate is usually kept at room temperature, and the deposition rate is controlled at 2–40 nm/min. The thin film is then annealed under vacuum at $<230°C$. Arrays of PR149 grow on the substrate; the whisker cross-sectional dimensions are ~27 × 55 nm, their lengths are 1–1.5 μm, and the real number densities reach ~50 per square micron. The morphology of the PR149 whiskers, the whisker arrays on the substrate and the Pt-nanostructured thin-film electrode are shown in Figure 3.11. The detailed morphology of the Pt nanostructure on the whisker can be seen in Figure 3.12 [100]. It appears that the nanostructured thin film is not originally film-like; instead, it resembles nanograins packed together. The crystallinity analysis indicates that the structure of the Pt grain tips depends greatly on the crystal orientation of the PR149 facets. However, this fine structure will disappear and will be replaced by a continuous thin film when the NSTF catalyst is applied under potential [101].

Pt-based catalysts can be deposited onto the whiskers by dry chemical and physical methods, specifically magnetron sputtering deposition. Much effort has been devoted to the development of binary and ternary Pt-based NSTF electrocatalysts such as $Pt_{1-x}M_x$ (M = Ru, Mo, Co, Ta, Au, Sn) as anode electrocatalysts [102], Pt and Pt–Co–Mn electrocatalysts for ORR [103], dealloyed

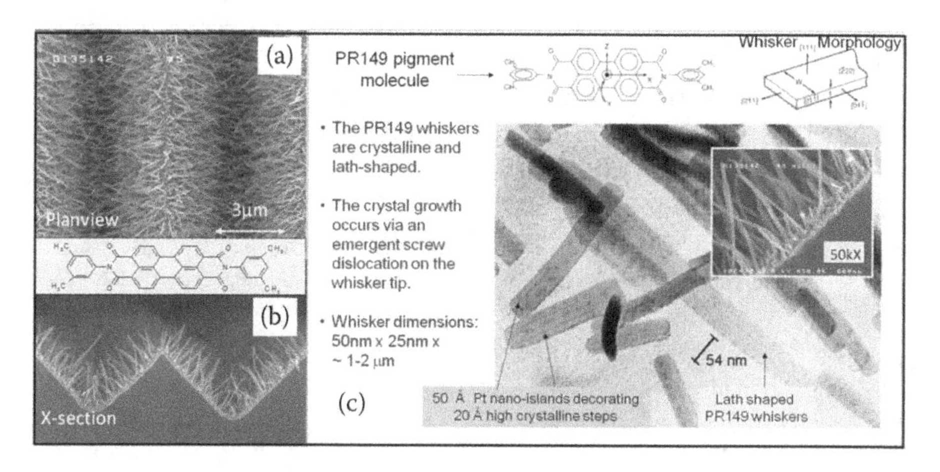

FIGURE 3.11 SEM and TEM images of NSTF catalyst support whiskers: (a and b) as a monolayer of oriented whiskers on a microstructured substrate web; (c) decorated with Pt nanoislands for TEM imaging. Reproduced by permission from ref. [97]; Copyright 2012, American Electrochemical Society.

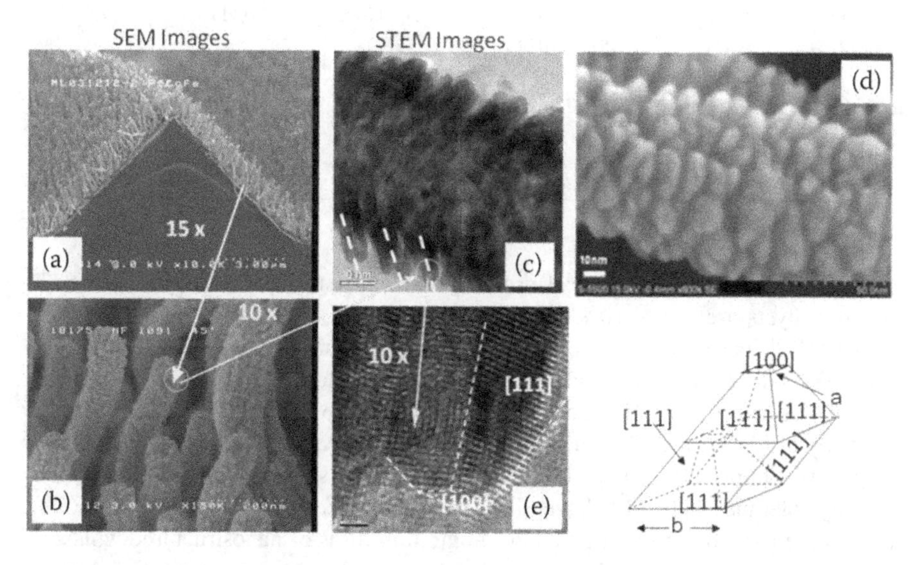

FIGURE 3.12 The morphology and crystallinity of Pt-decorated PR149 whiskers. Reproduced by permission from ref. [100]; Copyright 2008, American Chemical Society.

$Pt_{1-x}Ni_x$($0.65 < x < 0.75$) as cathode electrocatalysts [104], $Pt_x M_y$ and $Pt_x M_y N_z$ (M, N = Ni, Co, Zr, Hf, Fe, Mn; $0 \le x, y, z < 1$) for ORR [105], and to determination of their stability [106,107] and intrinsic activities [108–110]. The RDE mass activity and MEA mass activity are shown in Figure 3.1. Although most NSTF Pt alloys or intermixes can exceed the DOE 2020 RDE mass activity target, and some can even reach 7 mA/cm^2 PGM, only dealloyed Pt_3Ni_7 can meet the MEA

TABLE 3.1
Best-of-class MEAs. The PEM was either 3M 825 EW 24 μm thick (2012) or supported 3M 725 EW 14 μm PEM (2015). The cathode GDL was 3M 2979, and the anode GDL was either 3M 2979 (2012, 2015) or 3M experimental "X2" (cited from Ref. [99])

MEA ID	Anode (mg Pt/cm^2)	Cathode (mg Pt/cm^2)	Interlayer (mg Pt/cm^2)	Total PGM (mg Pt/cm^2)	Spec. Power @ 0.67 V (kW/g)
2012 (Mar.) (2)	Pt/ NSTF (0.03)	Baseline Dealloyed Pt_3Ni_7/ NSTF (0.121)	–	0.151	4.9
2015 (Jan.) (3)	PtCoMn/ NSTF, (0.015)	ChemD P1 Pt_3Ni_7/ NSTF (0.103)	–	0.118	8.4
2015 (Mar.) (2)	PtCoMn/ NSTF, (0.015)	ChemD P1 Pt_3Ni_7/ NSTF (0.103)	A (0.015)	0.133	7.8
2015 New "A" (2)	PtCoMn/ NSTF, (0.015)	ECD P1 Pt_3Ni_7 (0.087)	A (0.015)	0.117	7.5
2015 New "B" (1)	PtCoMn/ NSTF, (0.015)	ECD P1 Pt_3Ni_7 (0.075)	A (0.015)	0.105	7.5

mass activity target. For practical use, the best classes of MEAs based on NSTF electrocatalysts are summarized in Table 3.1.

The UTCL based on NSTF electrocatalysts is shown in Figure 3.13. The UTCL is only 0.25–0.67 μm, and the whiskers are not normal to the membrane; the latter is not really critical to the MEA performance according to the Toyota group [80]. The durability of MEA, including metal leaching from the catalysts [111,112], oxidative resistivity at high voltage [113,114], the endurance of start-stop cycling [115], and impurity tolerance [116,117] have all been investigated by 3M. $Pt_{1-x}M_x$ (M = Fe, Ni) was prepared by sputtering Pt and M onto thin films of nanostructured whisker-like supports; for $0<x<1.0$, the formation of randomly ordered substitutional solid solutions of $Pt_{1-x}Fe_x$ and $Pt_{1-x}Ni_x$ alloys was observed. For small values of x ($x<0.6$), no substantial changes in the lattice size were observed upon dissolution of Fe or Ni, suggesting that the dissolved transition metals originate from the surface. For $x>0.6$, the lattice constant expanded, indicating that transition metals also dissolve from the bulk [111]. The dissolution of transition metals (Co, Ni, Mn, Fe) from $Pt_{1-x-y}M_xM'_y$ electrocatalysts after operation in PEM hydrogen fuel cells or after treatment with 1 M H_2SO_4 at 80°C was reported. For the elements studied here, it was found that the catalyst with a composition of $x+y\geqslant0.25$ changed to $x+y\approx0.25$ after fuel cell or acid testing due to the dissolution of transition metals,

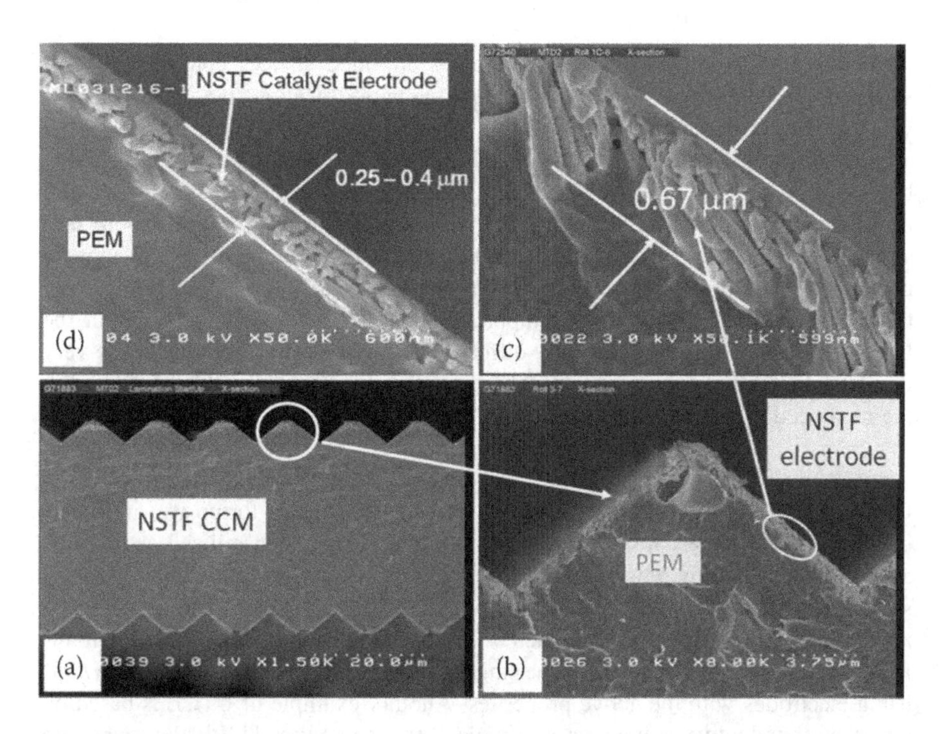

FIGURE 3.13 SEM cross-sections of NSTF electrodes with the whole CCM, expanded view of UTCL. Reproduced by permission from ref. [98]; Copyright 2013, American Electrochemical Society.

independent of the choice of M and M'. Therefore, the composition with $x+y \approx 0.25$ was near the percolation limit for diffusion of transition metals from the interior of the alloy [112]. Both for scanning between 0.6 and 1.2 volts at 20 mV/sec under H_2/N_2 and under steady-state conditions of 1.5 volts, NSTF demonstrated significantly greater durability. At 1.5 volts, NSTF catalysts lost no surface area, specific activity or fuel cell performance over periods as long as three hours, whereas the Pt/C-based electrodes lost large amounts of surface area, activity, and performance in just 30 minutes at 1.5 volts [113]. Continuous cycling between 0.6 and 1.2 V at various temperatures between 65 and 95°C with H_2/N_2 on the anode and cathode showed that the surface areas of the NSTF electrocatalysts were significantly more stable than those of the Pt/C electrocatalysts [114]. An automotive start/stop (S/S) testing protocol indicated that the NSTF catalyst showed much less loss of performance under the S/S cycling, and most of the loss was recoverable [115]. A ternary composition spread, $(Pt_{1-x}Ru_x)_{1-y}Mo_y$, $0<x<1$, $0<y<0.3$, was prepared through sputter deposition onto a nanostructured thin film support. The addition of either Mo or Ru to Pt led to a reduction in hydrogen oxidation overpotential for a simulated reformate gas stream containing up to 50 ppm CO, with a best composition of $Pt_{0.40}Ru_{0.35}Mo_{0.25}$ [116]. Chloride and sulfide ions were found to cause rapid but highly recoverable performance loss. It was interesting to note that the impact of chloride on the decay rate depended strongly on catalyst type, surface area, and cell temperature. The effects of these externally provided species could remain for several hours after they were removed from the humidification streams, and in this way the cell performance could be largely recovered [117]. It appears that due to the absence of carbon and ionomer materials in UTCL, the electrode is very resistant at high voltage and robust under potential cycling. However, due to its very thin electrode structure and limited ECA, the NSTF electrode appears to be more sensitive to the impurity. The metal leaching, which is a common phenomenon in binary and ternary electrocatalysts and is not specifically limited to NSTF electrodes, is another issue. Another important topic, water management within the NSTF electrode, which significantly differs from the traditional CL due to the very thin CL of NSTF and the very hydrophilic properties of NSTF, will be discussed in the next section.

3.3.3.1.5 Others

Due to their relatively large surface area, large voidage, low tortuosity, and interconnected macropores, inverse opal 3-D structures have been utilized as a robust and integrated configuration of catalyst layers [118]. Pt infiltration into a polystyrene (PS) template and galvanostatic-pulsed electrodeposition onto the template are performed. The PS matrix can then be removed, and a Pt inverse opal skeleton is formed, as shown in Figure 3.14. The single cell performance can reach 1.25 W/cm^2 using electrodes with the above properties. Another example of UTCL is based on electronic conductive polymer–polypyrrole (PPy) nanowires [119]. PPy nanowire arrays are synthesized using a typical electrochemical polymerization method and conducted through a workstation. Pd nanoparticles are then sputtered onto the PPy nanowire arrays by PVD in an argon atmosphere. Subsequently, Pt galvanic displacement is applied to form a PtPd catalyst thin layer with whiskerette shapes along the PPy nanowires. The single cell yields a maximum performance of 762.1 mW cm^{-2} with a low MEA Pt loading of 0.241 mg Pt cm^{-2}.

3.3.3.2 Proton Conductor Arrays

Since electron pathways and proton pathways physically coexist in the catalyst layer, proton conductor arrays using Nafion-based ionomers naturally come to mind. Two ideas have been advanced regarding the control of the morphologies of proton conductor arrays. The first is to grow ionomer pillars on the membrane and

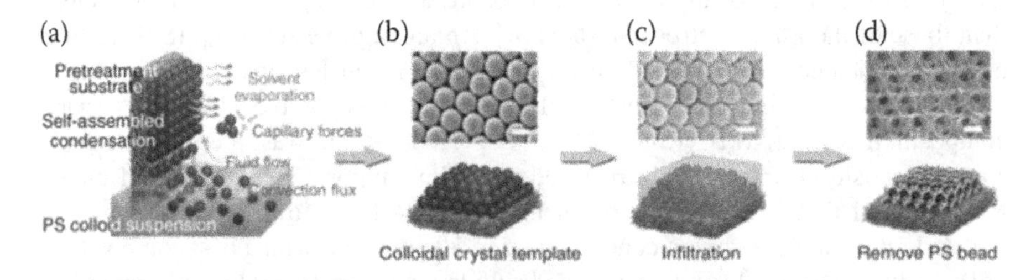

FIGURE 3.14 Processing of inverse opal 3-D structures with a Pt skeleton and MEA structure. Reproduced by permission from ref. [118]; Copyright 2013, Macmillan Publishers Limited.

to then fill the spaces between these pillars with traditional carbon-supported electrocatalysts. The second idea is to burst forth regular ionomer arrays and cover the ionomer wires with electrocatalysts, forming thin-film catalysts supported by ionomer. These two ideas have been implemented by the DOE National Laboratories, as shown in Figure 3.15 [120]. Although the single cell reaches 1.1 A/cm^2 @ 0.6 V, it appears that there is still a long way to go in this field. Regarding the filler, Pt/C, of course, is the candidate in the traditional fuel cell catalyst layer; recently, Volkswagen and Stanford collaborated to demonstrate the use of a passivation-gas-incorporated ALD (PALD) to reduce the thickness of nanoparticles deposited during ALD. The direct deposition of thinner Pt nanoparticles onto carbon-based catalyst supports enables greater Pt utilization due to the greater accessibility of the nanoparticles. At 1 μg/cm^2 Pt loading, the mass activity reaches 1.4 A/mg Pt [121].

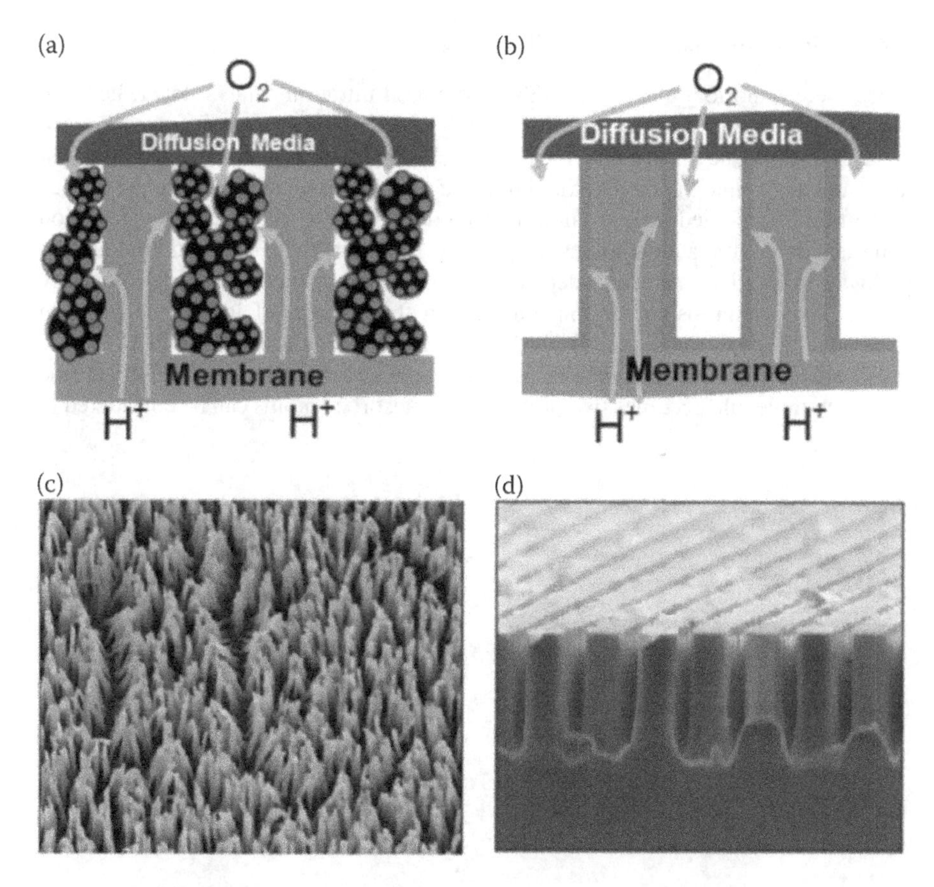

FIGURE 3.15 The idea and SEM images of proton conductor arrays for UTCL. Reproduced by permission from ref. [120]; Copyright 2018, US DOE.

3.4 CHALLENGES

Although much effort has been expended to revolutionize fuel cell electrode structure beyond the traditional ionomer-immersed agglomerate structure, which was a significant breakthrough contribution in the early 1990s that made it possible to reduce the PGM loading from 4 mg/cm^2 to 0.4 mg/cm^2, among the types of electrocatalysts discussed previously, the NSTF-structured UTCL based on Co–OH–CO_3 nanowire-assisted arrays or PR149 nanowire-assisted arrays seems most practical for real application to date. In both cases, Pt-based thin film electrocatalysts on nanowires form a single-phase, freestanding, ionomer-free, and array-like ultrathin catalyst layer. The ionomer-free and ultrathin characteristics of these electrodes differentiate them from agglomerate-aggregates-thin-film electrodes. The main challenges include charge transport, charge transfer, and mass transport; water management will be specifically discussed below.

3.4.1 Proton Transport and Transfer

Since there is no ionomer in the NSTF-structured ultrathin catalyst layer, how are protons transported through the whole catalyst layer? To answer this question, we need to investigate the proton transport mechanism in the ionomer-free catalyst layer and determine the factors that influence this process. Derived from the porous system and combined with a simplified model with two different Pt distribution patterns, one with a continuous Pt-pathway and one with a discontinuous Pt-pathway realized by selective deposition of Pt film on the rib-channel of glassy carbon, two scenarios for proton transport in the absence of ionomer have been generated, as shown in Figure 3.16 [122]. The first scenario involves Pt-H formed by H^+ adsorption either from the Pt/Nafion interface or from hydrogen spillover (the hydrogen molecule becomes two protons). The adsorbed atoms can be considered to

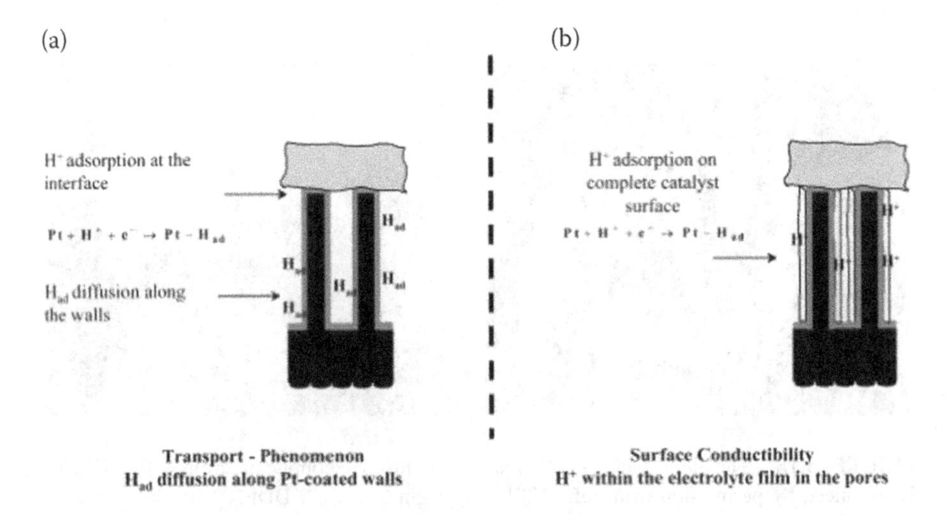

FIGURE 3.16 Two proton transport mechanisms in the absence of ionomer. Reproduced by permission from ref. [122]; Copyright 2003, Elsevier.

diffuse along the channel walls. This explanation is supported by the decrease in catalyst surface utilization with increasing sweep rate observed in experiments. The second possible scenario pertains to the thin water film on the wall, which actually determines the surface conductivity. It is the solvation of H^+ ions within the water film and diffusion along the water film that determines the whole proton transport process. The study is substantiated by the observation that when the thickness of the catalyst layer is decreased, the ECA or Pt utilization efficiency also decreases [123]. It is reported that the spillover hydrogen may bind, for example, at unsaturated edge sites or may form a layer of weakly bound, mobile hydrogen atoms. This is akin to the situation depicted in Figure 3.16a. Surface diffusion of the spilled-over hydrogen proceeds by a series of almost classical hops from carbon trap to carbon trap. The electrochemical utilization of Pt particles that are not in contact with the electrolyte will be strongly dependent upon the length scale of the surface spillover. Another experimental finding takes into account the existence of mesoporous Pt arrays filled by pure water and in direct contact with the Nafion membrane [124]. By comparing the electrochemical behaviors of the designed systems with that of mesoporous Pt arrays filled by H_2SO_4, it was found that the Pt surface in pores without Nafion in such a perpendicular mesoporous electrode was electrochemically active and that its response was very different from that of the Pt surface covered with Nafion. By electrochemical impedance spectroscopy (EIS), Pt-black electrodes with PTFE (bound, unbound, and ionomer-bound), as well as conventional Pt/Vulcan, were analyzed and compared [125]. The ionic conductivity of the ionomer-free electrode is 1–2 orders of magnitude lower than that of the ionomer-bound electrodes, but it is 2–3 orders of magnitude higher than that of bulk water. In effect, the interaction between the Pt surface and adsorbed water determines the proton conduction. By designing a layered structure, the ionic conductivity is directly measured in ionomer-free (and binder-free) Pt black electrodes [126]. The ionic conductivities of ionomer/binder-free Pt black respond positively to the relative humidity of the gases, and protons diffuse through the Pt-water pathway. A numerical model that describes the electrode in the above fashion is developed in which the Poisson-Nernst-Planck theory accounts for the proton transport [127]. The thickness of the catalyst layer plays a crucial role in determining the cathode performance with respect to mass transport and kinetics. A model that calculates the oxygen reduction in water-filled, cylindrical nanopores with platinum walls [128] supports the idea that interaction of the protons with the charged pore walls drives proton migration into the ionomer-free channels. This interaction highly determines the pore effectiveness and thus has a significant influence on the UTCL performance. A one-dimensional single-phase fuel cell model is deployed to estimate NSTF whisker proton conductivity as a function of RH [129]. It is found that the poor proton conductivity of NSTF at lower RHs leads to poor Pt utilization and thus to significant losses, especially at higher current densities. To make the best use of Pt without significant liquid water blockage in the electrode, a higher water vapor concentration in the NSTF electrode is required. Overall, the key factors that impact proton transport within the ionomer-free ultrathin catalyst layer are as follows: i) proton species that result from hydrogen spillover, the Pt/Nafion interface and water; ii) relative humidity or gas water from the humidity gas, resulting in a proton source or a solvation atmosphere; iii) the potential of the pore wall, which will interact with protons, impairing proton distribution and transport along the pore; iv) the pore size and pore length. Regarding the fourth factor, for example, the diffusion length of the double layer from the interface of the Nafion membrane can be estimated by the Gouy-Chapman law for the electrical double layer [124] according to the expression $\frac{1}{k} = \frac{1}{(3.29 \times 10^7)zC^{0.5}}$, where C is the bulk z electrolyte concentration in mol/L and κ is given in cm^{-1}. Thus, even for pure water (C = ~10^{-7} mol/L), the thickness of the diffuse layer is estimated to be ~961 nm, implying that the Nafion can still acidify the pure water providing the proton present in a pore of such length. Another issue concerns the proton transfer, since ORR is a proton-coupled electron transfer (PCET) process. Once proton transfer becomes the limiting step, H_2O_2 production might be preferred, especially at the tip of the tube, and this would negatively impact the fuel cell's durability. Therefore, the future research interesting might be i) the proton transport mechanisms for the ionomer-free ultrathin catalyst layer; ii) the role of water in supplying proton sources and carrying protons; iii) re-view of proton transport and transfer within the traditional catalyst layer.

3.4.2 Electron Conduction

Two main points should be emphasized regarding electron conduction at different scales of the ultrathin catalyst layer; these are the surface charge density at the pore walls of Pt-based electrocatalysts and electron transport or current flow in the plane of the catalyst layer. Based on the discussion in the section on proton transport and transfer, the relationship of the surface charge density at the pore walls to the electrode potential can be depicted by the Stern model, while potential distribution within the pores is delineated according to the Poisson–Nernst–Planck theory [128]. The electrostatic interaction between protons and the metal's surface charge determines the distribution of protons and the electrostatic potential in the pore. Thus, the deviation of the electrode potential from the potential of zero charge of the metal phase is the vital determinant of the effectiveness of platinum utilization. The lower the applied potential relative to the potential of zero charge is, the more negative the surface charge will be, and the higher will be the proton concentration and the current density produced at the pore walls. On the other hand, the oxygen depletion profile will depend highly on the interplay of electrostatic function from the overlap of the double layer with respect to the tube array structure, which is physically expressed by the radius and the thickness of the tube [128]. EIS coupled with modeling work further confirmed that kinetics, electrostatic interactions, and transport all contribute to the overall Faradaic current density in ionomer-free catalyst layers [130]. Unlike the agglomerate structure of the traditional catalyst layer, the electron conductor is randomly distributed in both the in-plane and through-plane directions; therefore, proton and electron gradients may form within the catalyst layer parallel to the membrane. In the case of the ionomer-free ordered array structure, in-plane current flow could theoretically be avoided, mitigating the possible degradation caused by the H_2-air interface. Regarding the future research areas, the followings might be worthy of exploring i) fundamental of modeling work on the potential distribution of the pore structure; ii) the current distribution in 3-D directions of the catalyst layer.

3.4.3 Mass Transport

According to 3M's surface collision frequency model, it is assumed that electrode surface area densities of at least 3×10^5 cm^2 Pt/cm^3 are most desirable because they contribute directly to the rate of successful chemisorption and oxygen reduction in the high current density region (iR corrected 0.7 V) [131]. While the NSTF electrode can fulfill the requirement for active surface area density, a sudden voltage drop is observed during a rapid potentiodynamic polarization scan [132] caused by flooding, as shown in Figure 3.17(a). Modeling work suggests that the first generation of GDL based on Toray carbon paper is dominated by gas phase diffusion driven by a gradient of water vapor saturation across the GDL. For ultrathin catalyst electrode layers, the most successful path to improving water management under cool conditions will be to improve the liquid water transport characteristics within the GDL. Another study addresses the failure start due to the flooding based on the transit behaviors; this problem can be solved by substituting a new anode backing layer, as shown in Figure 3.17(b) [133]. With respect to water management, the

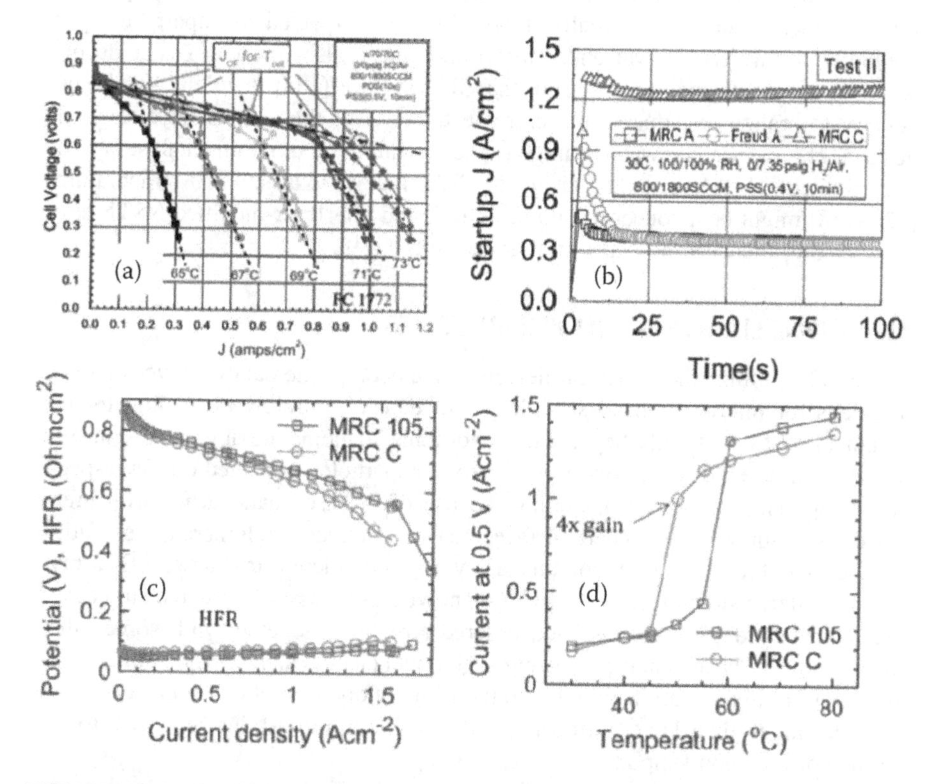

FIGURE 3.17 (a) Sudden voltage drop at the high current density region at different operating voltages. Reproduced by permission from ref. [132]; Copyright 2007, American Electrochemical Society. (b) Transit behaviors with different GDLs starting at 0.4 V with MRC C successful start-up. Reproduced by permission from ref. [133]; Copyright 2011, American Electrochemical Society. (c, d) Cell performance improvement with new anode backing layers. Reproduced by permission from ref. [135]; Copyright 2018, Cell Press.

properties of the anode GDL backing layer, which include factors such as backing treatment, MPL thickness and composition (adding ZrO_2), play the most important role in the response of MEAs with ultrathin electrodes to both low and high temperatures. In addition, when the operating temperature is lower, significant water condensation occurs within the fuel cell, resulting in very poor cell performance. By reducing the anode hydrogen reactant operating pressure from 150 kPa to 30 kPa, the limiting current density was found to increase by a factor of up to four [134]. This shows that the generated liquid water can be removed through the anode side by water back-diffusion. Recently, the current water-management limitations of these thin layers were overcome through the use of novel materials and, surprisingly, by the use of a novel GDL on the anode side [135]. This upgraded GDL functions differently under different operating conditions: it removes liquid water from the anode at low and moderate temperatures, and it has no significant impact at high temperature, where water exists and is transported mainly in the vapor phase; thus, it provides a better water management solution. It is found that the modulated structure of banded regions with high and low fiber densities creates high-porosity regions where water preferentially flows; these are expected to impart high permeability for water removal and, more importantly, also result in easier droplet water removal into the channel from the GDL surface [135]. It might also be necessary to realize that due to the extreme thinness of the catalyst layer and the absence of ionomer, the water storage capacity within the ultrathin catalyst layer is very limited. Therefore, cold starting of PEM fuel cells at subzero temperatures [136–138] might be a concern. Ionomer, SiO_2, and interlayer-modified NSTF have been used to increase the water storage capacity [139].

3.5 CONCLUSIONS AND PERSPECTIVES

To further optimize the structural design of nanocomposite catalyst layer of PEM fuel cells for improving mass-specific power density, ultralow Pt loading for the design of PEM fuel cells beyond the traditional agglomerate-aggregate-thin-film structure based on discrete low-aspect-ratio nanoparticles dispersed on low-aspect-ratio supports, is reviewed, including the use of Pt-based nanoparticles on high-aspect-ratio supports, extended surface area of confined agglomerates, extended surface area of nanowires, carbon tube array-assisted ordered structures, TiO_2 array-assisted ordered structures, Co-OH-CO_3 nanowire array-assisted ordered structures, and pigment red 149 array-assisted ordered structures, together with some other novel designs. The advantages of using such novel electrocatalysts and cutting-edge structures are high Pt utilization due to their highly open structures, enhanced mass transport due to their lower tortuosity, and possible improved durability due to the absence of a carbon support.

Catalysts based on extended surface area of confined agglomerates and nanowires exhibit high RDE mass activity, but their MEA mass activity can hardly reach the target set by the DOE. This implies that electrodes featuring fibers or nanowires packed parallel to membrane-like nonwoven fabrics might require further modification. With respect to real-world applications, electrodes with ordered structures are more practical compared to other novel electrode design discussed in this review. Electrodes with such structures are usually obtained by using ordered array substrates to assist the formation of either Pt-decorated or nanostructured thin films. Among them, carbon-tube-array-assisted structures developed by Toyota, Co-OH-CO_3 nanowire array-assisted structures developed by Dalian Institute of Chemical Physics, and pigment red 149 array-assisted structures developed by 3M have all achieved the DOE 2020 target for Pt mass-specific power density. The catalysts demonstrate excellent mass activity and show extraordinary cycling durability.

However, Pt-based thin film electrocatalysts on nanowires forms a single-phase, freestanding, ionomer-free, and array-like ultrathin catalyst layer structure. These characteristics make them quite different from the agglomerate-aggregate-thin film structure with respect to charge transfer, charge transport, mass transport, and water management. Five factors impact proton transport within the ionomer-free ultrathin catalyst layer: i) proton species originating from hydrogen spillover, reactions at the Pt/Nafion interface; ii) relative humidity or water, which can provide a direct proton source or solvation atmosphere; iii) the potential of the pore wall, which will interact with protons and thereby affect proton distribution and transport along the pore; iv) the scale of pore size and pore length. There are two main points regarding electron conduction at different scales within the ultrathin catalyst layer: i) surface charge density at the pore walls of Pt-based electrocatalysts; and ii) electron transport or current flow in the plane of the catalyst layer. The most challenging issues encountered in the use of electrocatalysts with ultrathin and ionomer-free structure are the water management-related issues in PEM fuel cells, i.e., flooding of the high current density region, poor transient behavior, and cold-start failure. Recent important findings show that the use of an appropriate anode gas diffusion layer can successfully remove water generated at the cathode from the anode side. The ultrathin and ionomer-free structure represents a novel model structure that opens a new path to achieving regulation and control of proton transport and transfer, electron conduction, and mass transport.

ACKNOWLEDGMENTS

This work was supported by the National Natural Science Foundation of China (21503134, 21406220), the National Key Research and Development Program of China (2016YFB0101201), and the Science and Technology Commission of Shanghai Municipality (15YF1406500).

REFERENCES

[1] B. C. H. Steele, A. Heinzel, Materials for fuel-cell technologies, *Mater. Renew. Sustain. Energy*, 2010, 224–231.

[2] A. J. Appleby, F.R. Foulkes, *Fuel cell handbook, United States: N. p.,. Web.*, 1988.

[3] T. E. Springer, T. A. Zawodzinski, S. Gottesfeld, *Polymer electrolyte fuel cell model, J. Electrochem. Soc.*, 1991, **138**, 2334.

[4] R. Borup, J. Meyers, B. Pivovar, Y. S. Kim, R. Mukundan, N. Garland, D. Myers, M. Wilson, F. Garzon, D. Wood, P. Zelenay, K. More, K. Stroh, T. Zawodzinski, J. Boncella, J. E. McGrath, M. Inaba, K. Miyatake, M. Hori, K. Ota, Z. Ogumi, S. Miyata, A. Nishikata, Z. Siroma, Y. Uchimoto, K. Yasuda, K.-i. Kimijima, N.

Iwashita, *Scientific aspects of polymer electrolyte fuel cell durability and degradation, Chem. Rev.*, 2007, **107**, 3904.
[5] J. Hou, M. Yang, C. Ke, G. Wei, C. Priest, Z. Qiao, G. Wu, J. Zhang, *Platinum-group-metal catalysts for proton exchange membrane fuel cells: From catalyst design to electrode structure optimization, EnergyChem*, 2020, **2**, 100023.
[6] R. M. Mark, F. Mathias, Hubert A. Gasteiger, Jason J. Conley, Timothy J. Fuller, Craig J. Gittleman, Shyam S. Kocha, Daniel P. Miller, Corky K. Mittelsteadt, Tao Xie, S. G. Van, Paul T. Yu, *Two fuel cell cars in every garage, Electrochem. Soc. Interface*, 2005, **14**, 24.
[7] F. T. Wagner, B. Lakshmanan, M. F. Mathias, *Electrochemistry and the Future of the Automobile, J. Phys. Chem. Lett.*, 2010, **1**, 2204.
[8] J. Hou, Y. Shao, M. W. Ellis, R. B. Moore, B. Yi, *Graphene-based electrochemical energy conversion and storage: fuel cells, supercapacitors and lithium ion batteries, PCCP*, 2011, **13**, 15384.
[9] H. Abe, *Current status and future of the car exhaust catalyst, Q. Rev.*, 2011, **39**.
[10] S. Rodatz, G. Paganelli, L. Guzzella, *Optimizing air supply control of a PEM fuel cell system, Proceedings of the 2003 American Control Conference*, 2003.
[11] V. Yarlagadda, M. K. Carpenter, T. E. Moylan, R. S. Kukreja, R. Koestner, W. Gu, L. Thompson, A. Kongkanand, *Boosting fuel cell performance with accessible carbon mesopores, ACS Energy Lett.*, 2018, **3**, 618.
[12] J. Zhang, M. B. Vukmirovic, Y. Xu, M. Mavrikakis, R. R. Adzic, *Controlling the catalytic activity of platinum-monolayer electrocatalysts for oxygen reduction with different substrates, Angew. Chem.*, 2005, **117**, 2170.
[13] H. A. Gasteiger, N. M. Marković, *Just a dream—or future reality? Science*, 2009, **324**, 48.
[14] H. A. Gasteiger, S. S. Kocha, B. Sompalli, F. T. Wagner, *Activity benchmarks and requirements for Pt, Pt-alloy, and non-Pt oxygen reduction catalysts for PEMFCs, Appl. Catal. B*, 2005, **56**, 9.
[15] *Fuel Cell R&D Annual Merit Review Proceedings*. 2018.
[16] P. Strasser, S. Koh, T. Anniyev, J. Greeley, K. More, C. Yu, Z. Liu, S. Kaya, D. Nordlund, H. Ogasawara, M. F. Toney, A. Nilsson, *Lattice-strain control of the activity in dealloyed core–shell fuel cell catalysts, Nat. Chem.*, 2010, **2**, 454.
[17] J. K. Nørskov, J. Rossmeisl, A. Logadottir, L. Lindqvist, J. R. Kitchin, T. Bligaard, H. Jónsson, *Origin of the overpotential for oxygen reduction at a fuel-cell cathode, J. Phy. Chem. B*, 2004, **108**, 17886.
[18] N. M. Marković, P. N. Ross, *Surface science studies of model fuel cell electrocatalysts, Surf. Sci. Rep.*, 2002, **45**, 117.
[19] R. R. Adzic, J. Zhang, K. Sasaki, M. B. Vukmirovic, M. Shao, J. X. Wang, A. U. Nilekar, M. Mavrikakis, J. A. Valerio, F. Uribe, *Platinum monolayer fuel cell electrocatalysts, Top. Catal.*, 2007, **46**, 249.
[20] L. Qu, Y. Liu, J.-B. Baek, L. Dai, *Nitrogen-doped graphene as efficient metal-free electrocatalyst for oxygen reduction in fuel cells, ACS Nano*, 2010, **4**, 1321.
[21] K. Gong, F. Du, Z. Xia, M. Durstock, L. Dai, *Nitrogen-doped carbon nanotube arrays with high electrocatalytic activity for oxygen reduction. Science*, 2009, **323**, 760.
[22] Y. Shao, G. Yin, Y. Gao, *Understanding and approaches for the durability issues of Pt-based catalysts for PEM fuel cell, J. Power Sources*, 2007, **171**, 558.
[23] M. K. Debe, *Electrocatalyst approaches and challenges for automotive fuel cells, Nature*, 2012, **486**, 43.
[24] L. Chong, J. Wen, J. Kubal, F. G. Sen, J. Zou, J. Greeley, M. Chan, H. Barkholtz, W. Ding, D.-J. Liu, *Ultralow-loading platinum-cobalt fuel cell catalysts derived from imidazolate frameworks, Science*, 2018.

[25] M. S. Wilson, S. Gottesfeld, *Thin-film catalyst layers for polymer electrolyte fuel cell electrodes*, *J. Appl. Electrochem.*, 1992, **22**, 1.
[26] M. S. Wilson, S. Gottesfeld, *High performance catalyzed membranes of ultra-low Pt loadings for polymer electrolyte fuel cells*, *J. Electrochem. Soc.*, 1992, **139**, L28.
[27] N. Nonoyama, S. Okazaki, A. Z. Weber, Y. Ikogi, T. Yoshida, *Analysis of oxygen-transport diffusion resistance in proton-exchange-membrane fuel cells*, *J. Electrochem. Soc.*, 2011, **158**, B416.
[28] A. Ohma, T. Mashio, K. Sato, H. Iden, Y. Ono, K. Sakai, K. Akizuki, S. Takaichi, K. Shinohara, *Analysis of proton exchange membrane fuel cell catalyst layers for reduction of platinum loading at Nissan*, *Electrochim. Acta*, 2011, **56**, 10832.
[29] A. Kongkanand, N. P. Subramanian, Y. Yu, Z. Liu, H. Igarashi, D. A. Muller, *Achieving high-power PEM fuel cell performance with an ultralow-Pt-content core–shell catalyst*, *ACS Catal.*, 2016, **6**, 1578.
[30] A. Kongkanand, M. F. Mathias, *The priority and challenge of high-power performance of low-platinum proton-exchange membrane fuel cells*, *J. Phys. Chem. Lett.*, 2016, **7**, 1127.
[31] T. A. Greszler, D. Caulk, P. Sinha, *The impact of platinum loading on oxygen transport resistance*, *J. Electrochem. Soc.*, 2012, **159**, F831.
[32] W. T. Grubb, L. W. Niedrach, *Batteries with solid ion-exchange membrane electrolytes: II. low-temperature hydrogen-oxygen fuel cells*, *J. Electrochem. Soc.*, 1960, **107**, 131.
[33] L. W. Niedrach, H. R. Alford, *A new high-performance fuel cell employing conducting-porous-Teflon electrodes and liquid electrolytes*, *J. Electrochem. Soc.*, 1965, **112**, 117.
[34] *Solid polymer electrolyte fuel cells (SPEFCs)*. *Energy*, 1986, **11**, 137.
[35] I. D. Raistrick, *Modified gas diffusion electrode for proton exchange membrane fuel cells. Proceedings of the symposium on diaphragms, separation, and ion-exchange membranes*, *Ponnington (NJ): Electrochemical Society*, 1986.
[36] X. H. Yan, P. Gao, G. Zhao, L. Shi, J. B. Xu, T. S. Zhao, *Transport of highly concentrated fuel in direct methanol fuel cells*, *Appl. Therm. Eng.*, 2017, **126**, 290.
[37] X. H. Yan, X. L. Zhou, T. S. Zhao, H. R. Jiang, L. Zeng, *A highly selective proton exchange membrane with highly ordered, vertically aligned, and subnanosized 1D channels for redox flow batteries*, *J. Power Sources*, 2018, **406**, 35.
[38] X. Yan, C. Guan, Y. Zhang, K. Jiang, G. Wei, X. Cheng, S. Shen, J. Zhang, *Flow field design with 3D geometry for proton exchange membrane fuel cells*, *Appl. Therm. Eng.*, 2019, **147**, 1107.
[39] T. E. Springer, I. D. Raistrick, *Electrical impedance of a pore wall for the flooded-agglomerate model of porous gas-diffusion electrodes*, *J. Electrochem. Soc.*, 1989, **136**, 1594.
[40] I. D. Raistrick, *Impedance studies of porous electrodes*, *Electrochim. Acta*, 1990, **35**, 1579.
[41] M. Ciureanu, R. Roberge, *Electrochemical impedance study of PEM fuel cells. experimental diagnostics and modeling of air cathodes*, *J. Phys. Chem. B*, 2001, **105**, 3531.
[42] N. P. Siegel, M. W. Ellis, D. J. Nelson, M. R. von Spakovsky, *Single domain PEMFC model based on agglomerate catalyst geometry*, *J. Power Sources*, 2003, **115**, 81.
[43] W. Sun, B. A. Peppley, K. Karan, *An improved two-dimensional agglomerate cathode model to study the influence of catalyst layer structural parameters*, *Electrochim. Acta*, 2005, **50**, 3359.
[44] E. Middelman, *Improved PEM fuel cell electrodes by controlled self-assembly*, *Fuel Cells Bull.*, 2002, **2002**, 9.

[45] C. Y. Du, X. Q. Cheng, T. Yang, G. P. Yin, P. F. Shi, *Numerical simulation of the ordered catalyst layer in cathode of proton exchange membrane fuel cells*, *Electrochem. Commun.*, 2005, **7**, 1411.

[46] M. Chisaka, H. Daiguji, *Design of ordered-catalyst layers for polymer electrolyte membrane fuel cell cathodes*, *Electrochem. Commun.*, 2006, **8**, 1304.

[47] C. Y. Du, T. Yang, P. F. Shi, G. P. Yin, X. Q. Cheng, *Performance analysis of the ordered and the conventional catalyst layers in proton exchange membrane fuel cells*, *Electrochim. Acta*, 2006, **51**, 4934.

[48] S. M. Rao, Y. Xing, *Simulation of nanostructured electrodes for polymer electrolyte membrane fuel cells*, *J. Power Sources*, 2008, **185**, 1094.

[49] M. M. Hussain, D. Song, Z. S. Liu, Z. Xie, *Modeling an ordered nanostructured cathode catalyst layer for proton exchange membrane fuel cells*, *J. Power Sources*, 2011, **196**, 4533.

[50] S. Sun, H. Zhang, M. Pan, *Dynamic simulation of oxygen transport rates in highly ordered electrodes for proton exchange membrane fuel cells*, *Fuel Cells*, 2015, **15**, 456.

[51] Y. Zeng, H. Zhang, Z. Wang, J. Jia, S. Miao, W. Song, Y. Xiao, H. Yu, Z. Shao, B. Yi, *Nano-engineering of a 3D-ordered membrane electrode assembly with ultrathin Pt skin on open-walled PdCo nanotube arrays for fuel cells*, *J. Mater. Chem. A*, 2018, **6**, 6521.

[52] M. Sudan Saha, R. Li, M. Cai, X. Sun, *High electrocatalytic activity of platinum nanoparticles on SnO2 nanowire-based electrodes*, *Electrochem. Solid-State Lett.*, 2007, **10**, B130.

[53] M. S. Saha, M. N. Banis, Y. Zhang, R. Li, X. Sun, M. Cai, F. T. Wagner, *Nanowire-based three-dimensional hierarchical core/shell heterostructured electrodes for high performance proton exchange membrane fuel cells*, *J. Power Sources*, 2009, **192**, 330.

[54] M. S. Saha, R. Li, M. Cai, X. Sun, *Tungsten oxide nanowires grown on carbon paper as Pt electrocatalyst support for high performance proton exchange membrane fuel cells*, *J. Power Sources*, 2008, **185**, 1079.

[55] M. S. Saha, R. Li, X. Sun, S. Ye, *3-D composite electrodes for high performance PEM fuel cells composed of Pt supported on nitrogen-doped carbon nanotubes grown on carbon paper*, *Electrochem. Commun.*, 2009, **11**, 438.

[56] S. Sun, G. Zhang, D. Geng, Y. Chen, M. N. Banis, R. Li, M. Cai, X. Sun, *Direct growth of single-crystal Pt nanowires on Sn@CNT nanocable: 3D electrodes for highly active electrocatalysts*, *Chem. Eur. J.*, 2010, **16**, 829.

[57] S. Jiang, B. Yi, C. Zhang, S. Liu, H. Yu, Z. Shao, *Vertically aligned carbon-coated titanium dioxide nanorod arrays on carbon paper with low platinum for proton exchange membrane fuel cells*, *J. Power Sources*, 2015, **276**, 80.

[58] S. Jiang, B. Yi, H. Zhang, W. Song, Y. Bai, H. Yu, Z. Shao, *Vertically aligned titanium nitride nanorod arrays as supports of Platinum–Palladium–Cobalt catalysts for thin-film proton exchange membrane fuel cell electrodes*, *ChemElectroChem*, 2016, **3**, 734.

[59] S. Cavaliere, S. Subianto, I. Savych, D. J. Jones, J. Rozière, *Electrospinning: designed architectures for energy conversion and storage devices*, *Energy Environ. Science*, 2011, **4**, 4761.

[60] Y. S. Kim, S. H. Nam, H.-S. Shim, H.-J. Ahn, M. Anand, W. B. Kim, *Electrospun bimetallic nanowires of PtRh and PtRu with compositional variation for methanol electrooxidation*, *Electrochem. Commun.*, 2008, **10**, 1016.

[61] R. Takemori, H. Kawakami, *Electrospun nanofibrous blend membranes for fuel cell electrolytes*, *J. Power Sources*, 2010, **195**, 5957.

[62] W. Zhang, P. N. Pintauro, *High-performance nanofiber fuel cell electrodes*, *ChemSusChem*, 2011, **4**, 1753.

[63] A. B. Suryamas, G. M. Anilkumar, S. Sago, T. Ogi, K. Okuyama, *Electrospun Pt/SnO2 nanofibers as an excellent electrocatalysts for hydrogen oxidation reaction with ORR-blocking characteristic*, *Catal. Commun.*, 2013, **33**, 11.

[64] P. Chen, H. Wu, T. Yuan, Z. Zou, H. Zhang, J. Zheng, H. Yang, *Electronspun nanofiber network anode for a passive direct methanol fuel cell*, *J. Power Sources*, 2014, **255**, 70.

[65] M. Brodt, R. Wycisk, P. N. Pintauro, T. Han, N. Dale, K. Adjemian, *Nanofiber fuel cell electrodes I. fabrication and performance with commercial Pt/C catalysts*, *ECS Trans.*, 2013, **58**, 381.

[66] M. Brodt, T. Han, N. Dale, E. Niangar, R. Wycisk, P. Pintauro, *In-situ performance, and durability of nanofiber fuel cell electrodes*, *J. Electrochem. Soc.*, 2015, **162**, F84.

[67] S. Hong, M. Hou, Y. Xiao, Z. Shao, B. Yi, *Investigation of a high-performance nanofiber cathode with ultralow Platinum loading for proton exchange membrane fuel cells*, *Energy Technol.*, 2017, **5**, 1457.

[68] S. Hong, M. Hou, H. Zhang, Y. Jiang, Z. Shao, B. Yi, *A high-performance PEM fuel cell with ultralow platinum electrode via electrospinning and underpotential deposition*, *Electrochim. Acta*, 2017, **245**, 403.

[69] L. Li, S. S. Wong, *Ultrathin metallic nanowire-based architectures as high-performing electrocatalysts*, *ACS Omega*, 2018, **3**, 3294.

[70] S. M. Alia, Y. S. Yan, B. S. Pivovar, *Galvanic displacement as a route to highly active and durable extended surface electrocatalysts*, *Catal. Sci. Technol.*, 2014, **4**, 3589.

[71] S. M. Alia, B. A. Larsen, S. Pylypenko, D. A. Cullen, D. R. Diercks, K. C. Neyerlin, S. S. Kocha, B. S. Pivovar, *Platinum-coated nickel nanowires as oxygen-reducing electrocatalysts*, *ACS Catal.*, 2014, **4**, 1114.

[72] S. M. Alia, S. Pylypenko, K. C. Neyerlin, D. A. Cullen, S. S. Kocha, B. S. Pivovar, *Platinum-coated cobalt nanowires as oxygen reduction reaction electrocatalysts*, *ACS Catal.*, 2014, **4**, 2680.

[73] Z. Baroud, M. Benmiloud, A. Benalia, C. Ocampo-Martinez, *Novel hybrid fuzzy-PID control scheme for air supply in PEM fuel-cell-based systems*, *Int. J. Hydrogen Energy*, 2017, **42**, 10435.

[74] S. A. Mauger, K. C. Neyerlin, S. M. Alia, C. Ngo, S. K. Babu, K. E. Hurst, S. Pylypenko, S. Litster, B. S. Pivovar, *Fuel cell performance implications of membrane electrode assembly fabrication with Platinum-Nickel nanowire catalysts*, *J. Electrochem. Soc.*, 2018, **165**, F238.

[75] W. W. McNeary, C. Ngo, A. E. Linico, J. W. Zack, A. M. Roman, K. M. Hurst, S. M. Alia, J. W. Medlin, S. Pylypenko, B. S. Pivovar, A. W. Weimer, *Extended thin-film electrocatalyst structures via Pt atomic layer deposition*, *ACS Appl. Nano Mater.*, 2018, **1**, 6150.

[76] M. D. Gasda, G. A. Eisman, D. Gall, *Nanorod PEM fuel cell cathodes with controlled porosity*, *J. Electrochem. Soc.*, 2010, **157**, B437.

[77] W. Zhang, J. Chen, A. I. Minett, G. F. Swiegers, C. O. Too, G. G. Wallace, *Novel ACNT arrays based MEA structure-nano-Pt loaded ACNT/Nafion/ACNT for fuel cell applications*, *Chem. Commun.*, 2010, **46**, 4824.

[78] W. Zhang, A. I. Minett, M. Gao, J. Zhao, J. M. Razal, G. G. Wallace, T. Romeo, J. Chen, *Integrated high-efficiency Pt/carbon nanotube arrays for PEM fuel cells*, *Adv. Energy Mater.*, 2011, **1**, 671.

[79] Z. Q. Tian, S. H. Lim, C. K. Poh, Z. Tang, Z. Xia, Z. Luo, P. K. Shen, D. Chua, Y. P. Feng, Z. Shen, J. Lin, *A highly order-structured membrane electrode assembly with vertically aligned carbon nanotubes for ultra-low Pt loading PEM fuel cells*, *Adv. Energy Mater.*, 2011, **1**, 1205.

[80] S. Murata, M. Imanishi, S. Hasegawa, R. Namba, *Vertically aligned carbon nanotube electrodes for high current density operating proton exchange membrane fuel cells*, *J. Power Sources*, 2014, **253**, 104.
[81] W.-J. Lee, M. Alhosan, S. L. Yohe, N. L. Macy, W. H. Smyrl, *Synthesis of Pt/TiO2 nanotube catalysts for cathodic oxygen reduction*, *J. Electrochem. Soc.*, 2008, **155**, B915.
[82] D.-H. Lim, W.-J. Lee, N. L. Macy, W. H. Smyrl, *Electrochemical durability investigation of Pt/TiO2 nanotube catalysts for polymer electrolyte membrane fuel cells*, *Electrochem. Solid-State Lett.*, 2009, **12**, B123.
[83] D.-H. Lim, W.-J. Lee, J. Wheldon, N. L. Macy, W. H. Smyrl, *Electrochemical characterization and curability of sputtered Pt catalysts on TiO2 nanotube arrays as a cathode material for PEFCs*, *J. Electrochem. Soc.*, 2010, **157**, B862.
[84] C. Zhang, H. Yu, Y. Li, Y. Gao, Y. Zhao, W. Song, Z. Shao, B. Yi, *Supported noble metals on hydrogen-treated TiO2 nanotube arrays as highly ordered electrodes for fuel cells*, *ChemSusChem*, 2013, **6**, 659.
[85] C. Zhang, H. Yu, Y. Li, W. Song, B. Yi, Z. Shao, *Preparation of Pt catalysts decorated TiO2 nanotube arrays by redox replacement of Ni precursors for proton exchange membrane fuel cells*, *Electrochim. Acta*, 2012, **80**, 1.
[86] S. Galbiati, A. Morin, N. Pauc, *Supportless platinum nanotubes array by atomic layer deposition as PEM fuel cell electrode*, *Electrochim. Acta*, 2014, **125**, 107.
[87] C. Zhang, H. Yu, L. Fu, Y. Xiao, Y. Gao, Y. Li, Y. Zeng, J. Jia, B. Yi, Z. Shao, *An oriented ultrathin catalyst layer derived from high conductive TiO2 nanotube for polymer electrolyte membrane fuel cell*, *Electrochim. Acta*, 2015, **153**, 361.
[88] C. Zhang, H. Yu, L. Fu, Y. Gao, J. Jia, S. Jiang, B. Yi, Z. Shao, *A novel ultra-thin catalyst layer based on wheat ear-like catalysts for polymer electrolyte membrane fuel cells*, *RSC Adv.*, 2014, **4**, 58591.
[89] Y. Zeng, X. Guo, Z. Wang, J. Geng, H. Zhang, W. Song, H. Yu, Z. Shao, B. Yi, *Highly stable nanostructured membrane electrode assembly based on Pt/Nb2O5 nanobelts with reduced platinum loading for proton exchange membrane fuel cells*, *Nanoscale*, 2017, **9**, 6910.
[90] Y. Zeng, Z. Shao, H. Zhang, Z. Wang, S. Hong, H. Yu, B. Yi, *Nanostructured ultrathin catalyst layer based on open-walled PtCo bimetallic nanotube arrays for proton exchange membrane fuel cells*, *Nano Energy*, 2017, **34**, 344.
[91] K. Sasaki, H. Naohara, Y. Cai, Y. M. Choi, P. Liu, M. B. Vukmirovic, J. X. Wang, R. R. Adzic, *Core-protected Platinum monolayer shell high-stability electrocatalysts for fuel-cell cathodes*, *Angew. Chem. Int. Ed.*, 2010, **49**, 8602.
[92] K. K. Kam, M. K. Debe, R. J. Poirier, A. R. Drube, *Summary Abstract: Dramatic variation of the physical microstructure of a vapor deposited organic thin film*, *J. Vac. Sci. Technol. A*, 1987, **5**, 1914.
[93] M. K. Debe, K. K. Kam, J. C. Liu, R. J. Poirier, *Vacuum vapor deposited thin films of a perylene dicarboximide derivative: Microstructure versus deposition parameters*, *J. Vac. Sci. Technol. A*, 1988, **6**, 1907.
[94] M. K. Debe, R. J. Poirier, *Postdeposition growth of a uniquely nanostructured organic film by vacuum annealing*, *J. Vac. Sci. Technol. A*, 1994, **12**, 2017.
[95] M. K. Debe, A. R. Drube, *Structural characteristics of a uniquely nanostructured organic thin film*, *J. Vac. Sci. Technol.*, 1995, **13**, 1236.
[96] M. K. Debe, R. T. Atanasoski, A. J. Steinbach, *Nanostructured thin film electrocatalysts - current status and future potential*, *ECS Trans.*, 2011, **41**, 937.
[97] M. K. Debe, *Nanostructured thin film electrocatalysts for PEM fuel cells - a tutorial on the fundamental characteristics and practical properties of NSTF catalysts*, *ECS Trans.*, 2012, **45**, 47.

[98] M. K. Debe, *Tutorial on the fundamental characteristics and practical properties of nanostructured thin film (NSTF) catalysts*, *J. Electrochem. Soc.*, 2013, **160**, F522.
[99] A. J. Steinbach, D. van der Vliet, A. E. Hester, J. Erlebacher, C. Duru, I. Davy, M. Kuznia, D. A. Cullen, *Recent progress in nanostructured thin film (NSTF) ORR electrocatalyst development for PEM fuel cells*, *ECS Trans.*, 2015, **69**, 291.
[100] L. Gancs, T. Kobayashi, M. K. Debe, R. Atanasoski, A. Wieckowski, *Crystallographic characteristics of nanostructured thin-film fuel cell electrocatalysts: a HRTEM study*, *Chem. Mater.*, 2008, **20**, 2444.
[101] A. Kongkanand, Z. Liu, I. Dutta, F. T. Wagner, *Electrochemical and microstructural evaluation of aged nanostructured thin film fuel cell electrocatalyst*, *J. Electrochem. Soc.*, 2011, **158**, B1286.
[102] D. A. Stevens, J. M. Rouleau, R. E. Mar, A. Bonakdarpour, R. T. Atanasoski, A. K. Schmoeckel, M. K. Debe, J. R. Dahn, *Characterization and PEMFC testing of Pt1 − x M x (M = Ru , Mo , Co , Ta , Au , Sn) anode electrocatalyst composition spreads*, *J. Electrochem. Soc.*, 2007, **154**, B566.
[103] A. Bonakdarpour, K. Stevens, G. D. Vernstrom, R. Atanasoski, A. K. Schmoeckel, M. K. Debe, J. R. Dahn, *Oxygen reduction activity of Pt and Pt–Mn–Co electrocatalysts sputtered on nano-structured thin film support*, *Electrochim. Acta*, 2007, **53**, 688.
[104] G. Chih-Kang Liu, D. A. Stevens, J. C. Burns, R. J. Sanderson, G. Vernstrom, R. T. Atanasoski, M. K. Debe, J. R. Dahn, *Oxygen reduction activity of dealloyed Pt1–xNix catalysts*, *J. Electrochem. Soc.*, 2011, **158**, B919.
[105] M. K. Debe, A. J. Steinbach, G. D. Vernstrom, S. M. Hendricks, M. J. Kurkowski, R. T. Atanasoski, P. Kadera, D. A. Stevens, R. J. Sanderson, E. Marvel, J. R. Dahn, *Extraordinary oxygen reduction activity of Pt3Ni7*, *J. Electrochem. Soc.*, 2011, **158**, B910.
[106] D. A. Stevens, R. Mehrotra, R. J. Sanderson, G. D. Vernstrom, R. T. Atanasoski, M. K. Debe, J. R. Dahn, *Dissolution of Ni from high Ni content Pt1−xNix alloys*, *J. Electrochem. Soc.*, 2011, **158**, B905.
[107] D. A. Stevens, S. Wang, R. J. Sanderson, G. C. K. Liu, G. D. Vernstrom, R. T. Atanasoski, M. K. Debe, J. R. Dahn, *A combined rotating disk electrode/X-Ray diffraction study of Co dissolution from Pt1−xCox alloys*, *J. Electrochem. Soc.*, 2011, **158**, B899.
[108] D. van der Vliet, C. Wang, M. Debe, R. Atanasoski, N. M. Markovic, V. R. Stamenkovic, *Platinum-alloy nanostructured thin film catalysts for the oxygen reduction reaction*, *Electrochim. Acta*, 2011, **56**, 8695.
[109] J. E. Harlow, D. A. Stevens, R. J. Sanderson, G. C.-K. Liu, L. B. Lohstreter, G. D. Vernstrom, R. T. Atanasoski, M. K. Debe, J. R. Dahn, *Structural changes induced by Mn mobility in a Pt1−xMnx binary composition-spread catalyst*, *J. Electrochem. Soc.*, 2012, **159**, B670.
[110] D. F. van der Vliet, C. Wang, D. Tripkovic, D. Strmcnik, X. F. Zhang, M. K. Debe, R. T. Atanasoski, N. M. Markovic, V. R. Stamenkovic, *Mesostructured thin films as electrocatalysts with tunable composition and surface morphology*, *Nat. Mater.*, 2012, **11**, 1051.
[111] A. Bonakdarpour, J. Wenzel, D. A. Stevens, S. Sheng, T. L. Monchesky, R. Löbel, R. T. Atanasoski, A. K. Schmoeckel, G. D. Vernstrom, M. K. Debe, J. R. Dahn, *Studies of transition metal dissolution from combinatorially sputtered, nanostructured Pt1 − x M x (M = Fe, Ni; 0 < x < 1) electrocatalysts for PEM fuel cells*, *J. Electrochem. Soc.*, 2005, **152**, A61.

[112] A. Bonakdarpour, R. Löbel, R. T. Atanasoski, G. D. Vernstrom, A. K. Schmoeckel, M. K. Debe, J. R. Dahn, *Dissolution of transition metals in combinatorially sputtered Pt1 − x − y M x M y ′ (M, M′ = Co, Ni, Mn, Fe) PEMFC electrocatalysts, J. Electrochem. Soc.*, 2006, **153**, A1835.

[113] M. K. Debe, A. Schmoeckel, S. Hendricks, G. Vernstrom, G. Haugen, R. Atanasoski, *Durability aspects of nanostructured thin film catalysts for PEM fuel cells, ECS Trans.*, 2006, **1**, 51.

[114] M. K. Debe, A. K. Schmoeckel, G. D. Vernstrom, R. Atanasoski, *High voltage stability of nanostructured thin film catalysts for PEM fuel cells, J. Power Sources*, 2006, **161**, 1002.

[115] M. K. Debe, A. J. Steinbach, K. Noda, *Stop-start and high-current durability testing of nanostructured thin film catalysts for PEM fuel cells, ECS Trans.*, 2006, **3**, 835.

[116] D. A. Stevens, J. M. Rouleau, R. E. Mar, R. T. Atanasoski, A. K. Schmoeckel, M. K. Debe, J. R. Dahn, *Enhanced CO-tolerance of Pt–Ru–Mo hydrogen oxidation catalysts, J. Electrochem. Soc.*, 2007, **154**, B1211.

[117] A. J. Steinbach, C. V. Hamilton, M. K. Debe, *Impact of micromolar concentrations of externally-provided chloride and sulfide contaminants on PEMFC reversible stability, ECS Trans.*, 2007, **11**, 889.

[118] O.-H. Kim, Y.-H. Cho, S. H. Kang, H.-Y. Park, M. Kim, J. W. Lim, D. Y. Chung, M. J. Lee, H. Choe, Y.-E. Sung, *Ordered macroporous platinum electrode and enhanced mass transfer in fuel cells using inverse opal structure, Nat. Commun.*, 2013, **4**, 2473.

[119] S. Jiang, B. Yi, L. Cao, W. Song, Q. Zhao, H. Yu, Z. Shao, *Development of advanced catalytic layer based on vertically aligned conductive polymer arrays for thin-film fuel cell electrodes, J. Power Sources*, 2016, **329**, 347.

[120] K. M. Rod Borup, Adam Weber, *FC135: FC-PAD: fuel cell performance and durability consortium, DOE 2018 Fuel Cell Annual Proceding*, 2018.

[121] S. Xu, Y. Kim, J. Park, D. Higgins, S.-J. Shen, P. Schindler, D. Thian, J. Provine, J. Torgersen, T. Graf, T. D. Schladt, M. Orazov, B. H. Liu, T. F. Jaramillo, F. B. Prinz, *Extending the limits of Pt/C catalysts with passivation-gas-incorporated atomic layer deposition, Nat. Catal.*, 2018, **1**, 624.

[122] U. A. Paulus, Z. Veziridis, B. Schnyder, M. Kuhnke, G. G. Scherer, A. Wokaun, *Fundamental investigation of catalyst utilization at the electrode/solid polymer electrolyte interface: Part I. Development of a model system, J. Electroanal. Chem.*, 2003, **541**, 77.

[123] J. Jiang, B. Yi, *Thickness effects of a carbon-supported platinum catalyst layer on the electrochemical reduction of oxygen in sulfuric acid solution, J. Electroanal. Chem.*, 2005, **577**, 107.

[124] S. Tominaka, C.-W. Wu, K. Kuroda, T. Osaka, *Electrochemical analysis of perpendicular mesoporous Pt electrode filled with pure water for clarifying the active region in fuel cell catalyst layers, J. Power Sources*, 2010, **195**, 2236.

[125] E. L. Thompson, D. Baker, *Proton conduction on ionomer-free Pt surfaces, ECS Trans.*, 2011, **41**, 709.

[126] S. J. An, S. Litster, *Ionic conductivity measurement of ionomer/binder-free Pt catalyst under fuel cell operating condition, ECS Trans.*, 2013, **58**, 831.

[127] Q. Wang, M. Eikerling, D. Song, Z.-S. Liu, *Modeling of ultrathin two-phase catalyst layers in PEFCs, J. Electrochem. Soc.*, 2007, **154**, F95.

[128] K. Chan, M. Eikerling, *A pore-scale model of oxygen reduction in ionomer-free catalyst layers of PEFCs, J. Electrochem. Soc.*, 2011, **158**, B18.

[129] P. K. Sinha, W. Gu, A. Kongkanand, E. Thompson, *Performance of nano structured*

thin film (NSTF) electrodes under partially-humidified conditions, *J. Electrochem. Soc.*, 2011, **158**, B831.

[130] K. Chan, M. Eikerling, *Impedance model of oxygen reduction in water-flooded pores of ionomer-free PEFC catalyst layers*, *J. Electrochem. Soc.*, 2011, **159**, B155.

[131] M. K. Debe, *Effect of electrode surface area distribution on high current density performance of PEM fuel cells*, *J. Electrochem. Soc.*, 2011, **159**, B53.

[132] M. K. Debe, A. J. Steinbach, *An empirical model for the flooding behavior of ultra-thin PEM fuel cell electrodes*, *ECS Trans.*, 2007, **11**, 659.

[133] A. J. Steinbach, M. K. Debe, M. J. Pejsa, D. M. Peppin, A. T. Haug, M. J. Kurkowski, S. M. Maier-Hendricks, *Influence of anode GDL on PEMFC ultra-thin electrode water management at low temperatures*, *ECS Trans.*, 2011, **41**, 449.

[134] A. J. Steinbach, M. K. Debe, J. Wong, M. J. Kurkowski, A. T. Haug, D. M. Peppin, S. K. Deppe, S. M. Hendricks, E. M. Fischer, *A new paradigm for PEMFC ultra-thin electrode water management at low temperatures*, *ECS Trans.*, 2010, **33**, 1179.

[135] A. J. Steinbach, J. S. Allen, R. L. Borup, D. S. Hussey, D. L. Jacobson, A. Komlev, A. Kwong, J. MacDonald, R. Mukundan, M. J. Pejsa, M. Roos, A. D. Santamaria, J. M. Sieracki, D. Spernjak, I. V. Zenyuk, A. Z. Weber, *Anode-design strategies for improved performance of polymer-electrolyte fuel cells with ultra-thin electrodes*, *Joule*, 2018, **2**, 1297.

[136] J. Hou, H. Yu, S. Zhang, S. Sun, H. Wang, B. Yi, P. Ming, *Analysis of PEMFC freeze degradation at −20°C after gas purging*, *J. Power Sources*, 2006, **162**, 513.

[137] J. Hou, B. Yi, H. Yu, L. Hao, W. Song, Y. Fu, Z. Shao, *Investigation of resided water effects on PEM fuel cell after cold start*, *Int. J. Hydrog. Energy*, 2007, **32**, 4503.

[138] J. Hou, W. Song, H. Yu, Y. Fu, Z. Shao, B. Yi, *Electrochemical impedance investigation of proton exchange membrane fuel cells experienced subzero temperature*, *J. Power Sources*, 2007, **171**, 610.

[139] A. Kongkanand, M. Dioguardi, C. Ji, E. L. Thompson, *Improving operational robustness of NSTF electrodes in PEM fuel cells*, *J. Electrochem. Soc.*, 2012, **159**, F405.

4 Outlines for the Next-Generation Cathode Materials Utilized in Lithium Batteries

Zhenlian Chen and Deyu Wang
Jianghan University, Wuhan, Hubei, People's Republic of China

CONTENTS

4.1 INTRODUCTION

Cathode materials exerted the crucial effects on determining energy density and safety of lithium ion batteries. Since the first proposed layer-structured $LiCoO_2$ by Professor J. B. Goodenough in the 1980s [1], various materials were successfully commercialized, such as $LiFePO_4$ and $LiNi_{1-x-y}Co_xMn_yO_2$, which was massively utilized in commercial lithium ion batteries to power all kinds of terminals. In the foreseeing future, LIBs are still the most important energy-storage technology with

DOI: 10.1201/9781003133971-4

the increasing ratio of ~20% per year due to the booming markets of electric vehicles and smart grids, implying the brilliant future of all major materials including cathodes.

As known, high reversible capacity and long durability are the persistent pursuits on cathode materials. The commercialized cathodes of layer-structured $LiMO_2$ (M = Mn, Ni, Co), olivine-structured $LMPO_4$ (M = Fe, Mn, Co), and spinel-structured LiM_2O_4 (M = Mn, Ni, Co) are based on the valence variation of transitional metals (TMs). Currently, their reversible capacity has gradually reached to its theoretical limitation after continuous optimization. Therefore, novel reaction mechanisms, especially on oxygen-anion involved reactions, were explored to further improve cathode's reversible capacity. Here we shall introduce Li-rich layer cathodes, which is the representative of possessing the oxygen reactions and then high reversible capacities.

Long durability has also been highly pursued since LIBs were launched. Plenty of remedies, such as passivating surface and improving carriers' transportation, were adopted to prolong LIBs' operation. Recently, researchers found that mechanical properties played the crucial role on particles' stability. The continuous Li^+ intercalation/deintercalation could deteriorate particles' mechanical strengths and eventually lacerated particles, resulting in the plummet of cyclic life. As an emerging multidisciplinary field, mechanical properties should attract much more attention to achieve a long durability in the future. Here we shall discuss the progress of mechanical investigation on cathode materials.

Furthermore, solid-state batteries are widely accepted as the most appropriate technology for large-scale utilization, such as electric vehicles and smart grids, due to their excellent safety. They also advanced the new requirements for cathode materials. This technology requires zero- or micro-strain materials to assure the physical contacts. Also, the appropriate modifications are necessary to stabilize the interfacial layer. In the final part, we shall discuss the progress of cathode materials in this technology.

In this chapter, we shall introduce the progresses of these promising fields, which should be the directions for the next generation of cathode materials.

4.2 LAYERED LI-RICH OXIDE CATHODE AND OXYGEN REDOX MECHANISM

4.2.1 PIONEERING FINDINGS

The beginning of the research of layered Li-rich oxide (LLO) cathode can be dated back to 1991, when Thackeray et al. from Argonne National Laboratory obtained electrochemically active Li-excess oxide $Li_{1.09}Mn_{0.91}O_2$ by leaching Li_2O from Li_2MnO_3 via acid treatment [2]. And years later, they developed a family of high capacity, Li_2MnO_3-stabilized $LiMO_2$ (M = Mn, Ni, Co) electrodes for lithium-ion batteries with three-step fabrication: 1) leaching Li_2O from Li_2MnO_3 by acid treatment; 2) subsequent re-lithiation, either electrochemically in a lithium cell or chemically with n-butyl-lithium; and 3) recognizing the structural compatibility between Li_2MO_3 (e.g., Mn, Ti, Zr) and layered $LiMO_2$ [3]. In 2001, Dahn et al.

found highly reversible capacities of 220 mA h g^{-1} can be delivered by the $Li[Ni_{1/3}Li_{1/9}Mn_{5/9}]O_2$ cathode at 55°C in the voltage range of 2.0–4.6 V [4].

4.2.2 Crystal Structure

The pristine structure of LLO materials could be a composite denoted as $(1-x)Li_2MnO_3 \cdot LiMO_2$ (M = Mn, Co, Ni) or a solid solution denoted as $Li_{1+x}M_{1-x}O_2$, of the rhombohedral $LiMO_2$ ($R\bar{3}m$ space group) and monoclinic Li_2MnO_3 structures (C2/m space group). The two structures are O_3-type where interslab octahedral sites are only occupied by lithium ions, and slab octahedral sites are occupied by M ions in rhombohedral $LiMO_2$, whereas lithium and manganese ions with the (1:2) ratio in Li_2MnO_3, c.f. Figure 4.1(a–c). The structure of Li_2MnO_3 is also described by the $Li[Li_{1/3}Mn_{2/3}]O_2$ formula in relation with its layered structure, and Li-Mn cationic ordering pattern looks like a honeycomb, as shown in Figure 4.1(b) and gives superlattice reflection in XRD (c.f., Figure 4.1(d)) [5]. While in rhombohedral $LiMO_2$, the atom scattering factors of Ni, Co, and Mn are very close, and no superlattice X-ray reflections observable (c.f., Figure 4.1(e)). Whether the pristine structure of LLO material is composite or solid solution and then the intensities of the superlattice reflections is largely dependent on the composition and synthesis conditions [3,7].

The migration of TM ion to Li vacancy site with a charging process is widely recognized according to the increase of mixed occupation on Li and TM sites in highly delithiated states respective to pristine materials suggested from the Rietveld refinement of XRD [8]. That disturbs the intra-layer honeycomb ordering. The intensity of the superlattice reflections in XRD pattern is reduced, and the two-atom TM–TM dumbbells characteristic of TM honeycomb ordering within each TM layer in annular dark field–scanning transmission electron microscopy (ADF-TEM) image along the – [100] zone axis disappeared [9]. In the bulk of the material, the migration of TM ions is partially reversible. For example, the occupancy of TM ions in a Li layer is estimated to be 2.8%, 9.0%, and 4.7% in the pristine, charged and discharged materials, respectively [10]. While near the surface, the migration of TM ions leads to layered to spinel structure transition, and the spinel phase may extend from the surface to the bulk during cycling [11]. In parallel, the delithiation may also lead to planar gliding and stacking fault as indicated by the singularities in the 3-D displacement field inside primary battery particles during battery operation reconstructed using the in-situ Bragg coherent diffractive imaging (BCDI) technique [12]. The migration of TM ions with delithiation can be accelerated with heating [13], while the structural degradation can also be partially recovered when the discharged sample is heated at 150 to 300, which pushes the charging and discharging plateaus upward to close to the initial cycle [12].

4.2.3 Electrochemical Activity, Redox Mechanism, and Characterization Techniques

During the charging process of LLO, there's an irreversible plateau above 4.4 V, which does not appear in conventional LMO_2 cathode and is also called activation plateau, as shown in Figure 4.2(a). Aided with this activation plateau, the initial

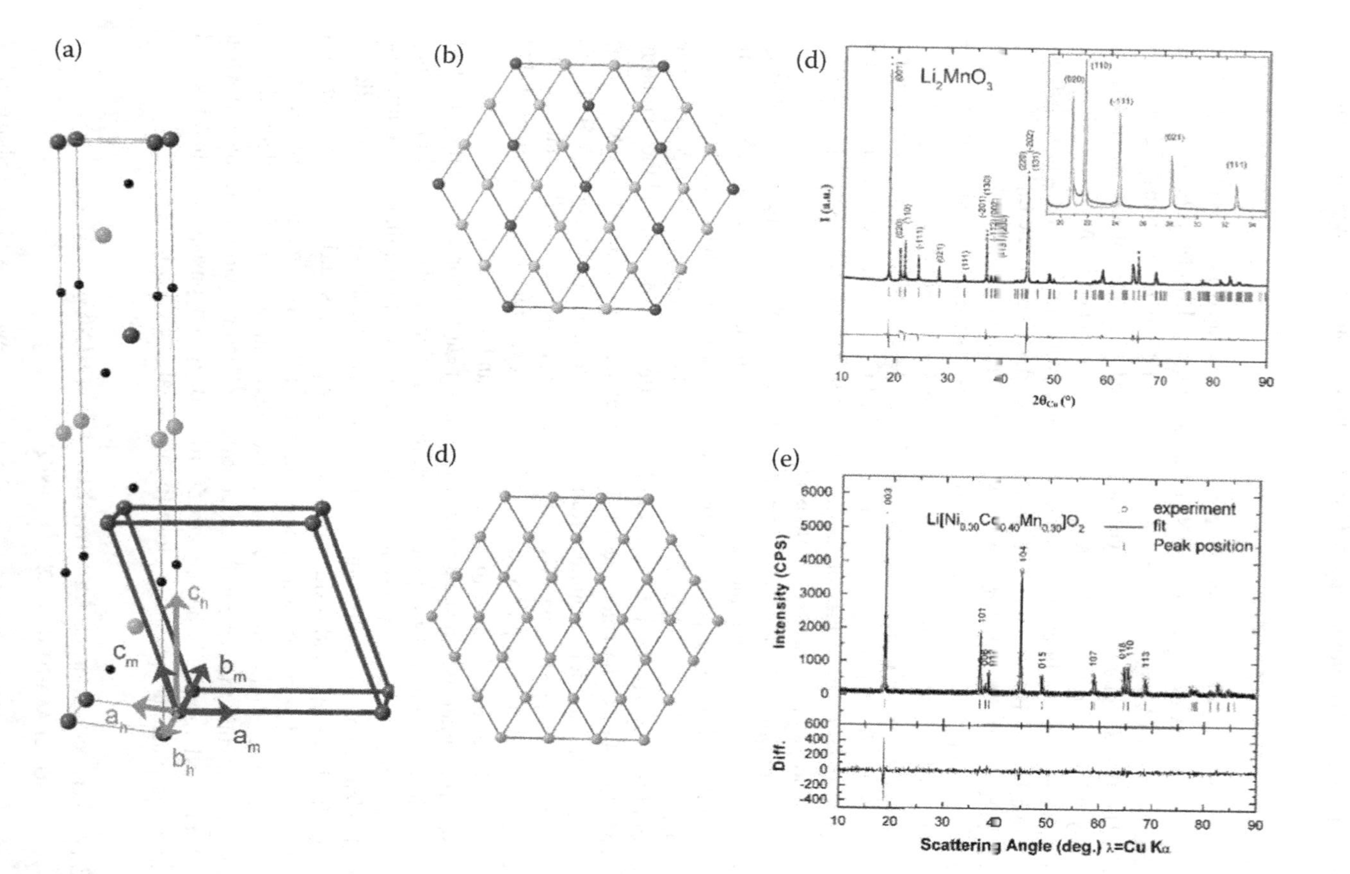

FIGURE 4.1 Crystallographic feaurure of LLO. (a) Crystallographic relationship between parent hexagonal cell of $LiMO_2$ and monclinic cell of Li_2MnO_3. Intra-layer atomistic distribution of (b), monclinic Li_2MnO_3 and (c), rhombohedral $LiMO_2$. The XRD patterns of (d), Li_2MnO_3 (Reproduced with permission of ref. [5], Copyright 2009 American Chemical Society), and (e), $LiMO_2$ (Reproduced with permission of ref. [6], Copyright 2002 IOP Publishing).

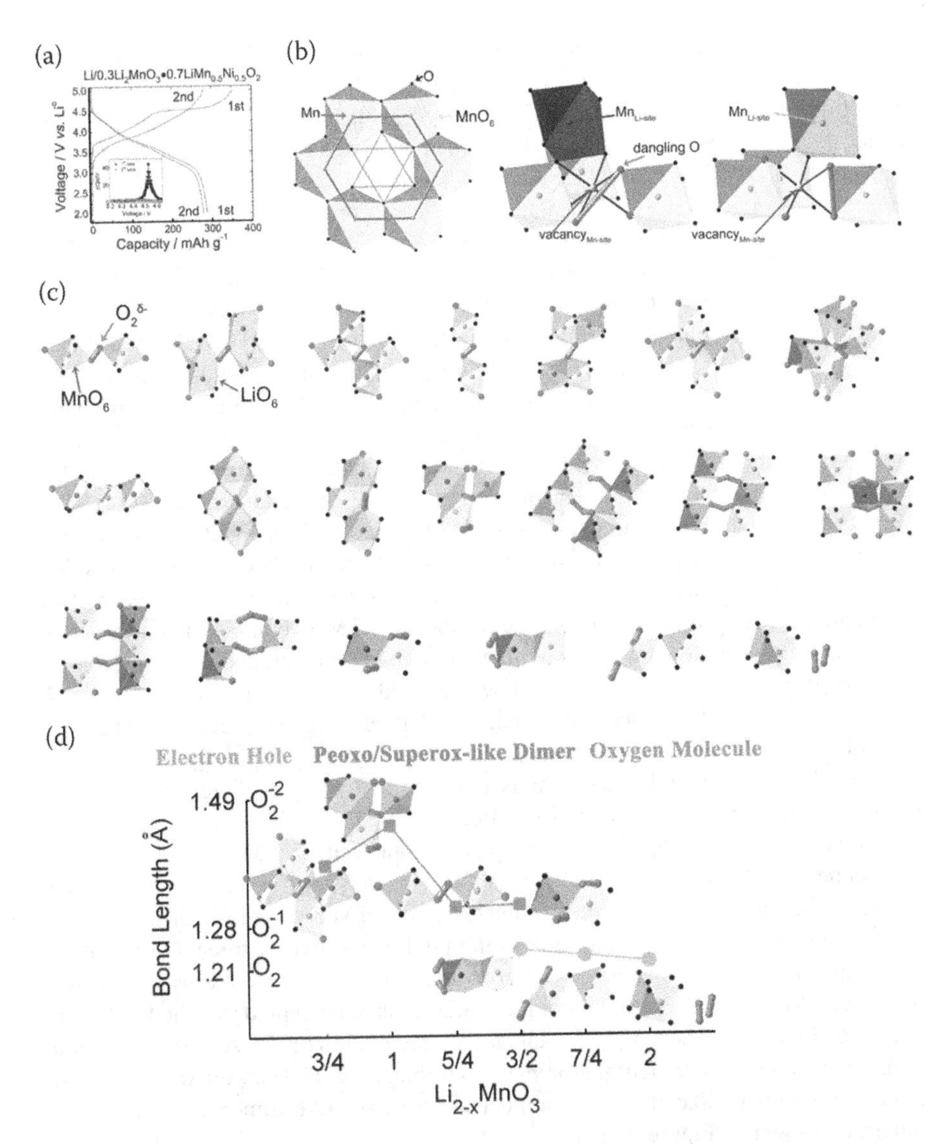

FIGURE 4.2 Electrochemical activity and oxygen dimers. (a) Initial two charge/discharge voltage profiles. Reproduced with permission of ref. [14], Copyright 2006 Elsevier. (b) Coordination condition of oxygen dimer formation. A top view of perfect Li_2MnO_3 and a side view of Mn ion migration to first and second neighboring Li site, respectively. (c) Five classes of binding modes on how oxygen dimers connected to Mn ions, including bridge dimers, edge dimers, v-bridge dimers, dangling dimers, and floating dimers. (d) The predicted evolution of oxygen dimers versus the degree of delithiation. (b), (c), and (d) are reproduced with permission of ref. [15], Copyright 2019 American Chemical Society.

charging and discharging capacities can be remarkably higher than the capacity that can be offered by the available TM redox pairs [14]. Several mechanisms such as extraction of lattice oxygen, Li^+-H^+ exchange [16], and oxidation of Mn^{4+} to a higher oxidation state such as Mn^{7+} have been proposed to contribute to the excess capacity beyond TM redox [17]. Another mechanism, oxygen redox, was firstly suggested by first-principles calculations [18], and prevailed after the concept of peroxo-like O_2^{2-} dimer formation proposed in $Li_2Ru_{3/4}Sn_{1/4}O_3$ [19]. However, O-O covalent bonds (<1.6 Å) are hardly detected. For example, only a small decrease of distance between an O-O pair from ~2.8 Å to ~2.4 Å measured by high-angle annular dark-field scanning transmission electron microscopy (HAADF-STEM) and annular bright field STEM(ABF-STEM) images [20], in line with first-principle calculations on conventional delithitation models [21,22]. Moreover, peroxo or superoxo stretching bands were also undetected in routine Raman spectroscopy [23]. Simultaneously, the dimer formation was predicted to be inhibited from the view of coordination environment of oxygen ion. In rhombohedral $LiMO_2$ (M = Mn, Ni, Co), each oxygen ion is coordinated to three TM ions and, in monoclinic Li_2MnO_3, each oxygen ion is coordinated to two Mn ions. Then in LLO with solid solution structure, some of the oxygen ions are coordinated to two TM ions and the others are coordinated to three ions. The two or three TM-O bonds from one oxygen ion are closely orthogonal with each other and prevent the neighboring oxygen ions to form peroxo species [21]. Then, it was argued that the form of the oxidized oxygen species should be electron depletion of lattice oxygen ion but without dimer formation [21,23].

Recently, first-principle calculations uncover that oxygen dimer can form in virtue of antisite-cation-vacancy formation (ACVF), which commonly emerges with mixed Li/TM occupation in the pristine material or TM migration during electrochemical process [8,15]. Given ACVF, the oxygen ion coordinated to the vacancy TM site becomes singly coordinated to one TM ion, and is called dangling oxygen, c.f. Figure 4.2(b). That unlocks the bind of the two or three TM-O dative bonds and facilitates oxygen dimer formation. Five binding patterns of oxygen dimer with Mn/Li ions, as shown in Figure 4.2(c), have been predicted in delithiated phases of Li_2MnO_3 when ACVF is taken into calculations [15]. And the evolution of an oxygen dimer with delithiation may go through a dinuclear peroxo-like dimer, dinuclear superoxo-like dimer, mononuclear superoxo-like dimer, and molecular dimer, as shown in Figure 4.2(d).

The organization of the O-O bonding and anti-bonding states for an oxygen dimer in delithiated Li_2MnO_3 is close to that in stand-alone alkali-peroxides/superoxides or oxygen molecule. It starts with the lowest-lying σ bonding states, then π bonding states followed by π^* anti-bonding states, and ends with the highest-lying σ^* states. However, the states, especially π and π^* states, are largely dispersed by the back-bonding of an oxygen dimer with TM ions and electrostatic field of the neighboring ions. With electronic structure different from oxygen dimer in standalone alkali-peroxides/superoxides and oxygen molecule, as well as normal lattice oxygen ion, the oxygen dimer in the bulk of cathode is expected to give unique features in X-ray absorption spectroscopy (XAS). For lattice oxygen ion, the pre-edge peaks are solely determined by the unoccupied $2p$ states that overlap with the

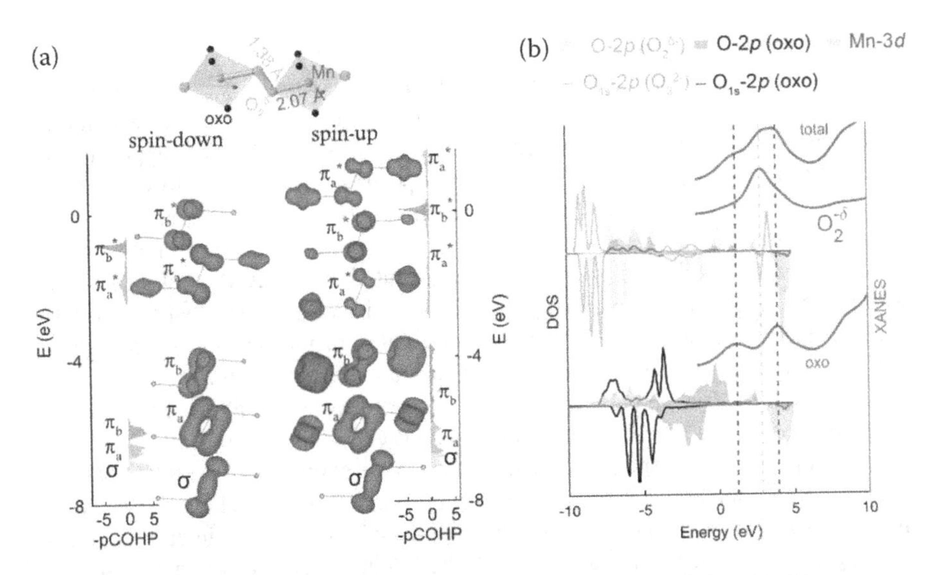

FIGURE 4.3 Electronic structure of oxygen dimer in $Li_{5/4}MnO_3$. (a) The energy-decomposed partial charge density plots (top view along ***b*** axis). Reproduced with permission of ref. [15], Copyright 2019 American Chemical Society. (b) Oxygen-K XANES predicted with first-principle calculations. Reproduced with permission of ref. [25], Copyright 2021 Wiley-VCH.

unoccupied TM-*d* states. For example, in $Li_{5/4}MnO_3$, there are two oxo pre-edge peaks with gap of ca. 2.5 eV associated with the unoccupied spin-up and spin-down Mn-3*d* states, respectively. And in Li_2O_2, only the σ^* orbital is unoccupied and excitation of the O-1*s* electron gives strong signal peak at ~531.0 eV; in LiO_2 and molecular O_2, one of the two π^* orbitals are half occupied and empty, respectively, and both the excitations to the unoccupied σ^* and π^* orbitals give strong signal peaks [24]. While in the bulk of $Li_{5/4}MnO_3$, the oxygen dimer bridging two MnO_6 octahedra, most of the π^* states are occupied but leaving a little spin-up π^* states unoccupied, gives only one signal peak. That signal peak is given by the excitation of O-1*s* electron to the σ^* orbital and the contribution from the excitation to the unoccupied π^* states is hardly observable associated with the dispersion of the π^* states. The signal peak is predicted to be located at an energy position between the two oxo peaks representing an increase of the left shoulder of the second oxo peak (Figure 4.3).

Experimentally, the increasement on the right (higher energy position) shoulder of the first oxo peak has been observed during the charging process of LLO [21]. And it has been taken as a signature of oxidized oxygen species in LLO. However, it is hard to tell the increasement origins whether from the oxidation of lattice oxygen ion or increased *d-p* hybridization with cationic oxidation. And it has proposed the spot with excitation at around 531.0 eV and emission at around 523.7 eV in mapping resonant inelastic X-ray scattering (RIXS) as the signature of oxidized oxygen species in LLO [8,26]. However, the excitation of O-1*s* electron to the unoccupied π^* state of molecular O_2 is also located at around 531.0 eV. And the fine structure of high-resolution RIXS with excitation energy at 531.0 eV for fully charged LLO is very similar to molecular O_2. The separation between the first two peaks shown in Figure 4.4(c) is 0.192 eV, equivalent to 1,550 cm^{-1}, corresponding to the vibrational spectrum of molecular O_2 [9]. So, molecular O_2 trapped in the bulk of LLO was proposed to account for that spectroscopic feature. However, from lattice oxygen ion to molecular O_2, there should be an intermediate state, i.e., peroxo/superoxo dimer. Interestingly, the peroxo species has found binding to Si ions in silicate cathode and giving an independent signature XAS peak below and well separated from the single oxo pre-edge. So, it is still puzzling that whether the peroxo/superoxo dimer formed in LLO is transient to be undetectable or the signals are only able to be resolved with more advanced characterization.

In addition to XAS signals expected to be given by the excitation of O-1s to the unoccupied σ^* orbital, the O-O stretching mode is expected to be detected in Raman spectroscopy. The frequency of O-O stretching mode could vary with the O-O bond length, e.g., around 800, 1,137, and 1,556 cm^{-1} in Li_2O_2, LiO_2 and O_2, respectively. It will also vary with particular binding mode of an oxygen dimer ligand and binding cations [27]. In the bulk of LLO, the binding modes of oxygen dimers to ligand TM ion could be dinuclear end-on, mononuclear side-on, mononuclear end-on, and double end-to-end ozone bridge as predicted in Li_2MnO_3 and shown in Figure 4.2(c). Usually the peroxo dimer binding to Co/Mn/Ni in dioxygen complexes gives single signature stretching band around 800 cm^{-1}. However, that vibration band had not been detected except with shell-isolated nanoparticle-enhanced Raman spectroscopy for LLO [21,28], whereas there are two bands in $Li_{4.15}Ni_{0.85}WO_6$ and $Li_2Co_{0.975}Mn_{0.0625}SiO_4$ detected with routine Raman measurements associated with the coupling of the peroxo stretch with W-O and Si-O stretching modes, respectively [25].

4.2.4 Degradation Issues and Mitigation Strategies

LLO cathode materials have high specific energy, low cost, and good thermal stability, but the inherent shortcomings hinder the practical applications. These include a low initial Coulombic efficiency (CE), voltage hysteresis, poor rate capability, serious voltage decay, and significant capacity loss upon cycling. Those are deeply involved with oxygen loss and structural transition. The oxygen loss was first measured with in-situ differential electrochemical mass spectrometry (DEMS) in $Li[Ni_{0.2}Li_{0.2}Mn_{0.6}]O_2$ [29]. So far, it is still unclear that the oxygen loss is proceeded with the diffusion of transition metal ions from surface to bulk where they occupy vacancies created by Li removal or the oxygen vacancies produced at the surfaces are continuously pumped into the bulk lattice [29,30]. Effective surface modification including atomic layer deposition, chemical vapor deposition, reactive gas treatment, and deposition of coating materials from liquid phases can avoid detrimental release of oxygen and avoid crack formation resulting from irreversible structural changes [31]. Using electrolyte solutions that contain additives that react on the active mass surface to form passivating surface films was also proved to effectively protect the surface. Besides a parasitic reaction on the surface and structural transition near the surface, the structural degradation in the bulk is vital to the electrochemical degradation. The structural transition could lead to

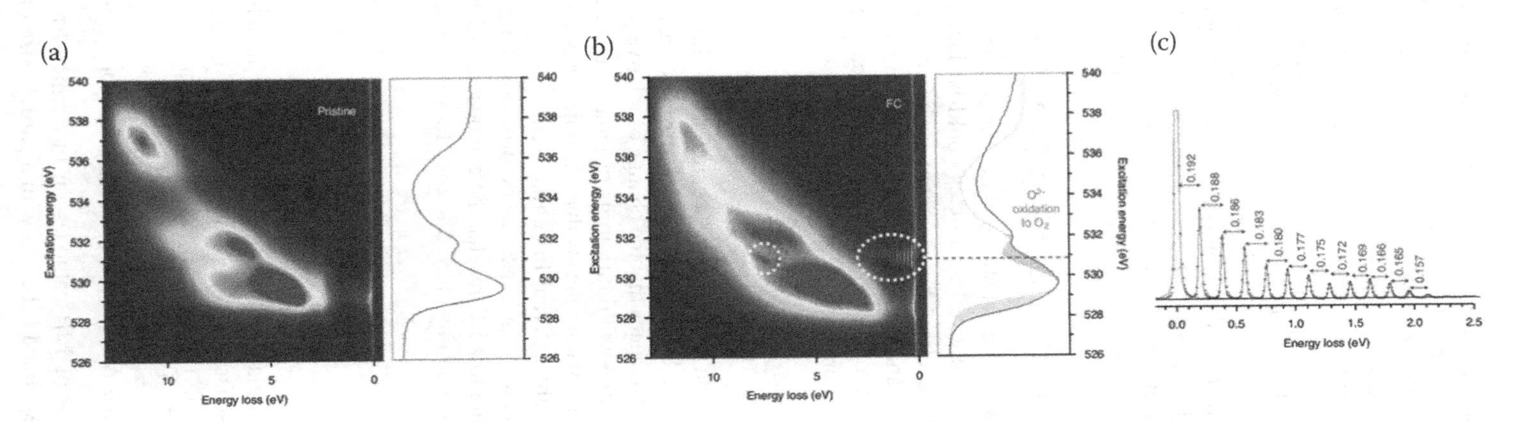

FIGURE 4.4 X-ray spectroscopic characterization of O. High-resolution mapping RIXS and O K-edge partial fluorescence yield XAS data for (a) pristine and (b) fully charged $Li_{1.2}Ni_{0.13}Co_{0.13}Mn_{0.54}O_2$. (c) The low-energy loss, which can be resolved into a progression of peaks that arise from transitions to different vibrational energy levels. Reproduced with permission of ref. [9], Copyright 2020 Springer Nature.

dislocations, inert domains, and decrease Li^+ kinetics. For example, the propagation of dislocations generates wider and more loosely packed inter-slabs, leading to progressive intragranular crack formation. And inter-crystalline cracks are formed when the particles are subjected to inelastic deformation in the course of fast lithiation/delithiation processes. Effective element substitution/doping are applied to stabilize the host framework, prevent TM migration, and expand Li layer spacing [31]. However, to resolve the issues associated with the bulk from the bottom, rational design of the pristine structure to achieve highly reversible oxygen redox is necessary. For example, recent researches show that the superlattice other than honeycomb ordering and stacking sequence other than O_3 can highly improve the reversibility of structural transition and mitigate voltage hysteresis and decay [26,32], and disordered rock-salt structure may avoid layer to spinel transition and improve the rate capability [33].

4.3 INVESTIGATIONS ON CATHODES' MECHANICS

The crystalline lattice of LIB cathode should endure the mechanical stress during Li+ intercalation and deintercalation. The particles could be lacerated by the internal strains after the specific cycles. The generated surface reacted with electrolytes to raise the polarization and then caused cyclic degradation. Besides Li+ interfacial inhabitation and electrolyte consumption, mechanical stability is one of the key reasons for cyclic deterioration.

4.3.1 Mechanics on LIB Cathodes

As we all know, a force acting on an object will change its motion state or make it deformed. The deformation can be divided into elastic deformation and plastic deformation. Elastic deformation means that the change of relative position between points caused by an external force will return to its original state when the external force is withdrawn. Plastic deformation means that the object's deformation cannot recover its original state after the external force removed.

For LIBs, the internal stress will cause material deformation during charging and discharging [34–36]. According to XRD results, the cathodes could be attributed to elastic materials, which internal strains obeyed to the Hook equation, as shown in Equation (4.1).

$$\sigma = E\varepsilon \tag{4.1}$$

where E is Young's Moduli and ε is particle stress.

It is worthy to note what cathodes endure the internal stain; namely, particles are in equilibrium state after discharge and relaxation process. In other words, the particles are under stress only in the state of Li^+ transportation.

4.3.2 Progress of Cathode Mechanics in LIB Cathodes

The first public proposed case of LIB cathodes on mechanics was proposed by Wang et al. in 2004 [37]. They found $LiFePO_4$ with a particle size of 300~500 nm

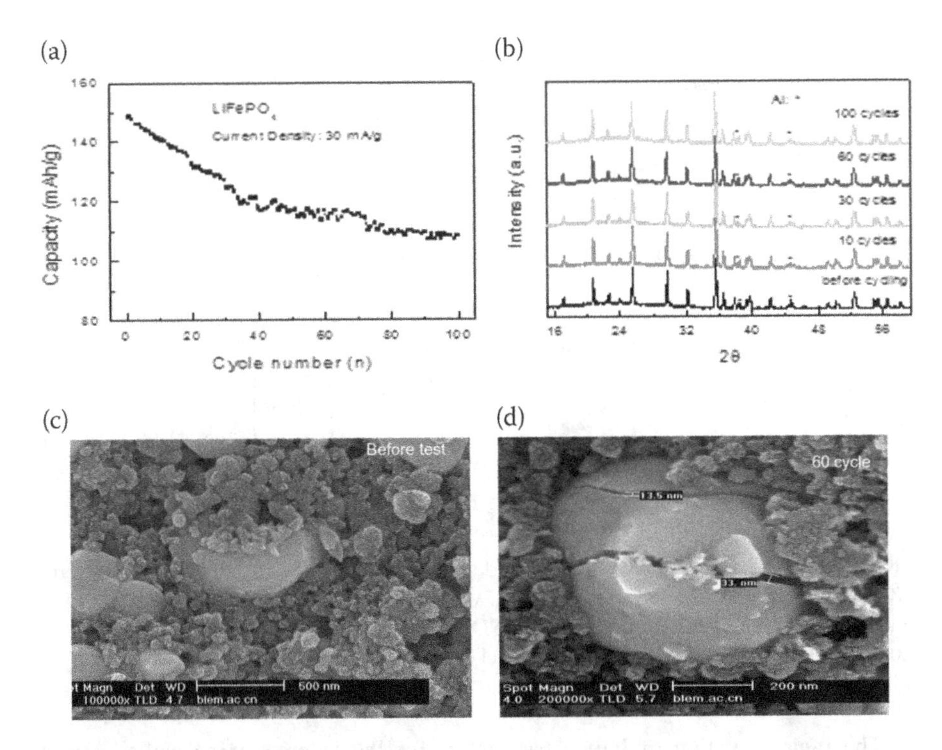

FIGURE 4.5 (a) Cycling performance of $LiFePO_4$; (b) the XRD patterns of LiFePO4 after different cycles; (c and d) the particle of $LiFePO_4$ before and after cycling. Reproduced from ref. [37], Copyright 2005 Elsevier.

was cracked and accompanied with cyclic degradation during operation (Figure 4.5).

Since the excellent electrochemical performances and low cost, the layered transition metal oxide is one of the most promising cathode materials for the next-generation energy storage devices [40]. Herein, some pioneering investigations have focused on the formation mechanism of the internal stress, the relationship between the stress and fatigue-damage of cathodes, and methods to address it.

Researchers generally believe that during charging, Li^+ ions extract from cathodes, the deformed lattice shall generate the mechanical stress, which exert on every single particle in the whole cycle [41]. It found that the change of lattice volume mainly resulted from the variation of the c-axis because of the same change trend, as shown in Figure 4.6(a,b) [43]. From it, the variation of the c-axis is much bigger than the a-axis, about 10 times. It indicates the variation of lattice parameters for layered oxide cathodes is anisotropic. From Figure 4.6(c), the change of lattice parameters (c-axis) can be divided into two stages at ~0.6 Li^+ removed during charging [44]. The change of the c-axis reaches the maximum when the delithiation reaches 0.6. After that, with the increase of charging depth, the c-axis decreases quickly with the twice variation ratio of that in the first stage [39,45,46].

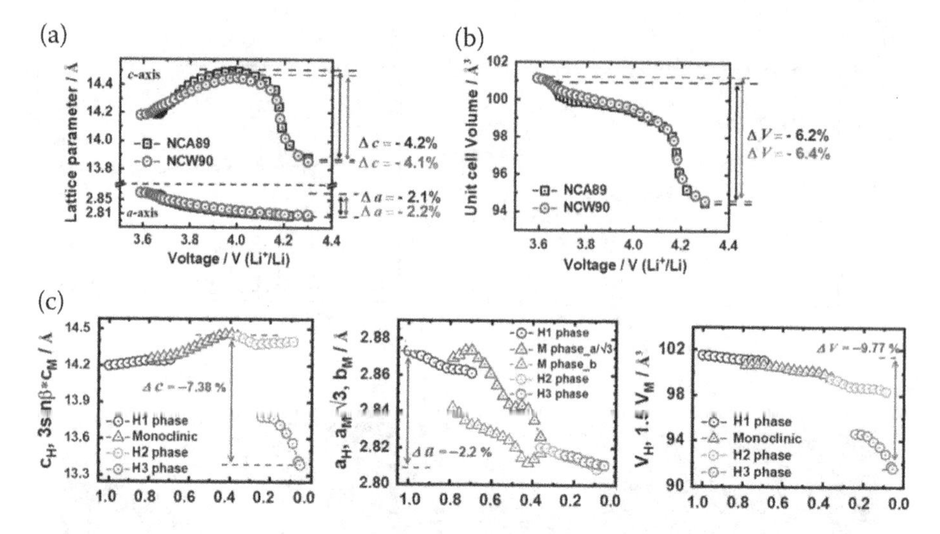

FIGURE 4.6 Changes in (a) a- and c-axis lattice parameters and (b) unit cell volumes for NCA89 and NCW90 as a function of the cell voltage; (c) the lattice parameters of the c-axis, a-axis, and unit cell volumes of LiNiO2 (LNO). (a-b) Reproduced from ref. [43], Copyright 2019 WILEY-VCH; (c) reproduced from ref. [44], Copyright 2019 The Royal Society of Chemistry.

The rapid mutation of lattice parameters for the layered oxide cathodes will produce internal stress [44]. To disclose the inherent mechanism, Ren et al. [47,48] took the commercial ~4 μm $LiNi_{1/3}Co_{1/3}Mn_{1/3}O_2$ (NCM111) single crystal as a sample material to explore the relationship between internal stress and electrochemical performance. The variation trend of lattice parameters measured by ex-situ XRD during charging was consistent with that reported in literature, as shown in Figure 4.7(a,b). According to Hook Equation, the change of lattice will produce internal stress. In order to simplify the analysis, only the change of the c-axis was considered to roughly estimate the internal stress [48].

From the mechanical point of view, particle cracking could result from tensile stress due to the non-transient delithiation and lithiation, leading to non-uniform distribution of Li^+ in different areas of one particle. Correspondingly the repeating volumetric contraction and expansion due to Li^+ extraction and insertion can lead to fatigue and crack growth. The stress originated from non-uniform distribution of Li^+ can be derived theoretically under the following assumptions: 1) it is a sharp interface between the fully lithiated region and the region without any Li^+ in the particle; 2) stress relaxation could be completed very quickly and the stress wave is not under consideration; 3) the deformation is small every cycle. Within these assumptions, the region with the largest Li^+ variation rate is account for the particle cracking. Ren et al. took the length along axis *c* of the particle with maximum Li^+ as the reference state, denoted as L_0, force balance at the interface perpendicular to axis *c* between the region fully occupied by Li^+, in which the stress is σ_{Li}, and without Li^+, in which the stress is σ, gives

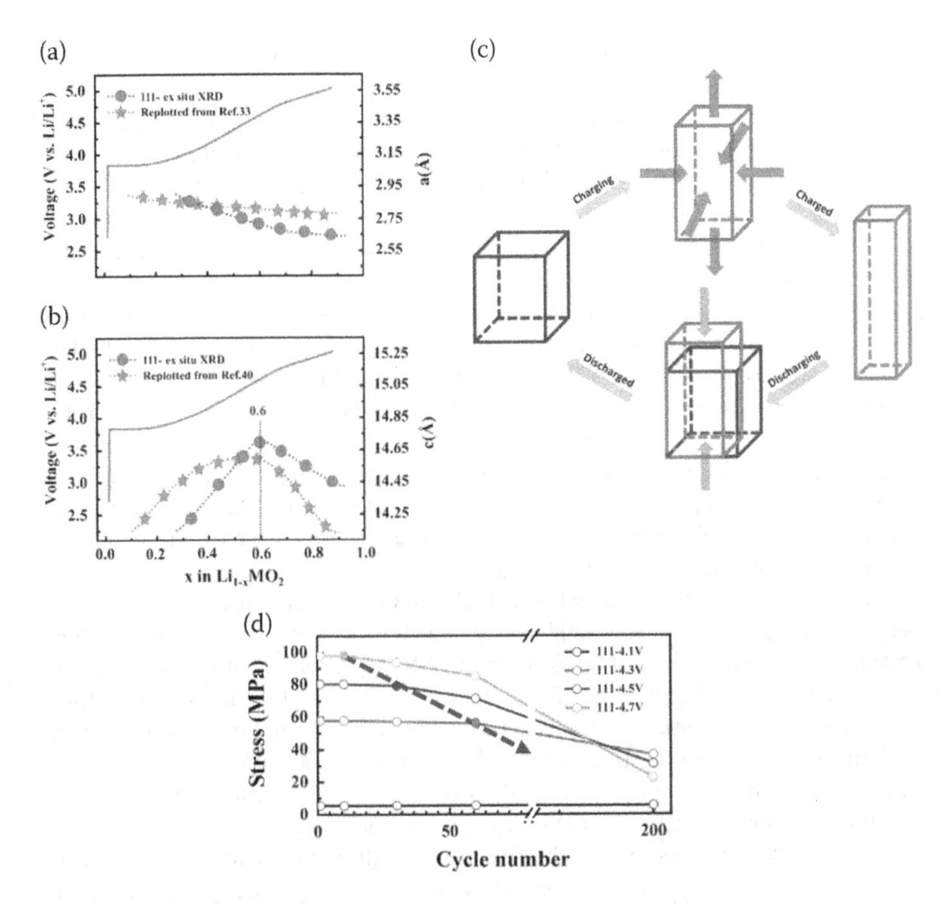

FIGURE 4.7 (a)–(b) The variations of axis a and c during the charging of the investigated sample and previously reported NCM111; under different cut-off voltages; (c) the schematic diagram of the particle stress for NCM111 in every cycle; (d) the tensile stresses of the electrodes after 10, 30, 60, and 200 cycles. (a–d) Reproduced from ref. [48], Copyright 2019 Elsevier.

$$\sigma(L_0-\Delta \mathbf{L}) + \sigma_{Li}(L_0-\Delta \mathbf{L}) = 0 \tag{4.2}$$

For a particle without any Li^+, the experimentally measured strains are eigen strains referred to a particle fully occupied by Li^+, denoted by $\boldsymbol{\varepsilon_0}$. The corresponding eigen stress σ_0 can be calculated by using $\sigma_0 = Y_{Li}\varepsilon_0$, where Y_{Li} is the Young's modulus of particle with Li insertion. Thus, $\sigma_{Li} = \sigma_0$-$Y_{Li}\varepsilon$ at the nominal strain ε. Both σ_{Li} and σ_0 are tensile stresses in our experiments. From Equation (4.2), the tensile stress can be expressed as

$$\sigma = Y_{Li}\varepsilon - \sigma_0 = Y_{Li}(\varepsilon - \varepsilon_0). \tag{4.3}$$

The schematic diagrams of the particle stresses for NCM111 in every cycle are shown in Figure 4.7(c), which is the removal Li^+ is no more than 0.6. Almost the same values of Young's modulus (Y_{Li}) with ~0.95 GPa were obtained for all particles with different charging states. From Figure 4.7(d), leaving the same constant, $Y_{Li}\varepsilon_0$, undetermined due to the unknown value of eigen strain ε_0. In these plots, $Y_{Li}\varepsilon_0 = 0$ was set due to the zero stress in the pristine state. As shown in Figure 4.7(d), the tensile stress for the particles is unchanged in the whole 200 cycles when cycled between 2.8~4.1 V vs. Li^+/Li. They also keep the similar values in the initial ~60, ~30, and ~10 cycles with the cut-off potentials of 4.3, 4.5, and 4.7 V vs. Li/Li^+, respectively.

From the mechanical point of view, the internal stress under repeated action, which results from tensile stress due to the non-transient delithiation and lithiation, can lead to fatigue and crack growth of the layered oxide cathodes [49–51]. The morphology of NCM111 after long cycling is consistent with that of the other layered oxide cathode materials; that is, the particles of layered oxide cathode material appeared as micro-cracks, shown in Figure 4.8(a–c) [48,53,54]. Obviously, stresses start to decrease when the particles are cracked. After particle cracking, the electrochemical performance also decayed rapidly, and the decaying trend was consistent with the degree of internal stress damage, as shown in Figure 4.9(a,b) [48].

According to the reported experimental results and analysis, the mechanical behavior of the investigated sample obeys the fracture-damage models. The mechanical evolved processes of cracking for NCM111 particles could be described as in Equation (4.4). If the particle's fracture strength is higher than the strains exerted on the particles, the particles' strains generated per cycle should be almost the same with the assumption of the unchanged Li^+ diffusion coefficient. In the initial stage of cell's operation, the fracture strength is fatigued by the particles' stress with a function of damage degree via the bulky defects, such as line dislocations.

As the cycles are proceeding, the fracture strength is gradually weakened, corresponding to the proper service period. When the fracture strength is unable to sustain the strains, the particles are lacerated and induced the plummet of cyclic stability due to the quickly aggrandized polarization. If the damage degrees in all cycles were assumed as the same value, the equation could be converted to Equation (4.5):

$$F_n = F_o * (1 - D_1) * (1 - D_2) * (1 - D_3) * \ldots * (1 - D_n) \tag{4.4}$$

$$F_n = F_o * (1 - D)^n \tag{4.5}$$

where F_n represents the fracture strength in the nth cycle; F_o represents the intrinsic fracture strength; and D_1, D_2, D_3, and D_n represent the damage degrees in the first, second, third, and Nth cycle.

This model could explain the cell's capacity fading related to particles' internal stress and estimate the probable working life in term of mechanics.

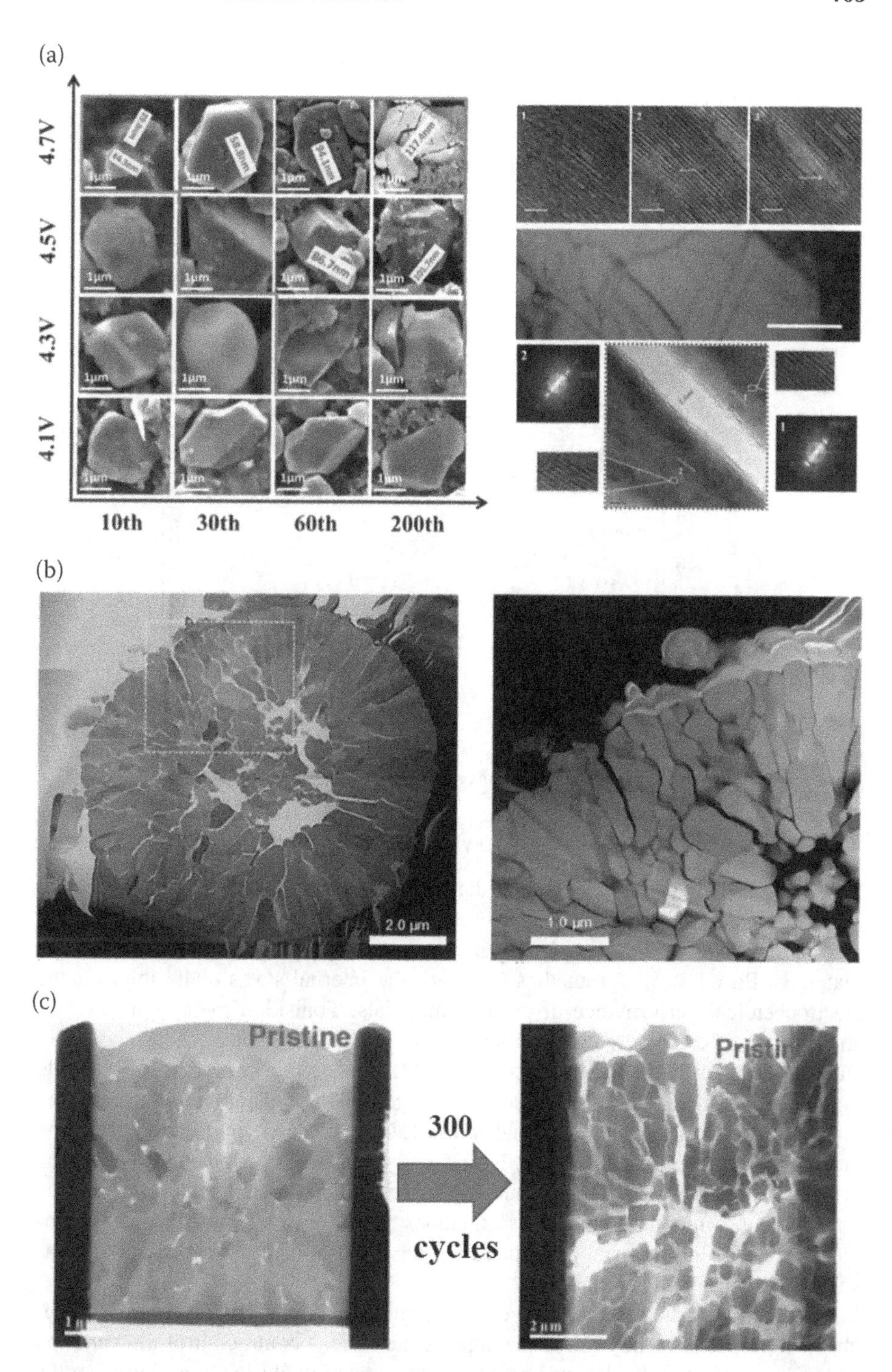

FIGURE 4.8 (a) SEM images of the electrode after 10, 30, 60, and 200 cycles; lattice images after 0 cycles; 10 cycles; 30 cycles, and TEM images and corresponding lattice images taken from the regions marked by the dash-red rectangular after 200 cycles; (b) mosaic scanning TEM image of the cycled NCA90 electrode after 1,000 cycles; (c) pristine sample before and after cycles. (a) Reproduced from ref. [48], (b) reproduced from ref. [54], Copyright 2018 Elsevier. (c) reproduced from ref. [53], Copyright 2016 WILEY-VCH.

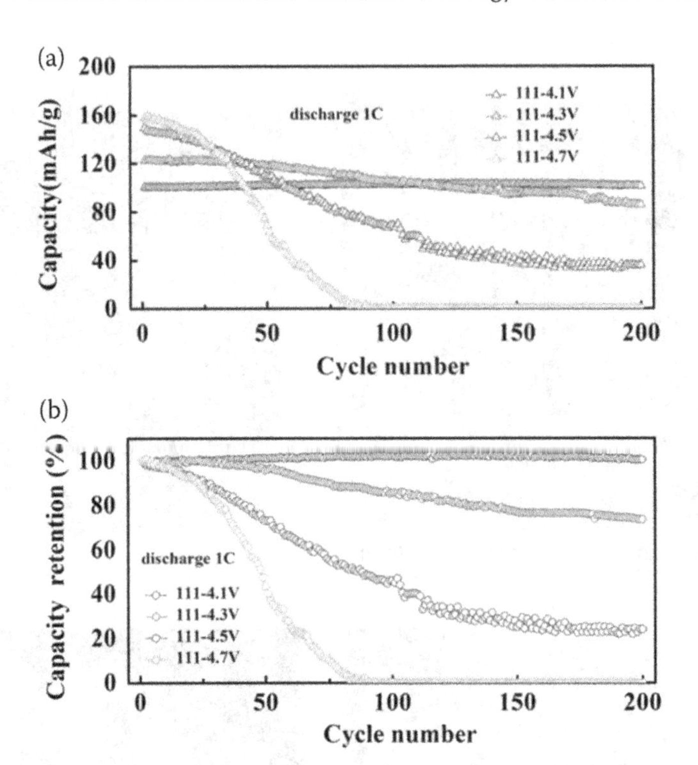

FIGURE 4.9 (a) Cycling performance at 1C during 200 cycles; (b) capacity retentions at 1C during 200 cycles. Reproduced from ref. [48], Copyright 2019 Elsevier.

4.3.3 Bulky Substitution to Alleviate Internal Stress

The previous analysis demonstrates that the internal stress of the layered oxide cathode materials during cycling causes mechanical damage, evolves to micro-cracks, and leads to the degradation of electrochemical properties of the electrode materials. Based on this, remedies to reduce the internal stress could improve the electrochemical performance of cathode materials. Transition metal element substitution, an effective strategy to alleviate the internal stress, has been intensively studied [55]. Arumugam Manthiram used Mg-dopant the layered oxide cathode material $LiNi_{0.94}Co_{0.06}O_2$ to reduce the change of lattice parameters during charging, so as to mitigate the internal stress [56]. As shown in Figure 4.10a, the variation of c-axic for $Li_{0.98}Mg_{0.02}Ni_{0.94}Co_{0.06}O_2$ was 3.7%, lower than that of a pristine sample (5.6%). The fatigue damage was ameliorated by decreasing the corresponding internal stress. These methods were successful in reducing the internal stress of the material and mitigating the damage of the cathode material, as shown in Figure 4.10(b–d).

Ren et al. also investigated the effect of substitution cations on materials' mechanical properties. They found the appropriate cations could control the variation of lattice parameters for layered oxide cathode materials during charging from the level of the electron cloud, so as to reduce the generation of internal stress [47]. The

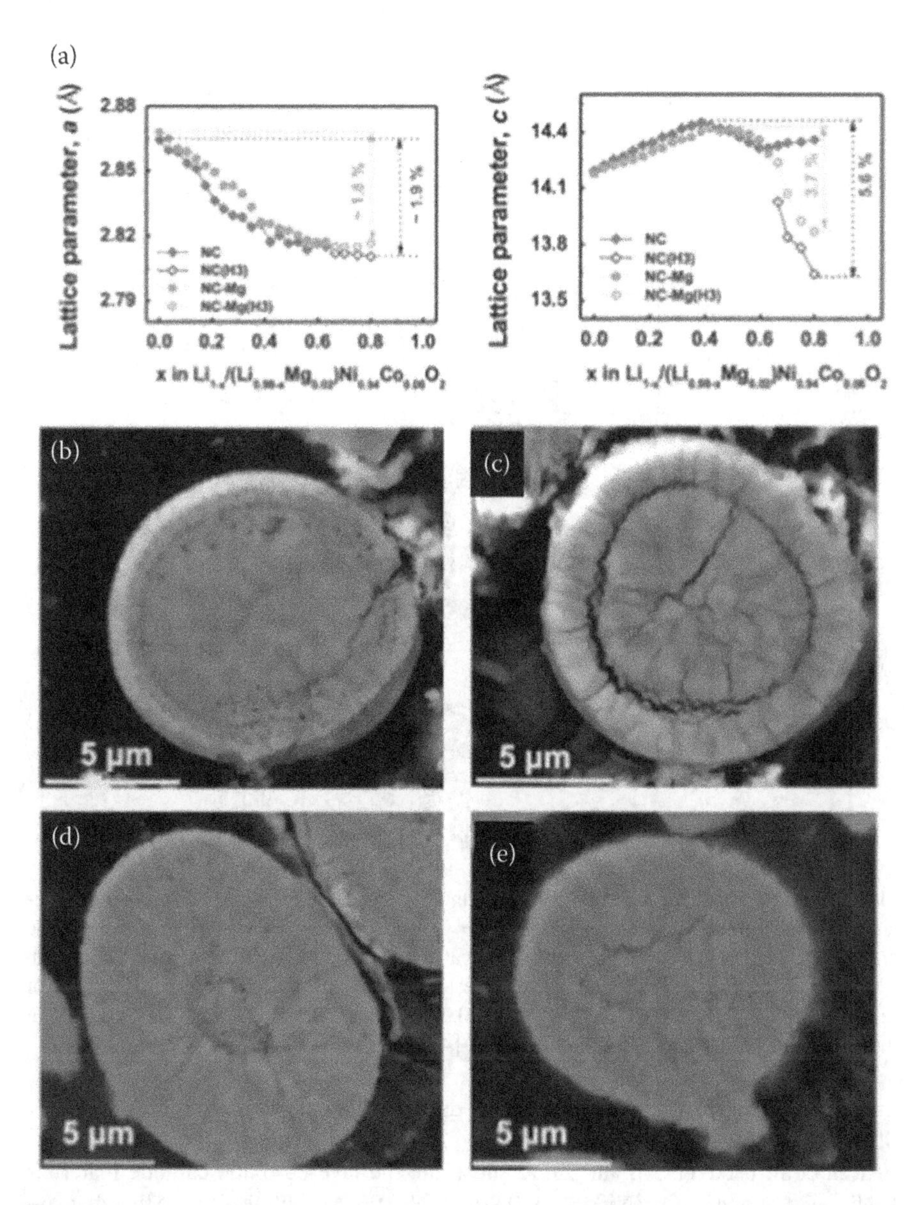

FIGURE 4.10 (a) The lattice parameters a and c of $LiNi_{0.94}Co_{0.06}O_2$; SEM cross section images of (b and c) $LiNi_{0.94}Co_{0.06}O_2$ and (d and e) $Li_{0.98}Mg_{0.02}Ni_{0.94}Co_{0.06}O_2$ after 200 cycles in the pouch full cell. Reproduced from ref. [56], Copyright 2019 American Chemical Society.

variations of the c-axis are mainly determined by the propelling force between O^{2-} anions in O-Li-O layer (Figure 4.11a). The elongation of the c-axis is a result of gradually increasing propelling force as x ≤~0.6 in $Li_{1-x}MO_2$, and the shrinkage as x ≥~0.6 in $Li_{1-x}MO_2$ should be attributed to the withdrawal of oxygen's charge to transition metal layer by Ni^{4+} ions. Therefore, the c-axis variations, namely the

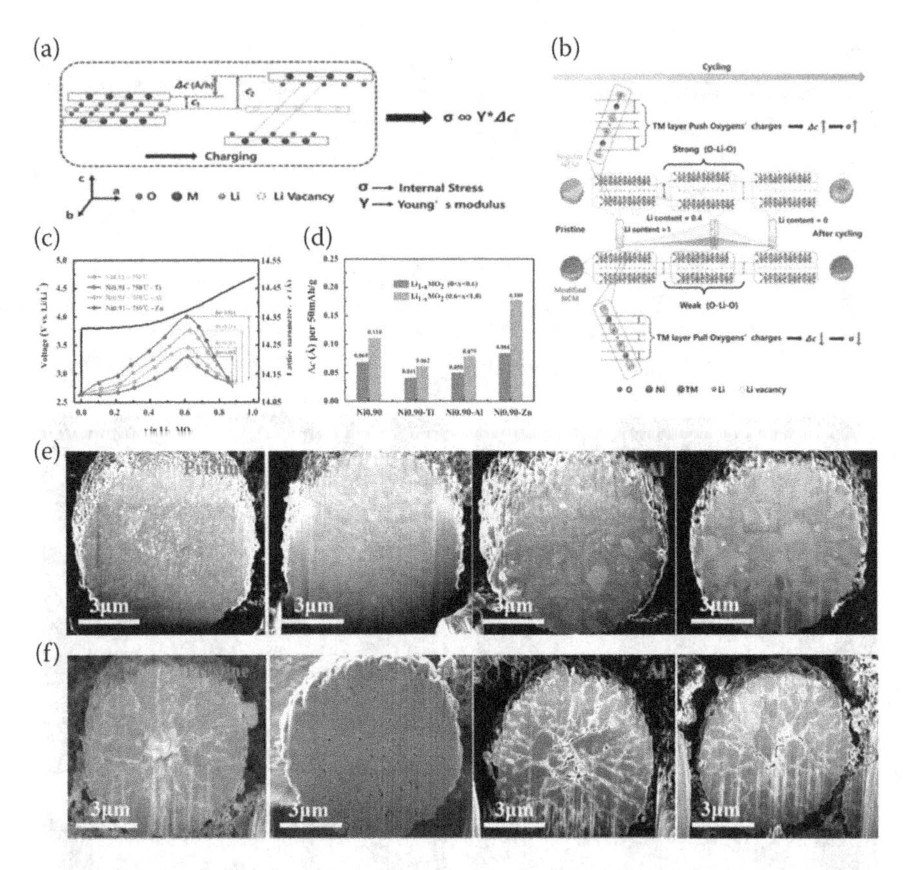

FIGURE 4.11 The schematic illustration on the mechanical damages: (a) mechanism of mechanical damage; (b) relationship among the O's charges, mechanical stress, and particle cracking; (c) variation of lattice c on all investigated samples during charge; (d) Δc per 50 mAh gl in two distinct phase transitions of all investigated samples; SEM cross-section images of (e) Ni0.90 and (f) Ni0.90-Ti, Ni0.90-Al, and Ni0.90-Zn dopant samples after 0 and 200 cycles in the pouch full cell. (a–f) Reproduced from ref. [47], Copyright 2020 Elsevier.

mechanical stress, could be weakened by drawing the oxygen's charges to the layer of TM, as shown in Figure 4.11(b).

Ren et al. used Ti, Al, and Zn to substitute the layered oxide cathode materials $LiNi_{0.90}Co_{0.02}Mn_{0.08}O_2$ (Ni0.90, Ni0.90-Ti, Ni0.90-Al, Ni0.90-Zn). After that, in-situ XRD was used to test the variation of the c-axis for a pristine sample and the other samples during charging and the results are shown in Figure 4.11(c,d). Among the investigated species, Ti^{4+} and Al^{3+} draw the oxygen charges to transition metal (TM) layers, and Zn^{2+} plays the pushing effect. Ti-dopant sample renders the largest reduction of lattice c variation, ~40% less than the pristine in both regions. According to Hooke's law, the internal stress of the Ti-dopant sample was also much lower than that of the pristine sample. This conclusion was proven by the fatigue damage of the samples that resulted from the repeated internal stress, which was shown in Figure 4.11(e,f). After 200 charge/discharge cycles, Ni0.90 primary

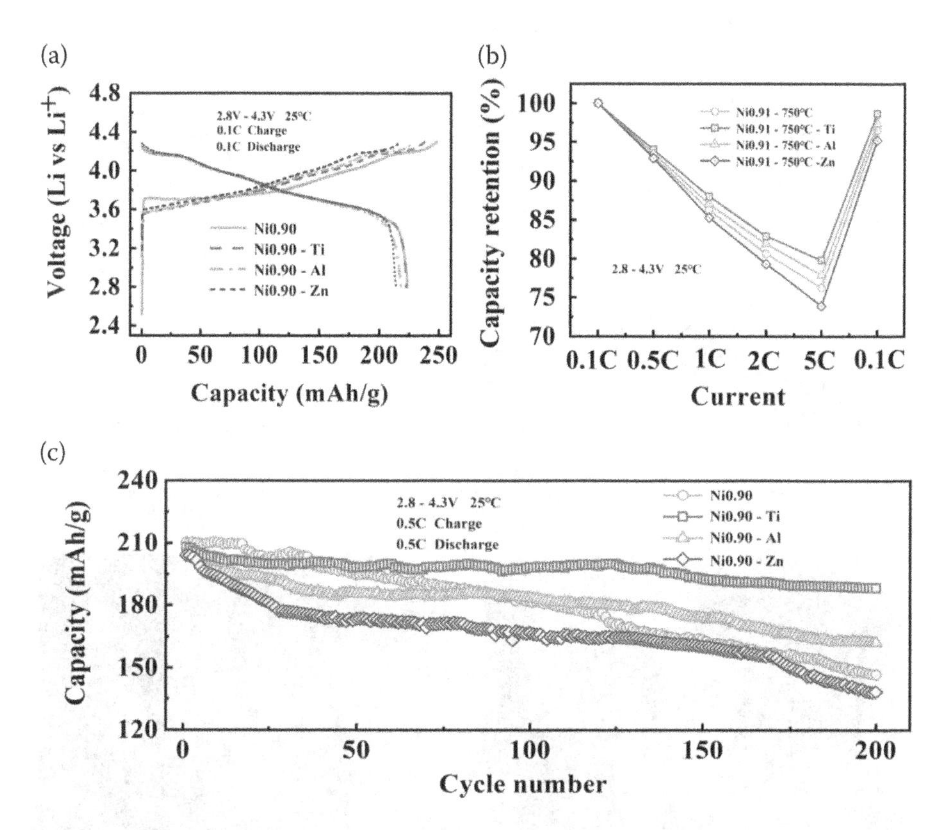

FIGURE 4.12 Electrochemical performance of Ni0.90 and Ni0.90-M cathodes: (a) first cycle charge/discharge profiles of Ni0.90 and Ni0.90-Ti, measured at 0.1C; (b) rate performance; and (c) cyclic stability. (a–c) Reproduced from ref. [47], Copyright 2020 Elsevier.

particles are obviously pulverized, with the cracks of 50 nm to 1 μm. On the other hand, Ni0.90-Ti particles retain the original boundary between the primary particles without obvious cracks. And the electrochemical performance for the Ti-dopant sample was greatly improved after decreasing the internal stress, as shown in Figure 4.12(a–c).

4.3.4 Surface Engineering to Suppress Particle Cracking

Apart from the bulky substitution, the surface engineering could also suppress the particle cracking for cathode materials [57–61]. Kim et al. proposed a novel method of surface coating that nano cobalt hydroxide and residual lithium on the surface of layered oxide cathode materials were used for an in-situ reaction [62]. After that, a powerful surface coating layer was formed on the surface of layered oxide cathode material. Through 500 cycles, the particle with a surface coating layered oxide cathode was kept integrated, while the pristine sample appeared as micro-cracks in the particle.

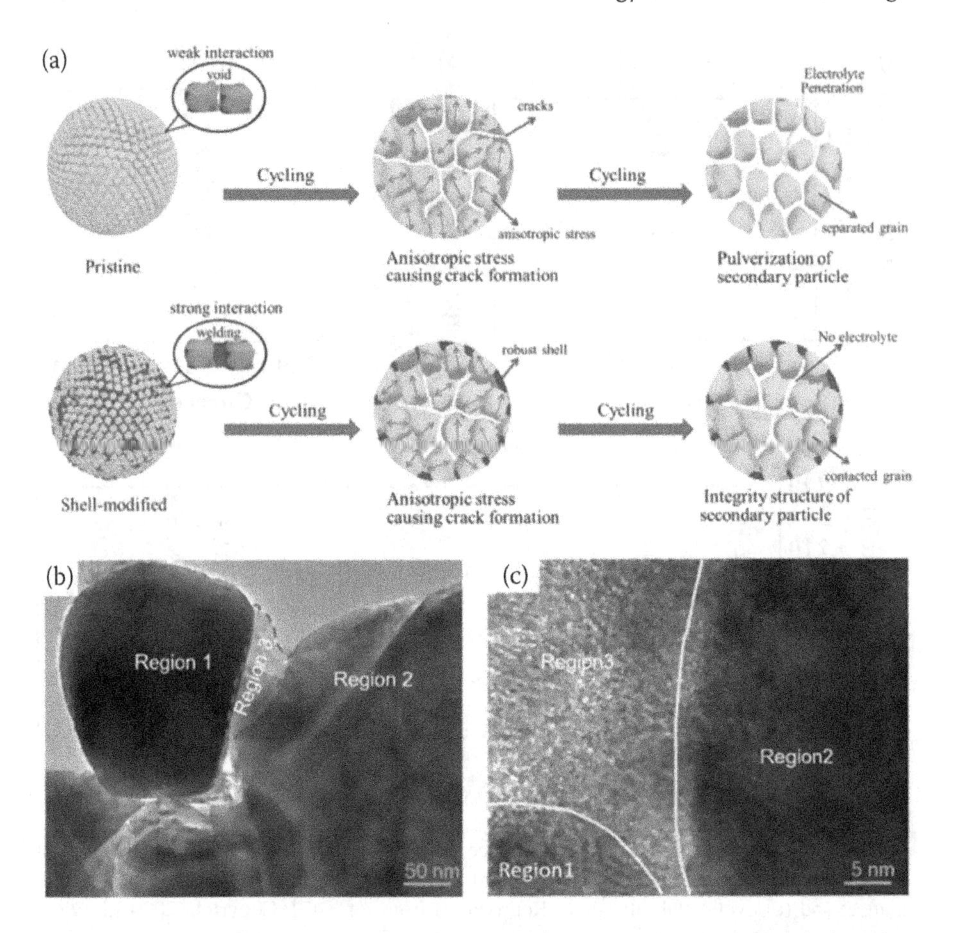

FIGURE 4.13 (a) Schematic of the shell-modified improve the electrochemical performance for Ni-rich materials during the cycling; (b) low-magnification images; (c) the corresponding HRTEM images. Reproduced from ref. [63], Copyright 2021 American Chemical Society. [63]

Liu et al. also constructed a strengthened shell layer on a polycrystalline secondary particle to address the unfavorable influence of particles' cracking instead of suppressing their bulky pulverization, which is shown in Figure 4.13(a). Through the solid state sintering method, the nanoscale Nb-based ceramic material was embedded in the gap of the primary particles on the surface (Figure 4.13(b,c)) Meanwhile, as shown in Figure 4.14, the Young's Moduli of particles for the modified cathode materials is 2.6 times higher than that of the original samples, which is measured by Atomic Force Microscopy (AFM). After 200 cycles, the modified sample appeared as a few smaller cracks due to the protection of the shell on the particle surface, while the fragmentation of the novel sample is more serious, which is shown in Figure 4.15.

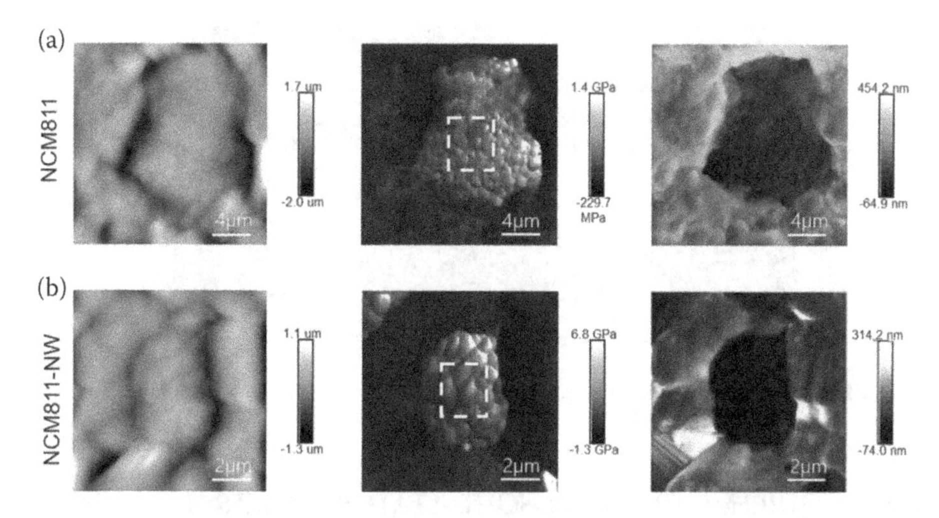

FIGURE 4.14 AFM images of (a) NCM811 and (b) NCM811-NW. Reproduced from ref. [63], Copyright 2021 American Chemical Society.

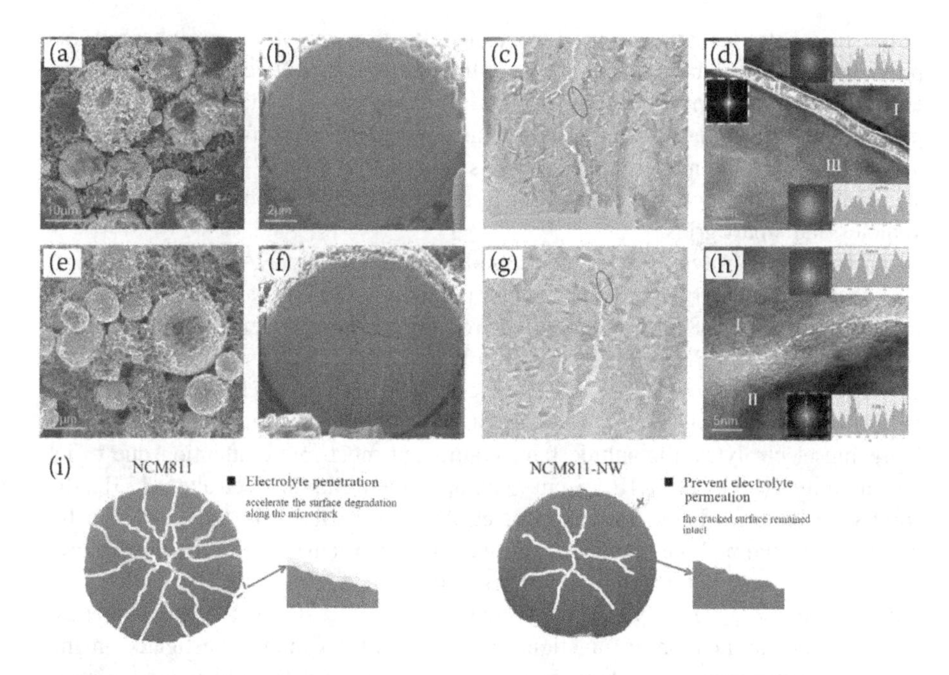

FIGURE 4.15 Post-analysis on electrodes and particles of the pristine and Nb-modified sample. (a, e) SEM images of NCM811 and NCM811-NW; (b, f) cross-section SEM images of NCM811 and NCM811-NW; (c, g) magnified TEM images of NCM811 and NCM811-NW; (d, h) high-resolution TEM images and FFT images (inset) of NCM811 and NCM811-NW the marked red oval regions in (c) and (g); (i) the schematic illustration of microcrack surface changes after cycling. The green dots represent cracks, the purple rings represent the shell, and the yellow arrow curve represents the electrolyte. Reproduced from ref. [63], Copyright 2021 American Chemical Society.

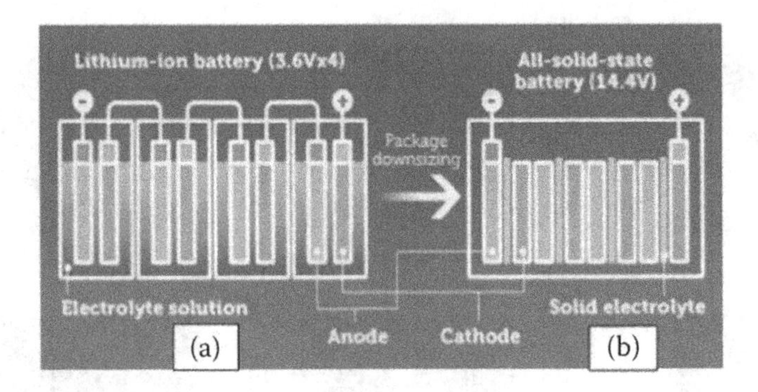

FIGURE 4.16 Diagram of (a) lithium ion battery; (b) all-solid-state battery. Source: Toyota (2016), All Solid State Batteries.

4.4 REQUIREMENTS OF ALL SOLID-STATE BATTERIES FOR CATHODES

4.4.1 Brief on Solid-State Batteries

Solid-state batteries attract ever-increasing attention due to their inherent improvement on safety by replacing organic-liquid electrolytes to ceramic or polymer electrolytes, as shown in Figure 4.16 [64,65]. Moreover, solid-state electrolytes possess the advantages of a higher mechanical property and no free solvents, which are capable of utilizing a lithium metal as an anode [52,66,67]. Therefore, solid-state batteries are highly preferable in large-format applications, such as electric vehicles and smart grids.

The solid electrolytes are mainly classified into sulfides, NASICON-structured phosphates, garnet-structured oxides, and polymers; their characteristics are compared in Table 4.1 [68–70]. Polymer electrolytes are the first generation of a solid-state electrolyte. A PEO-based lithium polymer battery was proposed in the last century and currently is utilized in electric vehicles in Paris [71,72]. Its operating temperature ranges at ~80°C, which limits its application scenarios. Among the inorganic electrolytes, the sulfide family attracted much more attention due to its conductivities (up to 2.5×10^{-2} S cm^{-1}) comparable to liquid electrolytes [73]. All-solid-state batteries based on sulfite electrolytes were listed in the road map by Toyota, who claimed the practical worldwide commercialization in the near future. The other ceramic solid-state electrolytes, such as LATP and LLZO, couldn't be utilized as the sole electrolytes because they are unable to sustain the cells' kinetics. Instead, their application as functional additives were widely investigated in the academic community. As is discussed previously, sulfide solid sulfites are the most promising solid-state electrolyte.

As one of the major materials, cathodes should process novel properties to fulfill the requirements of solid-state batteries. Since an organic polymer system is close to an organic electrolyte cell, only the requirements of ceramic solid-state batteries are discussed in this part. The challenge for cathodes is mainly related to the

TABLE 4.1
The most common solid electrolytes (SEs) and their examples

Type	Structure	Formula	Con. ($S.cm^{-1}$)	Development
Oxides	NASICON	$LiZr_2(PO4)_3$	2.1×10^{-5}	Ilika, Sony, Qingtao
		$Li_{1.4}Al_{0.4}Ti_{1.6}(PO_4)_3$	5.16×10^{-4}	
		$Li_{1.3}Ti_{1.7}Al_{0.3}(PO_4)_3$	7×10^{-4}	
		$Li_{1+x}Al_xGe_{2-x}(PO_4)_3$	6.2×10^{-4}	
	Garnet	$Li_5La_3M_2O_{12}$(M = Nb, Ta)	10^{-6}	
		$Li_6BaLa_2Ta_2O_{12}$	4×10^{-5}	
		$Li_7La_3Zr_2O_{12}$	8.1×10^{-4}	
Polymer	Polymer	PEO–$LiClO_4$	10^{-3}	Bollore
Sulfides	Li_2S-SiS_2	0.6(0.4SiS_2–0.6Li_2S)–0.4LiI	1.8×10^{-3}	Toyota, SumsungCATLGanfeng
		Li_2S–SiS_2–Li_3PO_4	6.9×10^{-4}	
	Li_2S-P_2S_5	80Li_2S–20P_2S_5	10^{-3}	
		$Li_7P_3S_{11}$	3.2×10^{-3}	
		$Li_{10}SnP_2S_{12}$	4×10^{-3}	
	Li_6PS_5X	Argyrodite Li_6PS_5Cl	1.9×10^{-3}	
		Argyrodite Li_6PS_5Br	6.8×10^{-3}	
	$Li_{11-x}M_{2-x}P_{1+x}S_{12}$	$Li_{10}GeP_2S_{12}$	1.2×10^{-2}	
		$Li_{9.54}Si_{1.7}4P_{1.44}S_{11.7}Cl_{0.3}$	2.5×10^{-2}	

Note: A polymer electrolyte was tested at 80°C; others were tested at room temperature.

electrolyte/cathode solid-solid contacts, which could be deteriorated by the unmatched phases and continuously volumetric variations.

4.4.2 Cathodes' Interfacial Modifications for All-Solid-State Batteries

The mechanism between cathodes and solid-state electrolytes could be understood with space charge layer (SCL). SCL refers to the area where the concentration of charge carriers varies between two phases, as shown in Figure 4.17 [74–76]. The concept originally came from explaining the conductive effect of the interface between two phases of semiconductors. In the process of forming a "junction" after the p-type and n-type semiconductor contact, a thin layer of positive and negative charges space is formed on both sides of the contact surface, as shown in Figure 4.18.

Based on the core-space charge mode [76,77], assume that the defect (a type of carrier) has the same continuous electrochemical potential, that the crystal structure at the interface between the two phases does not change (x = 0), and that the mobility of the defect in the space charge layer is the same as that in the bulk

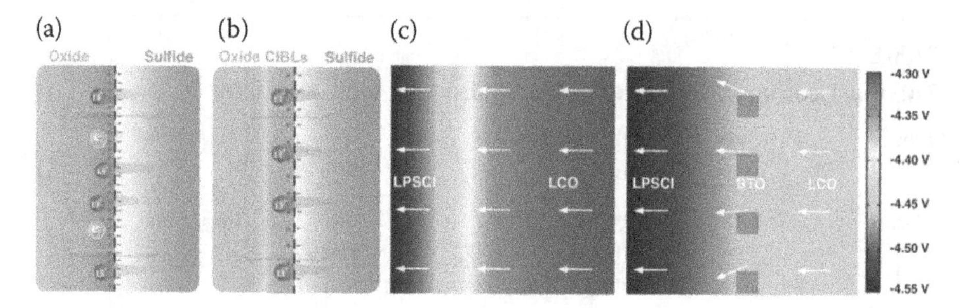

FIGURE 4.17 (a) a Initial interface; (b) double-phase interface with fast Li-ion conductor CIBLs; (c) simulation results of the internal electrical field for the LCO/LPSCl interfaces; (d) simulation results of the internal electrical field for the BTO–LCO/LPSCl interfaces. (Reproduced with permission of ref. [74], Copyright 2020 Springer Nature).

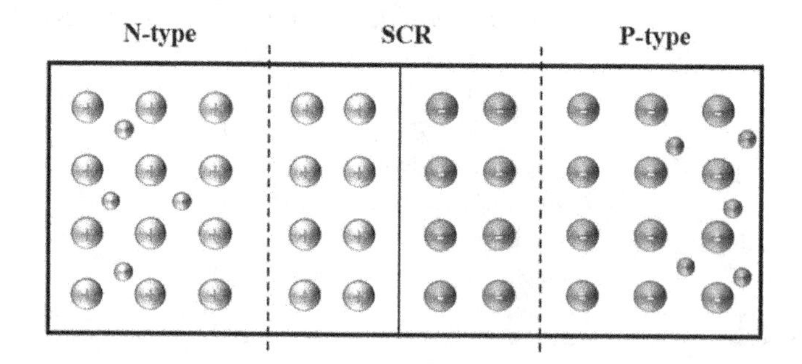

FIGURE 4.18 The SCL formed by a PN junction in a semiconductor (SCL:space charge layer).

phase (x = ∞), the concentration ζ at the equilibrium state of the migration carrier j can be expressed as

$$\varsigma_j^{\frac{1}{Z_j}}(\mathrm{x}) = \left[\frac{c_j(x)}{c_{j,\infty}}\right]^{\frac{1}{Z_j}} = \exp\left\{-\frac{[\varphi(\mathrm{x}) - \varphi_\infty]}{k_B T}\right\} \tag{4.6}$$

where c is the carrier concentration of the defect and φ is the electric potential.

It was then used to understand the interface phenomenon in the ionic conductor system of all-solid-sate batteries [81]. The formation of SCL in cathode and solid state interface is different from the semiconductor hetero-junctures. According to the typical model, the space charge should be enriched in the hetero-interface without species mitigation. In fact, in $LiCoO_2$/sulfide electrolytes, scanning transmission electron microscopy with energy-dispersive X-ray spectroscopy revealed Co diffusion from $LiCoO_2$ (LCO) to the sulfide electrolytes, even at a distance of 50 nm from the interface, as shown in Figure 4.19 [82,83].

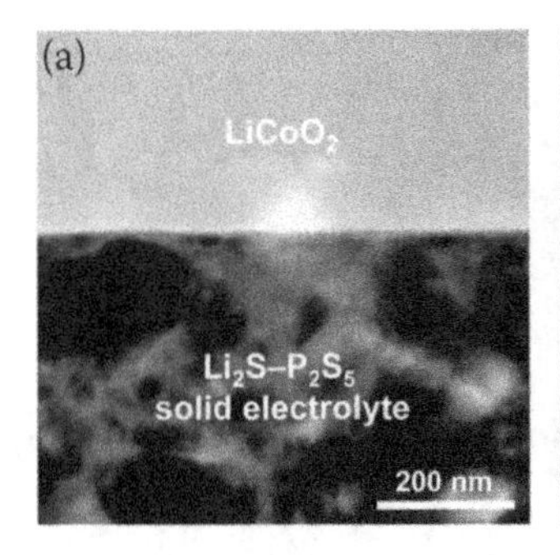

FIGURE 4.19 (a) Cross-sectional HAADF-STEM image and (b) the corresponding EDX mapping for the Co element near the $LiCoO_2$ electrode/Li_2S-P_2S_5 solid electrolyte interface after initial charging (Reproduced with permission of ref. [78], *Copyright © 2010, American Chemical Society*).

When a solid-state electrolyte contacts the mixture conductor of cathodes, such as $LiCoO_2$, the space charge layer will be formed in the interface. In the case of $LiCoO_2$ and sulfides, Li^+ ions are enriched in the side of a solid-state electrolyte and electrons are centralized in the side of a cathode. Ohta et al. [84–87] adopted the appropriate surface-treated layer to modify the interfacial space-charge layer to decrease the resistance and improve the high-rate capability. As a result, the rate capability of Li/sulfide/$LiCoO_2$ cells was significantly improved when cathode particles were coated with $Li_4Ti_5O_{12}$, as shown in Figure 4.20.

Tateyama et al. [73] theoretically elucidated the characteristics of the space-charge layer (SCL) at interfaces between oxide cathode and sulfide electrolyte in all-solid-state batteries and the effect of the buffer layer interposition via theoretical calculations with density functional theory (DFT) + U framework. The results revealed that LCO/LPS could form the stable interface, which possesses Li adsorption sites and a rather disordered structure. Moreover, the interposition of the LNO buffer layers between $LiCoO_2$ and sulfides could smooth the SCL's structure and facilitate Li transportation, as shown in Figure 4.21. The calculated energies of the Li-vacancy formation and Li migration revealed that the subsurface in the LPS side can transfer at the under-voltage condition through the interface, which suggests the SCL growth at the beginning of charge leads to the interfacial resistance.

The space charge layer effect, interfacial strain/stress concentration, and side reaction originating from the undesirable interfacial structure will lead to serious deterioration of electrochemical performance [73,86,87]. Thus, engineering the stable interfaces to improve the chemical/electrochemical compatibility are crucial for all-solid-state battery development. Surface decorating appropriate materials onto a cathode particle has been proved to be an effective strategy to passivate and stabilize the cathode–electrolyte interface to avoid the effect of space charge layer, thus inhibiting the side reactions [88–94].

In previous studies, Yubuchi et al. [95] have employed a pulsed laser deposition approach to uniformly coat the LNMO particle surface with amorphous Li_3PO_4 for blocking its direct contact with the sulfide-based SE. When working with

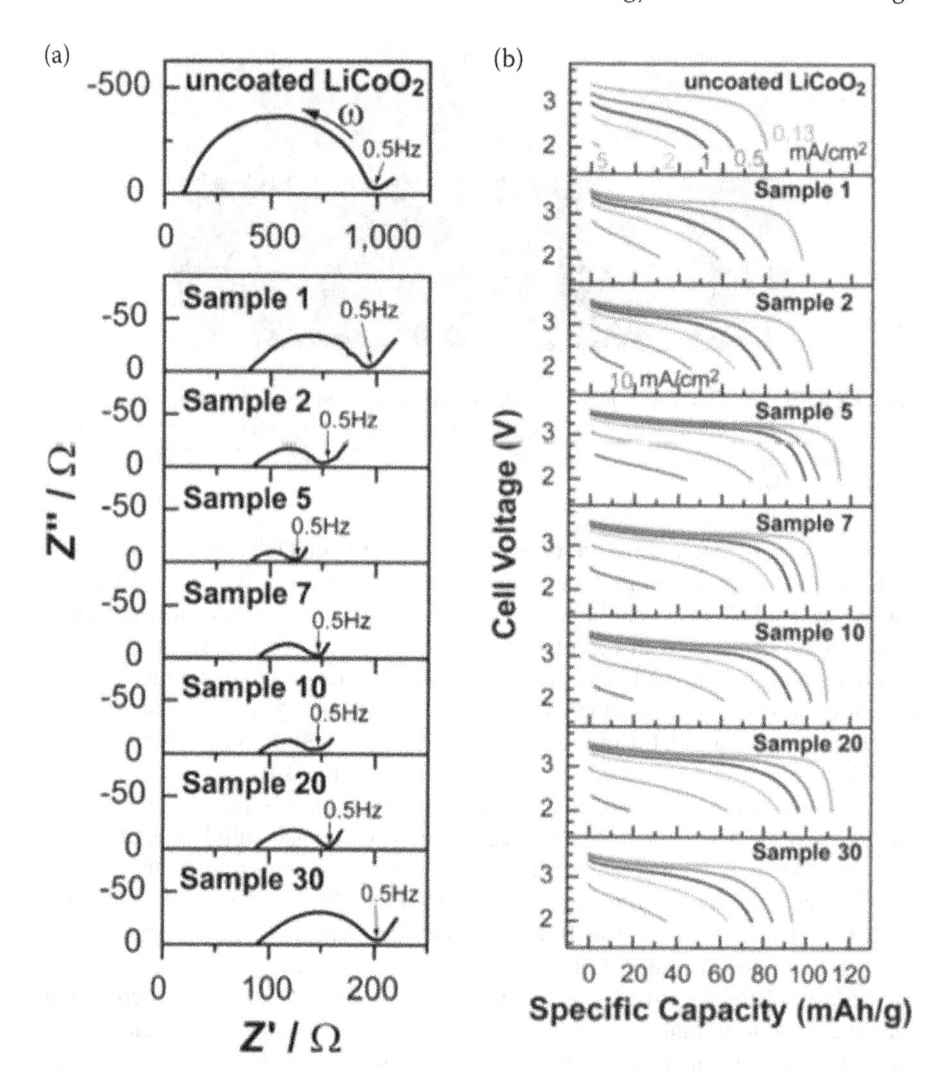

FIGURE 4.20 (a) Complex impedance (Z) plots of the In–Li/$LiCoO_2$ cells; (b) discharge curves of the In–Li/$LiCoO_2$ cells. The horizontal axes indicate the specific capacity calculated on the basis of the weight of the uncoated or coated $LiCoO_2$. (Reproduced with permission of ref. [84], Copyright 2013 Elsevier.)

$0.8Li_2S$-$0.2P_2S_5$ SEs, this Li_3PO_4-coated LNMO (28 wt% in the cathode layer) could demonstrate a markedly reduced interfacial resistance with an initial reversible capacity of 62 mA h g^{-1}, as is shown in Figure 4.22.

Jihui Yang et al. [96] used atomic layer deposition technology to coat the surface of $LiCoO_2$ with ultra-thin Li_3NbO_4. The interatomic diffusion narrowed the SCL structure and reduced the interface resistance, which greatly improved the cycle performance and rate performance. Electrochemical impedance spectroscopy results indicated a rapid growth of charge transfer resistance upon cycling for the treated

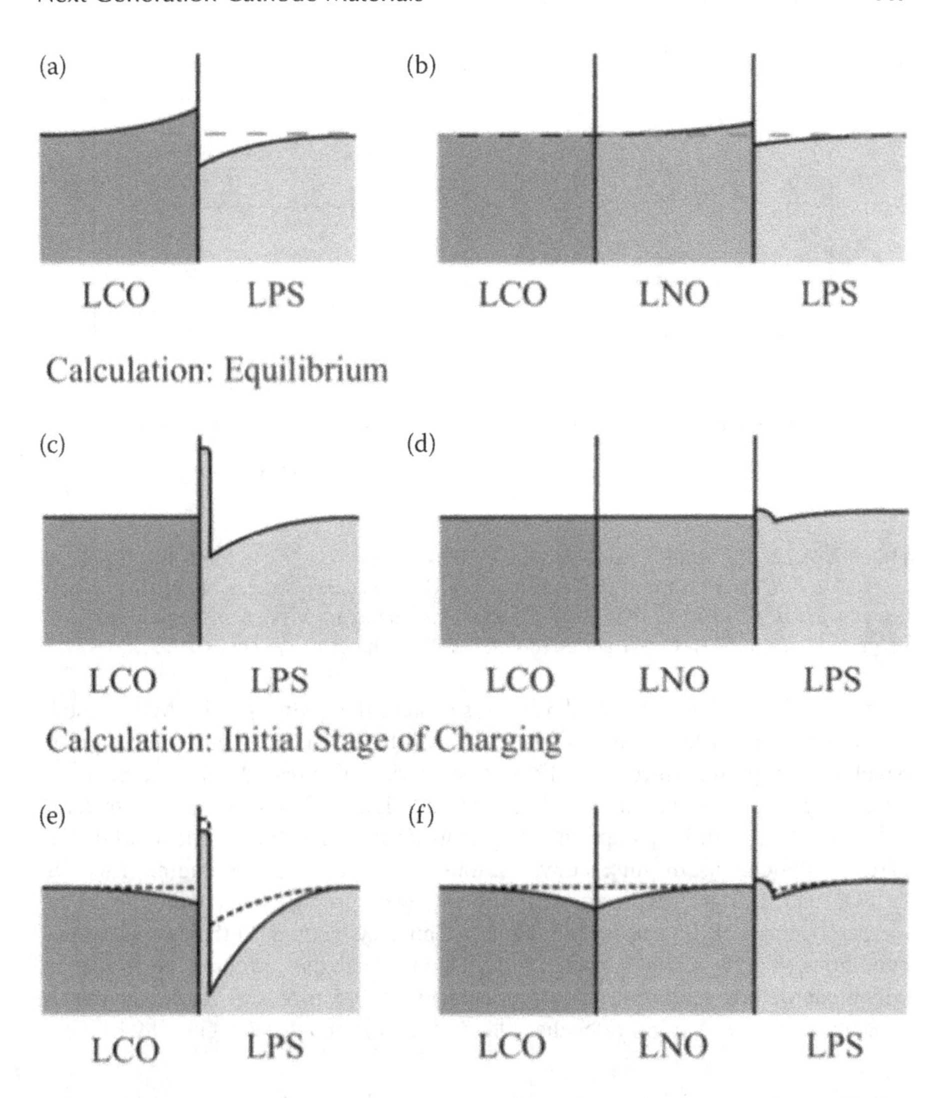

FIGURE 4.21 Schematic illustrations of the interfacial Li concentration. The equilibrium concentrations expected by the conventional model and indicated by the present calculations for the LCO/LPS interface (a and c) as well as the LCO/LNO/LPS (b and d). The Li concentrations in (e) and (f) describe the expected changes at the initial stage of charging for both interfaces, respectively, proposed in the present calculations. (Reproduced with permission of ref. [85], Copyright 2014 American Chemical Society.)

$LiCoO_2$. In contrast, Li_3NbO_4-coated $LiCoO_2$ showed a more stable interfacial impedance upon cycling, demonstrating the buffering and passivating roles of the Li_3NbO_4 coating layer. The specific discharge capacity of the c-LCO cell is ~118 mA h g^{-1} in initial cycles and gradually lifted to ~127 mA h/g by the fifth cycle, as shown in Figure 4.23.

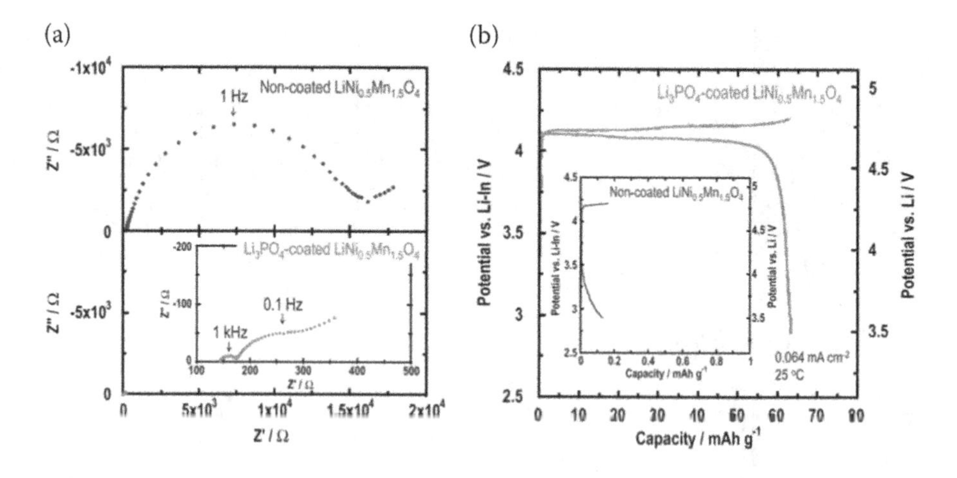

FIGURE 4.22 (a) Impedance spectra of cells using non-coated and Li_3PO_4-coated $LiNi_{0.5}Mn_{1.5}O_4$ after charging to 4.8 V (vs. Li); (b) impedance spectra of cells using non-coated and Li_3PO_4-coated $LiNi_{0.5}Mn_{1.5}O_4$ after charging to 4.8 V (vs. Li). (Reproduced with permission of ref. [95], Copyright 2016 Elsevier Publishing.)

Xiayin Yao et al. [97] adopted 5 V-class spinel $LiNi_{0.5}Mn_{1.5}O_4$ (LNMO) cathode to combine with a high-ionic-conductivity Li_6PS_5Cl (LPSCl) solid electrolyte to develop high-performance all-solid-state batteries. Compared with germanium-containing materials including LGPS, argyrodite Li_6PS_5Cl was a low-cost material with cheaper precursors, implying its promising potential for practical solid-state battery application. In this study, various oxide materials including $LiNbO_3$, Li_3PO_4, and $Li_4Ti_5O_{12}$ with different amounts were proposed as coating layers for the passivating LNMO surface via the wet-chemistry method. In their experimental conditions, 8 wt% $LiNbO_3$-coated LNMO presented the optimum performance, which can deliver an initial discharge capacity of 115 mA h g^{-1} and a reversible capacity of ~80 mA h g^{-1} after the 20th cycle, as is shown in Figure 4.24.

4.4.3 Concerns of Volumetric Variation

The volumetric changes of cathode materials were the inherent reason to deteriorate the physical contacts of cathodes and solid-state electrolytes. Currently, the commercial cathodes are layer-structured oxide cathodes, spinel-structured cathodes, and olivine phosphates [42,98,99]. Their lattice variations ranged from 2% to 7%, as shown in Table 4.2.

Significant dimensional and volume changes are associated with variations in lattice parameters and transformations of crystalline/amorphous phases, which occur during electrochemical cycling. These phenomena result in deformations and stress generation in the active cathode. Such stresses can cause fragmentation, disintegration, fracturing, and losing contact between electrolytes and cathodes. Cracks can occur between SE and active materials, or the SE may even locally be deformed such that its microstructure and local conductivity changes [98]. These

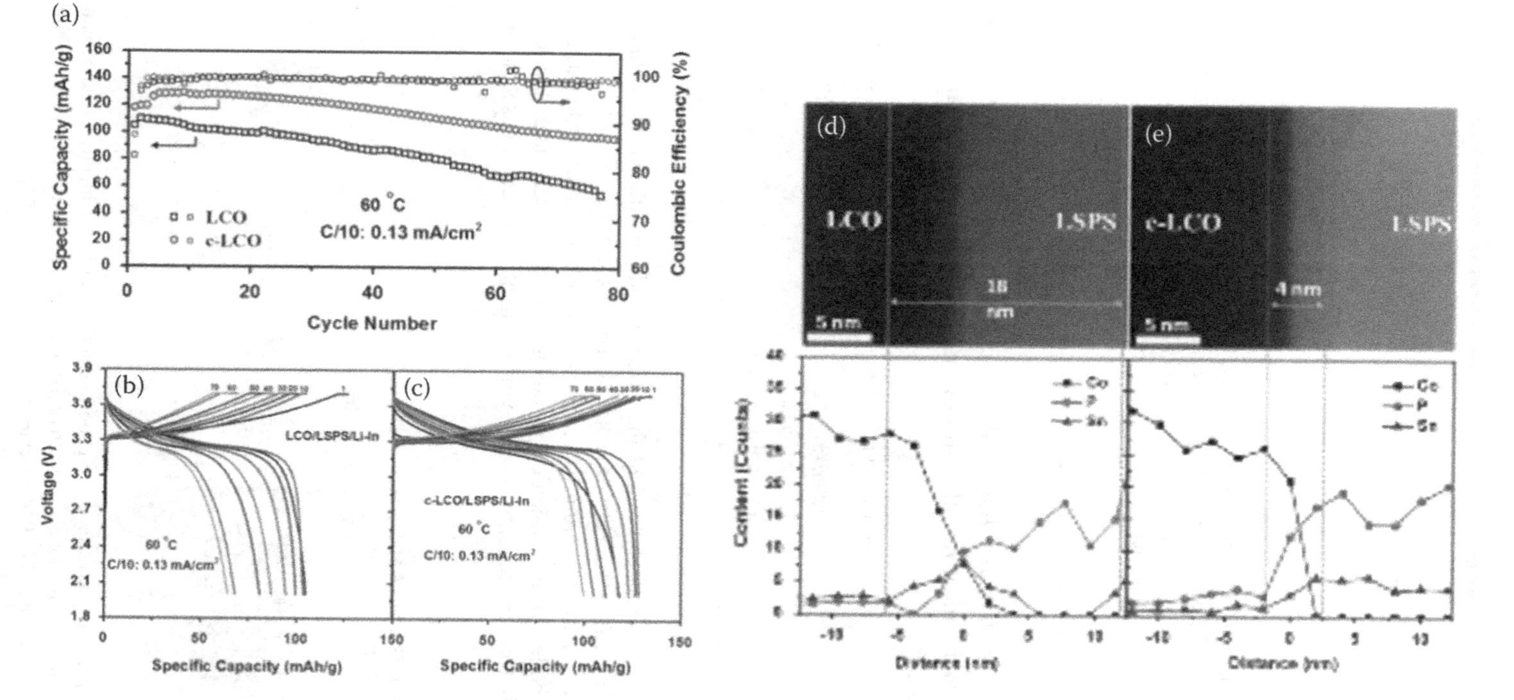

FIGURE 4.23 (a) Cycling performance of LCO/LSPS/Li-In and c-LCO/LSPS/Li-In between 2.0 and 3.7 V vs. Li-In (2.62–4.32 V vs Li^+/Li) at C/10 (0.13 mA/cm^2) and 60°C. Voltage profiles of (b) LCO/LSPS/Li-In and (c) c-LCO/LSPS/Li-In; TEM images and corresponding EDX line scans of (d) LCO/LSPS and (e) c-LCO/LSPS interfaces after 10 C/10 cycles at 60°C. (Reproduced with permission of ref. [96], Copyright 2018 Elsevier Publishing.)

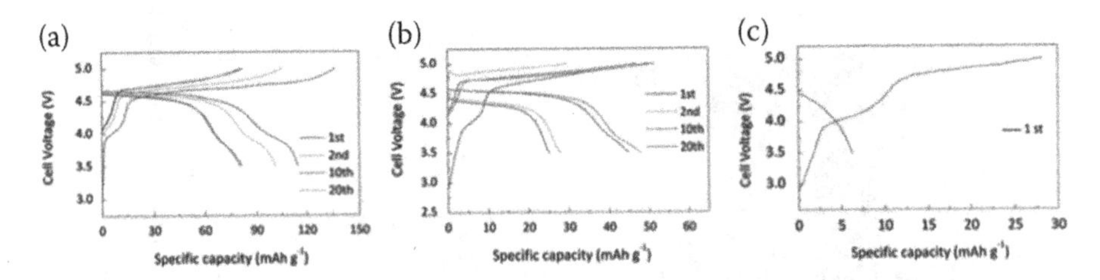

FIGURE 4.24 Discharge-charge profiles of batteries with different composite cathodes: (a) 8 wt% $LiNbO_3$-coated LNMO/LPSCl = 70/30; (b) 2 wt% Li_3PO_4-coated LNMO/LPSCl = 70/30; and (c) 8 wt% $Li_4Ti_5O_{12}$-coated LNMO/LPSCl = 70/30. (Reproduced with permission of ref. [97], Copyright 2020 American Chemical Society Publishing.)

TABLE 4.2
Crystal structures and crystal volumes (also volume changes) that form during progressive electrochemical de-lithiation of the various electrode materials for Li-ion batteries

Electrode material	Space group (structure)	Volume variation
$LiCoO_2$	R3m (trigonal)	3%
$LiMn_2O_4$	Fd3m (cubic)	2.5%
$LiFePO_4$	Pnma (orthorhombic)	6.8%

degradation processes ultimately caused the capacity fading with electrochemical cycling for nearly all cathode electrode materials (Figure 4.25).

Obviously, zero- or micro- strain cathodes are feasible for solid-state battery. However, the cathodes with non-lattice variations are absent till date. Particles engineering with buffer components to realize the volumetric variation in secondary particle level.

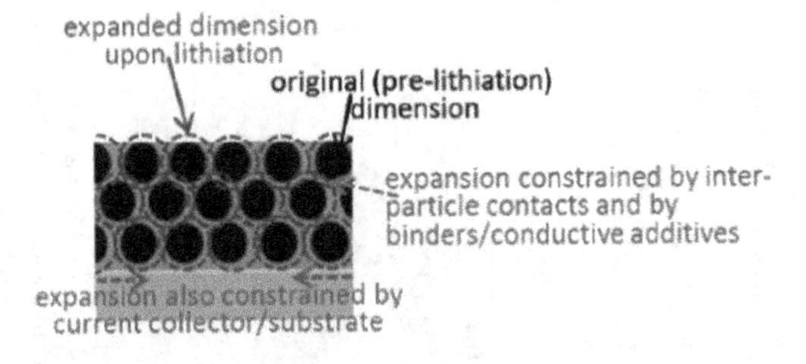

FIGURE 4.25 Constraining effects of neighboring active particles, inactive matrix, and current collector/substrate on the expanding active particles upon lithiation. Reproduced with permission of ref [98]., Copyright 2014 Elsevier.)

4.5 PROSPECTS

In this chapter, part of the outlines for next-generation cathodes are discussed to meet the novel requirements of lithium batteries in the future. The multidisciplinary collaborations are necessary to advance the real-world application of the next-generation LIB. Here, the utilizations of anionic redox chemistry, mechanics, and semiconductor physics are proposed to improve the reversible capacity, durability, and feasibility of cathodes. Finally, the engineering approaches from industry are also required to realize the real-world application of the next-generation cathode.

REFERENCES

[1] Mizushima, K., P. C. Jones, P. J. Wiseman, and J. B. Goodenough. 1980. "LixCoO2 (0<x<-1): A new cathode material for batteries of high energy density." *Materials Research Bulletin* 15 (6):783–789. 10.1016/0025-5408(80)90012-4.

[2] Rossouw, M. H., and M. M. Thackeray. 1991. "Lithium manganese oxides from Li2MnO3 for rechargeable lithium battery applications." *Materials Research Bulletin* 26 (6):463–473. 10.1016/0025-5408(91)90186-P.

[3] Thackeray, Michael M., Sun-Ho Kang, Christopher S. Johnson, John T. Vaughey, Roy Benedek, and S. A. Hackney. 2007. "Li2MnO3-stabilized LiMO2 (M = Mn, Ni, Co) electrodes for lithium-ion batteries." *Journal of Materials Chemistry* 17 (30):3112–3125.

[4] Lu, Zhonghua, D. D. MacNeil, and J. R. Dahn. 2001. "Layered cathode materials Li [Ni[sub x]Li[sub (1/3 - 2x/3)]Mn[sub (2/3 - x/3)]]O[sub 2] for lithium-ion batteries." *Electrochemical and Solid-State Letters* 4 (11):A191–A194.

[5] Boulineau, A., L. Croguennec, C. Delmas, and F. Weill. 2009. "Reinvestigation of Li2MnO3 structure: Electron diffraction and high resolution TEM." *Chemistry of Materials* 21 (18):4216–4222. 10.1021/cm900998n.

[6] MacNeil, D. D., Z. Lu, and J. R. Dahn. 2002. "Structure and electrochemistry of Li [$Ni_xCo_{1-2x}Mn_x$]O_2 (0 <= x <= 1/2)." *Journal of the Electrochemical Society* 149 (10):A1332–A1336. 10.1149/1.1505633.

[7] Shunmugasundaram, Ramesh, Rajalakshmi Senthil Arumugam, and J. R. Dahn. 2016. "A study of stacking faults and superlattice ordering in some Li-rich layered transition metal oxide positive electrode materials." *Journal of the Electrochemical Society* 163 (7):A1394–A1400. 10.1149/2.1221607jes.

[8] Hong, Jihyun, William E. Gent, Penghao Xiao, Kipil Lim, Dong-Hwa Seo, Jinpeng Wu, Peter M. Csernica, Christopher J. Takacs, Dennis Nordlund, Cheng-Jun Sun, Kevin H. Stone, Donata Passarello, Wanli Yang, David Prendergast, Gerbrand Ceder, Michael F. Toney, and William C. Chueh. 2019. "Metal–oxygen decoordination stabilizes anion redox in Li-rich oxides." *Nature Materials* 18 (3):256–265. 10.1038/s41563-018-0276-1.

[9] House, Robert A., Gregory J. Rees, Miguel A. Pérez-Osorio, John-Joseph Marie, Edouard Boivin, Alex W. Robertson, Abhishek Nag, Mirian Garcia-Fernandez, Ke-Jin Zhou, and Peter G. Bruce. 2020. "First-cycle voltage hysteresis in Li-rich 3d cathodes associated with molecular O2 trapped in the bulk." *Nature Energy* 5:777–785. 10.1038/s41560-020-00697-2.

[10] Gent, W. E., K. Lim, Y. F. Liang, Q. H. Li, T. Barnes, S. J. Ahn, K. H. Stone, M. McIntire, J. Y. Hong, J. H. Song, Y. Y. Li, A. Mehta, S. Ermon, T. Tyliszczak, D. Kilcoyne, D. Vine, J. H. Park, S. K. Doo, M. F. Toney, W. L. Yang, D. Prendergast, and W. C. Chueh. 2017. "Coupling between oxygen redox and cation migration

explains unusual electrochemistry in lithium-rich layered oxides." *Nature Communications* 8:2091. 10.1038/s41467-017-02041-x.

[11] Hu, Enyuan, Xiqian Yu, Ruoqian Lin, Xuanxuan Bi, Jun Lu, Seongmin Bak, Kyung-Wan Nam, Huolin L. Xin, Cherno Jaye, Daniel A. Fischer, Kahlil Amine, and Xiao-Qing Yang. 2018. "Evolution of redox couples in Li- and Mn-rich cathode materials and mitigation of voltage fade by reducing oxygen release." *Nature Energy* 3 (8):690–698. 10.1038/s41560-018-0207-z.

[12] Singer, A., M. Zhang, S. Hy, D. Cela, C. Fang, T. A. Wynn, B. Qiu, Y. Xia, Z. Liu, A. Ulvestad, N. Hua, J. Wingert, H. Liu, M. Sprung, A. V. Zozulya, E. Maxey, R. Harder, Y. S. Meng, and O. G. Shpyrko. 2018. "Nucleation of dislocations and their dynamics in layered oxide cathode materials during battery charging." *Nature Energy* 3 (8):641–647. 10.1038/s41560-018-0184-2.

[13] Yu, H. J., Y. G. So, Y. Ren, T. H. Wu, G. C. Guo, R. J. Xiao, J. Lu, H. Li, Y. B. Yang, H. S. Zhou, R. Z. Wang, K. Amine, and Y. Ikuhara. 2018. "Temperature-sensitive structure evolution of lithium-manganese-rich layered oxides for lithium-ion batteries." *Journal of the American Chemical Society* 140 (45):15279–15289. 10.1021/jacs.8b07858.

[14] Thackeray, M. M., S. H. Kang, C. S. Johnson, J. T. Vaughey, and S. A. Hackney. 2006. "Comments on the structural complexity of lithium-rich Li1+xM1-xO2 electrodes (M = Mn, Ni, Co) for lithium batteries." *Electrochemistry Communications* 8 (9):1531–1538. 10.1016/j.elecom.2006.06.030.

[15] Chen, Zhenlian, Jun Li, and Xiao Cheng Zeng. 2019. "Unraveling oxygen evolution in Li-rich oxides: A unified modeling of the intermediate peroxo/superoxo-like dimers." *Journal of the American Chemical Society* 141 (27):10751–10759. 10.1021/jacs.9b03710.

[16] Robertson, Alastair D., and Peter G. Bruce. 2003. "Mechanism of electrochemical activity in Li2MnO3." *Chemistry of Materials* 15 (10):1984–1992. 10.1021/cm030047u.

[17] Zuo, Wenhua, Mingzeng Luo, Xiangsi Liu, Jue Wu, Haodong Liu, Jie Li, Martin Winter, Riqiang Fu, Wanli Yang, and Yong Yang. 2020. "Li-rich cathodes for rechargeable Li-based batteries: Reaction mechanisms and advanced characterization techniques." *Energy & Environmental Science* 13 (12):4450–4497. 10.1039/D0EE01694B.

[18] Koyama, Yukinori, Isao Tanaka, Miki Nagao, and Ryoji Kanno. 2009. "First-principles study on lithium removal from Li2MnO3." *Journal of Power Sources* 189 (1):798–801. 10.1016/j.jpowsour.2008.07.073.

[19] Sathiya, M., G. Rousse, K. Ramesha, C. P. Laisa, H. Vezin, M. T. Sougrati, M. L. Doublet, D. Foix, D. Gonbeau, W. Walker, A. S. Prakash, M. Ben Hassine, L. Dupont, and J. M. Tarascon. 2013. "Reversible anionic redox chemistry in high-capacity layered-oxide electrodes." *Nature Materials* 12 (9):827–835. 10.1038/nmat3699. http://www.nature.com/nmat/journal/v12/n9/abs/nmat3699.html#supplementary-information.

[20] McCalla, Eric, Artem M. Abakumov, Matthieu Saubanere, Dominique Foix, Erik J. Berg, Gwenaelle Rousse, Marie-Liesse Doublet, Danielle Gonbeau, Petr Novak, Gustaaf Van Tendeloo, Robert Dominko, and Jean-Marie Tarascon. 2015. "Visualization of O-O peroxo-like dimers in high-capacity layered oxides for Li-ion batteries." *Science* 350 (6267):1516–1521. 10.1126/science.aac8260.

[21] Luo, K., M. R. Roberts, R. Hao, N. Guerrini, D. M. Pickup, Y. S. Liu, K. Edstrom, J. H. Guo, A. V. Chadwick, L. C. Duda, and P. G. Bruce. 2016. "Charge-compensation in 3d-transition-metal-oxide intercalation cathodes through the generation of localized electron holes on oxygen." *Nature Chemistry* 8 (7):684–691. 10.1038/nchem.2471.

[22] Saubanere, M., E. McCalla, J. M. Tarascon, and M. L. Doublet. 2016. "The intriguing question of anionic redox in high-energy density cathodes for Li-ion batteries." *Energy & Environmental Science* 9 (3):984–991. 10.1039/c5ee03048j.
[23] Seo, D. H., J. Lee, A. Urban, R. Malik, S. Kang, and G. Ceder. 2016. "The structural and chemical origin of the oxygen redox activity in layered and cation-disordered Li-excess cathode materials." *Nature Chemistry* 8 (7):692–697. 10.1038/nchem.2524.
[24] Frati, Federica, Myrtille O. J. Y. Hunault, and Frank M. F. de Groot. 2020. "Oxygen K-edge X-ray absorption spectra." *Chemical Reviews* 120 (9):4056–4110. 10.1021/acs.chemrev.9b00439.
[25] Chen, Zhenlian, Bjoern Schwarz, Xianhui Zhang, Wenqiang Du, Lirong Zheng, Ailing Tian, Ying Zhang, Zhiyong Zhang, Xiao Cheng Zeng, Zhifeng Zhang, Liyuan Huai, Jinlei Wu, Helmut Ehrenberg, Deyu Wang, and Jun Li. 2021. "Peroxo species formed in the bulk of silicate cathodes." *Angewandte Chemie International Edition* 61:370. 10.1002/anie.202100730.
[26] Eum, Donggun, Byunghoon Kim, Sung Joo Kim, Hyeokjun Park, Jinpeng Wu, Sung-Pyo Cho, Gabin Yoon, Myeong Hwan Lee, Sung-Kyun Jung, Wanli Yang, Won Mo Seong, Kyojin Ku, Orapa Tamwattana, Sung Kwan Park, Insang Hwang, and Kisuk Kang. 2020. "Voltage decay and redox asymmetry mitigation by reversible cation migration in lithium-rich layered oxide electrodes." *Nature Materials* 19:419–427. 10.1038/s41563-019-0572-4.
[27] Jasniewski, A. J., and L. Que. 2018. "Dioxygen activation by nonheme diiron enzymes: Diverse dioxygen adducts, high-valent intermediates, and related model complexes." *Chemical Reviews* 118 (5):2554–2592. 10.1021/acs.chemrev.7b00457.
[28] Li, X., Y. Qiao, S. Guo, Z. Xu, H. Zhu, X. Zhang, Y. Yuan, P. He, M. Ishida, and H. Zhou. 2018. "Direct visualization of the reversible O(2-) /O(-) redox process in Li-rich cathode materials." *Advanced Materials* 30 (14):1705197. 10.1002/adma.201705197.
[29] Armstrong, A. Robert, Michael Holzapfel, Petr Novák, Christopher S. Johnson, Sun-Ho Kang, Michael M. Thackeray, and Peter G. Bruce. 2006. "Demonstrating oxygen loss and associated structural reorganization in the lithium battery cathode Li[Ni0.2Li0.2Mn0.6]O2." *Journal of the American Chemical Society* 128 (26):8694–8698. 10.1021/ja062027+.
[30] Yan, Pengfei, Jianming Zheng, Zhen-Kun Tang, Arun Devaraj, Guoying Chen, Khalil Amine, Ji-Guang Zhang, Li-Min Liu, and Chongmin Wang. 2019. "Injection of oxygen vacancies in the bulk lattice of layered cathodes." *Nature Nanotechnology* 14 (6):602–608. 10.1038/s41565-019-0428-8.
[31] Lei, Yike, Jie Ni, Zijun Hu, Ziming Wang, Fukang Gui, Bing Li, Pingwen Ming, Cunman Zhang, Yuval Elias, Doron Aurbach, and Qiangfeng Xiao. 2020. "Surface modification of Li-rich Mn-based layered oxide cathodes: Challenges, materials, methods, and characterization." *Advanced Energy Materials* 10 (41):2002506. 10.1002/aenm.202002506.
[32] House, Robert A., Urmimala Maitra, Miguel A. Pérez-Osorio, Juan G. Lozano, Liyu Jin, James W. Somerville, Laurent C. Duda, Abhishek Nag, Andrew Walters, Ke-Jin Zhou, Matthew R. Roberts, and Peter G. Bruce. 2020. "Superstructure control of first-cycle voltage hysteresis in oxygen-redox cathodes." *Nature* 577 (7791):502–508. 10.1038/s41586-019-1854-3.
[33] Ji, Huiwen, Jinpeng Wu, Zijian Cai, Jue Liu, Deok-Hwang Kwon, Hyunchul Kim, Alexander Urban, Joseph K. Papp, Emily Foley, Yaosen Tian, Mahalingam Balasubramanian, Haegyeom Kim, Raphaële J. Clément, Bryan D. McCloskey, Wanli Yang, and Gerbrand Ceder. 2020. "Ultrahigh power and energy density in

partially ordered lithium-ion cathode materials." *Nature Energy* 5 (3):213–221. 10.1038/s41560-020-0573-1.

[34] Aurbach, D. 2003. "Electrode-solution interactions in Li-ion batteries: a short summary and new insights." *Journal of Power Sources* 119:497–503. 10.1016/S0378-7753(03)00273-8.

[35] Feng, X. N., X. M. He, M. G. Ouyang, L. G. Lu, P. Wu, C. Kulp, and S. Prasser. 2015. "Thermal runaway propagation model for designing a safer battery pack with 25 Ah LiNixCoyMnzO2 large format lithium ion battery." *Applied Energy* 154:74–91. 10.1016/j.apenergy.2015.04.118.

[36] Jiang, L. H., Q. S. Wang, and J. H. Sun. 2018. "Electrochemical performance and thermal stability analysis of LiNixCoyMnzO2 cathode based on a composite safety electrolyte." *Journal of Hazardous Materials* 351:260–269. 10.1016/j.jhazmat.2018.03.015.

[37] Wang, D. Y., X. D. Wu, Z. X. Wang, and L. Q. Chen. 2005. "Cracking causing cyclic instability of LiFePO4 cathode material." *Journal of Power Sources* 140 (1):125–128. 10.1016/j.jpowsour.2004.06.059.

[38] Wang, L., R. Xie, B. Chen, X. Yu, J. Ma, C. Li, Z. Hu, X. Sun, C. Xu, S. Dong, T. S. Chan, J. Luo, G. Cui, and L. Chen. 2020. "In-situ visualization of the space-charge-layer effect on interfacial lithium-ion transport in all-solid-state batteries." *Nature Communications* 11 (1):5889. 10.1038/s41467-020-19726-5.

[39] Bi, Y. J., W. C. Yang, R. Du, J. J. Zhou, M. Liu, Y. Liu, and D. Y. Wang. 2015. "Correlation of oxygen non-stoichiometry to the instabilities and electrochemical performance of LiNi0.8Co0.1Mn0.1O2 utilized in lithium ion battery." *Journal of Power Sources* 283:211–218. 10.1016/j.jpowsour.2015.02.095.

[40] Ren, F. H., Z. Peng, M. Q. Wang, Y. Xie, Z. D. Li, H. Wan, H. Lin, and D. Y. Wang. 2019. "Over-potential induced Li/Na filtrated depositions using stacked graphene coating on copper scaffold." *Energy Storage Materials* 16:364–373. 10.1016/j.ensm.2018.06.012.

[41] Lee, W., S. Muhammad, T. Kim, H. Kim, E. Lee, M. Jeong, S. Son, J. H. Ryou, and W. S. Yoon. 2018. "New insight into Ni-rich layered structure for next-generation Li rechargeable batteries." *Advanced Energy Materials* 8 (4). 10.1002/aenm.201701788.

[42] Bouwman, P. J. 2002. "Lithium intercalation in preferentially oriented submicron LiCoO2 films." *PhD Thesis. University of Twente, Enschede, Sweden.*

[43] Ryu, H. H., K. J. Park, D. R. Yoon, A. Aishova, C. S. Yoon, and Y. K. Sun. 2019. "Li[Ni0.9Co0.09W0.01]O-2: A new type of layered oxide cathode with high cycling stability." *Advanced Energy Materials* 9 (44). 10.1002/aenm.201902698.

[44] Ryu, H. H., G. T. Park, C. S. Yoon, and Y. K. Sun. 2019. "Suppressing detrimental phase transitions via tungsten doping of LiNiO2 cathode for next-generation lithium-ion batteries." *Journal of Materials Chemistry A* 7 (31):18580–18588. 10.1039/c9ta06402h.

[45] Jung, S. K., H. Gwon, J. Hong, K. Y. Park, D. H. Seo, H. Kim, J. Hyun, W. Yang, and K. Kang. 2014. "Understanding the degradation mechanisms of LiNi0.5Co0.2Mn0.3O2 cathode material in lithium ion batteries." *Advanced Energy Materials* 4 (1). 10.1002/aenm.201300787.

[46] Kim, B. R., K. S. Yun, H. J. Jung, S. T. Myung, S. C. Jung, W. Kang, and S. J. Kim. 2013. "Electrochemical properties of the TiO2(B) powders ball mill treated for lithium-ion battery application." *Chemistry Central Journal* 7. 10.1186/1752-153x-7-174.

[47] Ren, Z. M., C. Shen, M. Liu, J. Liu, S. Q. Zhang, G. Yang, L. Y. Huai, X. S. Liu, D. Y. Wang, and H. Li. 2020. "Improving LiNi0.9Co0.08Mn0.02O2's cyclic

stability via abating mechanical damages." *Energy Storage Materials* 28:1–9. 10.1016/j.ensm.2020.02.028.
[48] Ren, Z. M., X. H. Zhang, M. Liu, J. J. Zhou, S. Sun, H. Y. He, and D. Y. Wang. 2019. "Constant dripping wears away a stone: Fatigue damage causing particles' cracking." *Journal of Power Sources* 416:104–110. 10.1016/j.jpowsour.2019.01.084.
[49] Miller, D. J., C. Proff, J. G. Wen, D. P. Abraham, and J. Bareno. 2013. "Observation of microstructural evolution in Li battery cathode oxide particles by in situ electron microscopy." *Advanced Energy Materials* 3 (8):1098–1103. 10.1002/aenm.201300015.
[50] Watanabe, S., M. Kinoshita, T. Hosokawa, K. Morigaki, and K. Nakura. 2014. "Capacity fading of LiAlyNi1-x-yCoxO2 cathode for lithium-ion batteries during accelerated calendar and cycle life tests (effect of depth of discharge in charge-discharge cycling on the suppression of the micro-crack generation of LiAlyNi1-x-yCoxO2 particle)." *Journal of Power Sources* 260:50–56. 10.1016/j.jpowsour.2014.02.103.
[51] Xu, J., E. Y. Hu, D. Nordlund, A. Mehta, S. N. Ehrlich, X. Q. Yang, and W. Tong. 2016. "Understanding the degradation mechanism of lithium nickel oxide cathodes for Li-ion batteries." *ACS Applied Materials & Interfaces* 8 (46):31677–31683. 10.1021/acsami.6b11111.
[52] Janek, Jürgen, and Wolfgang G. Zeier. 2016. "A solid future for battery development." *Nature Energy* 1 (9). 10.1038/nenergy.2016.141.
[53] Kim, H., S. Lee, H. Cho, J. Kim, J. Lee, S. Park, S. H. Joo, S. H. Kim, Y. G. Cho, H. K. Song, S. K. Kwak, and J. Cho. 2016. "Enhancing interfacial bonding between anisotropically oriented grains using a glue-nanofiller for advanced Li-ion battery cathode." *Advanced Materials* 28 (23):4705–4712. 10.1002/adma.201506256.
[54] Kim, U. H., J. H. Kim, J. Y. Hwang, H. H. Ryu, C. S. Yoon, and Y. K. Sun. 2019. "Compositionally and structurally redesigned high-energy Ni-rich layered cathode for next-generation lithium batteries." *Materials Today* 23:26–36. 10.1016/j.mattod.2018.12.004.
[55] Xu, Z. R., M. M. Rahman, L. Q. Mu, Y. J. Liu, and F. Lin. 2018. "Chemomechanical behaviors of layered cathode materials in alkali metal ion batteries." *Journal of Materials Chemistry A* 6 (44):21859–21884. 10.1039/c8ta06875e.
[56] Xie, Q., W. D. Li, and A. Manthiram. 2019. "A Mg-doped high-nickel layered oxide cathode enabling safer, high-energy-density Li-ion batteries." *Chemistry of Materials* 31 (3):938–946. 10.1021/acs.chemmater.8b03900.
[57] Bianchini, M., M. Roca-Ayats, P. Hartmann, T. Brezesinski, and J. Janek. 2019. "There and back again-The journey of LiNiO2 as a cathode active material." *Angewandte Chemie International Edition* 58 (31):10434–10458. 10.1002/anie.201812472.
[58] Lai, C. H., D. S. Ashby, T. C. Lin, J. Lau, A. Dawson, S. H. Tolbert, and B. S. Dunn. 2018. "Application of Poly(3-hexylthiophene-2,5-diyl) as a protective coating for high rate cathode materials." *Chemistry of Materials* 30 (8):2589–2599. 10.1021/acs.chemmater.7b05116.
[59] Xie, J., A. D. Sendek, E. D. Cubuk, X. K. Zhang, Z. Y. Lu, Y. J. Gong, T. Wu, F. F. Shi, W. Liu, E. J. Reed, and Y. Cui. 2017. "Atomic layer deposition of stable LiAlF4 lithium ion conductive interfacial layer for stable cathode cycling." *ACS Nano* 11 (7):7019–7027. 10.1021/acsnano.7b02561.
[60] Yin, Shouyi, Wentao Deng, Jun Chen, Xu Gao, Guoqiang Zou, Hongshuai Hou, and Xiaobo Ji. 2021. "Fundamental and solutions of microcrack in Ni-rich layered oxide cathode materials of lithium-ion batteries." *Nano Energy* 83:105854. 10.1016/j.nanoen.2021.105854.

[61] Zhou, P. F., H. J. Meng, Z. Zhang, C. C. Chen, Y. Y. Lu, J. Cao, F. Y. Cheng, and J. Chen. 2017. "Stable layered Ni-rich LiNi0.9Co0.07Al0.03O2 microspheres assembled with nanoparticles as high-performance cathode materials for lithium-ion batteries." *Journal of Materials Chemistry A* 5 (6):2724–2731. 10.1039/c6ta09921a.

[62] Kim, J., H. Ma, H. Cha, H. Lee, J. Sung, M. Seo, P. Oh, M. Park, and J. Cho. 2018. "A highly stabilized nickel-rich cathode material by nanoscale epitaxy control for high-energy lithium-ion batteries." *Energy & Environmental Science* 11 (6):1449–1459. 10.1039/c8ee00155c.

[63] Liu, Meng, Ren, Zhongming, Wang, Deyu, Zhang, Haitao, Bi, Yujing, Shen, Cai, & Guo, Bingkun 2021. Addressing Unfavorable Influence of Particle Cracking with a Strengthened Shell Layer in Ni-Rich Cathodes. ACS Applied Materials & Interfaces, 13, 18954–1896010.1021/acsami.1c05535.

[64] Kartini, Evvy, and Carla Theresa Genardy. 2020. "The future of all solid state battery." *IOP Conference Series: Materials Science and Engineering* 924. 10.1088/1757-899x/924/1/012038.

[65] Xu, Lin, Shun Tang, Yu Cheng, Kangyan Wang, Jiyuan Liang, Cui Liu, Yuan-Cheng Cao, Feng Wei, and Liqiang Mai. 2018. "Interfaces in solid-state lithium batteries." *Joule* 2 (10):1991–2015. 10.1016/j.joule.2018.07.009.

[66] Famprikis, T., P. Canepa, J. A. Dawson, M. S. Islam, and C. Masquelier. 2019. "Fundamentals of inorganic solid-state electrolytes for batteries." *Nature Materials* 18 (12):1278–1291. 10.1038/s41563-019-0431-3.

[67] Pang, Yuepeng, Jinyu Pan, Junhe Yang, Shiyou Zheng, and Chunsheng Wang. 2021. "Electrolyte/electrode interfaces in all-solid-state lithium batteries: A review." *Electrochemical Energy Reviews* 4:169–193. 10.1007/s41918-020-00092-1.

[68] Yao, Xiayin, Bingxin Huang, Jingyun Yin, Gang Peng, Zhen Huang, Chao Gao, Deng Liu, and Xiaoxiong Xu. 2016. "All-solid-state lithium batteries with inorganic solid electrolytes: Review of fundamental science." *Chinese Physics B* 25 (1). 10.1088/1674-1056/25/1/018802.

[69] Zhang, Q., D. Cao, Y. Ma, A. Natan, P. Aurora, and H. Zhu. 2019. "Sulfide-based solid-state electrolytes: synthesis, stability, and potential for all-solid-state batteries." *Advanced Materials* 31 (44):e1901131. 10.1002/adma.201901131.

[70] Zhao, R., L. Li, T. Xu, D. Wang, D. Pan, G. He, H. Zhao, and Y. Bai. 2019. "One-step integrated surface modification to build a stable interface on high-voltage cathode for lithium-ion batteries." *ACS Applied Materials & Interfaces* 11 (17):16233–16242. 10.1021/acsami.9b02996.

[71] Agrawal, R. C., and G. P. Pandey. 2008. "Solid polymer electrolytes: materials designing and all-solid-state battery applications: An overview." *Journal of Physics D: Applied Physics* 41 (22). 10.1088/0022-3727/41/22/223001.

[72] Karabelli, Duygu, Kai Peter Birke, and Max Weeber. 2021. "A performance and cost overview of selected solid-state electrolytes: Race between polymer electrolytes and inorganic sulfide electrolytes." *Batteries* 7 (1). 10.3390/batteries7010018.

[73] Haruyama, Jun, Keitaro Sodeyama, Liyuan Han, Kazunori Takada, and Yoshitaka Tateyama. 2014. "Space–charge layer effect at interface between oxide cathode and sulfide electrolyte in all-solid-state lithium-ion battery." *Chemistry of Materials* 26 (14):4248–4255. 10.1021/cm5016959.

[74] Wang, L., R. Xie, B. Chen, et al. 2020. "In-situ visualization of the space-charge-layer effect on interfacial lithium-ion transport in all-solid-state batteries." *Nature Communications* 11(1):1–9. 10.1038/s41467-020-19726-5.

[75] Wagner, C. 1972. "The electrical conductivity of semi-conductors involving inclusions of another phase." *Journal of Physics and Chemistry of Solids* 33(5):1051–1059. 10.1016/S0022-3697(72)80265-8.

[76] Maier, J. 1987. "Defect chemistry and conductivity effects in heterogeneous solid electrolytes." *Journal of the Electrochemical Society* 13(6):1524–1535. 10.1149/1.2100703.
[77] Jamnik, J., J. Maier, and S. Pejovnik. 1995. "Interfaces in solid ionic conductors: Equilibrium and small signal picture." *Solid State Ionics* 75: 51–58. 10.1016/0167-2738(94)00184-T.
[78] Sakuda, A., A. Hayashi, and M. Tatsumisago. 2010. "Interfacial observation between $LiCoO_2$ electrode and $Li_2S{\cdot}P_2S_5$ solid electrolytes of all-solid-state lithium secondary batteries using transmission electron microscopy." *Chemistry of Materials* 22(3):949–956. 10.1021/cm901819c.
[79] Ohtomo, T., A. Hayashi, M. Tatsumisago, et al. 2013. "All-solid-state lithium secondary batteries using the $75Li_2S{\cdot}25P_2S_5$ glass and the $70Li_2S{\cdot}30P_2S_5$ glass–ceramic as solid electrolytes." *Journal of Power Sources* 233:231–235. 10.1016/j.jpowsour.2013.01.090.
[80] Woo, J. H., J. E. Trevey, A. S. Cavanagh, et al. 2012. "Nanoscale interface modification of $LiCoO_2$ by Al_2O_3 atomic layer deposition for solid-state Li batteries." *Journal of the Electrochemical Society* 159(7):A1120–A1124. 10.1149/2.085207jes.
[81] Jow, T., and J. B. Wagner. 1980. "The effect of dispersed alumina particles on the electrical conductivity of cuprous chloride." *Journal of the Electrochemical Society* 11 (12).
[82] Ohtomo, Takamasa, Akitoshi Hayashi, Masahiro Tatsumisago, Yasushi Tsuchida, Shigenori Hama, and Koji Kawamoto. 2013. "All-solid-state lithium secondary batteries using the 75Li2S·25P2S5 glass and the 70Li2S·30P2S5 glass–ceramic as solid electrolytes." *Journal of Power Sources* 233:231–235. 10.1016/j.jpowsour.2013.01.090.
[83] Woo, Jae Ha, James E. Trevey, Andrew S. Cavanagh, Yong Seok Choi, Seul Cham Kim, Steven M. George, Kyu Hwan Oh, and Se-Hee Lee. 2012. "Nanoscale interface modification of LiCoO2by Al2O3 atomic layer deposition for solid-state Li batteries." *Journal of the Electrochemical Society* 159 (7):A1120–A1124. 10.1149/2.085207jes.
[84] Ohta, N., K. Takada, L. Zhang, et al. 2006. "Enhancement of the high-rate capability of solid-state lithium batteries by nanoscale interfacial modification." *Advanced Materials* 18(17):2226–2229. 10.1149/2.085207jes.
[85] Haruyama, J., K. Sodeyama, L. Han, et al. 2014. "Space–charge layer effect at interface between oxide cathode and sulfide electrolyte in all-solid-state lithium-ion battery." *Chemistry of Materials* 26(14):4248–4255. 10.1021/cm5016959.
[86] Okumura, T., T. Nakatsutsumi, T. Ina, Y. Orikasa, H. Arai, T. Fukutsuka, Y. Iriyama, T. Uruga, H. Tanida, Y. Uchimoto, and Z. Ogumi. 2011. "Depth-resolved X-ray absorption spectroscopic study on nanoscale observation of the electrode-solid electrolyte interface for all solid state lithium ion batteries." *Journal of Materials Chemistry* 21:10051. 10.1039/C0JM04366D.
[87] Nie, K., Y. Hong, J. Qiu, Q. Li, X. Yu, H. Li, and L. Chen. 2018. "Interfaces between cathode and electrolyte in solid state lithium batteries: Challenges and perspectives." *Frontiers in Chemistry* 6:616. 10.3389/fchem.2018.00616.
[88] Yang, Q., J. Huang, Y. Li, Y. Wang, J. Qiu, J. Zhang, H. Yu, X. Yu, H. Li, and L. Chen. 2018. "Surface-protected $LiCoO_2$ with ultrathin solid oxide electrolyte film for high-voltage lithium ion batteries and lithium polymer batteries." *Journal of Power Sources* 388:65–70. 10.1016/j.jpowsour.2018.03.076.
[89] Choi, J.-w., and J.-w. Lee. 2016. "Improved electrochemical properties of Li $(Ni_{0.6}Mn_{0.2}Co_{0.2})O_2$ by surface coating with $Li_{1.3}Al_{0.3}Ti_{1.7}(PO_4)_3$." *Journal of Power Sources* 307:63–68. 10.1016/j.jpowsour.2015.12.055.
[90] Zhao, R., L. Li, T. Xu, D. Wang, D. Pan, G. He, H. Zhao, and Y. Bai. 2019. "One-step integrated surface modification to build a stable interface on high-voltage cathode for

lithium-ion batteries." *ACS Applied Materials & Interfaces*11:16233–16242. 10.1021/acsami.9b02996.
[91] Huang, J., H. Liu, N. Zhou, K. An, Y. S. Meng, and J. Luo. 2017. "Enhancing the ion transport in $LiMn_{1.5}Ni_{0.5}O_4$ by altering the particle wulff shape via anisotropic surface segregation." *ACS Applied Materials & Interfaces* 9:36745–36754. 10.1021/acsami.7b09903.
[92] Chen, M., W. Li, X. Shen, and G. Diao. 2014. "Fabrication of core-shell A-Fe_2O_3@$Li_4Ti_5O_{12}$ composite and its application in the lithium ion batteries." *ACS Applied Materials & Interfaces* 6:4514–4523. 10.1021/am500294m.
[93] Liang, J.-Y., X. Zhang, X. Zeng, M. Yan, Y. Yin, S. Xin, W. Wang, X. Wu, J. Shi, and Y. Guo. 2020. "Enabling durable electrochemical interface via artificial amorphous cathode electrolyte interphase for hybrid solid/liquid lithium-metal batteries". Angewandte Chemie 132:6647. 10.1002/ange.201916301.
[94] Liang, J.-Y., X.-X. Zeng, X.-D. Zhang, P.-F. Wang, J.-Y. Ma, Y.-X. Yin, X.-W. Wu, Y.-G. Guo, and L.-J. Wan. 2018. "Mitigating interfacial potential drop of cathode-solid electrolyte via ionic conductor layer to enhance interface dynamics for solid batteries." *Journal of the American Chemical Society* 140:6767–6770. 10.1021/jacs.8b03319.
[95] Yubuchi, S., Y. Ito, T. Matsuyama, A. Hayashi, and M. Tatsumisago. 2016. "5V class $LiNi_{0.5}Mn_{1.5}O_4$ positive electrode coated with Li_3PO_4 thin film for all-solid-state batteries using sulfide solid electrolyte." *Solid State Ionics* 285:79–82. 10.1016/j.ssi.2015.08.001.
[96] Vinado, C., S. Wang, Y. He, et al. 2018. "Electrochemical and interfacial behavior of all solid state batteries using $Li_{10}SnP_2S_{12}$ solid electrolyte." *Journal of Power Sources* 396:824–830. 10.1016/j.jpowsour.2018.06.038.
[97] Liu, G., Y. Lu, H. Wan, et al. 2020. "Passivation of the cathode–electrolyte interface for 5V-class all-solid-state batteries." *ACS Applied Materials & Interfaces* 12(25):28083–28090. 10.1021/acsami.0c03610.
[98] Mukhopadhyay, Amartya, and Brian W. Sheldon. 2014. "Deformation and stress in electrode materials for Li-ion batteries." *Progress in Materials Science* 63:58–116. 10.1016/j.pmatsci.2014.02.001.
[99] Shadow Huang, Hsiao-Ying, and Yi-Xu Wang. 2012. "Dislocation based stress developments in lithium-ion batteries." *Journal of the Electrochemical Society* 159(6):A815–A821. 10.1149/2.090206jes.

5 Cathode Materials for Lithium-Sulfur Batteries

Zhenhua Wang
School of Chemistry and Chemical Engineering, Beijing Institute of Technology, Beijing, People's Republic of China

CONTENTS

5.1 INTRODUCTION

With the consumption of fossil fuels and overwhelming environmental pollution, developing clean energy is imperative under the situation. Lithium ion batteries (LIBs) have attracted people's attention because of portability, high specific capacity, and power density after their commercialization. However, when applied in electric vehicles (EV), LIBs cannot dispel consumer' mile and cost anxiety to some extent due to limited energy density. It is urgently needed to introduce a new battery system enabling high energy density and low cost. Lithium-sulfur (Li-S) batteries consist of sulfur cathode, separator, and lithium metal anode. By virtue of natural abundance, environmental friendliness, as well as high specific capacity (1,675

DOI: 10.1201/9781003133971-5

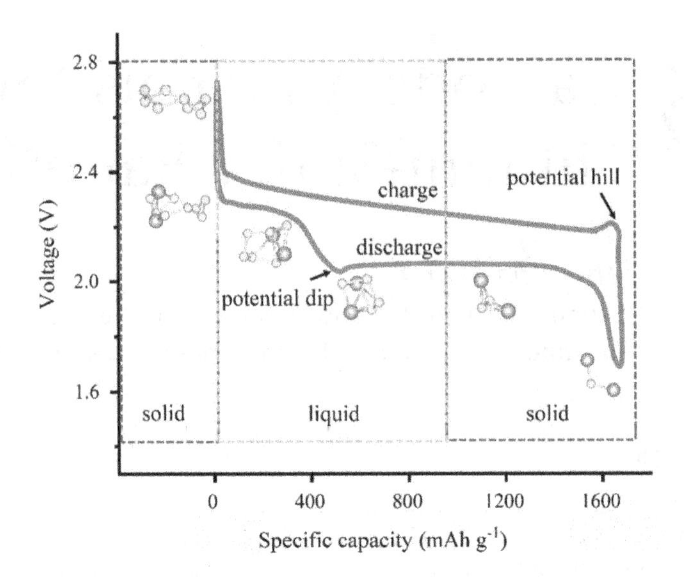

FIGURE 5.1 Discharge-charge profile of Li-S batteries.

mAh $g^{-1}{}_{sulfur}$), sulfur cathode exhibits huge advantages over conventional cathodes of LIBs [1]. Li-S batteries are expected to be the next generation of an energy storage system. Despite high energy density (2,600 wh $kg^{-1}{}_{sulfur}$ or 2,800 wh $L^{-1}{}_{sulfur}$), Li-S batteries suffer from severe lithium polysulfides (LiPSs) shuttle and sluggish kinetics [2].

The reaction mechanism of Li-S batteries is shown in Equations (5.1) and (5.2) [3]:

$$S_8 + Li^+ + \mathrm{e} \rightarrow Li_2S_n\,(2.4\mathrm{V}\sim 2.1\mathrm{V}) \tag{5.1}$$

$$Li_2S_n + e \rightarrow Li_2S\ \ and/or\ \ Li_2S_2(<2.1V) \tag{5.2}$$

Actually, the reaction process of Li-S batteries goes through solid-liquid-solid conversion (Figure 5.1). For the discharge process, the element sulfur (S_8) is reduced to a soluble Li_2S_n ($2<n\le 8$), LiPSs can easily dissolve in an ether-based electrolyte as binding energy of LiPSs; ether is stronger than that of aggregation of LiPSs clusters [4]. Subsequently, LiPSs can be further reduced to insoluble Li_2S or Li_2S_2. A potential dip can be clearly observed at approximately 2.05 V, attributing to overcome a high-energy barrier (usually 10 mV) to nucleate [3]. For the charge process, the oxidation of Li_2S is faced with a large overpotential due to the strong interaction of lithium and sulfur atoms in the crystalline; therefore, a potential hill appears to conquer energy barrier.

In practical applications, there are four main obstacles in a sulfur cathode, which hamper the commercialization of Li-S batteries (Figure 5.2).

(1) Due to the insulating nature of Li_2S (1×10^{-14} S cm^{-1}) [5] and S_8 (5×10^{-30} S cm^{-1}) [6], it puts forward more requirements on the cathode design. Much more proportion

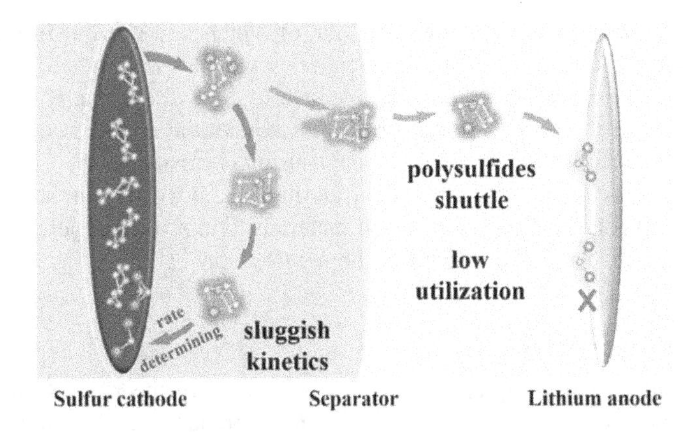

FIGURE 5.2 Schematic illustration of major challenges in Li-S batteries.

of the conducting agent should be added in cathode materials to construct a good electron and ion network and decrease polarization. (2) The density of S_8 (2.07 g cm^{-3}) [7] shows significant variance with Li_2S (1.66 g cm^{-3}) [1]; 80% of volume expansion can lead to poor contact of active material and conducting agent and eventually electrode pulverization. (3) The conversion of Li_2S_2 to Li_2S basically depends on Li^+ ion transfer in the solid phase, which thus becomes a rate-determining step and causes sluggish kinetics [8]. (4) Soluble LiPSs can dissolve in electrolytes and transfer to a Li anode; the low chemical potential of Li metal makes LiPSs easily reduced to terminal product Li_2S, resulting in corrosion of Li anode and low sulfur utilization. This phenomenon is called the shuttle effect.

For the sake of immobilizing LiPSs, buffering volume expansion, constructing better conductive composite sulfur cathode, and boosting LiPSs redox kinetics, researchers made every effort to design a sulfur cathode. The following part will focus on active substances and host materials of a sulfur cathode in Li-S batteries.

5.2 ACTIVE SUBSTANCE IN SULFUR CATHODE

As one of most abundant natural resources in the world, sulfur can be easily found in natural gas desulfurization; the price of the element sulfur is merely \$150 per ton [9]. Therefore, the application of sulfur in Li-S batteries is advantageous in the cost of raw materials. The most stable state of the element sulfur is the orthorhombic α-S_8 molecule, which is a crown-like structure. α-S_8 becomes metastable from 95°C and gradually transforms to monoclinic β-S_8. It starts to melt at 113°C and the viscosity is the smallest at 155°C [10]. Therefore, the melting method is often used to infiltrate sulfur into porous host materials at 155°C under inert atmosphere [11]. The conductivity of sulfur is 5.0×10^{-30} S cm^{-1} at room temperature, which causes large polarization and retards electrochemical redox kinetics of Li-S batteries. To solve this tough problem, conducting agents such as carbon black, acetylene black, and Ketjenblack are chosen to enhance the conductivity of sulfur cathode. In spite of this, the dispersion of sulfur particles still needs to be solved. With the proceeding of the discharge-charge process, the terminal product of Li_2S precipitates on the cover of sulfur particles, the insulation of which will prevent further reactions. The reactant in the core of sulfur particles will not be utilized, leading to deterioration of capacity. Moreover, intermediate Li_2S_n will constantly dissolve in the electrolyte, due to poor LiPSs immobilization. Under the effect of concentration gradient, LiPSs diffuse to the lithium anode, causing corrosion of the Li metal and irreversible loss of active materials. Considering all above, researchers tried to confine sulfur in polar and conductive host materials to improve the performance of Li-S batteries, which we will discuss in the next section.

5.2.1 Organic Sulfur

Sulfur is found in many organic compounds. Some play the role of supporting structural frameworks, and some sulfur atoms in the branched chain can participate the electrochemical reaction. With this idea, Wang et al. [12] first designed and fabricated a sulfurized pyrolyzed poly(acrylonitrile) cathode (S@pPAN) in 2002. Polymer-sulfur composites were synthesized by heating a mixture of sulfur and polyacrylonitrile (PAN) at 280 to 300°C under inert atmosphere for 6 hours (Equation (5.3)) [12]. Sulfur dehydrogenated the main chain of PAN to form a conductive polyacetylene structure, while the side chain cyano groups were cyclized. The remaining sulfur was complexed with the pyrolyzed PAN in the form of a single molecule [13]. This procedure achieved molecular-level contact of PAN and sulfur, and nearly 100% of active material utilization. Besides, almost all the S was reduced to S^{2-} via the reaction $2Li + S \rightarrow Li_2S$; soluble LiPSs will not be formed during such solid-state conversion, which can avoid the occurrence of the shuttle effect.

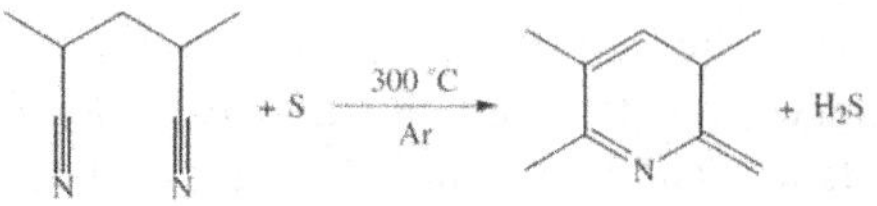

$$+ S \xrightarrow[Ar]{300\,°C} + H_2S \quad (5.3)$$

Fu et al. [14] reported a series of 1,3,5-benzenetrithiol (BTT) polymers, which can be synthesized by mixing BTT and a certain proportion of sulfur in carbon disulfide at room temperature. Generally, the reaction of organosulfide and lithium is as Figure 5.3 [14], the shuttle effect of LiPSs can be greatly inhibited. Meanwhile, the

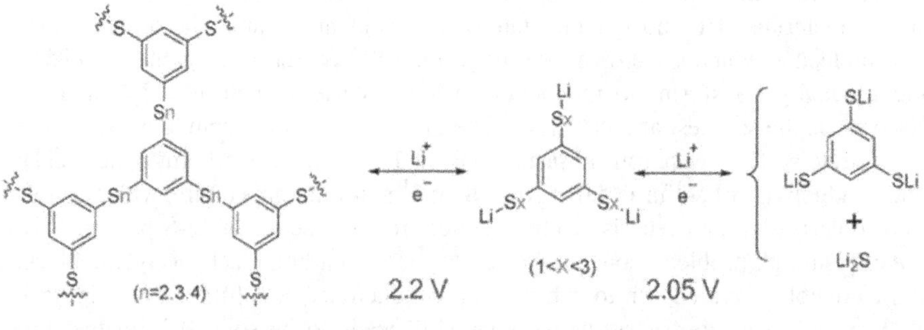

FIGURE 5.3 Reaction path by organosulfide BTT and lithium during charge and discharge process [14].

sulfur content of BTT polymers can be ~72%, guaranteeing relatively high specific capacity in practical applications.

$$R - S_n - R + (2n - 2)e^- + (2n - 2)Li^+ \leftrightarrow 2R - SLi + (n - 2)\ Li_2S, \quad \text{where } n \geq 2. \tag{5.4}$$

5.2.2 Li_2S_n Catholyte

Compared with insulating sulfur, replacing sulfur with a Li_2S_n catholyte seems a good choice by making full use of active materials. In a Li-polysulfide battery, as Li_2S_n dissolves easily in the electrolyte, the reaction kinetics will be faster than solid sulfur particles. Cui et al. [15] studied PVP-modified carbon paper in Li-polysulfides batteries. In this work, PVP can help prevent LiPSs from diffusing into a blank electrolyte, leading to improved cycling stability. The improvement of reaction kinetics is accompanied by a more serious shuttle and diffusion of LiPSs, which puts forward higher requirements for the design of the cathode host materials of Li-S batteries.

5.2.3 Li_2S

Due to the tough insulating nature of sulfur and dendrite of lithium anode, researchers put forward a non-lithium anode in Li-S batteries, which applies Li_2S (1166 mA h g^{-1}) [16] in the cathode simultaneously as a lithium and sulfur sources. By doing this, the electronic and ionic conductivity of a cathode improves to some extent. Moreover, the fabrication of Li-S batteries can be operated in a dry room and it indeed reduces the manufacture cost. However, there are several obstacles in application. The cost of Li_2S is expensive, and Li_2S can be unstable in humid air, resulting toxic H_2S gas formation; it is easy to cause safety hazards in the production process. More seriously, accounting for a big reactive energy barrier in the oxidation of Li_2S, the first cycle faces a huge onset potential in practical application.

5.2.4 S_2~S_4

In order to completely solve the problem of LiPSs dissolution and shuttle, a solid-state conversion mechanism is proposed by physically confining small sulfur molecules (S_2~S_4) in sub-micro pores. Gao et al. [6] first conducted sub-microporous carbon material to impregnate a small sulfur molecule. As the reaction processes, S_2~S_4 are directly converted to insoluble Li_2S_2 or Li_2S, avoiding the production of soluble intermediate products. The common method is to heat the sulfur at a high temperature to volatilize and decompose it into small molecules, and then make it penetrate into the sub-micron pores. Due to the steric hindrance effect, a large Li_2S_n will not be produced, so only a single platform will be observed during the discharge process (Figure 5.4(a,b)) [6,17]. Nevertheless, there is no silver bullet; this will drastically reduce the discharge capacity of the battery, and the preparation process of the material is complicated, which is difficult to meet the commercial demand.

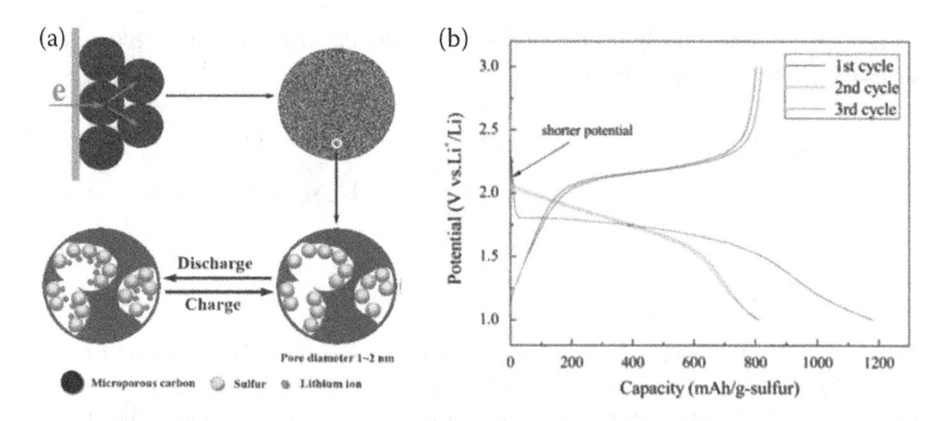

FIGURE 5.4 (a) Illustration of reaction pathway confined in sub-microporous carbon; (b) charge-discharge profile of sub-microporous carbon/S cathode [6].

5.3 SULFUR HOST MATERIALS

With the aim of enhancing poor electrical conductivity of sulfur, confining soluble LiPSs and boost redox kinetics, researchers reported various sulfur host materials to physically/chemically adsorb LiPSs, the performance of Li-S batteries was thus improved. Sulfur host materials can be divided by carbonaceous matrix, metal compounds, and heterostructures.

5.3.1 Carbonaceous Matrix

5.3.1.1 Porous Carbon Materials

As one of the most extensive materials on earth, carbon materials have the characteristics of high conductivity, low cost, and are lightweight. Therefore, carbonaceous frameworks with different dimensions have been used to host the element sulfur. There are mainly three dimensions of carbon materials from 1-dimension to 3-dimensions with various properties; for example, carbon nanofibers [18], carbon nanotubes [19,20], reduced graphene oxide [21–23], and porous biomass-derived carbon materials [24,25]. They not only construct excellent conductive framework, but also physically adsorb LiPSs, resulting in improved performance. Nazar et al. [26] reported highly ordered mesoporous nanostructure CMK-3/sulfur cathode to help retard the diffusion of LiPSs and reduce the loss of active substance (Figure 5.5(a)). Although carbon materials have a certain effect in inhibiting the dissolution of LiPSs and alleviating volume expansion, the non-polarity of carbon materials makes it difficult to effectively inhibit the shuttle of LiPSs. Initially, the researchers thought of doping carbon materials with heteroatoms and adsorbing LiPSs by polarity of heteroatoms (Figure 5.5(b,c)) [27–29]. Manthiram et al. [30] reported a lightweight three-dimensional nitrogen/sulphur-codoped graphene sponge as sulfur host. The doping elements of N and S increase active sites for LiPSs binding, leading to improved Li-S battery performance. According to the density functional theory (DFT) calculation, the Li atoms in LiPSs tend to bind with

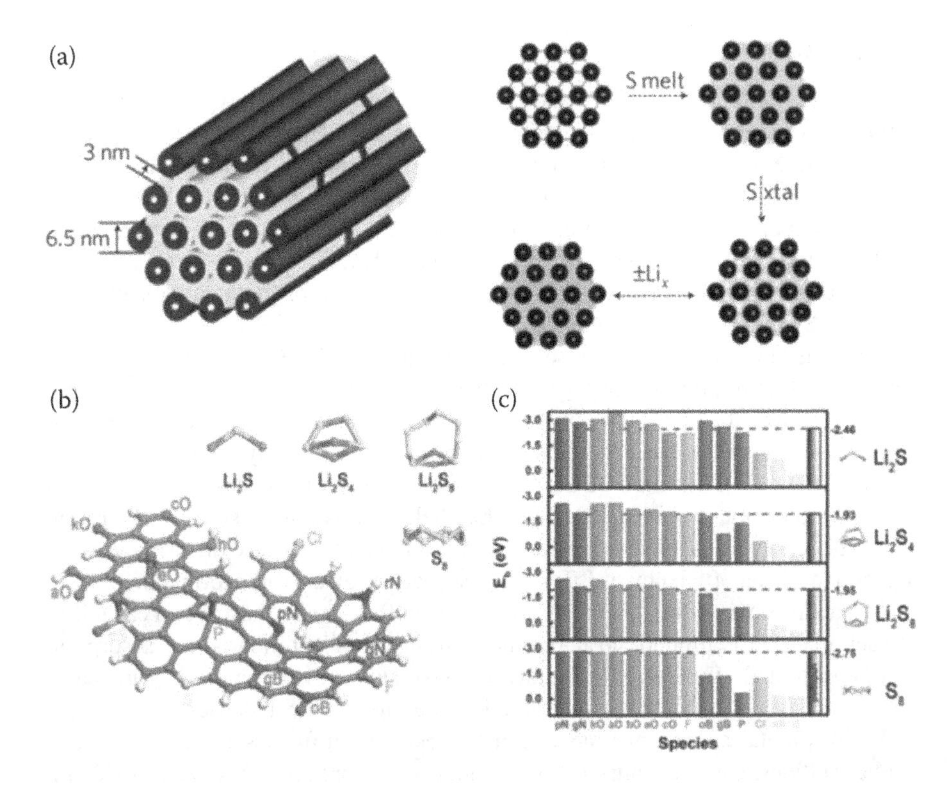

FIGURE 5.5 (a) Schematic illustration of structure and redox process of CMK-3/S cathode [26]. (b) Hetero atom doping in graphene; (c) binding energy of LiPSs and various hetero-atom doped graphene [32].

N, O, S, etc. [31,32]. However, the binding energy is still small, only by heteroatoms with Li, and it is not conducive to the migration of Li^+ ions in subsequent redox reactions. Therefore, more work began to focus on increasing the adsorption of sulfur atoms and accelerating the diffusion of Li^+ ions and metal compounds show obvious advantages.

5.3.1.2 Conductive Polymers

Common conductive polymers are polypyrrole (PPy), polyaniline (PANI), polythiophene (PTh), polyacrylonitrile (PAN), poly(2,2-dithiodiphenylamine) (PDTDA), polystyrene sulfonate (PSS), polyaniline nanotubes (PANI-NT), poly(3,4-ethylenedioxythiophene) (PEDOT), and polyimide (PI). Conductive polymer has excellent electrical conductivity, flexibility, and processability. The flexible polymer skeleton can effectively alleviate the volume expansion during the charge and discharge process. Under an electric field, the carriers move with the polymer molecule segments or jump between chains to form a current [33]. The introduction of a conductive polymer not only enhances the conductivity of sulfur cathode, but also adsorbs LiPSs by chemical or electrostatic interaction of polar functional group.

5.3.2 Metal Compounds

Compared with nonpolar carbon and heteroatoms in carbonaceous materials, metal compounds exhibit a strong anchoring ability to LiPSs. According to the Lewis acid-base interaction, a transition metal with partially occupied d orbital serves as Lewis acid and LiPSs with lone-pair electrons serve as Lewis base. DFT calculations show that sulfiphilic and lithiophilic sites can be verified, which form M-S (M for transition metals like Ni, Co, Ti, etc.) and/or Li...Y (Y for N, O, P, S, etc.) interaction and lengthen the S-S bond in LiPSs, favorable for fragmentation of long-chain LiPSs in a subsequent process.

With the deepening of the research mechanism of Li-S batteries, researchers have found that improving battery performance not only depends on the adsorption of LiPSs, but also on the timely diffusion of adsorbed LiPSs to the conductive surface to achieve the electron transfer of LiPSs and ultimately convert to lithium sulfide. Cui et al. [34] proposed adsorption and diffusion understanding based on a series of nonconductive metal oxides. They believe a balanced surface adsorption and diffusion will boost redox kinetics of LiPSs and better surface diffusion leads to higher deposition efficiency of LiPSs on electrodes, which avoids formation of dead sulfur and improve utilization of active substance. Because the adsorption capability of each material is limited, no matter how strong the adsorption is, the adsorption will be saturated. Therefore, the catalytic conversion will become a key indicator for evaluating the performance of cathode materials in Li-S batteries.

Various metal compounds have been proposed as sulfur hosts, including metal oxides, sulfides, carbides, nitrides, phosphides, metal organic frameworks (MOFs), etc. These compounds are designed as a hollow, core-shell, multi-layer, porous structure. Exposed active sites are prone to achieve strong affinity to LiPSs, which enhances the obstacle of LiPSs shuttle.

5.3.2.1 Metal Oxides

Metal oxides are easy to synthesize and achieve large-scale production without inert atmosphere. Metal oxides usually show strong adsorption capacity for LiPSs, while the electronic and ionic conductivity is relatively low due to strong electronegativity of oxygen. It was once generally recognized that metal oxides can effectively adsorb LiPSs, but the understanding of their structure-activity relationship on LiPSs was not clear. In 2015, Nazar et al. [35] reported manganese dioxide nanosheets can react with LiPSs to form intermediate polythionates and thiosulfates, and further disproportionate in the reaction to produce insoluble lithium sulfide. On one hand, MnO_2 nanosheets prohibit the shuttle of LiPSs; on the other hand, they accelerate the redox kinetics during the discharge-charge process. The Nazar group [36] further proposed the "Goldilocks principle", which shows the thiosulfate formation of metal oxide is selectively motivated with redox potential between $2.4\ V < E^o < 3.05\ V$. Other redox voltages either cause over-oxidization to the sulfate group or do not react with LiPSs only by polar interaction (Figure 5.6). Inspired by the "tip effect", Sun et al. [37] demonstrate that hollow cupric oxide spheres (HCOS) can be used as effective hosts for Li–S batteries. The HCOS host can effectively adsorb and promote the conversion of LiPSs to the discharge product Li_2S; meanwhile, it

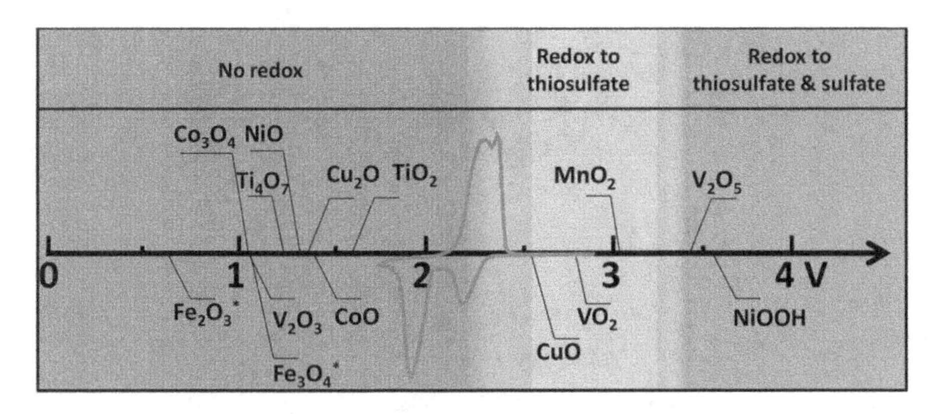

FIGURE 5.6 Chemical reactivity of different metal oxides with various redox potential [35].

provides a large number of reactive sites for the rapid and uniform deposition, maintaining the structural integrity of the host.

Perovskite, as a polar metal oxide with both chemical adsorption and conductive surfaces, can effectively inhibit the shuttle effect of LiPSs. The Wei group [38] conducted the ferroelectric perovskite $BaTiO_3$ as a sulfur host for the first time; the asymmetrical lattice structure causes the spontaneous polarization of the ferroelectric material, generating non-overlapping charges in the center of a single crystal unit, forming an electric dipole moment, and inducing polarization on the surface of the ferroelectric material, which has a strong adsorption effect on polar LiPSs. After that, $La_{0.6}Sr_{0.4}CoO_{3-\delta}$(LSC) [39], $La_{0.56}Li_{0.33}TiO_3$ (LLTO) [40], $Ba_{0.5}Sr_{0.5}Co_{0.8}Fe_{0.2}O_{3-\delta}$ (PrNP) [41] were subsequently proposed, which has broad application prospects in Li-S batteries.

Although metal oxides take advantage in sulfur host materials, the low conductivity limits their application. Increasing the ratio of the conducting agent should be added to the cathode to construct a good conductive framework, which doubtless decreases the energy density of Li-S batteries. There is an urgent need to find a sulfur host with both conductivity and LiPSs adsorption capacity.

5.3.2.2 Metal Sulfides

Since sulfur is less electronegative than oxygen, metal sulfides exhibit better conductivity than metal oxides. Various metal sulfides such as Co_9S_8 [11,42], MoS_2 [43,44], Co_3S_4 [45,46], TiS_2 [47], and WS_2 [48] have been synthesized to serve as a sulfur host. Metal sulfides not only show excellent LiPSs adsorption, but also help to build a good conductive framework to reduce polarization of cathode, so they have been widely studied. Zhang et al. [49] reported pyrite-type CoS_2 incorporated into carbon/sulfur cathode has sulfiphilic sites, which can effectively adsorb LiPSs and accelerate redox kinetics (Figure 5.7). A slow-capacity fading rate of 0.034% at 2.0°C was achieved after 2,000 cycles.

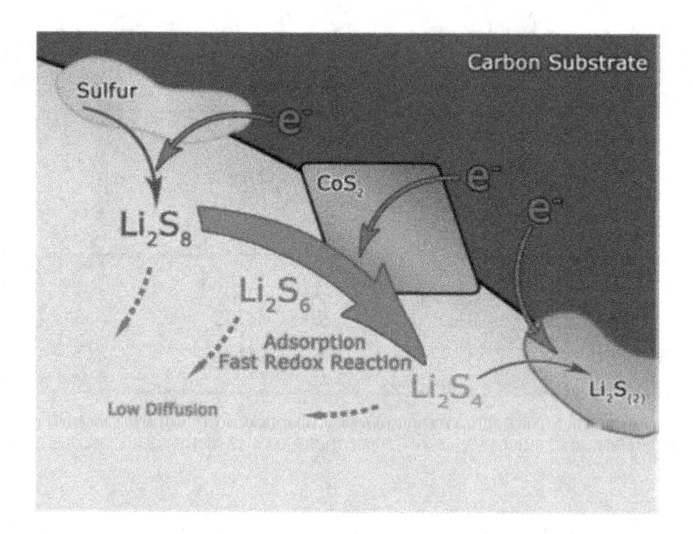

FIGURE 5.7 Scheme of discharge process in the CoS_2-based cathode [49].

5.3.2.3 Metal Carbides

Metal carbide owns both high conductivity and LiPSs adsorption capability. MXene, as an emerging two-dimensional nanomaterial, has received wide attention from researchers in recent years. It is currently mainly prepared by selective etching of the A atomic layer in the precursor $M_{n+1}AX_n$ phase, where M stands for pre-transition metal elements (Ti, V, Nb, etc.), A is mainly 13–16 group elements (Al, Si, etc.), and X is carbon and/or nitrogen. MXene provides more channels for the movement of ions, greatly increasing the speed of ion movement. Nazar et al. [50] presented S/Ti_2C composites, which own inherent high conductivity and surface functional groups. There is a strong interaction between Ti atoms and LiPSs; thus, high initial capacity and high capacity-retention was obtained. However, the fabrication of metal carbide can be complicated due to HF etching, unstable nature (easy to be oxidized and easy to stack), and expensive cost, and it is not conducive to achieve large-scale production.

5.3.2.4 Metal Nitrides

Metal nitrides are generally binary compounds where nitrogen atoms occupy the interstitial positions in the metal lattice. This compound is similar to metal with high hardness, high melting point, stable chemical properties, and conductivity [51]. Generally, metal nitrides have an ionic, covalent, and metallic nature, which endows metal nitrides' high conductivity; for instance, the conductivity of VN is 1.17×10^6 S m^{-1} [52]. The metal-N interaction in the metal nitride makes the parent metal lattice expansion and compression of metal d-band, and the higher density of states at the Fermi level renders precious metal-like catalytic activity [51,53]. When applied in sulfur cathode, metal nitrides exhibit effective adsorption capability and accelerate catalytic conversion of LiPSs.

5.3.2.5 Metal Phosphides

Metal phosphides show metallic nature and even superconductivity and can be prepared by a facile and gentle process [54]. Phosphides present intrinsic polarity as metal and phosphorous ions bind with S and Li, respectively. Such strong adsorption immobilizes LiPSs and alleviates the shuttle effect. The P atom has a strong electronegativity and more electrons in the metal of metal phosphides will be delocalized; thus showing good conductivity. Zhou et al. [55] systematically studied Co-based compounds by theoretical calculation and experiments and proposed that p bands originated from the non-metal anions can benefit the interfacial charge interaction by tuning the electron energy of the valence band. Therefore, a superior p-band level enables CoP excellent interfacial charge transfer capability, resulting in high catalytic activity of CoP. In addition, Sun et al. [56] employed amorphous cobalt phosphide grown on a reduced graphene oxide-multiwalled carbon nanotube (rGO-CNT-CoP(A)) as a sulfur host. The amorphous CoP strongly anchors LiPSs and significantly alleviates the LiPSs' shuttling. Moreover, the amorphous CoP can promote the diffusion of LiPSs on the surface together with accelerating the redox kinetics of LiPSs' conversion.

5.3.2.6 MOFs

MOFs have attracted much attention because of large pore volume, high specific surface area, and tunable pore structure [57]. Applications in Li-S batteries are mainly divided into two parts: pristine MOFs and MOFs-derived metal compound. MOFs are commonly used as precursors to prepare various metal compounds with controllable porosity and shape. By infiltrating sulfur into the micropores and mesopores of MOFs by high-temperature melting, effective encapsulation of sulfur and LiPSs intermediates can be achieved. Transition metal ions in the center can serve as active sites to polarly adsorb LiPSs and boost redox kinetics. Despite that, the conductivity of MOFs is generally poor, easily resulting in large overpotential, and a conductive skeleton is necessary when applied in sulfur host. Wang et al. [58] reported a 3-D heterogeneous sulfur host fabricated by an in-situ deposition of nickel-doped zeolitic imidazolate framework-8 (Ni-ZIF-8) on carbon cloth. Ni-ZIF-8 can effectively immobilize LiPSs due to the strong confining effect of cage-like pores. Meanwhile, the catalytic nickel species could not only adsorb the LiPSs strongly, but could also accelerate the LiPSs' redox kinetics effectively.

5.3.3 Heterostructures

Generally, it is difficult for most sulfur hosts to have both high conductivity and strong LiPSs adsorption performance. The researchers constructed heterostructures in-situ to combine material with strong LiPSs adsorption and material with catalytic capability together. Such a unique structure shows excellent electrochemical performance. Yu et al. [59] designed a freestanding carbon scaffold embedded with VN-TiN heterostructure nanoparticles as a bifunctional host of Li and S. A twinborn interface can be clearly observed by HRTEM (Figure 5.8(a–c)), and with the help of DFT calculation, the synergistic effect of VN and TiN is confirmed. The binding

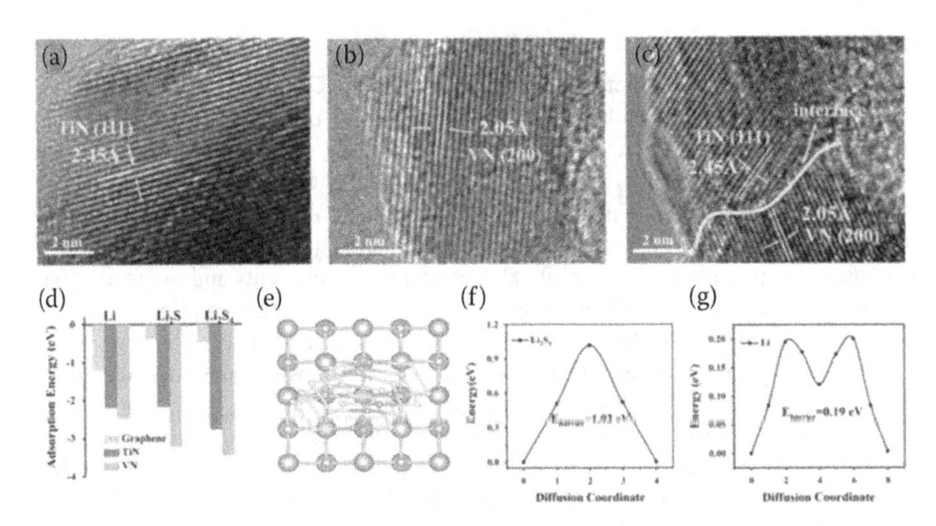

FIGURE 5.8 HRTEM images of (a) TiN(111), (b) VN(200), (c) lattice interface of TiN(111) and VN(200), (d) binding energy of LiPSs with various substrates, (e) diffusion pathway, (f) diffusion barrier of Li_2S_4 on TiN substrate, and (g) diffusion barrier of Li^+ ion on TiN substrates [59].

energy of VN-LiPSs is higher than TiN-LiPSs, while Li_2S_4 migration and Li^+ diffusion barrier on TiN surface is smaller than VN (Figure 5.8(d–g)). The results demonstrated better adsorption of VN and better reaction kinetics of TiN. Combining the advantages of perovskite and transition metal catalysts, Wang et al. [60] designed a bifunctional STO@Co with adsorption and catalysis. Perovskite effectively adsorbs LiPSs and inhibits the shuttle effect by a ferroelectric effect of perovskite. The in-situ exsolved Co nanometal enhances the conductivity and exhibits excellent LiPSs catalytic conversion capability.

Despite that, the construction of the heterostructure cannot be achieved casually. The lattice mismatch of the two crystals should be less than a certain value (usually <5%). At the same time, the processing time and temperature will affect the heterostructure, so it is necessary to find a balance among them.

5.4 SUMMARY

In recent years, cathode materials have been the main research direction of Li-S batteries. Reasonable sulfur cathode design can promote the application process of Li-S batteries. This requires the design of the positive electrode to meet the following demands:

1. The uniform distribution of polar metal compounds to provide adsorption and catalytic conversion sites for LiPSs, and alleviate LiPSs shuttles in Li-S batteries;
2. The conductive scaffold with high specific surface area connects active material sulfur and electrocatalyst, so as to realize rapid LiPSs diffusion and catalytic conversion at the three-phase interface;

3. In order to obtain a more stable sulfur cathode, it is of necessity to continuously deepen the understanding of the charge and discharge behavior of a sulfur cathode;
4. Advanced in-situ characterization and theoretical calculations can provide new insights for understanding the reaction mechanism and proposing solutions for Li-S batteries;
5. In order to realize the commercialization of Li-S batteries, further suppressing self-discharge, broadening the operating temperature range, increasing areal sulfur loading, and decreasing electrolyte/sulfur (E/S) ratio and negative/positive (N/P) capacity ratio are crucial.

Only when these key elements meet the requirements can Li-S batteries be able to compete with lithium-ion batteries. The road is tortuous, but the future is bright. With the joint efforts of researchers, Li-S batteries will eventually overcome numerous difficulties and realize commercialization and serve the world.

REFERENCES

[1] Ji X., Lee K. T., Nazar L. F. A highly ordered nanostructured carbon–sulphur cathode for lithium–sulphur batteries[J]. *Nature Materials*, 2009, 8(6): 500–506.
[2] Yuan Z., Peng H. J., Hou T. Z., et al. Powering lithium-sulfur battery performance by propelling polysulfide redox at sulfiphilic hosts[J]. *Nano Letters*, 2016, 16(1): 519–527.
[3] Zhang G., Zhang Z.-W., Peng H.-J., et al. A toolbox for lithium-sulfur battery research: Methods and protocols[J]. *Small Methods*, 2017, 1(7): 1700134.
[4] Wang B., Alhassan S. M., Pantelides S. T. Formation of large polysulfide complexes during the lithium-sulfur battery discharge[J]. *Physical Review Applied*, 2014, 2(3): 034004.
[5] Ren Y. X., Zhao T. S., Liu M., et al. Modeling of lithium-sulfur batteries incorporating the effect of Li2S precipitation[J]. *Journal of Power Sources*, 2016, 336: 115–125.
[6] Zhang B., Qin X., Li G. R., et al. Enhancement of long stability of sulfur cathode by encapsulating sulfur into micropores of carbon spheres[J]. *Energy & Environmental Science*, 2010, 3(10): 1531–1537.
[7] Evers S., Nazar L. F. New approaches for high energy density lithium-sulfur battery cathodes[J]. *Accounts of Chemical Research*, 2013, 46(5): 1135–1143.
[8] Wang C. G., Song H. W., Yu C. C., et al. Iron single-atom catalyst anchored on nitrogen-rich mof-derived carbon nanocage to accelerate polysulfide redox conversion for lithium sulfur batteries[J]. *Journal of Materials Chemistry A*, 2020, 8(6): 3421–3430.
[9] Chung S.-H., Manthiram A. Current status and future prospects of metal–sulfur batteries[J]. *Advanced Materials*, 2019, 31(27): 1901125.
[10] Seh Z. W., Sun Y. M., Zhang Q. F., et al. Designing high-energy lithium-sulfur batteries[J]. *Chemical Society Reviews*, 2016, 45(20): 5605–5634.
[11] Chen T., Ma L. B., Cheng B. R., et al. Metallic and polar Co9S8 inlaid carbon hollow nanopolyhedra as efficient polysulfide mediator for lithium-sulfur batteries [J]. *Nano Energy*, 2017, 38: 239–248.
[12] Wang J., Yang J., Xie J., et al. A novel conductive polymer–sulfur composite cathode material for rechargeable lithium batteries[J]. *Advanced Materials*, 2002, 14(13–14): 963–965.

[13] Yang H., Chen J., Yang J., et al. Prospect of sulfurized pyrolyzed poly(acrylonitrile) (S@pPAN) cathode materials for rechargeable lithium batteries[J]. *Angewandte Chemie International Edition*, 2020, 59(19): 7306–7318.

[14] Sang P., Song J., Guo W., et al. Hyperbranched organosulfur polymer cathode materials for Li-S battery[J]. *Chemical Engineering Journal*, 2021, 415: 129043.

[15] Cui Y., Fu Y. Enhanced cyclability of Li/polysulfide batteries by a polymer-modified carbon paper current collector[J]. *ACS Applied Materials & Interfaces*, 2015, 7(36): 20369–20376.

[16] Wang J., Jia L., Duan S., et al. Single atomic cobalt catalyst significantly accelerates lithium ion diffusion in high mass loading Li2S cathode[J]. *Energy Storage Materials*, 2020, 28: 375–382.

[17] Xin S., Gu L., Zhao N. H., et al. Smaller sulfur molecules promise better lithium-sulfur batteries[J]. *Journal of the American Chemical Society*, 2012, 134(45): 18510–18513.

[18] Ji L. W., Rao M. M., Aloni S., et al. Porous carbon nanofiber-sulfur composite electrodes for lithium/sulfur cells[J]. *Energy & Environmental Science*, 2011, 4(12): 5053–5059.

[19] Xu T., Song J., Gordin M. L., et al. Mesoporous carbon-carbon nanotube-sulfur composite microspheres for high-areal-capacity lithium-sulfur battery cathodes[J]. *ACS Applied Materials & Interfaces*, 2013, 5(21): 11355–11362.

[20] Hu G., Sun Z., Shi C., et al. A sulfur-rich copolymer@cnt hybrid cathode with dual-confinement of polysulfides for high-performance lithium-sulfur batteries[J]. *Advanced Materials*, 2017, 29(11).

[21] Kalaiappan K., Rengapillai S., Marimuthu S., et al. Kombucha scoby-based carbon and graphene oxide wrapped sulfur/polyacrylonitrile as a high-capacity cathode in lithium-sulfur batteries[J]. *Frontiers of Chemical Science and Engineering*, 2020, 14: 976–987.

[22] Yin L. C., Wang J. L., Lin F. J., et al. Polyacrylonitrile/graphene composite as a precursor to a sulfur-based cathode material for high-rate rechargeable Li-S batteries [J]. *Energy & Environmental Science*, 2012, 5(5): 6966–6972.

[23] Wu H., Xia L., Ren J., et al. A multidimensional and nitrogen-doped graphene/hierarchical porous carbon as a sulfur scaffold for high performance lithium sulfur batteries[J]. *Electrochimica Acta*, 2018, 278: 83–92.

[24] Li B., Xie M., Yi G. H., et al. Biomass-derived activated carbon/sulfur composites as cathode electrodes for Li-S batteries by reducing the oxygen content[J]. *RSC Advances*, 2020, 10(5): 2823–2829.

[25] Fan L., Li Z., Kang W., et al. Biomass-derived tube-like nitrogen and oxygen dual-doped porous carbon in the sulfur cathode for lithium sulfur battery[J]. *Renewable Energy*, 2020, 155: 309–316.

[26] Ji X. L., Lee K. T., Nazar L. F. A highly ordered nanostructured carbon-sulphur cathode for lithium-sulphur batteries[J]. *Nature Materials*, 2009, 8(6): 500–506.

[27] Ma Z. L., Dou S., Shen A. L., et al. Sulfur-doped graphene derived from cycled lithium-sulfur batteries as a metal-free electrocatalyst for the oxygen reduction reaction[J]. *Angewandte Chemie International Edition*, 2015, 54(6): 1888–1892.

[28] Wang X., Zhang Z., Qu Y., et al. Nitrogen-doped graphene/sulfur composite as cathode material for high capacity lithium–sulfur batteries[J]. *Journal of Power Sources*, 2014, 256: 361–368.

[29] Tang C., Zhang Q., Zhao M. Q., et al. Nitrogen-doped aligned carbon nanotube/graphene sandwiches: Facile catalytic growth on bifunctional natural catalysts and their applications as scaffolds for high-rate lithium-sulfur batteries[J]. *Advanced Materials*, 2014, 26(35): 6100–+.

[30] Zhou G. M., Paek E., Hwang G. S., et al. Long-life Li/polysulphide batteries with high sulphur loading enabled by lightweight three-dimensional nitrogen/sulphur-codoped graphene sponge[J]. *Nature Communications*, 2015, 6(1): 7760.
[31] Hou T.-Z., Xu W.-T., Chen X., et al. Lithium bond chemistry in lithium–sulfur batteries[J]. *Angewandte Chemie International Edition*, 2017, 56(28): 8178–8182.
[32] Hou T.-Z., Chen X., Peng H.-J., et al. Design principles for heteroatom-doped nanocarbon to achieve strong anchoring of polysulfides for lithium–sulfur batteries [J]. *Small*, 2016, 12(24): 3283–3291.
[33] Ouyang J. Recent advances of intrinsically conductive polymers[J]. *Acta Physica Sinica*, 2018, 34: 1211–1220.
[34] Tao X., Wang J., Liu C., et al. Balancing surface adsorption and diffusion of lithium-polysulfides on nonconductive oxides for lithium–sulfur battery design[J]. *Nature Communications*, 2016, 7(1): 11203.
[35] Liang X., Hart C., Pang Q., et al. A highly efficient polysulfide mediator for lithium–sulfur batteries[J]. *Nature Communications*, 2015, 6(1): 1–8.
[36] Liang X., Kwok C. Y., Lodi-Marzano F., et al. Tuning transition metal oxide-sulfur interactions for long life lithium sulfur batteries: The "goldilocks" principle[J]. *Advanced Energy Materials*, 2016, 6(6): 1501636.
[37] Yang Y. X., Wang Z. H., Li G. D., et al. Inspired by the "tip effect": A novel structural design strategy for the cathode in advanced lithium–sulfur batteries[J]. *Journal of Materials Chemistry A*, 2017, 5(7): 3140–3144.
[38] Xie K. Y., You Y., Yuan K., et al. Ferroelectric-enhanced polysulfide trapping for lithium–sulfur battery improvement[J]. *Advanced Materials*, 2017, 29(6): 1604724.
[39] Hao Z. X., Zeng R., Yuan L. X., et al. Perovskite $La_{0.6}Sr_{0.4}CoO_{3-\delta}$ as a new polysulfide immobilizer for high-energy lithium-sulfur batteries[J]. *Nano Energy*, 2017, 40: 360–368.
[40] Chen M. F., Huang C., Li Y. F., et al. Perovskite-type $La_{0.56}Li_{0.33}TiO_3$ as an effective polysulfide promoter for stable lithium–sulfur batteries in lean electrolyte conditions[J]. *Journal of Materials Chemistry A*, 2019, 7(17): 10293–10302.
[41] Kong L., Chen X., Li B. Q., et al. A bifunctional perovskite promoter for polysulfide regulation toward stable lithium–sulfur batteries[J]. *Advanced Materials*, 2018, 30(2): 1705219.
[42] Pang Q., Kundu D., Nazar L. F. A graphene-like metallic cathode host for long-life and high-loading lithium-sulfur batteries[J]. *Materials Horizons*, 2016, 3(2): 130–136.
[43] Lin H. B., Yang L. Q., Jiang X., et al. Electrocatalysis of polysulfide conversion by sulfur-deficient MoS2 nanoflakes for lithium-sulfur batteries[J]. *Energy & Environmental Science*, 2017, 10(6): 1476–1486.
[44] He J. R., Hartmann G., Lee M., et al. Freestanding 1T MoS2/graphene heterostructures as a highly efficient electrocatalyst for lithium polysulfides in Li–S batteries[J]. *Energy & Environmental Science*, 2019, 12(1): 344–350.
[45] Chen T., Zhang Z., Cheng B., et al. Self-templated formation of interlaced carbon nanotubes threaded hollow Co3S4 nanoboxes for high-rate and heat-resistant lithium-sulfur batteries[J]. *Journal of the American Chemical Society*, 2017, 139(36): 12710–12715.
[46] Pu J., Shen Z. H., Zheng J. X., et al. Multifunctional Co3S4@sulfur nanotubes for enhanced lithium-sulfur battery performance[J]. *Nano Energy*, 2017, 37: 7–14.
[47] Huang X., Tang J., Luo B., et al. Sandwich-like ultrathin TS_2 nanosheets confined within N, S codoped porous carbon as an effective polysulfide promoter in lithium-sulfur batteries[J]. Advanced Energy Materials, 2019, 9(32): 1901872.

[48] Huang S. Z., Wang Y., Hu J. P., et al. Mechanism investigation of high-performance li-polysulfide batteries enabled by tungsten disulfide nanopetals[J]. *ACS Nano*, 2018, 12(9): 9504–9512.

[49] Yuan Z., Peng H. J., Hou T. Z., et al. Powering lithium-sulfur battery performance by propelling polysulfide redox at sulfiphilic hosts[J]. *Nano Letters*, 2016, 16(1): 519–527.

[50] Liang X., Garsuch A., Nazar L. F. Sulfur cathodes based on conductive mxene nanosheets for high-performance lithium-sulfur batteries[J]. *Angewandte Chemie International Edition*, 2015, 54(13): 3907–3911.

[51] Gao B., Li X., Ding K., et al. Recent progress in nanostructured transition metal nitrides for advanced electrochemical energy storage[J]. *Journal of Materials Chemistry A*, 2019, 7(1): 14–37.

[52] Sun Z., Zhang J., Yin L., et al. Conductive porous vanadium nitride/graphene composite as chemical anchor of polysulfides for lithium-sulfur batteries[J]. *Nature Communications*, 2017, 8(1): 14627.

[53] Zhong Y., Xia X., Shi F., et al. Transition metal carbides and nitrides in energy storage and conversion[J]. *Advanced Science*, 2016, 3(5): 1500286.

[54] Yu S., Cai W., Chen L., et al. Recent advances of metal phosphides for Li–S chemistry[J]. *Journal of Energy Chemistry*, 2021, 55: 533–548.

[55] Zhou J., Liu X., Zhu L., et al. Deciphering the modulation essence of p bands in co-based compounds on li-s chemistry[J]. *Joule*, 2018, 2(12): 2681–2693.

[56] Sun R., Bai Y., Luo M., et al. Enhancing polysulfide confinement and electrochemical kinetics by amorphous cobalt phosphide for highly efficient lithium–sulfur batteries[J]. *ACS Nano*, 2021, 15(1): 739–750.

[57] Zhang X., Chen A., Zhong M., et al. Metal–organic frameworks (MOFs) and mof-derived materials for energy storage and conversion[J]. *Electrochemical Energy Reviews*, 2019, 2(1): 29–104.

[58] Yang Y., Wang Z., Jiang T., et al. A heterogenized ni-doped zeolitic imidazolate framework to guide efficient trapping and catalytic conversion of polysulfides for greatly improved lithium–sulfur batteries[J]. *Journal of Materials Chemistry A*, 2018, 6(28): 13593–13598.

[59] Yao Y., Wang H., Yang H., et al. A dual-functional conductive framework embedded with tin-vn heterostructures for highly efficient polysulfide and lithium regulation toward stable Li–S full batteries[J]. *Advanced Materials*, 2019, 32(6).

[60] Hou W., Yang Y., Fang L., et al. Perovskite with in situ exsolved cobalt nanometal heterostructures for high rate and stable lithium-sulfur batteries[J]. *Chemical Engineering Journal*, 2021, 409: 128079.

6 Anode Materials for Lithium-Sulfur Batteries

Zhenhua Wang
School of Chemistry and Chemical Engineering, Beijing Institute of Technology, Beijing, People's Republic of China

CONTENTS

6.1 INTRODUCTION

The anode is one of the indispensable components of Li-S batteries. However, there are few studies on anode materials in Li-S batteries. Unlike intercalation cathodes, the electrode potential and shuttle effect of sulfur cathodes make them have different requirements for anode materials. Among the various anode materials for Li-S batteries since they were discovered in 1960s [1–3], metallic lithium has become the best potential candidate with its unparalleled theoretical specific energy. However, as a popular anode material, lithium metal has not yet achieved real application due to its inherent properties. Cycle life and safety issues have become

DOI: 10.1201/9781003133971-6

the main factors restricting its development. In addition to lithium metal, many materials have also been tried as the anodes of Li-S batteries. This chapter introduces the reaction principle and research status of lithium metal anodes, as well as the application of carbon, silicon, and other alloy materials as anodes in Li-S batteries.

6.2 STATUS OF LITHIUM METAL ANODE

Lithium metal with a high theoretical specific capacity (3860 mAh g^{-1} or 2061 mAh cm^{-3}) and low chemical potential (–3.04 V versus the standard hydrogen electrode) is an ideal anode material for lithium batteries [4–7]. The biggest advantage of Li-S batteries is their extremely high theoretical specific capacity, which complements each other even more when using lithium metal as an anode. The theoretical specific energy and specific capacity of Li-S batteries using lithium metal anodes can reach 2,500 Wh kg^{-1} and 1,675 mAh g^{-1}, which is much higher than that of currently commercialized lithium-ion batteries (Figure 6.1) [8–11].

Unfortunately, the popular lithium metal anode has not been commercialized since the 1960s [12,13]. This is because the extremely low chemical electrode potential of lithium metal makes it easy to react with other components in the battery and cause irreversible capacity degradation. However, during the electroplating process of lithium, a phenomenon of dendrite growth due to uneven charge distribution occurs. When the dendrite reaches the cathode, a short circuit occurs in the battery, causing problems such as fire and explosion. However, the sulfur cathode has a lower chemical potential than other cathode materials, and the chemical potential difference with lithium metal is also smaller. The combination of the two electrodes has lower requirements on the electrolyte potential window. After a lot of experiments, researchers found that ether electrolytes can better match sulfur cathodes and lithium metal anodes at the same time [14–16]. This is also because lithium LiPSs will irreversibly react with the carbonate electrolyte and cause the loss of active materials [17–19]. Therefore, lithium metal is the best choice for Li-S

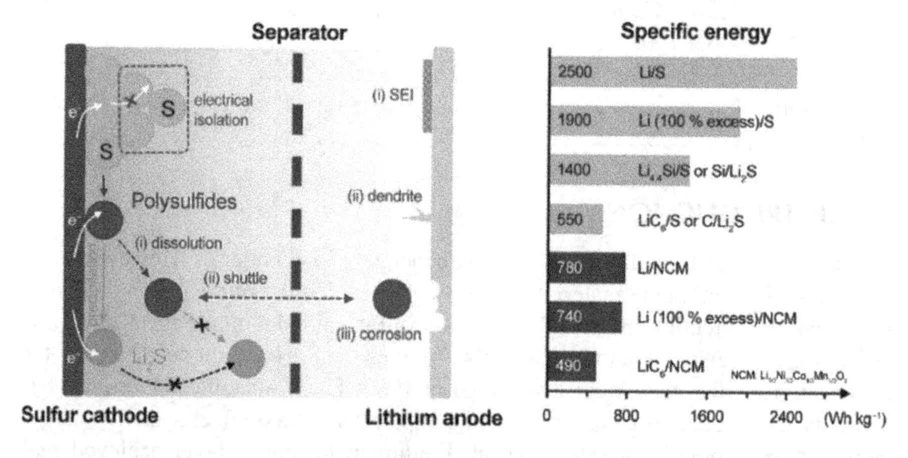

FIGURE 6.1 Schematic illustration and energy density of a Li-S battery with lithium metal as the anode [11].

battery anode materials. However, in the complex environment of Li-S batteries, lithium metal anodes are facing more problems and greater challenges.

6.3 PROBLEMS

During the charging and discharging of the Li-S battery, electroplating and stripping of lithium occurred on the anode side. In the process of alternate oxidation and reduction, lithium, which is highly reactive, can easily undergo irreversible chemical reactions with the electrolyte, lithium salt, and LiPSs dissolved in the electrolyte. In turn, the active material (lithium) is consumed in this side reaction, and the capacity of the battery is continuously reduced. In addition, in the electroplating process of lithium, the uneven electric fields and current density will cause the uneven nucleation and deposition of lithium. Negative charges are more likely to accumulate in the lithium-rich area to produce a "tip effect" [20], which leads to a higher enhanced electric and ionic fields in the lithium-rich area, which causes lithium to be continuously reduced and deposited in this part, and finally grows into lithium dendrites. In addition to the spontaneous formation of lithium dendrites under ideal conditions, the rupture of SEI is also one of the important reasons leading to the growth of dendrites [21], which will be introduced in Section 6.3.1.

As shown in Figure 6.2, the emergence of lithium dendrites will bring about two extremely serious consequences: (1) Lithium dendrites fracture and fall off from the current collector, causing them to lose electron conductivity and become an invalid "dead lithium", resulting in the loss of battery-active materials. (2) Dendrites continue to grow and pierce the separator to reach the cathode side, causing a short

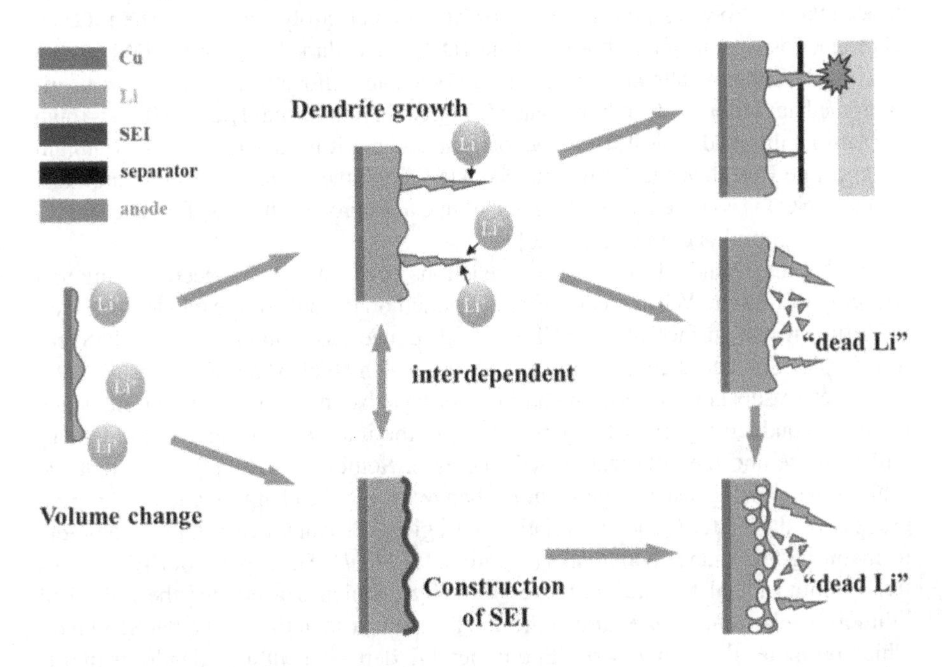

FIGURE 6.2 Failure mechanism of lithium metal anode.

circuit inside the battery. In a flammable organic electrolyte environment, thermal runaway caused by a short circuit can easily develop into a fire or explosion.

The two major issues of cycle life and safety have greatly restricted the application of lithium metal as an anode material. These problems are affected by many factors. The following will explain the problems existing in the working process of the lithium metal anode. It is worth noting that these factors influence and coordinate with each other. At present, there is no perfect formula or model that can perfectly explain this process.

6.3.1 Failure Caused by SEI

Lithium metals have a very low electrochemical potential, which allows them to react with almost all of the electrolytes to form a dense passivation layer between the lithium metal and the electrolyte, called the solid electrolyte interface (SEI) layer. SEI was proposed by Peled [22] in 1979 and has received extensive attention and research as an important part of the anode of lithium batteries. The formation process and chemical composition of SEIs are affected by the environment and working conditions of the battery, and play an important role in the cycle life of the lithium metal anode [23,24].

The formation of a SEI is mainly divided into chemical reaction and electrochemical reaction [23]. The chemical reaction is mainly formed by the instant reaction of lithium metal in contact with the electrolyte. The result of this reaction is that the passivated SEI layer limits the further reaction of the lithium metal with the electrolyte and protects the lithium metal from further corrosion. The composition of SEI has a lot to do with the electrolyte. At present, the most common electrolyte for Li-S batteries is an ether electrolyte containing 1,3-dioxolane (DOL) and dimethoxyethane (DME), and most of the lithium salts are lithium trifluoromethanesulfonate ($LiCF_3SO_3$) or bistrifluoromethanesulfonimide lithium salt (LiTFSI). In this electrolyte, DME has high LiPSs solubility and rapid LiPSs reaction kinetics, but it is easy to react with lithium metal, while DOL helps to form a more stable SEI on the surface of the lithium metal anode. LiNO3 is considered an effective additive to improve battery performance and is widely recognized and applied [25,26].

On the other hand, electrochemical reactions occur during the electroplating and stripping of lithium. When electrons are enriched on the side of the anode, they tend to diffuse to the surface of the SEI and will reduce the lithium ions and the electrolyte. The SEI keeps getting thicker and looser, which also causes the inactivation of the active substance to become an ineffective substance. At the same time, since the ionic conductivity of SEI is generally low, the internal resistance of the battery will increase and the polarization will increase. Some researchers believe that the lithium metal will cause a huge volume change during the charging and discharging process of the battery, and the relatively fragile SEI cannot maintain a complete form during this change and breaks (Figure 6.3) [21,27]. The rupture of SEI exposes fresh lithium metal to react with the electrolyte, which exacerbates the failure of lithium. In addition, the pressure at the cracks is lower than that of the flat SEI layer, which promotes the tendency of lithium metal to deposit as lithium dendrites in this area [5,21].

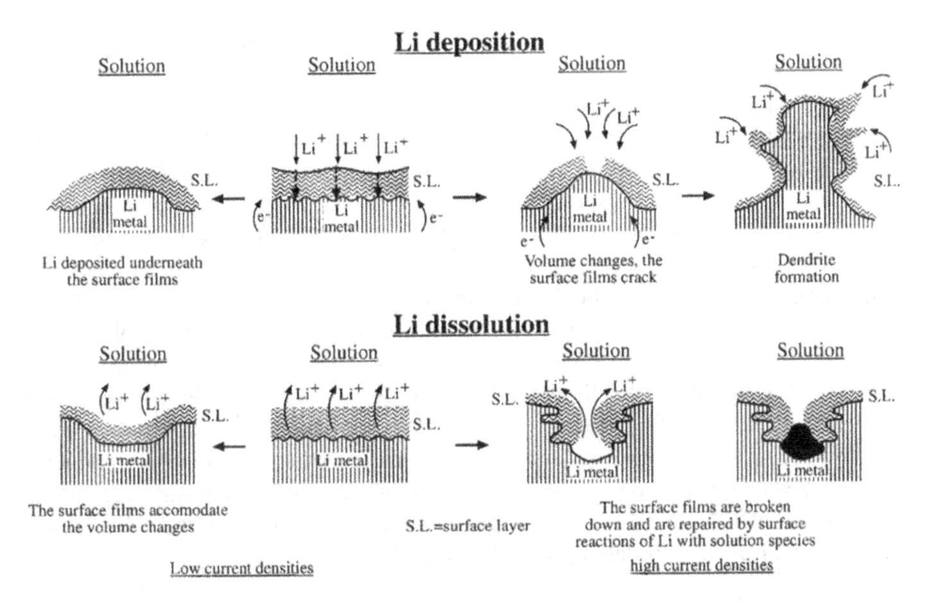

FIGURE 6.3 An illustration of the morphological phenomena developed on Li electrodes during Li deposition and dissolution [21].

6.3.2 Lithium Dendrites

Lithium metals with lower surface energy tend to accumulate in places where the current density and electric field are higher during the electroplating process [28,29]. The result of this uneven deposition is that a lithium metal grows in the form of particles or needles on the anode side, collectively referred to herein as "lithium dendrites". The formation of lithium dendrites is divided into two stages: nucleation and growth.

During the charging process of a Li-S battery, lithium ions obtain electrons and deposit on the current collector or lithium metal. In order to obtain a longer battery life, the current Li-S battery system mostly uses excess lithium metal as the anode, which means that the nucleation process of lithium deposition mainly occurs on the primary lithium metal. The lower surface energy and higher migration energy of a lithium metal make it have a higher surface diffusion barrier, which thermodynamically explains the reason why a lithium metal is more likely to form dendrites [30]. In addition, the nucleation of lithium dendrites is also affected by the space electric field and SEI. When lithium ions are deposited at a high speed, the concentration of anions on the electrode surface is low, resulting in a strong spatial electric field at the interface between the electrode and the electrolyte, accelerating the dendritic growth of lithium ions [31–33]. At the same time, since the volume of a lithium metal will change during the process of electroplating and stripping, this will inevitably bring about changes in the stress by SEI. In this process, the SEI is easily broken, which exposes fresh lithium, and lithium ions tend to deposit in cracks with lower resistance and thinner SEI to form lithium dendrites. After nucleation, lithium dendrites will continue to grow. Unlike the nucleation stage, which is affected by thermodynamics and interface properties, the growth of lithium dendrites is mainly controlled by external factors such as electrolyte and current density, charging capacity, temperature, and pressure [34–39].

In addition to the reaction consumption of lithium metal with electrolytes and LiPSs, the formation of dendrites is also an important reason for the degradation of Li-S battery capacity and the increase of polarization. When the lithium dendrite is connected to the collector or bulk lithium, it can still contribute to the battery's capacity through the charge-discharge reaction of obtaining and losing electrons. But when the SEI becomes thicker and looser, the dendrites will break and fall off the lithium electrode. This part of the lithium metal loses the ability to obtain and lose electrons, and becomes an invalid part that is freed in the battery, which is called "dead lithium" in the literature [40–42]. The existence of "dead lithium" not only greatly reduces the capacity of the battery, but also affects the conduction of lithium ions in the electrolyte and continues to consume electrolyte components through reduction reactions. When "dead lithium" and SEI residues accumulate on the surface of lithium metal, it will further restrict mass transfer and promote the rapid growth of lithium dendrites in the voids [43].

6.3.3 Volume Changes

Unlike the intercalated graphite anode, the lithium metal anode exhibits electroplating and stripping during battery discharging and charging processes. Compared with the volume change of a graphite anode and silicon anode (graphite: 10%, silicon: 300%), the volume of lithium metal anode shows a process from scratch [8]. The result of this is a huge change in the volume and pressure of the anode of the battery. During the charging process of the battery, lithium ions are electroplated onto the negative electrode current collector, and a larger capacity means a thicker plating layer. Correspondingly, during the discharge of the battery, lithium ions are stripped from the anode. In such a cycle, the negative electrode undergoes a change in volume, which not only causes a change in the internal pressure of the battery, but also ruptures and regenerates of the fragile SEI. As mentioned earlier, it will eventually lead to a series of problems such as electrolyte depletion and dendrite growth.

6.3.4 The Effect of Shuttle on Anode

LiPSs, an intermediate product formed during the charging and discharging of Li-S batteries, can be dissolved in the electrolyte. The specific mechanism has been introduced in Chapter 5. LiPSs will be dissolved in the electrolyte and shuttle to the anode to react with lithium metal. After being reduced, it will form low-level LiPSs and shuttle back to the positive electrode [44]. The shuttle behavior of LiPSs seriously affects the coulombic efficiency and battery capacity of Li-S batteries. It is worth noting that the addition of LiPSs changes the composition of SEI, and also makes the chemical composition and structure of the anode interface more complicated [24,45,46]. When the LiPSs shuttles to the side of the lithium metal electrode, it will react with the lithium metal and be reduced to Li_2S. Li_2S with better chemical stability is more likely to become a component of the inner layer of SEI. Although studies have shown that Li_2S with a higher ionic conductivity has a positive effect on the stability of SEI [47], the reaction and deposition of the positive electrode active material on the anode seriously affect the capacity of the Li-S battery.

6.4 STRATEGIES

6.4.1 ELECTROLYTE DESIGN

As mentioned earlier, dendrite growth is affected by multiple factors such as current density, SEI, and thermodynamics. In Li-S batteries, the presence of sulfur makes the internal environment of the battery more complicated. And in the battery system, these effects are often not a single existence, but mutual restriction and influence. In order to obtain a stable and safe lithium metal anode, research in various directions has been proposed for a long time. However, these strategies often only made improvements and enhancements to one of the factors, and did not perfectly solve the problems of lithium metal anodes.

6.4.1.1 Electrolyte Additives

As an important component in the electrolyte, a very small amount of additives can stabilize the SEI, adjust the solubility of LiPSs, and change the solvation structure. Additives acting on lithium metal anodes can be divided into two categories. The first category is the film-forming additives that react with lithium metal and participate in the construction of SEI, such as $LiNO_3$ [46], fluoroethylene carbonate [48,49], lithium bis(oxalato)borate (LiBOB) [50], and so on. This type of additive forms an in-situ SEI protective layer by reacting with lithium metal, thereby preventing the lithium metal from further reacting with the electrolyte. $LiNO_3$, as the most common additive in ether electrolyte, can spontaneously react to form relatively stable Li_xNO_y compounds in Li-S batteries to participate in the construction of SEI. Studies have shown that $LiNO_3$ can also change the solvation structure of LiTFSI in the electrolyte, allowing the anion $TFSI^-$ to be separated from the solvation shell to participate in the construction of SEI, thereby forming a more stable LiF-rich SEI layer [51]. In addition, when LiPSs and $LiNO_3$ exist at the same time, LiPSs will be oxidized to Li_xSO_y, which makes the composition of SEI more complicated (Figure 6.4) [26,46]. The other type is lithium ion plating additives, such as alkali metal ions and halide ions. Such additives do not participate in the reaction during the battery cycle, but stabilize the lithium metal anode by changing the electroplating and diffusion behavior of lithium ions. Cs^+ as a typical lithium ion plating additive has also been found to inhibit the growth of lithium dendrites (Figure 6.5) [52]. At low concentrations, Cs^+ has a lower effective reduction potential than Li^+. During the lithium deposition process, Cs^+ will preferentially gather at the tip of the dendrite to form an electrostatic shielding layer, thereby inhibiting the further growth of the dendrite.

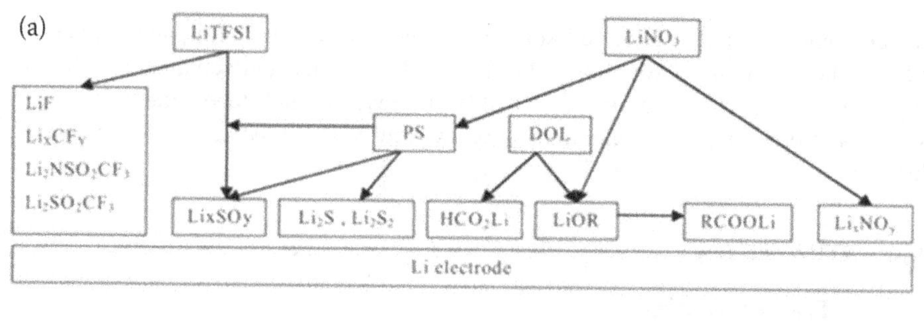

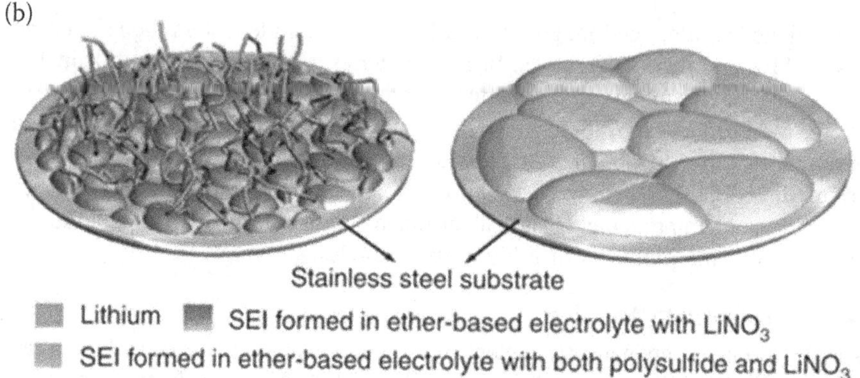

FIGURE 6.4 (a) The chemical composition of the anode surface of a Li-S battery containing $LiNO_3$ [26], (b) schematic illustration of the synergistic effect of LiNO3 and LiPSs [46].

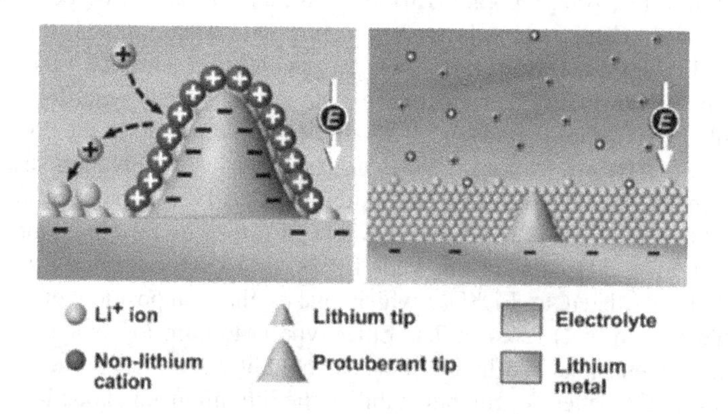

FIGURE 6.5 Schematic illustration of Li deposition process based on the self-healing electrostatic shield mechanism [52].

6.4.1.2 Superconcentrated Electrolyte

Limited by conditions such as solubility, ion conductivity, viscosity, and cost, the concentration of lithium salt in a general liquid electrolyte is ~1 M. A super concentrated electrolyte refers to an electrolyte system with a lithium salt concentration exceeding 3 M. Studies have shown that as the concentration of lithium salt increases, the ion conductivity will decrease and the viscosity of the electrolyte will increase, which greatly reduces its ion conductivity [14,53]. However, as the number of lithium ion migration increases, the solubility of LiPSs decreases, making the SEI on the lithium metal electrode in the high-salt electrolyte thinner and more stable. The high-salt electrolyte can effectively inhibit the formation and growth of lithium dendrites and reduce the occurrence of side reactions.

6.4.2 Current Collector Design

Different from the intercalation reaction of graphite anode, the electroplating and stripping of lithium metal on the current collector will bring about huge volume changes. By changing the shape of the current collector, the effect of dispersing the current density and limiting the volume change can be achieved. Three-position current collectors can be simply divided into two categories: metal (Figure 6.6(a)) and non-metal (Figure 6.6(b–d)). Among them, the metal current collector is mainly Cu [54–56] and Ni [57]. The non-metallic current collector is mainly carbon [58–62]. In addition, non-conductive materials such as glass fibers [63] and polymer fibers [64] are also used as the current collectors to achieve the purpose of restricting the deposition of lithium metal. The results show that in a current collector with a large specific surface area, lithium metal can circulate with a more uniform and smaller current density, which has a good effect on dendrite suppression and volume adaptability. However, it is worth noting that a higher specific surface area means more lithium is involved in the formation of a primary SEI, causing more capacity loss. In addition, the three-dimensional current collector often has a larger mass and volume, which reduces the overall energy density of the battery.

6.4.3 Artificial SEI

The composition and structure of the original SEI cannot match the environment and long-term use requirements of the negative electrode in the Li-S battery. Therefore, an artificial SEI controlled by materials and processes is constructed through in-situ and ex-situ methods. The composition and function of artificial SEI are similar to thin, solid electrolytes, which play the role of ion conduction and inhibit side reactions. In-situ SEI (Figure 6.7(a,b)) mainly forms a stable protective layer on the surface of a lithium metal through physical, chemical, and electrochemical reactions. Generally, the method of stabilizing SEI by changing the electrolyte composition is in-situ SEI. Further, the lithium metal negative electrode can also form in-situ SEI by a gas phase reaction [65,66], liquid phase reaction [67,68], or solid phase reaction.

The ex-situ SEI (Figure 6.7(c,d)) is mainly composed of polymers [69–71] or inorganic nanomaterials [72,73]. A general coating method is difficult to obtain a sufficiently thin interface layer, so atomic/molecular layer deposition [74] and magnetron sputtering [75] are used to prepare thinner and more stable SEIs. The artificial SEI layer can effectively inhibit dendrites' growth and reduce the efficacy of dendrite growth and reduction of sub-reactions from different mechanisms.

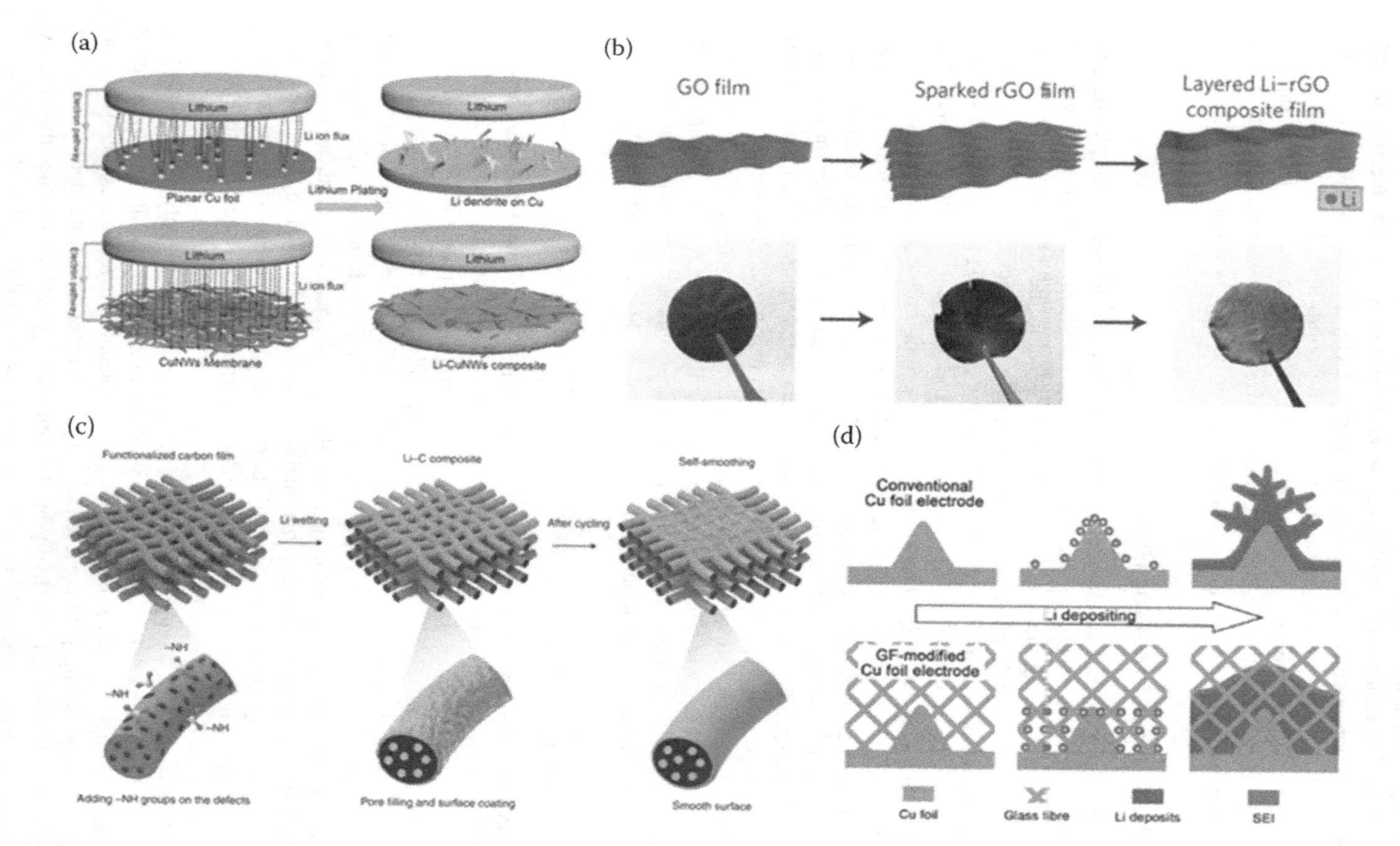

FIGURE 6.6 Schematic illustration of various current collector designs [54,60,62,63].

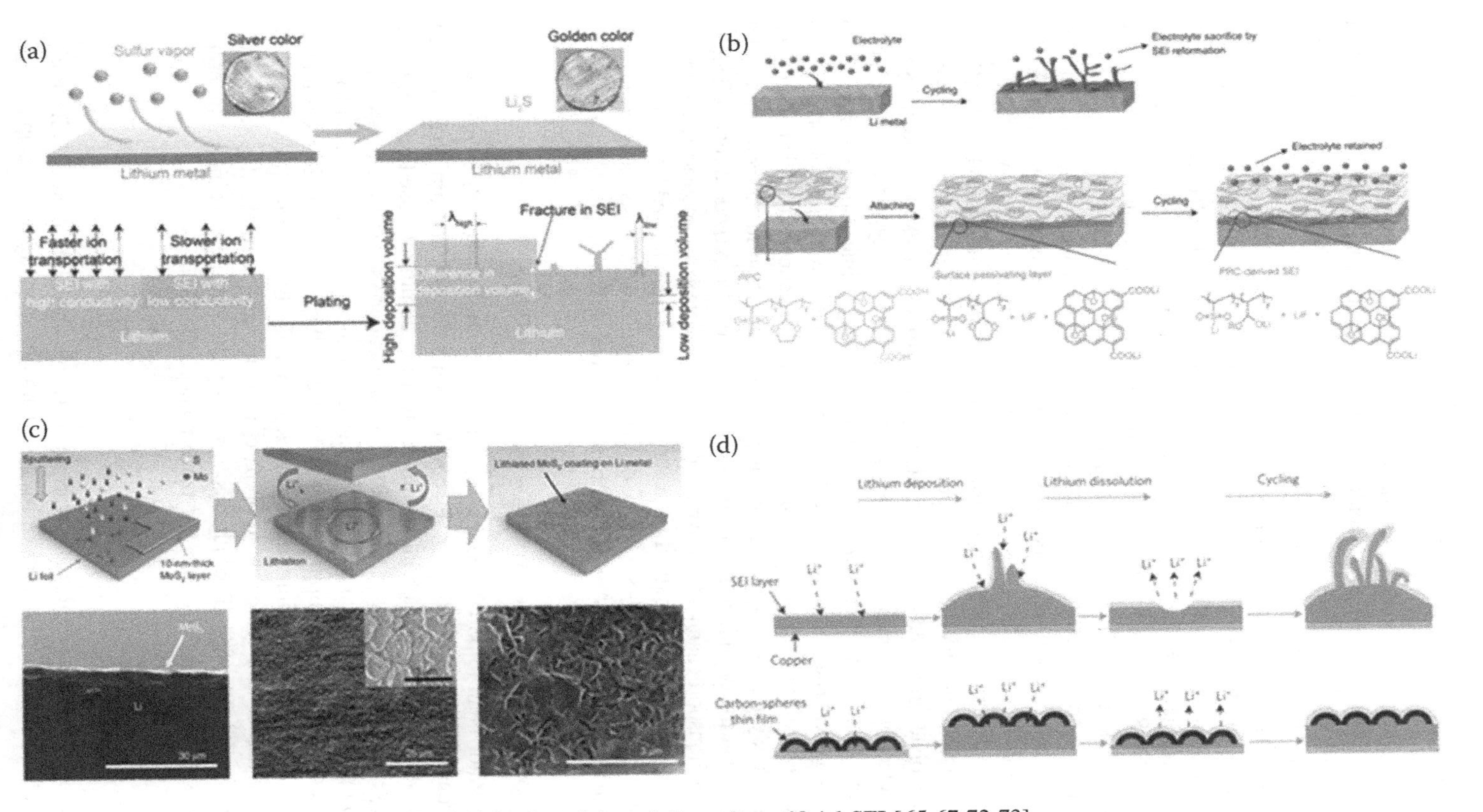

FIGURE 6.7 Schematic illustration of (a and b) in-situand (c and d) ex-situ artificial SEI [65,67,72,73].

6.4.4 Solid-State Electrolytes

Because a solid electrolyte does not contain flammable organic electrolyte components, it has extremely high safety and stability. At the same time, solid electrolytes with high modulus and stable interfaces are considered to be able to fundamentally solve the problems of dendrite growth and SEI instability. Solid electrolytes are mainly divided into solid polymer electrolytes (SPEs), inorganic ceramic electrolytes (ICEs), and solid composite electrolytes (SCEs). Among them, SPEs have good flexibility and workability, and their sticky surface also has good wettability with the electrodes. However, due to the limitation of crystallinity at room temperature, the ionic conductivity of SPEs is low ($<10^{-4}$ S cm^{-1}), and its narrow electrochemical window cannot be adapted to high-voltage electrodes. In contrast, ICEs have an ionic conductivity (10^{-3}–10^{-2} S cm^{-1}) comparable to that of liquid electrolytes, a wide electrochemical window, and high mechanical strength [76–78]. The most common ICEs are NASICON, garnet, perovskite, LISICON, LiPON, Li_3N, sulfide, garnet, etc. The SPEs mainly include polyethylene oxide (PEO), polyacrylonitrile (PAN), polymethylmethacrylate (PMMA), polyvinylpyrrolidone (PVP), polyvinylidene fluoride (PVDF), and so on. The composite solid electrolyte composed of organic and inorganic hybrids can exert the advantages of the two electrolytes, while ensuring high electrical conductivity, while having good mechanical strength and chemical/electrochemical stability, while also maintaining good interface contact. But it will also weaken the respective advantages of organic and inorganic solid electrolytes. It is worth noting that studies believe that it is not only the electrolyte body that participates in the lithium ion conduction in the composite solid electrolyte, but the organic-inorganic phase interface will also provide a more effective lithium ion transport channel [79]. Although the solid electrolyte is considered to match the lithium metal electrode, in the Li-S battery, the study is also limited by the interfacial contactability. The SPEs [80–82] and ICEs [83–85] applicable to Li-S batteries are also lacking in the in-depth study of lithium metal negative dendrites.

6.4.5 Membrane Modification

The separator is an important component of the lithium battery, which plays a role in preventing short-circuiting of electrodes and allowing lithium ions to pass through. At present, commercial separators that are commonly used are single or multi-layer separators made of PP or PE. For Li-S battery separators, in addition to good ion permeability, thermal stability, and mechanical flexibility, they also need to have a certain dendrite suppression ability and LiPSs blocking ability. Currently, separators designed for lithium metal anodes are mainly divided into two categories. One is the use of new materials to prepare high-strength, high-stability, and functional separators [86–88]. The other is to modify or coat the existing commercial separators to make them have better performance [89–93].

6.4.6 Alternatives

In addition to lithium metal, many carbon materials and alloy materials have also been extensively studied as anodes for Li-S batteries. Among them, graphite, as a mature anode material for lithium-ion batteries, has also been tried in Li-S batteries. However, the lower specific energy of graphite and the higher electrode potential relative to lithium metal are contrary to the original intention of the high specific energy of Li-S batteries. Therefore, it has not been promoted and researched vigorously. Beside carbon materials, silicon has also been tried as a representative of high specific energy battery anode materials. It is worth noting that these intercalation or alloy anode materials do not contain lithium themselves, and a cathode with lithium needs to be matched during the battery assembly process.

6.4.6.1 Carbon

Since it was developed by SONY in 1901 as an intercalation anode material for lithium-ion batteries [94], carbon-based materials, such as graphite, hard carbon, and soft carbon, have become the best candidates for lithium-ion battery anodes [45]. Graphite has been researched and applied for more than 180 years with its excellent electrochemical stability [13,95]. As a mature anode material, graphite has also been tried as an anode for Li-S batteries [96]. Although compared to lithium-ion batteries, Li-S batteries with graphite as the anode still have a high energy density. But this is far less than lithium metal, which greatly weakens the advantages of Li-S batteries.

6.4.6.2 Si

Si and carbon belong to the same group of elements, but unlike graphite's intercalation lithium storage, silicon and lithium form an alloy. Each silicon atom can hold 4.4 lithium atoms to form a $Li_{22}Si_5$ alloy, which makes the silicon anode have a considerable specific energy (4200 mAh g^{-1}) [97,98]. However, silicon as an anode has problems such as volume expansion and electrode pulverization. Before and after lithium storage, the volume of silicon will expand to 420% [99]. In a Li-S battery, the volume change will greatly reduce the interface and overall stability.

6.4.6.3 Else

There are many other materials used as anodes for Li-S batteries, such as Sn [100,101], Li-B alloy [102,103], and so on. Although these attempts have brought new opportunities for the advancement of Li-S batteries, they have not been further studied and applied due to the nature of the materials themselves.

6.5 SUMMARY

Many materials and strategies have been proposed to solve the problem of the negative electrode side in Li-S batteries, but there is still no method to achieve industrialization. More importantly, in cheap and high specific energy Li-S batteries, expensive source materials and materials with large mass and volume will weaken their advantages to a certain extent. Therefore, in the research of Li-S batteries, progress and cooperation of all parts are indispensable.

REFERENCES

[1] Danuta H, Juliusz U. Electric dry cells and storage batteries. [P]. US, US3043896 A, 1962.

[2] Rao M. Organic electrolyte cells[D]. *US*, 1968.

[3] Nole D A, Moss V. Battery employing lithium - sulphur electrodes with non-aqueous electrolyte [P]. US, US3532543A, 1970.

[4] Tarascon J M, Armand M. Issues and challenges facing rechargeable lithium batteries [J]. *Nature*, 2001, 414(6861): 359–367.

[5] Xu W, Wang J L, Ding F, et al. Lithium metal anodes for rechargeable batteries [J]. *Energy & Environmental Science*, 2014, 7(2): 513–537.

[6] Lin D, Liu Y, Cui Y. Reviving the lithium metal anode for high-energy batteries [J]. *Nature Nanotechnology*, 2017, 12(3): 194–206.

[7] Albertus P, Babinec S, Litzelman S, et al. Status and challenges in enabling the lithium metal electrode for high-energy and low-cost rechargeable batteries [J]. *Nature Energy*, 2018, 3(1): 16–21.

[8] Tao T, Lu S, Fan Y, et al. Anode improvement in rechargeable lithium-sulfur batteries [J]. *Advanced Materials*, 2017, 29(48).

[9] Zhao M, Li B-Q, Zhang X-Q, et al. A perspective toward practical lithium-sulfur batteries [J]. ACS Central *Science*, 2020, 6(7): 1095–1104.

[10] Bruce P G, Freunberger S A, Hardwick L J, et al. Li-O-2 and Li-S batteries with high energy storage [J]. *Nature Materials*, 2012, 11(1): 19–29.

[11] Peng H J, Huang J Q, Cheng X B, et al. Review on high-loading and high-energy lithium-sulfur batteries [J]. Advanced Energy *Materials*, 2017, 7(24).

[12] Xie J, Lu Y C. A retrospective on lithium-ion batteries [J]. *Nature Communications*, 2020, 11(1): 2499.

[13] Zhang H, Li C, Eshetu G G, et al. From solid-solution electrodes and the rocking-chair concept to today's batteries [J]. *Angewandte Chemie International Edition*, 2020, 59(2): 534–538.

[14] Suo L, Hu Y-S, Li H, et al. A new class of solvent-in-salt electrolyte for high-energy rechargeable metallic lithium batteries [J]. Nature *Communications*, 2013, 4.

[15] Peled E, Sternberg Y, Gorenshtein A, et al. Lithium-sulfur battery - evaluation of dioxolane-based electrolytes [J]. *Journal of the Electrochemical Society*, 1989, 136(6): 1621–1625.

[16] Kolosnitsyn V S, Karaseva E V. Lithium-sulfur batteries: Problems and solutions [J]. *Russian Journal of Electrochemistry*, 2008, 44(5): 506–509.

[17] Yim T, Park M-S, Yu J-S, et al. Effect of chemical reactivity of polysulfide toward carbonate-based electrolyte on the electrochemical performance of Li-S batteries [J]. *Electrochimica Acta*, 2013, 107: 454–460.

[18] Gao J, Lowe M A, Kiya Y, et al. Effects of liquid electrolytes on the charge-discharge performance of rechargeable lithium/sulfur batteries: Electrochemical and in-situ X-ray absorption spectroscopic studies [J]. *Journal of Physical Chemistry C*, 2011, 115(50): 25132–25137.

[19] Barchasz C, Lepretre J-C, Patoux S, et al. Electrochemical properties of ether-based electrolytes for lithium/sulfur rechargeable batteries [J]. *Electrochimica Acta*, 2013, 89: 737–743.

[20] Monroe C, Newman J. Dendrite growth in lithium/polymer systems - A propagation model for liquid electrolytes under galvanostatic conditions [J]. *Journal of the Electrochemical Society*, 2003, 150(10): A1377–A1384.

[21] Cohen Y S, Cohen Y, Aurbach D. Micromorphological studies of lithium electrodes in alkyl carbonate solutions using in situ atomic force microscopy [J]. *The Journal of Physical Chemistry B*, 2000, 104(51): 12282–12291.

[22] Paled E. The electrochemical behavior of alkali and alkaline earth metals in non-aqueous battery systems—The solid electrolyte interphase model [J]. *Journal of The Electrochemical Society*, 1979, 126(12).

[23] Wu H, Jia H, Wang C, et al. Recent progress in understanding solid electrolyte interphase on lithium metal anodes [J]. *Advanced Energy Materials*, 2020, 11(5).

[24] Yu X, Manthiram A. Electrode-electrolyte interfaces in lithium-sulfur batteries with liquid or inorganic solid electrolytes [J]. *Accounts of Chemical Research*, 2017, 50(11): 2653–2660.

[25] Zhang S S. Role of LiNO3 in rechargeable lithium/sulfur battery [J]. *Electrochimica Acta*, 2012, 70: 344–348.

[26] Aurbach D, Pollak E, Elazari R, et al. On the surface chemical aspects of very high energy density, rechargeable Li-sulfur batteries [J]. *Journal of the Electrochemical Society*, 2009, 156(8): A694–A702.

[27] Chen Hou J H, Pan Liu, Chuchu Yang, et al. Operando observations of SEI film evolution by mass-sensitive scanning transmission electron microscopy [J]. *Advanced Energy Materials*, 2019, 9(45).

[28] Jackle M, Gross A. Microscopic properties of lithium, sodium, and magnesium battery anode materials related to possible dendrite growth [J]. *Journal of Chemical Physics*, 2014, 141(17): 174710.

[29] Ling C, Banerjee D, Matsui M. Study of the electrochemical deposition of Mg in the atomic level: Why it prefers the non-dendritic morphology [J]. *Electrochimica Acta*, 2012, 76: 270–274.

[30] Cheng X B, Zhang R, Zhao C Z, et al. Toward safe lithium metal anode in rechargeable batteries: A review [J]. *Chemical Reviews*, 2017, 117(15): 10403–10473.

[31] Fleury V, Chazalviel J N, Rosso M, et al. The role of the anions in the growth speed of fractal electrodeposits [J]. *Journal of Electroanalytical Chemistry*, 1990, 290(1–2): 249–255.

[32] Chazalviel J N. Electrochemical aspects of the generation of ramified metallic electrodeposits [J]. *Physical Review A*, 1990, 42(12): 7355–7367.

[33] Brissot C, Rosso M, Chazalviel J N, et al. In situ study of dendritic growth in lithium/PEO-salt/lithium cells [J]. *Electrochimica Acta*, 1998, 43(10–11): 1569–1574.

[34] Zhang R, Cheng X-B, Zhao C-Z, et al. Conductive nanostructured scaffolds render low local current density to inhibit lithium dendrite growth [J]. *Advanced Materials*, 2016, 28(11): 2155–2162.

[35] Love C T, Baturina O A, Swider-Lyons K E. Observation of lithium dendrites at ambient temperature and below [J]. *ECS Electrochemistry Letters*, 2015, 4(2): A24–A27.

[36] Ota H, Shima K, Ue M, et al. Effect of vinylene carbonate as additive to electrolyte for lithium metal anode [J]. *Electrochimica Acta*, 2004, 49(4): 565–572.

[37] Akolkar R. Modeling dendrite growth during lithium electrodeposition at sub-ambient temperature [J]. *Journal of Power Sources*, 2014, 246: 84–89.

[38] Seong I W, Hong C H, Kim B K, et al. The effects of current density and amount of discharge on dendrite formation in the lithium powder anode electrode [J]. *Journal of Power Sources*, 2008, 178(2): 769–773.

[39] Barai P, Higa K, Srinivasan V. Effect of initial state of lithium on the propensity for dendrite formation: A theoretical study [J]. *Journal of the Electrochemical Society*, 2017, 164(2): A180–A189.

[40] Yoshimatsu I, Hirai T, Yamaki J. Lithium electrode morphology during cycling in lithium cells [J]. *Journal of the Electrochemical Society*, 1988, 135(10): 2422–2427.

[41] Steiger J, Kramer D, Moenig R. Microscopic observations of the formation, growth and shrinkage of lithium moss during electrodeposition and dissolution [J]. *Electrochimica Acta*, 2014, 136: 529–536.
[42] Arakawa M, Tobishima S, Nemoto Y, et al. Lithium electrode cycleability and morphology dependence on current-density [J]. *Journal of Power Sources*, 1993, 43(1–3): 27–35.
[43] Zhang J-G, Xu W, Henderson WA. Lithium metal anodes and rechargeable lithium metal batteries [M]. 2017, 249: 25–26.
[44] Cheon S E, Ko K S, Cho J H, et al. Rechargeable lithium sulfur battery - II. Rate capability and cycle characteristics [J]. *Journal of the Electrochemical Society*, 2003, 150(6): A800–A805.
[45] Cao R, Xu W, Lv D, et al. Anodes for rechargeable lithium-sulfur batteries [J]. Advanced Energy *Materials*, 2015, 5(16).
[46] Li W, Yao H, Yan K, et al. The synergetic effect of lithium polysulfide and lithium nitrate to prevent lithium dendrite growth [J]. *Nature Communications*, 2015, 6.
[47] Xu R, Belharouak I, Li J C M, et al. Role of polysulfides in self-healing lithium-sulfur batteries [J]. Advanced Energy *Materials*, 2013, 3(7): 833–838.
[48] Zhang X-Q, Cheng X-B, Chen X, et al. Fluoroethylene carbonate additives to render uniform Li deposits in lithium metal batteries [J]. *Advanced Functional Materials*, 2017, 27(10).
[49] Liu Q-C, Xu J-J, Yuan S, et al. Artificial protection film on lithium metal anode toward long-cycle-life lithium-oxygen batteries [J]. *Advanced Materials*, 2015, 27(35): 5241–5247.
[50] Xiong S, Kai X, Hong X, et al. Effect of LiBOB as additive on electrochemical properties of lithium–sulfur batteries [J]. *Ionics*, 2011, 18(3): 249–254.
[51] Fu J, Ji X, Chen J, et al. Lithium nitrate regulated sulfone electrolytes for lithium metal batteries [J]. *Angewandte Chemie International Edition*, 2020, 59(49): 22194–22201.
[52] Ding F, Xu W, Graff G L, et al. Dendrite-free lithium deposition via self-healing electrostatic shield mechanism [J]. *Journal of the American Chemical Society*, 2013, 135(11): 4450–4456.
[53] Qian J, Henderson W A, Xu W, et al. High rate and stable cycling of lithium metal anode [J]. *Nature Communications*, 2015, 6.
[54] Lu L L, Ge J, Yang J-N, et al. Free-standing copper nanowire network current collector for improving lithium anode performance [J]. *Nano Letters*, 2016, 16(7): 4431–4437.
[55] Yun Q, He Y-B, Lv W, et al. Chemical dealloying derived 3D porous current collector for Li metal anodes [J]. *Advanced Materials*, 2016, 28(32): 6932–+.
[56] Li Q, Zhu S, Lu Y. 3D porous cu current collector/Li-metal composite anode for stable lithium-metal batteries [J]. *Advanced Functional Materials*, 2017, 27(18).
[57] Chi S-S, Liu Y, Song W-L, et al. Prestoring lithium into stable 3D nickel foam host as dendrite-free lithium metal anode [J]. *Advanced Functional Materials*, 2017, 27(24).
[58] Yang C, Yao Y, He S, et al. Ultrafine silver nanoparticles for seeded lithium deposition toward stable lithium metal anode [J]. *Advanced Materials*, 2017, 29(38).
[59] Jin C, Sheng O, Luo J, et al. 3D lithium metal embedded within lithiophilic porous matrix for stable lithium metal batteries [J]. *Nano Energy*, 2017, 37: 177–186.
[60] Lin D, Liu Y, Liang Z, et al. Layered reduced graphene oxide with nanoscale interlayer gaps as a stable host for lithium metal anodes [J]. *Nature Nanotechnology*, 2016, 11(7): 626–+.

[61] Wang Z H, Wang X D, Sun W, et al. Dendrite-free lithium metal anodes in high performance lithium-sulfur batteries with bifunctional carbon nanofiber interlayers [J]. *Electrochimica Acta*, 2017, 252: 127–137.
[62] Niu C J, Pan H L, Xu W, et al. Self-smoothing anode for achieving high-energy lithium metal batteries under realistic conditions [J]. *Nature Nanotechnology*, 2019, 14(6): 594–+.
[63] Cheng X-B, Hou T-Z, Zhang R, et al. Dendrite-free lithium deposition induced by uniformly distributed lithium ions for efficient lithium metal batteries [J]. *Advanced Materials*, 2016, 28(15): 2888–2895.
[64] Liang Z, Zheng G, Liu C, et al. Polymer nanofiber-guided uniform lithium deposition for battery electrodes [J]. *Nano Letters*, 2015, 15(5): 2910–2916.
[65] Chen H, Pei A, Lin D, et al. Uniform high ionic conducting lithium sulfide protection layer for stable lithium metal anode [J]. *Advanced Energy Materials*, 2019, 9(22).
[66] Zhao J, Liao L, Shi F, et al. Surface fluorination of reactive battery anode materials for enhanced stability [J]. *Journal of the American Chemical Society*, 2017, 139(33): 11550–11558.
[67] Gao Y, Yan Z F, Gray J L, et al. Polymer-inorganic solid-electrolyte interphase for stable lithium metal batteries under lean electrolyte conditions [J]. *Nature Materials*, 2019, 18(4): 384–+.
[68] Gao Y, Zhao Y M, Li Y G C, et al. Interfacial chemistry regulation via a skin-grafting strategy enables high-performance lithium-metal batteries [J]. *Journal of the American Chemical Society*, 2017, 139(43): 15288–15291.
[69] Liu K, Pei A, Lee H R, et al. Lithium metal anodes with an adaptive "solid-liquid" interfacial protective layer [J]. *Journal of the American Chemical Society*, 2017, 139(13): 4815–4820.
[70] Zhu B, Jin Y, Hu X, et al. Poly(dimethylsiloxane) thin film as a stable interfacial layer for high-performance lithium-metal battery anodes [J]. *Advanced Materials*, 2017, 29(2).
[71] Liu Y Y, Lin D C, Yuen P Y, et al. An artificial solid electrolyte interphase with high Li-Ion conductivity, mechanical strength, and flexibility for stable lithium metal anodes [J]. *Advanced Materials*, 2017, 29(10).
[72] Zheng G, Lee S W, Liang Z, et al. Interconnected hollow carbon nanospheres for stable lithium metal anodes [J]. *Nature Nanotechnology*, 2014, 9(8): 618–623.
[73] Cha E, Patel M D, Park J, et al. 2D MoS2 as an efficient protective layer for lithium metal anodes in high-performance Li-S batteries [J]. *Nature Nanotechnology*, 2018, 13(4): 337–+.
[74] Jones J-P, Hennessy J, Billings K J, et al. Communication-atomic layer deposition of aluminum fluoride for lithium metal anodes [J]. *Journal of the Electrochemical Society*, 2020, 167(6).
[75] Wang L, Zhang L, Wang Q, et al. Long lifespan lithium metal anodes enabled by Al2O3 sputter coating [J]. *Energy Storage Materials*, 2018, 10: 16–23.
[76] Zhou D, Shanmukaraj D, Tkacheva A, et al. Polymer electrolytes for lithium-based batteries: advances and prospects [J]. *Chem*, 2019, 5(9): 2326–2352.
[77] Li S, Zhang S-Q, Shen L, et al. Progress and perspective of ceramic/polymer composite solid electrolytes for lithium batteries [J]. *Advanced Science*, 2020, 7(5).
[78] Gao Z, Sun H, Fu L, et al. Promises, challenges, and recent progress of inorganic solid-state electrolytes for all-solid-state lithium batteries [J]. *Advanced Materials*, 2018, 30(17).
[79] Hu C, Shen Y, Shen M, et al. Superionic conductors via bulk interfacial conduction [J]. *Journal of the American Chemical Society*, 2020, 142(42): 18035–18041.

[80] Yu X G, Xie J Y, Yang J, et al. All solid-state rechargeable lithium cells based on nano-sulfur composite cathodes [J]. *Journal of Power Sources*, 2004, 132(1–2): 181–186.

[81] Hassoun J, Scrosati B. A high-performance polymer Tin sulfur lithium ion battery [J]. *Angewandte Chemie International Edition*, 2010, 49(13): 2371–2374.

[82] Marmorstein D, Yu T H, Striebel K A, et al. Electrochemical performance of lithium/sulfur cells with three different polymer electrolytes [J]. *Journal of Power Sources*, 2000, 89(2): 219–226.

[83] Lin Z, Liu Z C, Dudney N J, et al. Lithium superionic sulfide cathode for all-solid lithium-sulfur batteries [J]. *ACS Nano*, 2013, 7(3): 2829–2833.

[84] Nagao M, Imade Y, Narisawa H, et al. All-solid-state Li-sulfur batteries with mesoporous electrode and thio-LISICON solid electrolyte [J]. *Journal of Power Sources*, 2013, 222: 237–242.

[85] Nagao M, Hayashi A, Tatsumisago M. High-capacity Li2S-nanocarbon composite electrode for all-solid-state rechargeable lithium batteries [J]. *Journal of Materials Chemistry*, 2012, 22(19): 10015–10020.

[86] Lin D C, Zhuo D, Liu Y Y, et al. All-integrated bifunctional separator for Li dendrite detection via novel solution synthesis of a thermostable polyimide separator [J]. *Journal of the American Chemical Society*, 2016, 138(34): 11044–11050.

[87] Hao X M, Zhu J, Jiang X, et al. Ultrastrong polyoxyzole nanofiber membranes for dendrite-proof and heat-resistant battery separators [J]. *Nano Letters*, 2016, 16(5): 2981–2987.

[88] Li J, Niu X H, Song J F, et al. Harvesting vapor by hygroscopic acid to create pore: Morphology, crystallinity and performance of poly (ether ether ketone) lithium ion battery separator [J]. *Journal of Membrane Science*, 2019, 577: 1–11.

[89] Ryou M-H, Lee D J, Lee J-N, et al. Excellent cycle life of lithium-metal anodes in lithium-ion batteries with mussel-inspired polydopamine-coated separators [J]. Advanced Energy *Materials*, 2012, 2(6): 645–650.

[90] Li C F, Liu S H, Shi C G, et al. Two-dimensional molecular brush-functionalized porous bilayer composite separators toward ultrastable high-current density lithium metal anodes [J]. *Nature Communications*, 2019, 10.

[91] Chi M M, Shi L Y, Wang Z Y, et al. Excellent rate capability and cycle life of Li metal batteries with ZrO2/POSS multilayer-assembled PE separators [J]. *Nano Energy*, 2016, 28: 1–11.

[92] Yang F, Sun W, Bai Y, et al. Rational design of sandwich-like "gel–liquid–gel" electrolytes for dendrite-free lithium metal batteries [J]. *Industrial & Engineering Chemistry Research*, 2020, 59(32): 14207–14216.

[93] Liu H, Peng D, Xu T, et al. Porous conductive interlayer for dendrite-free lithium metal battery [J]. *Journal of Energy Chemistry*, 2021, 53: 412–418.

[94] Brandt K. Historical development of secondary lithium batteries [J]. *Solid State Ionics*, 1994, 69(3–4): 173–183.

[95] Schafhaeutl C. Ueber die Verbindungen des Kohlenstoffes mit Silicium, Eisen und anderen Metallen, welche die verschiedenen Gallungen von Roheisen, Stahl und Schmiedeeisen bilden [J]. *Journal für Praktische Chemie*, 1840.

[96] Zheng S Y, Chen Y, Xu Y H, et al. In situ formed lithium sulfide/microporous carbon cathodes for lithium-ion batteries [J]. *ACS Nano*, 2013, 7(12): 10995–11003.

[97] Su X, Wu Q L, Li J C, et al. Silicon-based nanomaterials for lithium-ion batteries: A review [J]. *Advanced Energy Materials*, 2014, 4(1).

[98] Kasavajjula U, Wang C S, Appleby A J. Nano- and bulk-silicon-based insertion anodes for lithium-ion secondary cells [J]. *Journal of Power Sources*, 2007, 163(2): 1003–1039.

[99] Wu H, Cui Y. Designing nanostructured Si anodes for high energy lithium ion batteries [J]. *Nano Today*, 2012, 7(5): 414–429.
[100] Duan B C, Wang W K, Wang A B, et al. A new lithium secondary battery system: The sulfur/lithium-ion battery [J]. *Journal of Materials Chemistry A*, 2014, 2(2): 308–314.
[101] Hassoun J, Sun Y K, Scrosati B. Rechargeable lithium sulfide electrode for a polymer tin/sulfur lithium-ion battery [J]. *Journal of Power Sources*, 2011, 196(1): 343–348.
[102] Zhang X L, Wang W K, Wang A B, et al. Improved cycle stability and high security of Li-B alloy anode for lithium-sulfur battery [J]. *Journal of Materials Chemistry A*, 2014, 2(30): 11660–11665.
[103] Duan B C, Wang W K, Zhao H L, et al. Li-B alloy as anode material for lithium/sulfur battery [J]. *ECS Electrochemistry Letters*, 2013, 2(6): A47–A51.

7 Interlayer of Lithium-Sulfur Batteries

Zhenhua Wang
School of Chemistry and Chemical Engineering, Beijing Institute of Technology, Beijing, People's Republic of China

CONTENTS

7.1 INTRODUCTION

Separator plays an important role in a Li-S battery. It not only prevents the direct contact of the anode and cathode from causing a short circuit in the battery, but also has a rich pore structure that can be infiltrated by the electrolyte and allows lithium ions to migrate. However, these pore structures also provide convenience for the shuttle of LiPSs since sulfur will be transformed into a series of soluble LiPSs intermediates (L_2S_X, $x > 2$) during the discharge process.

In order to alleviate the shuttle effect, the main research of Li-S batteries was still focused on the modification of the positive electrode and the electrolyte at that time. But the strategy on the cathode has an inherent limiting factor in fully utilizing the merits of advanced lithium batteries. The weight of the unique host materials actually hampers the energy density of the sulfur cathode, as evaluated in the entire battery system. To fundamentally overcome the inherent limiting factors of cathodes, efforts should focus beyond the cathode side with the design transition from the closed structure to the open structure. The concept of the "interlayer", proposed by the Manthriam's research group for the first time [1], broke this limitation in 2012. They introduced porous carbon paper between the positive electrode and the separator to build a conductive network in the positive electrode area. The carbon paper acts as an upper current collector to inhibit the shuttle of LiPSs and improve the utilization of sulfur species. This idea expands the focus of researchers from the "inside" modification for cathode and electrolyte

DOI: 10.1201/9781003133971-7

to the "outside" design. Various carbon materials [2,4], conductive polymer [4,5], metal compounds [6,7], and their composite materials [8–12] used as positive electrode frameworks have been used in the construction of interlayers. Modifying the separator or introducing a functional interlayer between the positive electrode and the separator proved to be an effective method to improve the electrochemical performance of the Li-S battery. The interlayer has gradually attracted the attention of many researchers.

7.2 PRINCIPLES

The Li-S battery is a kind of lithium metal battery. It uses sulfur species as active material for a cathode and metal lithium as the anode with polymer separator and non-aqueous electrolyte in the middle. It offers a high theoretical specific capacity of 1,675 mAh g^{-1} and a high specific energy of 2,600 Wh kg^{-1}. The charging and discharging process of Li-S batteries is a complex redox reaction involving multiple phase changes and multiple steps. During the discharge process, the cyclic S8 molecule will first undergo a ring-opening reaction and combine with lithium ions and electrons to become a soluble intermediate product of LiPSs (Li_2S_8, Li_2S_6, Li_2S_4), and finally convert to insoluble Li_2S_2/Li_2S on the cathode. The charging process is just the opposite [13,14]. When charging or discharging, the LiPSs dissolved in the electrolyte will shuttle to the negative electrode side due to the influence of the concentration gradient. These LiPSs directly react with the lithium negative electrode, causing corrosion of the metal lithium [15]. This situation causing self-discharge and lowering the coulombic efficiency of the battery is called the "shuttle effect".

The interlayer can effectively solve this problem and improve the performance of Li-S batteries [16]. The modified separator and functional interlayer with good electronic conductivity can be used as the upper current collector of the sulfur cathode, reducing the battery interface impedance, promoting electron transmission, and improving the utilization of active materials. The interlayer mainly acts through three ways: physical barrier, chemical adsorption, and catalytic conversion.

7.3 WORKING MECHANISMS

7.3.1 Physical Barrier

Soluble LiPSs migrate to the negative electrode through the pores on the separators, causing irreversible capacity loss and self-discharge of the Li-S battery (Figure 7.1).

Inserting an interlayer between the positive electrode and separator can effectively inhibit the shuttle of LiPSs by a physical barrier. In 2012, the Manthiram's research group [1] introduced porous carbon paper and multi-walled carbon nanotubes as an interlayer into Li-S batteries, improving the specific capacity and cycle stability of the battery by capturing the LiPSs through their porous structure. Since then, there have been a large number of research reports on the physical barrier interlayer [3,17–22]. The interlayer based on physical barrier is an early type

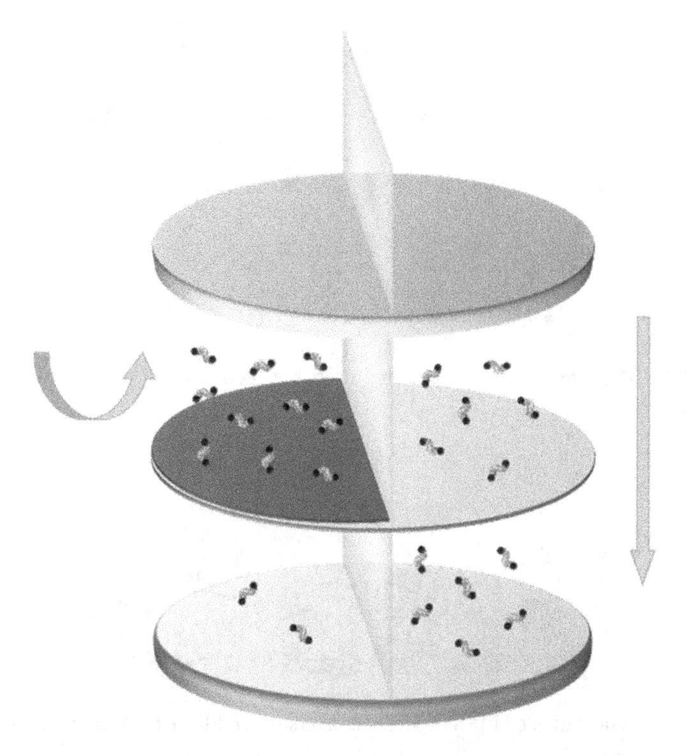

FIGURE 7.1 Schematic of physical blocking.

of functional interlayer. It can generally be achieved by restricting the migration path of LiPSs, porous adsorption, electrostatic repulsion, and so on.

7.3.2 Chemical Adsorption

Although the physical barrier effect can inhibit the shuttle of LiPSs to a certain extent, this kind of interlayer cannot achieve efficient adsorption of LiPSs by physical adsorption alone because the interaction between the barrier layer and LiPSs is too weak to inhibit the shuttle of polysulfides efficiently.

Chemical adsorption is a kind of adsorption, which forms the chemical bonds by the electrons transferring, exchanging, or sharing between adsorbed molecules and solid surface atoms (or molecules). Cui Yi's research group pointed out that having polar function groups is important to increase the interaction between Li_2S_n species and material [23]. Lithium atoms in LiPSs tend to bond with the host containing heteroatoms (such as oxygen, nitrogen, etc.) through lithium bonds under a dipole–dipole interaction [24,25]. At the same time, sulfur in LiPSs tends to combine with the metal atoms in some polar materials, such as metal oxides and metal sulfides [26–28]. Following that, extensive work has been done on introducing oxygen functional groups on carbon that can better anchor Li_2S_n species, or replacing carbon with other polar materials. Introducing nanoscale anchoring material is viewed as one of the most important methods to confine the Li_2S_n species

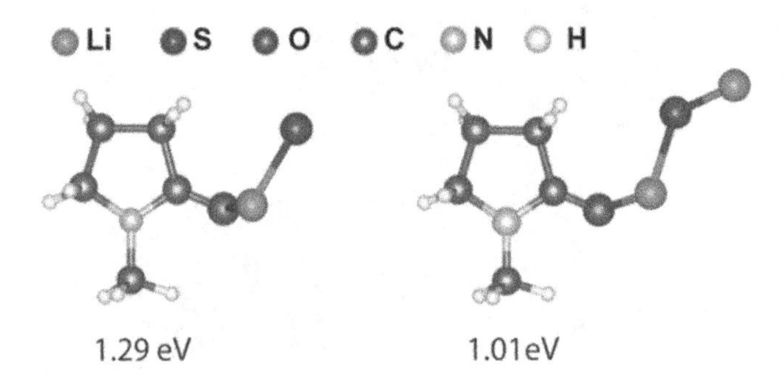

FIGURE 7.2 The interaction between the discharge products and the functional group on the polymer [23].

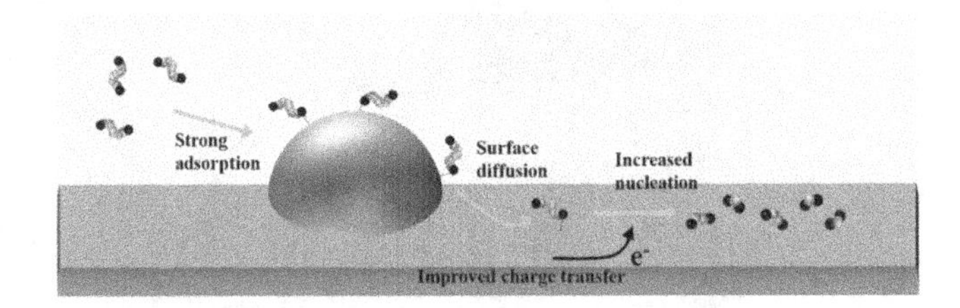

FIGURE 7.3 Schematics of LiPSs species conversion and Li_2S nucleation process through catalytic conversion.

and avoid their dissolution [29]. Therefore, researchers use metal oxides [8,30], metal sulfides [6,31,32], polymers [33], and other polar materials as interlayers to further slow down the shuttle of LiPSs by chemically adsorbing LiPSs to improve the electrochemical performance of Li-S batteries effectively (Figure 7.2).

7.3.3 Catalytic Conversion

With the furthering of research, it is found that only relying on a single physical barrier or chemical adsorption cannot solve the problem of slow conversion of LiPSs. Increasing the thickness of the interlayer can improve the inhibition of LiPSs shuttle. But this method will lose its advantages of high energy density and affect its commercial development of the Li-S battery. At the same time, a thicker interlayer will also reduce ion diffusion. Although polar compounds such as metal compounds, conductive polymers, and metal organic frameworks have a strong ability to bind LiPSs, their low electronic conductivity makes the conversion efficiency of LiPSs still unsatisfactory. A large amount of LiPSs cannot get further use but accumulate at the interlayer, causing a lot of capacity loss.

At that time, researchers proposed that by combining the physical barrier effect of conductive non-polar materials and the chemical adsorption and catalytic conversion capabilities of polar materials, Li-S batteries with better electrochemical performance can be obtained [11,34–37]. As shown in Figure 7.3, the LiPSs is first adsorbed on the surface by the polar material, and then diffuses to the conductive material to obtain electrons. Finally, they convert into Li_2S_2 or Li_2S. Professor Cui Yi's research group also specifically studied the relationship between the adsorption and diffusion of sulfides on a variety of metal oxides [38]. Through calculations, they found that although single-layer chemisorption occupies a major role in the capture of LiPSs, these non-conductive oxides are only a "transmission station". They can transport LiPSs from the oxide surface with poor conductivity to the carbon matrix with high conductivity, and ensure complete electrochemical conversion. Therefore, the adsorption amount of LiPSs depends on the surface area of the polar material, and a higher surface diffusion rate will increase the nucleation efficiency of LiPSs.

7.4 SUMMARY

The research and development of the functional interlayer provides a new direction for high-performance Li-S batteries, promoting the development and application of large-scale engineering. However, the introduction of the interlayer inevitably brings some new problems and challenges: (1) Increase the weight and reduce the energy density of the battery, losing its original advantage; (2) increase the transmission impedance of lithium ions and reduce the battery reaction rate; (3) the interlayer will absorb electrolytes, increase the amount of electrolytes, etc.

Therefore, in the process of engineering development, it is also necessary to design and optimize the interlayer rationally to reduce adverse effects and give full play to its role in improving and promoting the performance of Li-S batteries. The interlayer of Li-S batteries must meet the following points: (1) It can inhibit the shuttle of LiPSs, restrict the LiPSs to the positive side through physical barrier or chemical adsorption. (2) It must have high Li ion permeability, stability over cycling, and robust mechanical properties for continuous roll-to-roll operation. (3) Possess a high surface area and present excellent conductive nature to provide electron pathways and enable the reutilization of intercepted active materials during repeat cycles. The future development of the interlayer should also consider: (1) further research on the interlayer material system with catalytic conversion function; (2) the development of simple and low-cost preparation technology is also very important; (3) lightweight and thin layer are not to be ignored of material selection and design principles of interlayers in the future. With excellent progress and various efforts achieved thus far, future innovations of advanced interlayer systems may play an intricate role in any practical advancements of Li-S batteries.

REFERENCES

[1] Su Y-S, Manthiram A. Lithium-sulphur batteries with a microporous carbon paper as a bifunctional interlayer [J]. *Nature Communications*, 2012, 3.

[2] Zhang S, Qin X, Liu Y, et al. A conductive/ferroelectric hybrid interlayer for highly improved trapping of polysulfides in lithium-sulfur batteries [J]. *Advanced Materials Interfaces*, 2019, 6(22).

[3] Wang X, Wang Z, Chen L. Reduced graphene oxide film as a shuttle-inhibiting interlayer in a lithium–sulfur battery [J]. *Journal of Power Sources*, 2013, 2442(65–69).

[4] Tu S, Chen X, Zhao X, et al. A polysulfide-immobilizing polymer retards the shuttling of polysulfide intermediates in lithium-sulfur batteries [J]. *Advanced Materials*, 2018, 30(45).
[5] Li Y, Lin S, Wang D, et al. Single atom array mimic on ultrathin MOF nanosheets boosts the safety and life of lithium-sulfur batteries [J]. *Advanced Materials*, 2020, 32(8): e1906722.
[6] Han P, Chung S-H, Manthiram A. Thin-layered molybdenum disulfide nanoparticles as an effective polysulfide mediator in lithium-sulfur batteries [J]. *ACS Applied Materials & Interfaces*, 2018, 10(27): 23122–23130.
[7] Luo Y, Luo N, Kong W, et al. Multifunctional interlayer based on molybdenum diphosphide catalyst and carbon nanotube film for lithium-sulfur batteries [J]. *Small*, 2018, 14(8).
[8] Li Z, Zhou C, Hua J, et al. Engineering oxygen vacancies in a polysulfide-blocking layer with enhanced catalytic ability [J]. *Advanced Materials*, 2020, 1907444.
[9] Deng D-R, Bai C-D, Xue F, et al. Multifunctional ion-sieve constructed by 2D materials as an interlayer for Li-S batteries [J]. *ACS Applied Materials & Interfaces*, 2019, 11(12): 11474–11480.
[10] Tian D, Song X, Wang M, et al. MoN supported on graphene as a bifunctional interlayer for advanced Li-S batteries [J]. *Advanced Energy Materials*, 2019, 9(46): 1901940.
[11] Fan Y, Yang Z, Hua W, et al. Functionalized boron nitride nanosheets/graphene interlayer for fast and long-life lithium-sulfur batteries [J]. *Advanced Energy Materials*, 2017, 7(13).
[12] Wang M, Fan L, Qiu Y, et al. Electrochemically active separators with excellent catalytic ability toward high-performance Li–S batteries [J]. *Journal of Materials Chemistry A*, 2018, 6(25): 11694–11699.
[13] Pang Q, Liang X, Kwok CY, et al. Advances in lithium-sulfur batteries based on multifunctional cathodes and electrolytes [J]. *Nature Energy*, 2016, 1.
[14] Cui Z, Zu C, Zhou W, et al. Mesoporous titanium nitride-enabled highly stable lithium-sulfur batteries [J]. *Advanced Materials*, 2016, 28(32): 6926–+.
[15] Chung S-H, Manthiram A. Rational design of statically and dynamically stable lithium-sulfur batteries with high sulfur loading and low electrolyte/sulfur ratio [J]. *Advanced Materials*, 2018, 30(6).
[16] Fan L, Li M, Li X, et al. Interlayer material selection for lithium-sulfur batteries [J]. *Joule*, 2019, 3(2): 361–386.
[17] Chung SH, Manthiram A. High-performance Li-S batteries with an ultra-lightweight MWCNT-coated separator [J]. *Journal of Physical Chemistry Letters*, 2014, 5(11): 1978–1983.
[18] Kim HM, Hwang J-Y, Manthiram A, et al. High-performance lithium-sulfur batteries with a self-assembled multiwall carbon nanotube interlayer and a robust electrode electrolyte interface [J]. *ACS Applied Materials & Interfaces*, 2016, 8(1): 983–987.
[19] Huang J-Q, Xu Z-L, Abouali S, et al. Porous graphene oxide/carbon nanotube hybrid films as interlayer for lithium-sulfur batteries [J]. *Carbon*, 2016, 99: 624–632.
[20] Gu X, Lai C, Liu F, et al. A conductive interwoven bamboo carbon fiber membrane for Li–S batteries [J]. *Journal of Materials Chemistry A*, 2015, 3(18): 9502–9509.
[21] Zhai P-Y, Peng H-J, Cheng X-B, et al. Scaled-up fabrication of porous-graphene-modified separators for high-capacity lithium-sulfur batteries [J]. *Energy Storage Materials*, 2017, 7: 56–63.
[22] Wang Z, Wang X, Sun W, et al. Dendrite-free lithium metal anodes in high performance lithium-sulfur batteries with bifunctional carbon nanofiber interlayers [J]. *Electrochimica Acta*, 2017, 252: 127–137.

[23] Zheng G, Zhang Q, Cha JJ, et al. Amphiphilic surface modification of hollow carbon nanofibers for improved cycle life of lithium sulfur batteries [J]. *Nano Letters*, 2013, 13(3): 1265–1270.

[24] Hou TZ, Xu WT, Chen X, et al. Lithium bond chemistry in lithium-sulfur batteries [J]. *Angewandte Chemie International Edition*, 2017, 56(28): 8178–8182.

[25] Sun J, Sun Y, Pasta M, et al. Entrapment of polysulfides by a black-phosphorus-modified separator for lithium-sulfur batteries [J]. *Advanced Materials*, 2016, 28(44): 9797–9803.

[26] Cai D, Liu B, Zhu D, et al. Ultrafine Co3Se4 nanoparticles in nitrogen-doped 3D carbon matrix for high-stable and long-cycle-life lithium sulfur batteries [J]. *Advanced Energy Materials*, 2020, 10(19).

[27] Li Y, Chen G, Mou J, et al. Cobalt single atoms supported on N-doped carbon as an active and resilient sulfur host for lithium–sulfur batteries [J]. *Energy Storage Materials*, 2020, 28: 196–204.

[28] Sun Z, Vijay S, Heenen HH, et al. Catalytic polysulfide conversion and physiochemical confinement for lithium–sulfur batteries [J]. *Advanced Energy Materials*, 2020, 10(22): 1904010.

[29] Zhang Q, Wang Y, Seh ZW, et al. Understanding the anchoring effect of two-dimensional layered materials for lithium-sulfur batteries [J]. *Nano Lett*, 2015, 15(6): 3780–3786.

[30] Zhang Z, Lai Y, Zhang Z, et al. Al2O3-coated porous separator for enhanced electrochemical performance of lithium sulfur batteries [J]. *Electrochimica Acta*, 2014, 129: 55–61.

[31] Ghazi ZA, He X, Khattak AM, et al. MoS2/Celgard separator as efficient polysulfide barrier for long-life lithium-sulfur batteries [J]. *Advanced Materials*, 2017, 29(21).

[32] Yu X, Zhou G, Cui Y. Mitigation of shuttle effect in Li-S battery using a self-assembled ultrathin molybdenum disulfide interlayer [J]. *ACS Applied Materials & Interfaces*, 2019, 11(3): 3080–3086.

[33] Ma G, Wen Z, Wang Q, et al. Enhanced performance of lithium sulfur battery with self-assembly polypyrrole nanotube film as the functional interlayer [J]. *Journal of Power Sources*, 2015, 273: 511–516.

[34] Hu N, Lv X, Dai Y, et al. SnO2/Reduced graphene oxide interlayer mitigating the shuttle effect of Li-S batteries [J]. *ACS Applied Materials & Interfaces*, 2018, 10(22): 18665–18674.

[35] Wang N, Chen B, Qin KQ, et al. Rational design of Co9S8/CoO heterostructures with well-defined interfaces for lithium sulfur batteries: A study of synergistic adsorption-electrocatalysis function [J]. *Nano Energy*, 2019, 60: 332–339.

[36] Guo P, Liu D, Liu Z, et al. Dual functional MoS2/graphene interlayer as an efficient polysulfide barrier for advanced lithium-sulfur batteries [J]. *Electrochimica Acta*, 2017, 256: 28–36.

[37] Guan B, Zhang Y, Fan L, et al. Blocking polysulfide with Co2B@CNT via "synergetic adsorptive effect" toward ultrahigh-rate capability and robust lithium-sulfur battery [J]. *ACS Nano*, 2019, 13(6): 67442–6750.

[38] Tao X, Wang J, Liu C, et al. Balancing surface adsorption and diffusion of lithium-polysulfides on nonconductive oxides for lithium-sulfur battery design [J]. *Nature Communications*, 2016, 7: 11203.

8 Principles and Status of Lithium-Sulfur Batteries

Yi Wei
College of Materials Science and Engineering, Zhengzhou University, Zhengzhou, People's Republic of China

Huiyang Ma
College of Chemistry, Zhengzhou University, Zhengzhou, People's Republic of China

College of Materials Science and Engineering, Zhengzhou University, Zhengzhou, People's Republic of China

Wei Guo and Yongzhu Fu
College of Chemistry, Zhengzhou University, Zhengzhou, People's Republic of China

CONTENTS

DOI: 10.1201/9781003133971-8

8.1 INTRODUCTION

With the continuous development of our society, people's demand for energy is growing dramatically. The extensive use of energy causes the exhaustion of primary energy sources, i.e., fossil fuels. In addition, too much CO_2 release accompanied with the consumption of fossil fuels has polluted the atmosphere and elevated the temperature on the earth. To enable a renewable planet and future, alternative green energies such as solar and wind need to be adopted in the energy category. However, most of them are intermittent and limited by their locations. Therefore, energy storage systems are needed. Among them, electrochemical energy devices such as rechargeable batteries are one of the most promising solutions. Lithium-ion (Li-ion) batteries have dominated the power supply markets of portable electronics and electric vehicles and they are promising to play an important role for grid energy storage as well. The working principle of the Li-ion battery lies on the reversible intercalation/deintercalation of lithium ions in a graphite anode and transition metal oxide cathodes such as $LiCoO_2$, $LiMn_2O_4$, and $LiFePO_4$, as shown in Figure 8.1.

With the demand of increasing energy densities of Li-ion batteries, alternative battery systems are being actively pursued, such as lithium-sulfur (Li-S), lithium-oxygen, and zinc-ion batteries. Among them, rechargeable Li-S batteries have shown great promise because of the high capacities of a lithium metal anode and sulfur cathode. The charge storage mechanism in Li-S batteries is different from that in the Li-ion battery and conversion reactions vs. intercalation reactions. In addition, sulfur is an abundant element that is beneficial for widespread applications. In the early 1980s, Li-S batteries were studied; however, their limited cycling performance did not attract much attention. In the past two decades, significant progress was made, which has advanced the Li-S battery technology to an unprecedented frontier.

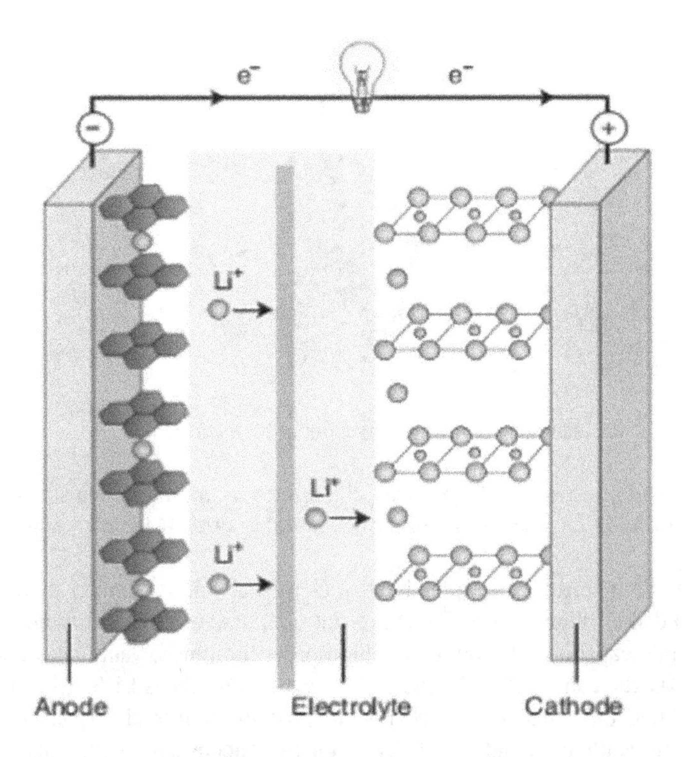

FIGURE 8.1 An illustration of the charge process and discharge process of a lithium-ion cell [1].

8.2 OPERATING PRINCIPLE AND CHALLENGES OF THE LI-S BATTERY

A Li-S battery usually consists of a lithium metal anode, liquid electrolyte, and sulfur cathode, as shown in Figure 8.2. The lithium metal anode has a theoretical specific capacity of 3,862 mAh g^{-1} and sulfur has a theoretical specific capacity of 1,675 mAh g^{-1}, thus the specific energy of Li-S battery is about 2,600 Wh kg^{-1}. In the discharge, lithium metal is oxidized to release lithium ions and electrons ($Li \rightarrow Li^+ + e^-$), both of which are conducted to the sulfur cathode and then sulfur cathode is reduced to form lithium sulfide ($S + 2Li^+ + e^- \rightarrow Li_2S$). In the charge process, both reactions are reversed.

At room temperature, sulfur is in the form of S_8. At the beginning of discharge in an ether electrolyte, a short voltage plateau at 2.3 V appears, which is due to the conversion of S_8 to high-order lithium polysulfides (e.g., Li_2S_8, Li_2S_6, and Li_2S_4). These lithium polysulfides are soluble in the electrolyte. When the discharge continues, Li_2S_4 is converted to low-order lithium polysulfides such as Li_2S_3, and then Li_2S_2 and Li_2S. Some of them start to precipitate, leading to the voltage plateau at 2.1 V. At the end, all sulfur is converted to Li_2S. The complete discharge process is shown in Figure 8.3. In the recharge process, Li_2S is oxidized, leading to the formation of low-order polysulfides, high-order polysulfides, and then elemental sulfur. During the whole cycle process, about 80% volume change occurs in the

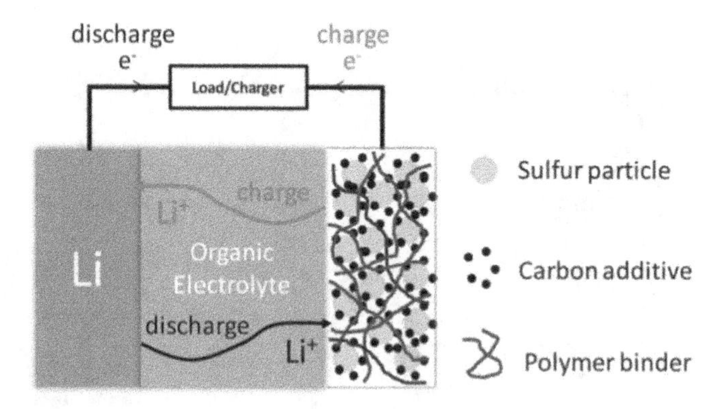

FIGURE 8.2 Schematic diagram of a conventional Li-S cell [2].

sulfur electrode, resulting in significant challenges in maintaining a stable electrode structure. In addition, the following issues are also critical for achieving favorable performance.

First of all, the electrical conductivity of elemental sulfur is only 5×10^{-30} S cm^{-1} at 25°C and the lithium polysulfides are not conductive either [4], which seriously affect the rate capability. Secondly, the high-order lithium polysulfides are soluble in the ether-based electrolyte, while the end discharged products Li_2S_2/Li_2S are not [5], posing a huge challenge on the reversibility of the sulfur electrode. Thirdly, the dissolved polysulfides could shuttle between two electrodes in the charge process, resulting in low Coulombic efficiency and short cycle life, which is called "shuttle effect" [6]. On the anode side, the repeated deposition and stripping of lithium cause passivation and dendrite growth on the surface of the anode, which is not conducive to the formation of solid-electrolyte interphase (SEI) film [7]. Moreover, self-discharge of Li-S battery is severe, resulting in short shelf life. To make matters worse, self-discharge may also promote the diffusion of dissolved polysulfide materials from the cathode region, which results in high active material loss, poor cycle life, and low Coulombic efficiency [8,9]. To overcome these issues, huge efforts have been focused on the sulfur cathode, lithium metal anode, and electrolyte. In the following, some of them are described.

8.3 CATHODES

To improve the utilization of active material in the sulfur cathode, conductive additives are needed, like carbon and conductive polymer [10]. In addition, hollow structures of TiO_2 that can retard the loss of active material have been developed [11]. Recently, some inorganic materials that can physically adsorb lithium polysulfides and catalyze the conversion of these materials have been paid particular attention. These approaches have shown advantages in improving the utilization of sulfur, electrode kinetics, and cycling stability. Moreover, organosulfides based on reversible breakage/formation of sulfur-sulfur (S-S) bonds have been developed as a new class of cathode materials. They have shown unique electrochemical behavior and redox mechanism.

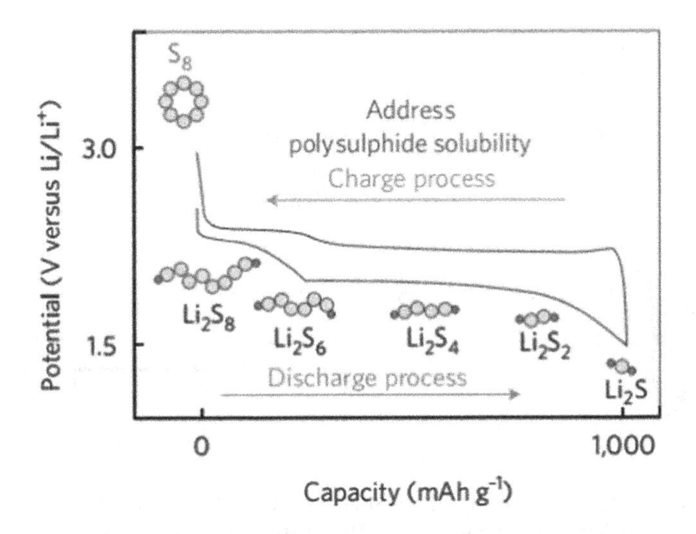

FIGURE 8.3 Schematic and voltage profiles of a Li-S cell [3].

8.3.1 Sulfur-Carbon Composite Cathodes

8.3.1.1 Sulfur-Porous Carbon Composites

A variety of porous carbons have been developed as hosts for sulfur, including microporous carbon (<2 nm), mesoporous carbon (2–50 nm), and macroporous carbon (>50 nm) [12]. Sulfur can be encapsulated in the pore structure of porous carbon, the pore size distribution of porous carbon greatly affects the redox reaction of sulfur, and the pore volume determines the sulfur content and material properties of the composite cathode [13].

Due to the high specific surface area and small pore size, microporous carbon can effectively limit the shuttle of polysulfides, so the Coulombic efficiency is usually high and the cycling performance is stable. For example, porous carbon with a micropore size of 0.6~0.7 nm, a specific surface area of 844 m^2 g^{-1}, and a pore volume of 0.474 cm^3 g^{-1} was used. After heat treatment with sulfur at 149°C and then 300°C, 42 wt% of sulfur can be encapsulated. The Li-S battery with such sulfur-porous carbon composite can be cycled 500 times with a capacity retention of 80% [14–16]. Wang et al. proposed that the high cycling stability of sulfur in micropores is due to an ionic desolvation mechanism through a two-step adsorption extraction process. Ions are often dissolved in micropores close to the size of ions. Micropores below 1 nm can lead to the desolvation of electrolyte ions, thus preventing or at least slowing down the dissolution of polysulfides, effectively eliminating the dissolution of polysulfides and ensuring good cycling stability [17]. This process and cycling performance are shown in Figure 8.4 [18]. However, due to the small pore size, the penetration and flow of electrolyte are affected, thus reducing the transport rate of lithium ions, and the reaction kinetics is slow [13].

On the contrary, due to the larger pore size of mesoporous and macroporous materials, the electrolyte permeation is better, the lithium-ion transport rate is higher, and the proportion of sulfur content is also increased, but the shuttle of

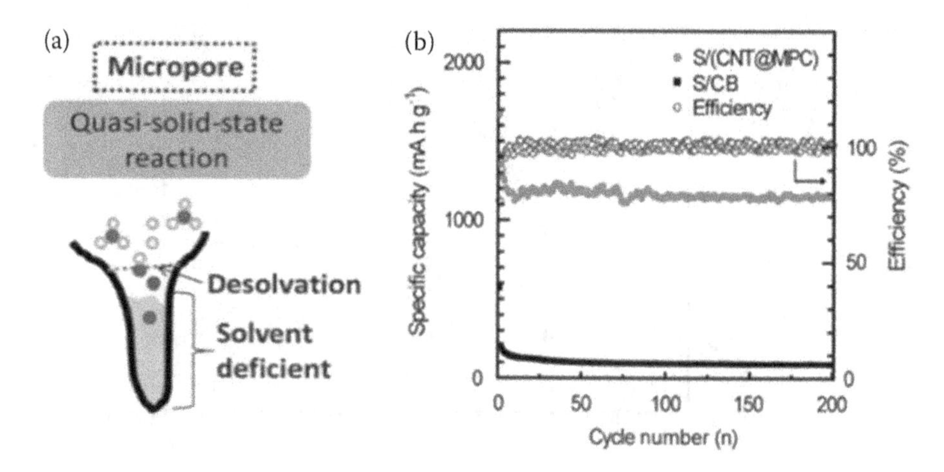

FIGURE 8.4 (a) Illustration of the ion-desolvation mechanism for the quasi-solid-state reaction of sulfur confined in micropores [18]. (b) Cycling performance and Coulombic efficiency of S/(CNT@MPC) and S/CB at 0.1C rate [19].

polysulfides cannot be effectively restricted [13]. The earliest study on sulfur-mesoporous carbon composite cathodes was based on highly ordered mesoporous carbon (CMK-3). Sulfur was loaded into CMK-3 by a melt diffusion method at 155°C. The sulfur nanoparticles grow in the narrow pore size of 3.3 nm, and the sulfur content is 70 wt%. The prepared Li-S cell can achieve a high Coulombic efficiency of 99.94% [20]. A diagram of the composite is shown in Figure 8.5 [20]. However, the size of different polysulfides varies, hierarchical porous carbon materials have been developed in order to make full use of the advantages of different pores [17,21–23]. Among them, hollow porous carbon has a large hole inside and a mesopore and micro-pore structure in the outer shell, which greatly reduces the dissolution and shuttle of polysulfides and the volume expansion upon cycling. Therefore, it has received widespread attention [24].

In addition, hollow porous carbon has become an active research material in recent years because it can be used as a good sulfur reservoir due to its large volume of internal cavity and porous shell. This material was originally reported by Archer et al. [25]. Using the hard-template method, hollow porous carbon spheres with inner pore size of 200 nm can be made. The specific surface area is 648 $cm^2\ g^{-1}$ and the average pore size on the carbon wall is 3 nm. A high sulfur content up to 70 wt % can be achieved in the hollow porous carbon. The performance of the prepared lithium-sulfur battery is shown in Figure 8.6. The capacity of the battery is extremely reversible and the cycling performance is also excellent.

8.3.1.2 Sulfur-Graphene Composites

Graphene is a carbon material with a honeycomb crystal structure formed by the sp^2 hybridization of a single layer of carbon atoms. It has rich pore structure, good electrical conductivity, and high chemical stability [27]. In a Li-S battery, the conductivity of sulfur electrode can be enhanced by adding conductive graphene,

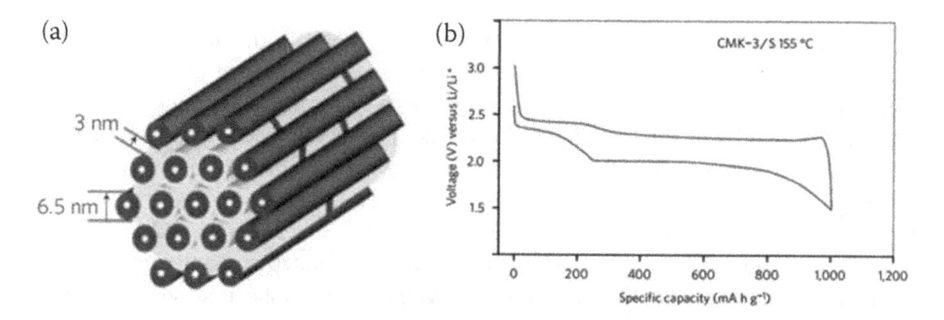

FIGURE 8.5 (a) A schematic diagram of the sulfur (yellow) confined in the interconnected pore structure of mesoporous carbon, CMK-3, formed from carbon tubes that are propped apart by carbon nanofibers. (b) Charge/discharge profiles and specific capacity of sulfur-CMK-3/5 composite cathode at 155°C [20].

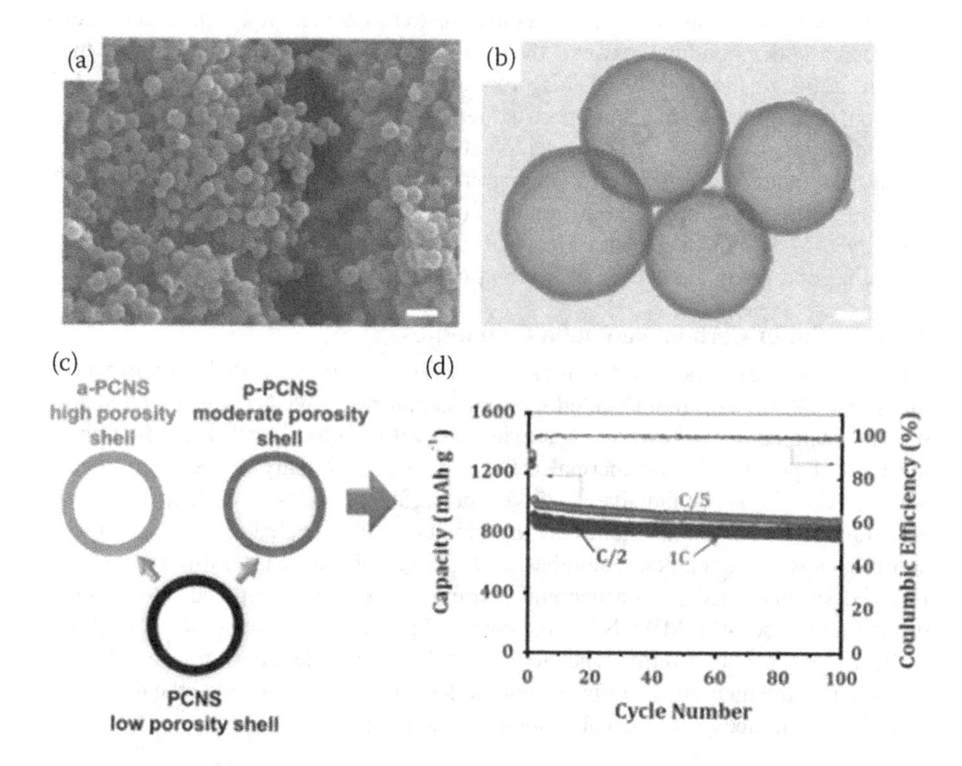

FIGURE 8.6 (a) SEM image and (b) TEM image of the hollow porous carbon nanospheres. Scale bars: 400 nm (a) and 50 nm (b). (c) Illustrations of hollow porous carbon nanospheres with low porosity, moderate porosity, and high porosity shells. (d) Cycling performance of the sulfur electrode using PCNS with moderate shell porosity at different current densities [26].

chemically modified graphene, and porous graphene [28–30]. In addition, graphene can help restrain the diffusion of soluble polysulfides and relieve the volume expansion of the electrode [30]. Therefore, graphene-based materials have been extensively studied for use in sulfur electrodes. Dai et al. used graphene sheets decorated with carbon black nanoparticles and sulfur particles to synthesize a sulfur-graphene composite cathode that shows a specific capacity of 600 mAh g^{-1} over 100 cycles [31]. Later, in order to further improve the electrical conductivity of the sulfur-graphene composites, so as to reduce the conductive agent content and increase the utilization of sulfur, Lin et al. synthesized graphene sheets with low defects through sulfur-assisted graphite peeling. When sulfur active materials are attached on the surface and edge of graphene, they show good performance under 2C [32].

Chemically modification can change the properties of graphene [33]. Studies have shown that graphene oxide can effectively immobilize polysulfides therefore improve the cycling stability. This is mainly due to functional group-induced ripples that increase the bonding between the sulfur atoms in the polysulfides and the carbon atoms in the graphene. This process is illustrated in Figure 8.7 [34]. In addition, manipulating the nano-pore structure of graphene is also an effective method to improve the performance of sulfur-graphene composite cathode by taking advantage of the size effect. Ding et al. prepared a high-density nanopore material with a pore size of 3.8 nm. Its composite with sulfur shows an initial capacity of 1,380 mAh g^{-1} and a capacity retention rate of 76 wt%. It also can maintain good cycling stability at different cycling rates [35].

8.3.1.3 Sulfur-Carbon Nanotube Composites

Carbon nanotubes generally fall into two categories: single-walled carbon nanotubes (SWCNTs) and multi-walled carbon nanotubes (MWCNTs) [36]. They are one-dimensional (1-D) carbon nanomaterials with high electrical conductivity, mechanical properties, and thermal stability [37–39]. Initially, carbon nanotubes were used only as conductive additives in sulfur cathodes [37]. The common structures when carbon nanotubes are used include sulfur-coated carbon nanotubes, sulfur-encapsulating carbon nanotubes, and tube-in-tube structure [40]. Yuan et al. reported sulfur-coated carbon nanotubes composite cathode prepared by the melt diffusion method [41]. MWCNT is the core and provides the electronic conduction pathway for the coated sulfur. The electrode exhibits excellent performance [42]. It is found that the theoretical sulfur content in the composite is related to the diameter of carbon nanotubes, and the calculation formula is [43]:

$$S_{wt\%} = \frac{\rho_s * (4a^2 + 4ad)}{\rho_{CNT} * \mathrm{d}^2 + \rho_s * (4a^2 + 4ad)}$$

In the formula, d represents the CNT diameter and a represents sulfur coating thickness [43]. ρCNT represents the density of CNTs and ρ_s is the density of sulfur [44].

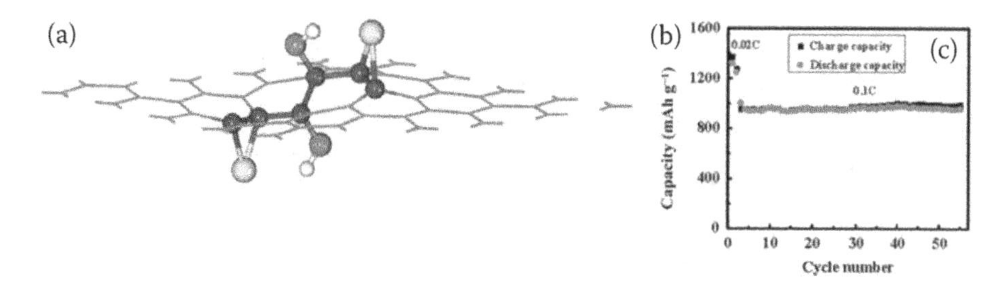

FIGURE 8.7 Representative pattern of graphene oxide immobilizing sulfur. The hydroxyl enhances the binding of S to the C-C bond due to the induced ripples by epoxy or hydroxyl group. Yellow, red, and white balls denote S, O, and H atoms, respectively, while the others are C atoms. The C atoms bonding to S or O are highlighted as blue balls. (b) Charge and discharge capacity at a current rate of 0.1C after initial activation processes at 0.02C for two cycles [34].

Based on the above theory, an ultra-light SWCNT conductive matrix with interconnection network structure is designed, and its specific structure is shown in Figure 8.8. The thickness of the sulfur coating is 6 nm, the sulfur content of the film is 95 wt%, and the electrochemical performance is excellent [43]. Sulfur can also be encapsulated in the hollow interior space of carbon nanotubes [45–47]. Moon et al. developed an encapsulated composite cathode of sulfur and carbon nanotubes by gas phase injection; its structure is shown in Figure 8.9 [45]. The sulfur content is 81 wt% and the composite shows a high specific capacity of 1,520 mAh g^{-1}.

Tube-in-tube structure is that carbon nanotubes with small diameters (e.g., 20 nm) are grown in those with large diameters (e.g., 200 nm). The hollow structure is used to encapsulate sulfur, and its structure is shown in Figure 8.10 [40]. The network structure formed in the tube provides an effective way for electron transport, effectively inhibits the shuttle of polysulfides, and adapts to the volume change of the active material upon cycling [48].

8.3.2 Sulfur-Conductive Polymers

Polymers generally do not conduct electricity, doping with anions or cations could make them conductive. Doping can be classified as n-type and p-type. The representative materials are polyacetylene (PAC), polyacrylonitrile (PAN), polythiobenzene (PTH), polypyrrole (PPy), poly(3,4-(ethylenedioxy)thiophene) (PEDOT), and polyaniline (PANI) [49]. With the development of conductive polymers, they can not only improve the conductivity, but also reduce the dissolution of polysulfides and inhibit the shuttle effect [50].

Sulfur-conductive polymers are usually divided into sulfur-PAN composites, sulfur-PP composites, sulfur-PANI composites, and other sulfur-conductive polymer composites. Sulfur-PAN composites [51] and sulfur-PP [52] composites for Li-S battery were both prepared by Wang et al. They heated elemental sulfur and PAN (mass ratio of 8:1) at 280–300°C in an argon atmosphere to form S@pPAN with a pore size of 200 nm and sulfur content of 53.4 wt% [51]. The Li-S cell

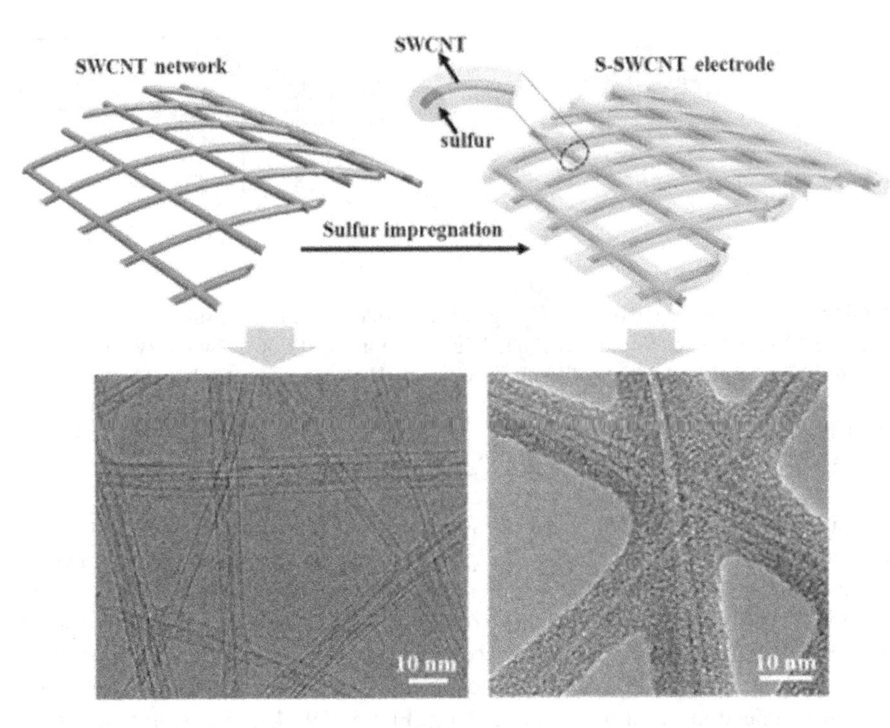

FIGURE 8.8 Illustrations of the SWCNT network and the sulfur-SWCNT electrode with TEM images of the SWCNTs and the sulfur-SWCNT composite, respectively [43].

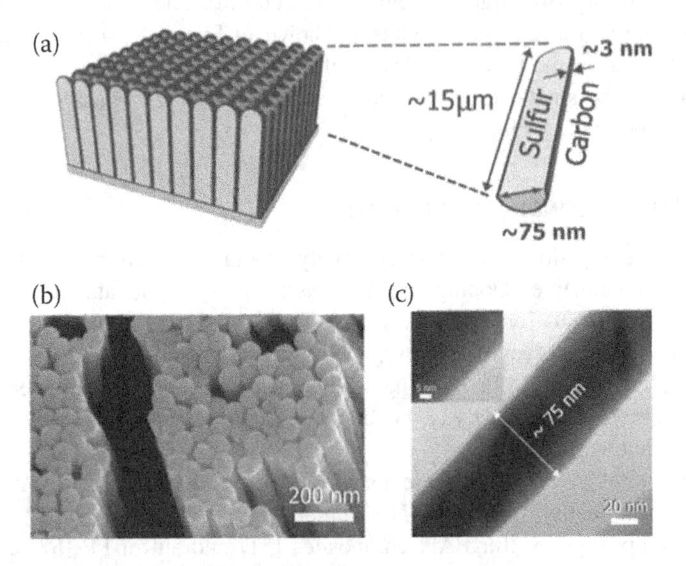

FIGURE 8.9 (a) Illustration of the encapsulated sulfur-CNT composite; (b) SEM; and (c) TEM images of the encapsulated sulfur-CNT composite [45].

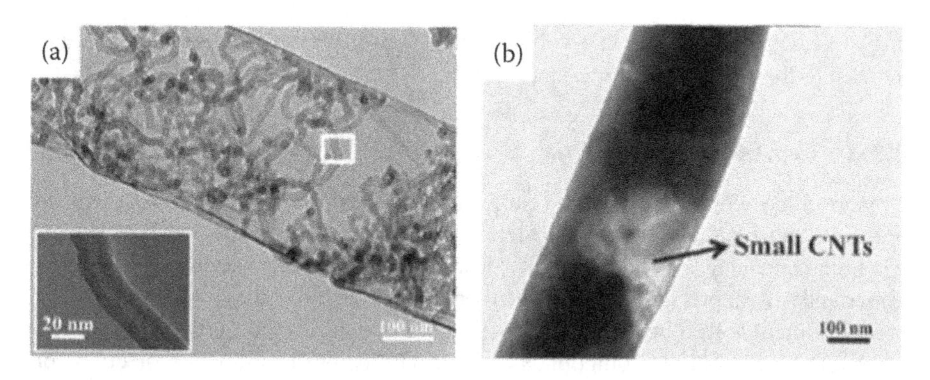

FIGURE 8.10 TEM images of (a) the tube-in-tube CNTs and (b) the sulfur-CNT composite [40].

with S@pPAN as cathode can maintain a high capacity of 600 mAh g^{-1} after 50 cycles. Later, people changed the heating time and temperature to optimize the sulfur content of S@pPAN composites or improve the material properties [53]. To make sulfur-PPy composite, sulfur powder was dispersed in the solution of sodium p-toluene sulfonate as the dopant, and pyrrole was polymerized to form the S-PPy composite [52]. Sulfur-PANI composite was studied by Wu et al. [54]. In this study, ball-milled and then-heated S-MWCNT composites were coated with PANI by an in-situ polymerization method. The discharge capacity of the composite cathode can reach 1,334.4 mAh g^{-1}. Other conductive polymers, like PTH [55] and PEDOT [56], have also been used in sulfur-polymer composites.

8.3.3 Sulfur-Inorganic Compounds

Inorganic compounds are often used as inherently polar substrates instead of carbon and polymers. In Li-S batteries, sulfur-inorganic compound composites are divided into three categories: metal oxide additives/composites, metal oxide coatings, and intercalation compounds/chalcogenide composites [57]. Studies have shown that metal oxide is used as an absorbent in metal oxide additives/composites to reduce the dissolution and shuttle of polysulfides. The absorption capacity of the absorbent is related to the radius of the metal particles [57]. At first, Gorkovenko et al. used vanadium oxide, silicate, alumina, and transition metal chalcogenides as absorbents [58]. Due to their large radius, the adsorption effect was not good and the electron transport was seriously affected. Later, Song et al. added nickel oxide nanoparticles (30–50 nm) as an adsorbent [59]. The resulting composite cathode shows good electrochemical performance and high capacity retention. Seh et al. developed a sulfur-TiO_2 egg yolk shell nanostructure for use in Li-S batteries [11]. The internal void of TiO_2 can accommodate the volume expansion of sulfur, reduce the dissolution of polysulfides, and realize long-term stable cycling performance of the battery. Intercalation compounds/chalcogenide composites are made of inorganic compounds with a voltage range similar to that of sulfur, which can be used as secondary active materials in the composite cathodes. For example, TiS_2 can be discharged/charged in the same voltage range (1.5–3.0 V) as that of sulfur, thereby increasing the overall specific capacity [60].

8.3.4 Organosulfides

Organosulfides consisting of sulfur-sulfur (S-S) bonds (e.g., $R\text{-}S_n\text{-}R$, $n \geq 2$, R is an organic group) can act as active materials for Li-S batteries. They have the overwhelming advantages of high capacity, low cost, and versatile structures [61]. Most importantly, with precise lithiation sites, organosulfide molecules, together with the organic functional groups connected with them, could provide unique properties. Therefore, organosulfide compounds can not only be used as model molecules for fundamental understanding of electrochemistry processes, but also provide a variety of opportunities for them to be electrodes and functional materials for Li-S batteries and beyond. Generally, organosulfide cathodes can be classified into two categories, i.e., polymers and small molecules, with three main transformation paths based on their solubility in the electrolytes, as shown in Figure 8.11 [62]. The three paths have apparent discrepancies in their redox potential, electrochemical kinetics, specific capacity, solubility in electrolytes, and electrochemical process [62].

8.3.4.1 Electrochemical Processes of Organosulfides

According to the solubility of the initial reactants, charged/discharged products and intermediates, the electrochemical processes of the organosulfide polymers could be classified into solid-liquid-solid state path (P-SLS) and solid-solid path (P-SS) [62]. The polymers which belong to different classes tend to show variant

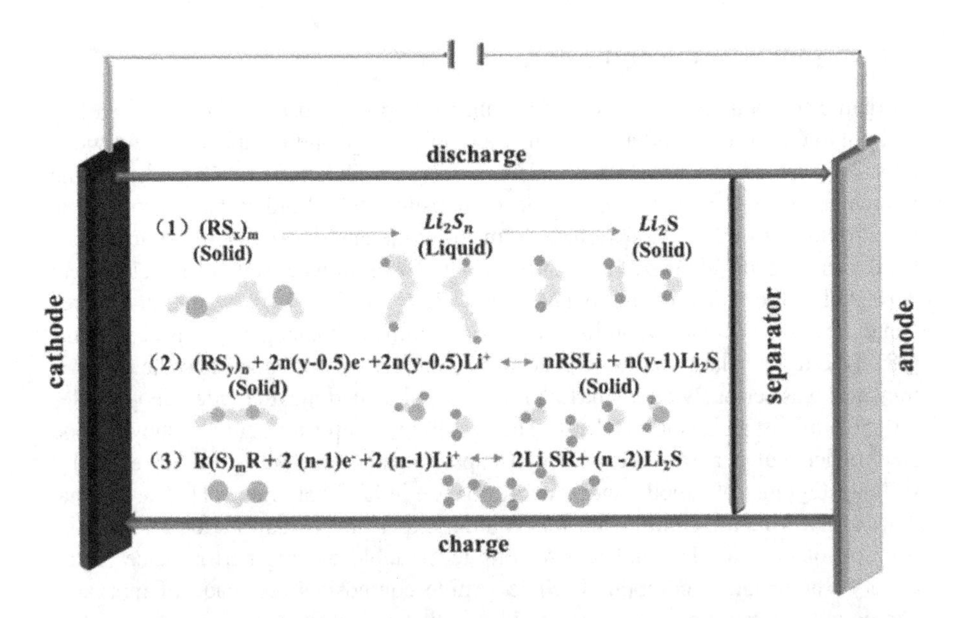

FIGURE 8.11 Transformation processes of P-SLS, P-SS, and small organosulfur molecules.

electrochemical processes. For the first path, ($(RS_x)_n \rightarrow Li_2S_n \rightarrow Li_2S$), it is found that P-SLS tends to lead to the formation of lithium polysulfides and then lithium sulfide through a series of reduction reactions during discharge, which is typically similar to the electrochemical process of S_8. However, for the second path, ($(RS_y)_n+2n(y-0.5)e^-+2n(y-0.5)Li^+ \leftrightarrow nRSLi+n(y-1)Li_2S$), without the formation of lithium polysulfides during discharge, both products and intermediates have limited solubilities in the electrolyte in the electrochemical process. And it only shows one potential stage in the discharge curves. Finally, the transformation path of the linear organosulfides can be concluded as $R\text{-}S_n\text{-}R+(2n-2)e^-+(2n-2)Li^+ \Leftrightarrow 2R\text{-}SLi+(n-2)Li_2S$ ($3 \leq n \leq 6$).

8.3.4.2 Organosulfide Polymers

During the past decade, research on the organosulfide cathodes have been conducted. As for the P-SS, the conjugated structure of the polymers plays a key role in the electrochemical properties of the cathodes. Taking the research of Wang's group as an example, it is illustrated that sulfurized polyacrylonitrile (SPAN) with insulating acrylonitrile is turned into a conjugated structure during sulfurization, which shows high sulfur utilization and superb rate performance in batteries. Furthermore, produced from a C_6S_6 monomer, a cross-linked disulfide polymer consisting of benzene rings and disulfide bonds possesses more than 70 wt% sulfur and shows great capacity retention of 98% after 100 cycles [63]. Moreover, changing the atoms in the sulfur-rich polymers will also bring out great properties, such as polyphenylene tetrasulfide (PPTS) [64], which show both great stable cycling performance and low electrochemical performance loss (Table 8.1).

For one hand, five kinds of organosulfide polymers which tend to transform in a SLS redox pathway have been studied and compared. Especially they contain different conjugated units, such as benzene rings, thiophene, and thiazine. It is found that polymers with low bandgap energy would have favorable performance based on density functional theory (DFT) analysis. For example, sulfur-linked tetra (allyloxy)-1,4-benzoquinone (S-TABQ) [67] with the lowest bandgap energy of

TABLE 8.1
Properties of P-SS polymers

Cathode	Theoretical capacity (mA h g^{-1})	Specific capacity (mA h g^{-1})	Cycle retention (%)
PPTS [64]	788	714 (0.05C)	77 (350 cycles)
PAQS [65]	225	Over 200 (0.25 mA)	Over 75 (125 cycles)
PAQnS [65]	976	600 (0.25 mA)	38.3 (125 cycles)
PDATtSSe [63]	742	700 (200 mA g^{-1})	92 (400 cycles)
Crosslinked disulfide polymer [66]	609	150 (0.1C)	98 (200 cycles)

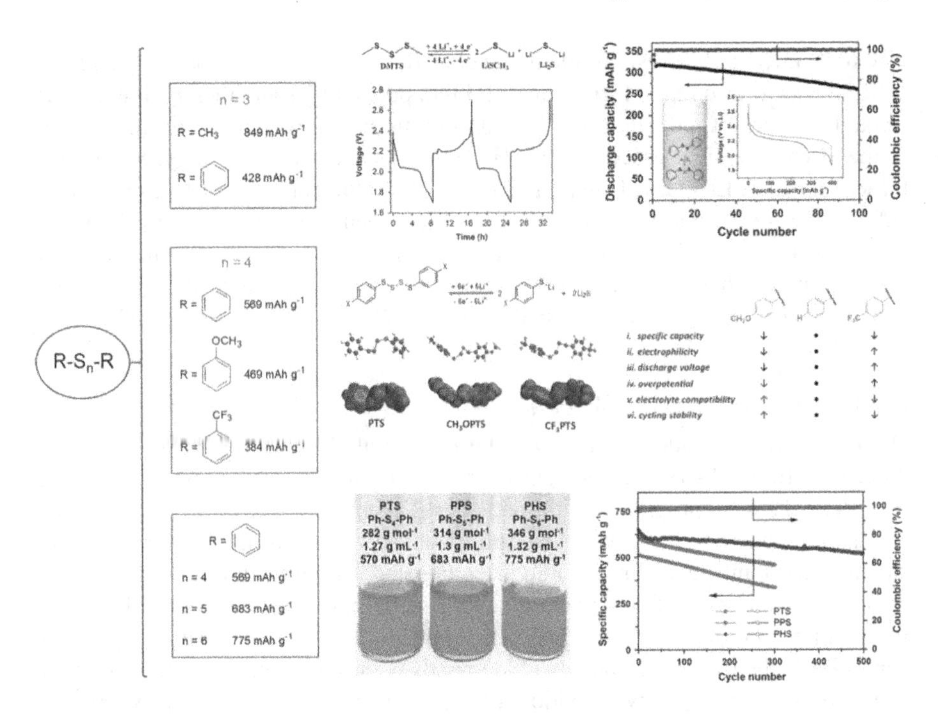

FIGURE 8.12 Organosulfide molecules (R-Sn-R, 6 ≥ n ≥ 3) that have been developed for lithium batteries: dimethyl trisulfide [69], diphenyl trisulfide [70], diphenyl tetrasulfides with different functional groups [71], and diphenyl polysulfides (tetrasulfide, pentasulfide, and hexasulfide) [64].

2.27 eV shows an outstanding rate performance of 833 mAh g^{-1} at 10 C and a stable cycling performance of 74% retention after 500 cycles. While for the other hand, sulfur contents in the P-SLS may also affect the properties. Pyun's group reported the poly(sulfur-random-1,3-diisopropenylbenzene) (S-DIB) [68] with the sulfur content up to 90 wt% presents high sulfur utilization and a promising cycling performance. The charging and discharging curve of S-DIB shows similarity with that of S_8, both of which have two typical discharge stages.

8.3.4.3 Small Organosulfide Molecules

Our group has focused on the redox reactions of small organosulfide molecules ($R\text{-}S_n\text{-}R$, $3 \leq n \leq 6$) in Li-S batteries, which are basically related to the cleavage and formation of S-S bonds during the charge and discharge processes [69]. With a low molecular weight and high electron storage, organosulfide molecules show high theoretical specific capacities, which could be ascribed to the specific R groups and the length of the sulfur chains (Figure 8.12).

In respect of the effect of the sulfur chains in the organosulfides, it is assumed that the longer chains of the sulfur may lead to high specific capacity, specific energies, and energy densities, which could be demonstrated through the comparison between variant liquid diphenyl polysulfides, e.g., diphenyl tetrasulfide (PhS_4Ph, PTS), diphenyl pentasulfide (PhS_5Ph, PPS), and diphenyl hexasulfide

TABLE 8.2
Properties of small organosulfide molecules

Cathode	Theoretical capacity (mA h g^{-1})	Areal loading (mg cm^{-2})	Specific capacity (mA h g^{-1})	Cycle retention (%)
DMTS [69]	849	6.7	597 (0.1C)	81 (50 cycles)
DPTS [72]	428	3.9	260 (0.5C)	79 (100 cycles)
PMTT [73]	418	6.7	353.4 (0.2C)	87 (100 cycles)
PMTH [73]	597	6	458 (0.2C)	85 (100 cycles)
PTS [74]	569.3	2.8	486 (0.5C)	74 (100 cycles)
CH_3OPTS [74]	469.3	3.4	324 (0.5C)	81 (100 cycles)
CF_3PTS [74]	384.3	4.2	272 (0.5C)	47 (100 cycles)
PHS [64]	775	Null	650 (1C)	80 (500 cycles)
PDSe-S [75]	311.4	3.51	194 (0.2C)	77 (200 cycles)
PDSe-S_2 [75]	427.4	3.92	241 (0.2C)	73 (200 cycles)
PhS-SePh [76]	202.1	Null	152 (0.2C)	52 (200 cycles)
1,2-LBDT [77]	347.8	0.7	340 (0.5C)	84 (100 cycles)

(PhS_6Ph, PHS) [64]. Among them, PhS_6Ph with the highest specific capacity of 650 mAh g^{-1}, specific energy of 1,302 Wh kg^{-1}, and energy density of 1,720 Wh L^{-1} shows great cycling stability of 80% capacity retention after 500 cycles [64]. Concerning the effect of different R groups in the organosulfides, it has been illustrated that the phenyl group of DPTS [70] shows greater properties of discharge voltage, energy efficiency, and cycling performance than those of the methyl group in DMTS [69]. A series of bis(aryl) tetrasulfides (PTS, CH_3OPTS, and CF_3PTS) have also been studied [71], which convey the effects of the functional groups on specific capacity, overpotential, electrolyte compatibility, and cycling stability (Table 8.2).

8.4 MODIFIED SEPARATORS

During the charge and discharge processes, the active materials in Li-S batteries always tend to go through a "solid-liquid-solid" electrochemical transformation path to form high-order lithium polysulfides and then the insoluble Li_2S_2/Li_2S [78]. This will lead to the low capacities during the discharge process and the constant loss of the active materials of the anode. In order to overcome the obstacles, the methods of modifying the separators/interface with functional coatings will be introduced as follows. According to the effect of the modified separators to the polysulfides, modified separators could be classified into three different types as modified separators based on adsorption effect, separation effect, and catalytic effect [79] (Figure 8.13).

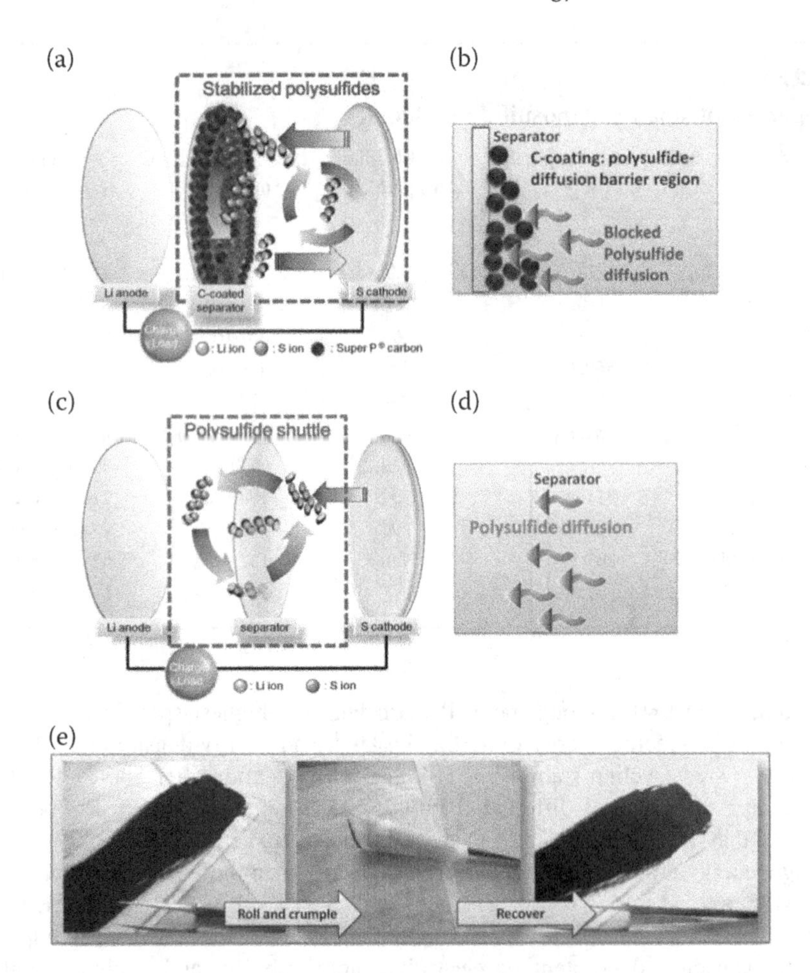

FIGURE 8.13 Schematic cell configuration modification of Li-S cells. (a) Schematic configuration of a Li-S cell with the C-coated separator and (b) the polysulfide-diffusion barrier region. (c) Schematic configuration of a Li-S cell with the Celgard separator and (d) the typical severe polysulfide diffusion. (e) Demonstration of the flexibility and mechanical strength of the C-coated separator [80].

8.4.1 Modified Separators Based on the Adsorption Effect

As to the modified materials based on the adsorption effect, it is found that long chain sulfides could adhere to the coatings of the separators. Accordingly, it results in that the production of the polysulfides could only happen in the cathode side including the cathode surface, electrolytes of the cathode side, and coating materials. According to the basic binding forces and patterns, basically the modified separators could be classified as a physical-adsorption-based one and a chemical-adsorption-based one.

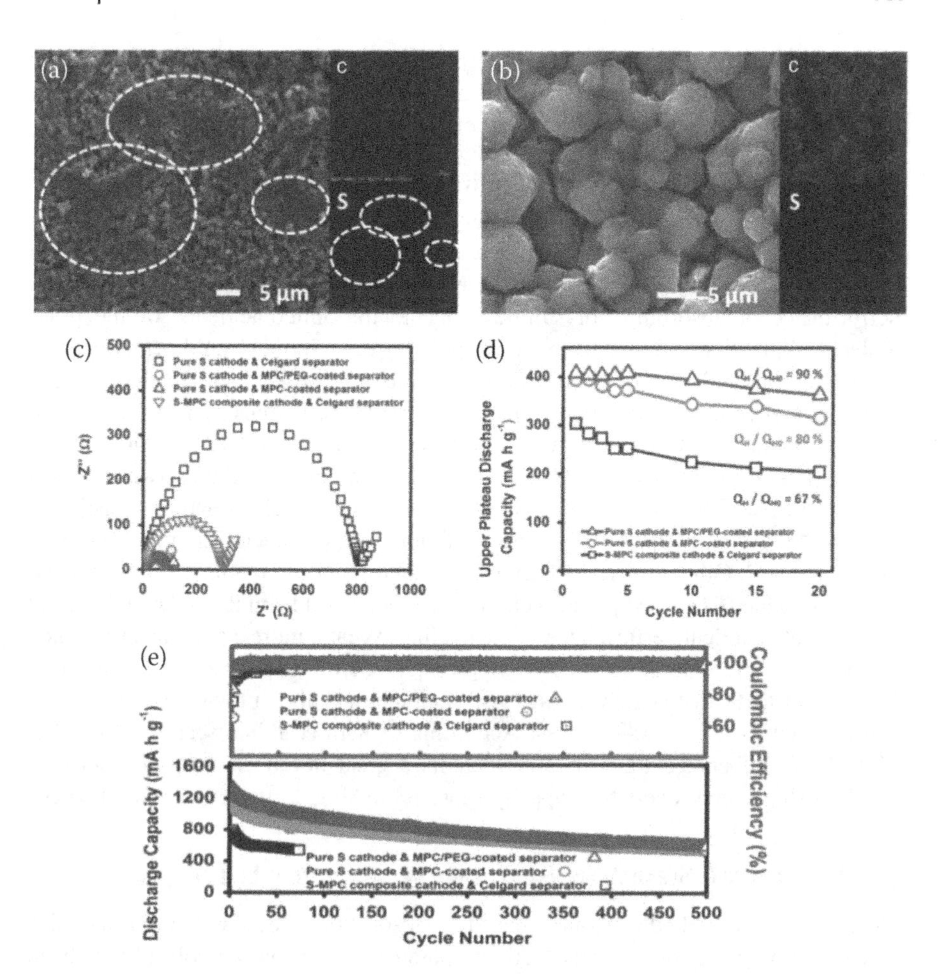

FIGURE 8.14 Comparison of the Li-S cell configurations utilizing a composite cathode, a MPC-coated separator, and a MPC/PEG-coated separator. SEM observation and elemental mapping: (a) cycled composite cathode and (b) the MPC-coated separator. Electrochemical analyses: (c) EIS data of the fresh cathode, (d) upper-plateau discharge capacities (Q_H), and (e) long-term cycle life at a C/5 rate [81].

8.4.1.1 Modified Separators Based on Physical Adsorption Effect

With high porosity and specific surface, nonpolar carbon materials, such as CNTs, carbon nanofibers (CNFs), graphene, carbon spheres, carbon flakes, carbon paper, carbon aerogel, conductive carbon, and activated carbon, are all able to form the intermolecular forces to encapsulate sulfur. In recent years, Manthiram's group has concentrated on finding the proper materials to trigger the adsorption effect [81]. The microporous carbon (MPC) interlayer, with microporous < 2 nm, shows excellent properties for inhibiting the lithium polysulfides from shuttling to the anode side. Moreover, composite coatings like MPC/polyethylene glycol (PEG) are synthesized to render a high reversible capacity and low fading rate of only 0.11% per cycle (Figure 8.14). In addition, conductive carbon (Super P) [80], coated with an ultra-lightweight layer of MWCNTs [82], flower-like hierarchical carbon spheres (FHCS) [83], and ultralight carbon flakes are used to modify the separators one after another, all exhibiting relatively superb effects of adsorbing polysulfides.

8.4.1.2 Modified Separators Based on Chemical Adsorption Effect

Different from physical adsorption, the chemical one takes effect by means of transferring, exchanging, or sharing an electron of the adsorbate molecule or atom and then successfully forming the adsorption chemical bond, resulting in great performance of absorbing polysulfides. Due to the anticoincidence of the polar molecules' positive and negative centers, polar covalent bonds could firmly capture the polysulfides. To generate the polar-polar bonds, researchers tend to design nonpolar carbon materials doped with electron-rich heteroatoms such as N, O, S, B, and P. For example, Chen et al. have developed a honeycomb-like N, P dual-doped carbon (HNPC) [84] modified separator to form the hard bonding between N and Li, P and S, which enables the Li-S cell with a high initial discharge capacity of 1,387 mAh g^{-1} and only 0.06% capacity fading rate per cycle for more than 900 cycles at 1.0 C. Other materials such as combined graphene oxide (GO) or reduced graphene oxide (RGO) [85] with metal oxides also could form the polar-polar bond and present excellent performance while cycling. What's more, some materials such as metal-organic frameworks (MOFs) [86] and MXenes [87] are often used as Lewis acid to form the Lewis acid-base interaction with a Lewis base that polysulfide anions (S_x^{2-}, $4 \leq x \leq 8$) act as. According to Kim et al.'s research [88], a Ni-MOF/MWCNT-coated PE separator inhabits a great initial discharge capacity of 1,358 mAh g^{-1} and reversible capacity of 1,183 mAh g^{-1} after 300 cycles at 0.2C.

8.4.2 Modified Separators Based on the Separation Effect

Other than the modified separators based on adsorption effect, separation may also take effect by using ionic permselective separators to prevent the polysulfides from moving, which otherwise could eliminate the shuttle effect without the loss of active materials. Generally, the electrostatic repulsive force and the steric hindrance are two main forms of the separation effect, exhibiting an alluring prospective for further study.

For one thing, through two molecules with the same charge such as the sulfonated derivatives and polysulfides anions (S_x^{2-}), electrostatic repulsive forces will be generated to perform distinctively during cycling. For example, Kim's group [89] proposed a UiO-66-S/Nafion hybrid-coated separator that enables the Li-S cell with an initial discharge capacity of 1,127.4 mAh g^{-1} at 0.1 C, with OCV remaining 98.4% after 48 hours. And it indicates the effective prohibition of the shuttle effect. For the other, the steric hindrance may otherwise obscure the high-order polysulfides relying on the specific molecules' spatial configuration. For example, materials like 2-D exfoliated [90] vermiculite sheets on separators have been exhibited dramatically on blocking the polysulfides, which enable the Li-S cell with an initial discharge capacity of 1,000 mAh g^{-1} and a Coulombic efficiency of 90.3% for 50 cycles.

8.4.3 Modified Separators Based on the Catalyst Effect

Compared with the other two modified separators that have been mentioned previously, the catalyst-based ones are able to transfer the high-order polysulfides into Li_2S more promptly. It will fundamentally lever the utilization of the sulfides. In addition, the coating materials that are mainly nitrides and nitrogen doped materials, ceramics, and some single atom catalysts (SACs) on modified separators, are able to adsorb the polysulfides as well. Then the coating materials will speed up the reduction of the polysulfides to the molecules. Huang's group [91] has developed a mixture of metal oxides made up of Al, Mg, Fe, Si, and O, which is called attapulgite, to suppress the shuttle effect of polysulfides by producing mounted polar sites to lock polysulfides. Statistically, the cell with modified separators shows great cycling stability after the 300 cycles at 1C and an excellent Coulombic efficiency of 99.4%.

8.5 ANODES

Herein, specific bottlenecks of the lithium metal anode have been listed as follows. For the first one, resulting from the lack of a host for a lithium ion, problems of the lithium metal's unstable composition and infinite bulk change often baffle the researchers. For the second one, the continuous formation of the lithium dendrites tends to penetrate into the separators and then may result in the short circuit of the batteries, or even the explosion of the cell. Thirdly, negative formation of the natural solid-electrolyte interface (SEI), which is generally electric-insulated and ionically conductive, usually leads to the electrochemical efficiency loss of the batteries [92]. In a bid to improve the properties of the lithium anode and then achieve the better performance of the full batteries, two mainstream methods of upgrading the lithium metal anode will be introduced.

8.5.1 Designing the Structured 3-D Lithium Hosts

8.5.1.1 Theory of Anode Interfaces' Kinetics

To effectively deal with the problem of actual areal current density due to the increase of the areal sulfur loading, one of the effective strategies is to design the structured 3-D lithium hosts. Based on the empirical diffusion model, the universally recognized theory of anode interfaces' kinetics, the fundamental principle of the lithium ions and electrons' transportation, could be depicted as:

$$Sand's\ time\ \tau = \pi \cdot D \cdot \left[\frac{eC_0(\mu_a + \mu_{Li^+})}{2 \cdot J \cdot \mu_a} \right]^2$$

τ: time that lithium dendrites begin to grow
D: the diffusion coefficient
e: the elementary charge

C_0: the initial concentration of lithium ion
μ_a: the anionic mobilities
μ_{Li^+}: lithium ion mobilities
J: absolute current density

8.5.1.2 Four Categories of Designing Hosts

Accordingly, the absolute current density in the function above is assumably related with the specific surface area (SSA) of the anode. Until now, the materials designed for the hosts could be generally concluded into four main categories: 3-D copper-based current collectors, high SSA current collectors (Figure 8.16) Figure 8.16, "lithiophilic" 3-D hosts, and the 3-D hosts with functionalized surface.

Naturally inspired by the nodes of lithium-ion batteries (LIBs) and lithium metal batteries (LMBs), copper (Cu) foils were first and applied as the hosts of lithium ions in the Li-S battery systems. Due to the porous 3-D structure, Yang's group once designed a submicron 3-D Cu skeleton and proved that the structure could offer enough space for the sediments of lithium. It is able to effectively prevent the formation of the lithium dendrites. Likewise, other than by the means of in-situ dehydration and reduction of $Cu(OH)_2$ fibers grown on a Cu foil [93] (Figure 8.15) Figure 8.15, the methods such as leaching Zn out of commercial Cu-Zn alloy [92], accumulating Cu nanowires into free-standing networks [94] have also been proven to be feasible to synthesize the Cu-based current collectors. Additionally, 3-D hosts for the lithium anode of the L-S batteries have been further studied by Huang's and Zhang's group [95]. They found that the 3-D fibrous boron matrix derived from irreversible electrochemical dealloying of a Li-B alloy showed great performance with no lithium dendrites even at 10.0 mA cm^{-2} and had a life span of over 2,000 cycles with a high Coulombic efficiency of over 90%.

8.5.2 Construction of Solid Electrolyte Interfaces

Besides the structured 3-D hosts (Figure 8.16) for the lithium deposition, another superb strategy is the formation of stable solid electrolytes interface (SEI). And it has been proved to be both electrically insulative and ionically conductive. To achieve the SEI with excellent properties of high lithium ionic conductivity, proper thickness, compact structure, high mechanical strength, and great chemical stability, researchers have proposed several effective strategies.

8.5.2.1 In-Situ Stabilization of Solid Electrolyte Interfaces

First, to deal with the problem of the shuttle effect of mobile polysulfides through in-situ stabilization of SEI, Aurbach's group has developed lithium nitrate ($LiNO_3$) to be the additive of the electrolytes in order to construct the $LiSO_y/Li_xNO_y$ species and eliminate the destruction of the anode implemented by polysulfides [96]. Mentioned with the unique agents, especially for the Li-S batteries, Liang's group first introduced phosphorous pentasulfide (P_2S_5) [97] in Li-S batteries. It is indicated that soluble additives in the electrolytes could effectively propel the formation of protective film on the anode side in the environment of lithium superionic

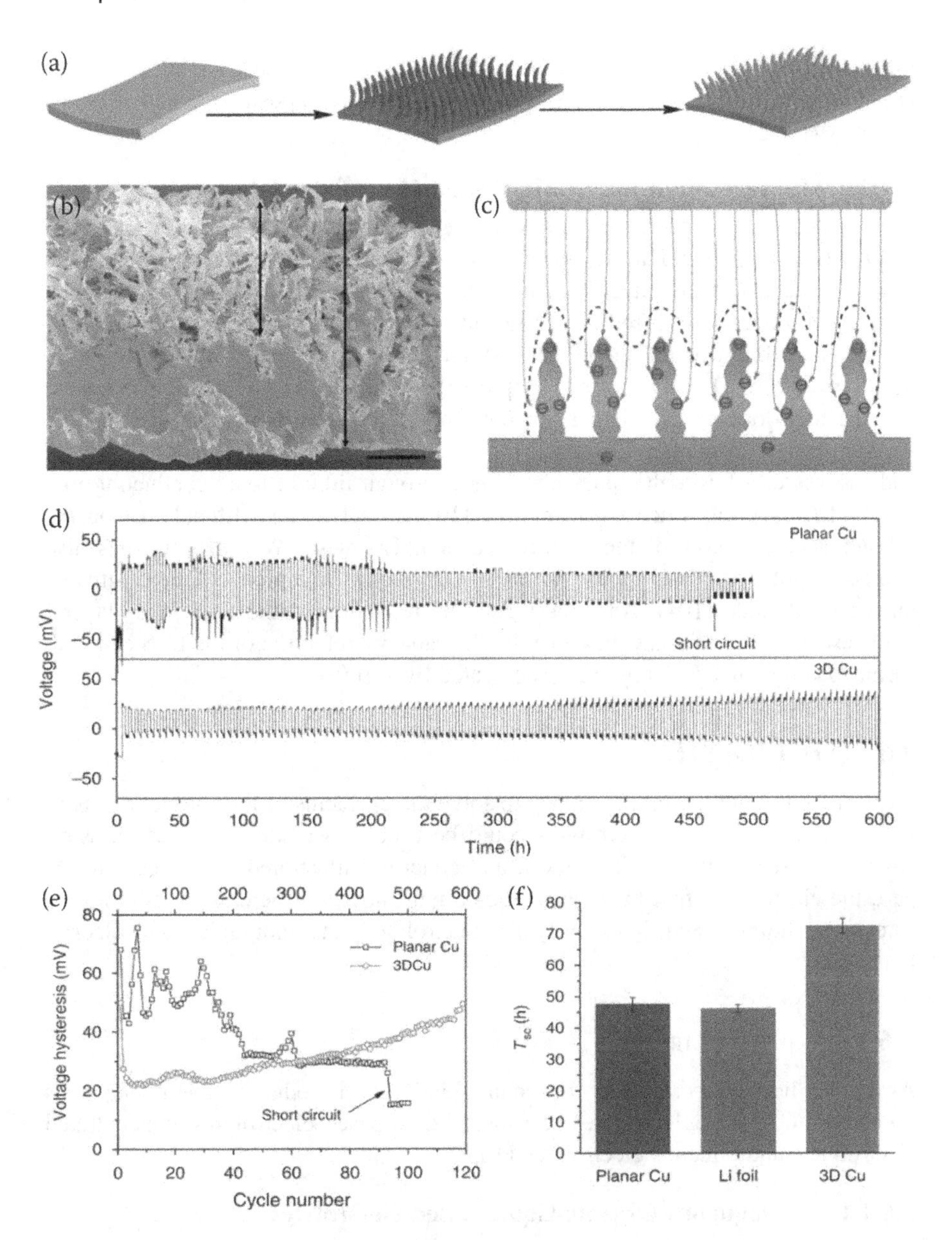

FIGURE 8.15 Preparation and characterization for 3-D Cu foil. (a) Schematic presentation of the procedures to prepare a 3-D porous Cu foil from a planar Cu foil. (b) X-ray diffraction profiles of $Cu(OH)_2$ on the Cu foil and the final 3-D Cu foil. The insets show digital images of the corresponding samples. (c) SEM images of the porous Cu (scale bar, 50 mm). The inset shows the high-magnification image (scale bar, 2 mm) [93].

Li_3PS_4 and counteract the capacity loss. At the same time, Manthiram and his co-workers once proposed the use of additives containing a metal ion with lower reactivity with sulfur than lithium such as copper acetate ($Cu(Ac)_2$) [98] to stabilize the solid electrolyte interface. It basically took effect by fabricating a surface passivation layer consisting of a sulfide matrix and decomposition products of the electrolyte.

8.5.2.2 Fabrication of the Artificial Solid Electrolytes Interface

For the alternative way, the formation of an exotic protecting layer or an "artificial" SEI on the lithium metal anode side shows great prospects in preventing the growth of lithium dendrites and intervening the side reaction from happening. For example, Jing's group once introduced the Al_2O_3 layer with porous structures to effectively protect the lithium metal anode [99]. Moreover, Kozen and his co-workers have developed the ALD Al_2O_3 layer to suppress the lithium corrosion and it is indicated that the high-loading Li-S battery (5 mg cm^{-2} sulfur) exhibits a capacity of over 1,000 mA h g^{-1} even after 100 cycles [114]. To compel the transportation of the Li^+ and successfully lower the inner impedance through lithium ionic conducting materials, Liu' s group once used graphite [100] to act as an artificial SEI on the lithium anode to control the surface reaction. Likewise, Wen's group has also composed polycrystalline lithium nitrides (Li_3N) to act as a proactive layer situated on the anode side [101]. And this led to the reduction of the Li_2S/Li_2S_2 layers' thickness down to 10 μm. Consequently, the capacity retention of the Li-S cell with sulfur loadings of 2.5–3 mg cm^{-2} is elevated over 100%.

8.6 ELECTROLYTES

In order to find out the basic strategy to solve the problems of the shuttle effects of Li-S batteries and how electrolytes could be used to wrestle with these hassles, advanced research will be introduced and concisely illustrated. So herein, in this part, the electrolytes that have been studied until now will be classified into three categories: liquid electrolytes, polymer electrolytes, and inorganic solid electrolytes [103].

8.6.1 Liquid Electrolytes

As to the liquid electrolytes, four main kinds are introduced: the nonaqueous organic-liquid-based electrolytes, the ionic liquid-based electrolytes, concentrated electrolytes, and aqueous electrolytes [103].

8.6.1.1 Nonaqueous Organic-Liquid-Based Electrolytes

First of all, the nonaqueous organic-liquid-based electrolytes are commonly achieved by dissolving the lithium salts into a matrix of one to two different organic solvents. Specifically, the lithium salts that have been found feasible are Li (SO_3CF_3), LiTf, LiTFSI, $LiPF_6$, and $LiClO_4$ and the excellent solvents are carbonates, ethers, tetrahydrofuran (THF), DME, DIOX, sulphones, and glymes [104]. They could be basically divided into the single ones and mixed ones. According to the recent researches, the first strategy to synthetize the useful solvents could be achieved by the utilization of the sulphones such as the methyl isopropyl sulphone (MiPS) [105]. Guo and his co-workers once combined the $LiPF_6$ with MiPS to

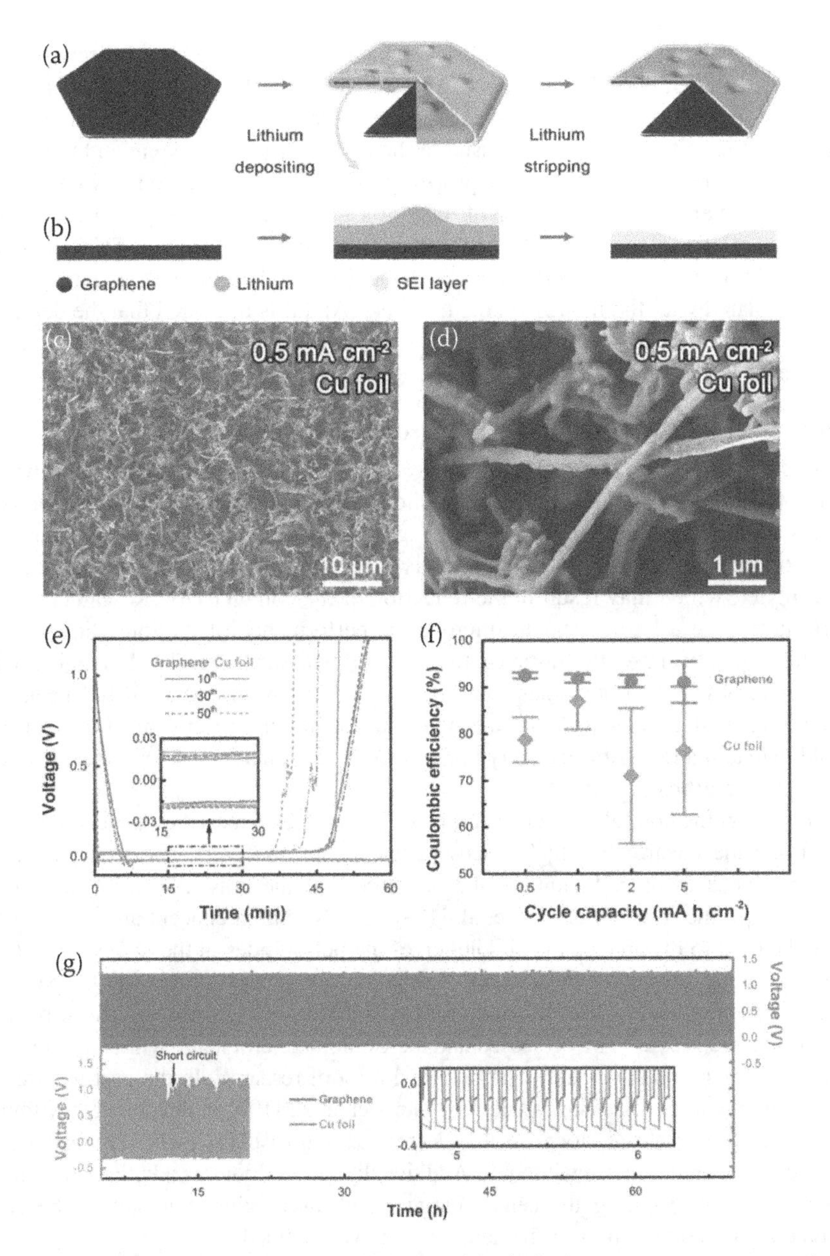

FIGURE 8.16 Schematic diagrams of Li depositing/stripping process on one graphene flake. The diagrams before cycles, after Li depositing, after Li stripping of (a) one graphene flake and (b) its sectional view. (c, d) Li dendrites on Cu foil-based anode at an increased current density of 0.5 mA cm^{-2}. (e) Voltage profiles of the 10th, 30th, and 50th cycle of graphene- and Cu foil-based anode with a cycling capacity of 0.5 mA h cm^{-2} at 0.5 mA cm^{-2}. (f) Average Coulombic efficiency and its variance in the cycles of graphene- and Cu foil-based anode with different cycling capacities at 0.5 mA cm^{-2}. Inset in (g) is an expanded view of the voltage–time curves at the time from 5 to 6 hours [102].

compose a solvent for the Li-S batteries, which shows a high initial discharge capacity of 1,080 mAh g^{-1} and superb rate capability [105]. However, the performance of the cell began to go downside when the temperature changed, which may ascribe to the salts that have been used in the test. As for the solvents in binary and ternary mixtures, they show great properties due to the combined advantages of different solvents. For instance, Kolosnitsyn's group [106] once proposed to mix the tetramethylene sulfone (TMS) with ethers like DIOX, DME, and THF at a ratio of 1:1. Accordingly, the performance of the Li-S cell with such electrolytes is closely related with the molar volume of ethers. And it is indicated that the depth of discharge and the sulfur utilization would increase with the decrease of the molar volume of the ethers.

0.6.1.2 The Ionic Liquid-Based Electrolytes

Secondly, because of their inflammable and involatile nature, ionic liquid-based electrolytes have also received much publicity and are assumed to be used for safe energy storage systems without any leakage or gassing. The most significant problem of the electrolytes is their high viscosities compared with organic liquids electrolytes, which may result in the reduction of the conductivity. Researchers like Park have done a lot of work to improve the performance of the ionic liquid electrolytes in practical use. For instance, the Li-S cell with 0.64 M LITFSI in ionic liquid electrolyte based on a quaternary ammonia cation (DEME) and TFSI has been explored [107]. It is indicated that the cell with an initial capacity of 800 mAh g^{-1} could maintain a reversible capacity of 576 mAh g^{-1} even after 100 cycles and shows a Coulombic efficiency of 98%.

Other than the normal salt concentration of less than 2 M, based on the common ion effect and the electrolytes' high viscosity, the highly concentrated electrolytes have been produced in order to impede the dissolution of the polysulfides intermediates. According to the research of Shin et al. [108], LiTFSI with a concentration up to 5 M could be used to manipulate the dissolution of the polysulfides in the DIOX:DME (1:1 v/v). The study shows that the reduced dissolution of the polysulfides will happen at a higher concentration and the diffusion of the polysulfides will slow down. Accordingly, the overcharge is sure to be decreased and the Coulombic efficiency will be levered up. Furthermore, the additives have also incurred a lot of research. In the past few years, $LiNO_3$ as an additive has been studied by Aurbach et al. [109] in detail and they found that the Li-foils that are stored in the electrolytes with $LiNO_3$ owned a smaller impedance compared to those without it. Additionally, polysulfides as additives also show great performance during the cells' charging and discharging, but only if the concentration and composition of the additives are well defined.

To achieve the highly safe and affordable Li-S batteries, there are also some cheap and noncombustible aqueous electrolytes that have been invented in recent years. By means of dissolving the Li_2S_4/Li_2S redox couple into the aqueous solution to construct a cathode and combining the Li metal with organic electrolytes made up of 1 M $LiClO_4$ in EC/DMC, Zou's group [110] has successfully produced a hybrid Li-S cell to achieve a high charge and discharge capacities of 1,030 mA h g^{-1} and 1,129 mA h g^{-1}, coupled with a high voltage plateau at 2.5–2.7 V vs Li^+/Li

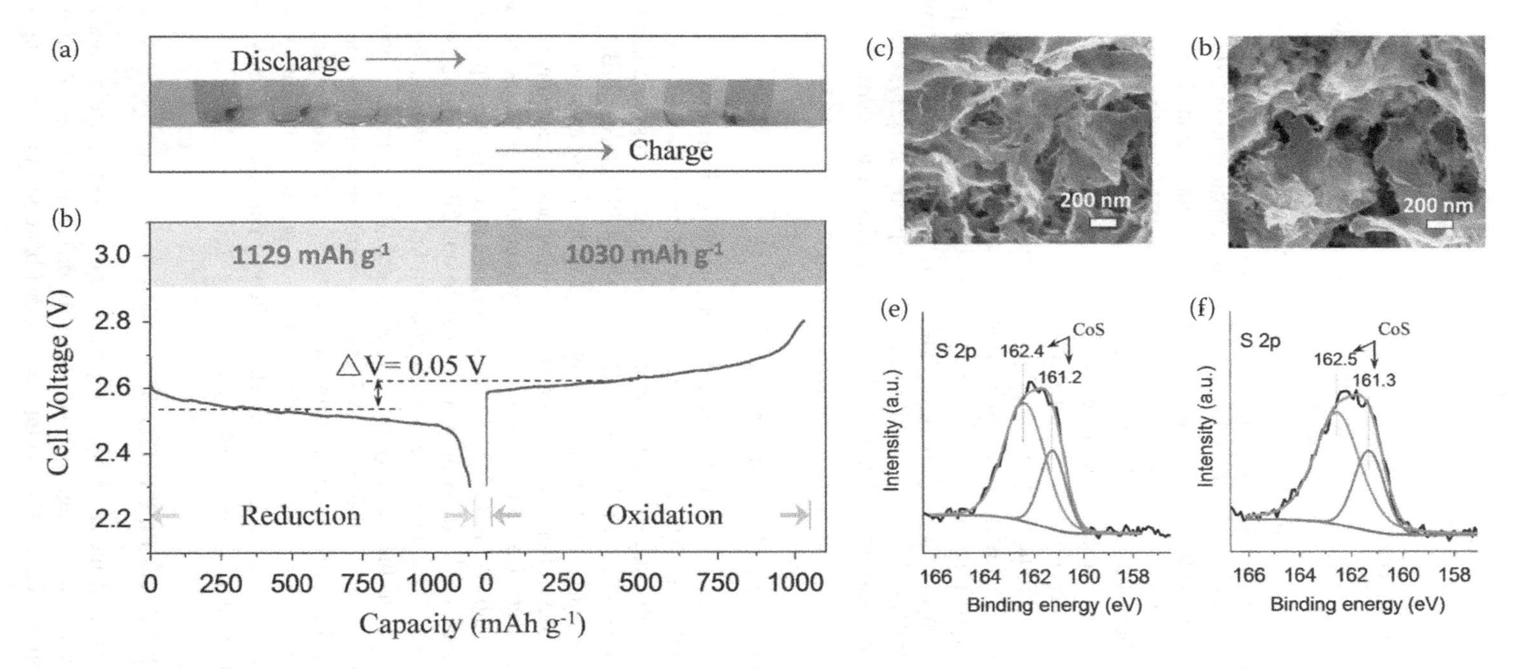

FIGURE 8.17 (a) An image of color gradients of the Li_2S_4/Li_2S electrolyte, from yellow to transparent with increasing S^{2-} concentration. (b) Initial discharge/charge curve of the aqueous lithium–polysulfide battery using 0.1 M Li_2S_4 solution at 0.2 mA cm^{-2}. SEM images of CoS deposited on a brass mesh before (c) and after (d) the initial discharge/charge process. S 2p XPS spectra of CoS deposited on a brass mesh before (e) and after (f) the initial discharge/charge process [110].

(Figure 8.17). From then on, cells using saturated aqueous electrolytes with 2.3 M Li_2S_4 are able to exhibit an excellent specific energy of 654 Wh kg^{-1}.

8.6.1.3 Polymer Electrolytes

Nowadays, the polymer electrolytes including the solid polymer electrolytes (SPEs) and the plasticized or gel polymer electrolytes (GPEs) have attracted much attention, serving as one of the most promising alternative materials to work out the problems of the shuttle effect. One of the surprising breakthrough of the utilization of SPEs in the Li-S cells could be the employment of nanostructured ZrO_2 as the additives [111] in the electrolytes. The use of SPEs with no more than 10 wt% ZrO_2 has been proved to effectively promote both the conductivity and the lithium-ion transference number and the stability of the electrode-electrolyte interface by coating a greatly dispersed layer on the anode. Compared to the SPEs, GPEs could perform with acceptably high ionic conductivities ($>10^{-3}$ S cm^{-1}) due to the special gel structures, which could be produced by plasticizing polymer matrixes with organic carbonates. For instance, researchers once fabricated a PVdF-HFP-based GEP [112], incorporating ionic-liquid-based electrolytes (0.5M LiTFSI-[P14][TFSI]). It shows an ionic conductivity of 2.54×10^{-4} S cm^{-1} at room temperature and enables the Li-S cell with an initial discharge capacity of 1,217.7 mAh g^{-1} and a reversible capacity of 818 mAh g^{-1} after 20 cycles at the current density of 50 mA g^{-1}.

8.6.2 Inorganic Solid Electrolytes

To completely eliminate the shuttle effect, the appearance of inorganic solid electrolytes has brought a lime light to the research. With the limited Li-ion transference numbers close to unity, i.e., only the lithium ion could come across, the inorganic solid electrolytes such as Li_2S-SiS_2, Li_2S-P_2S_5, thio-LISI-CONs, $LiBH_4$, and Li_3PS_4 show superb properties of suppressing the growth of lithium dendrites and impeding both the dissolution and diffusion of the polysulfides. Accordingly, some sulfide gases that were first proposed in the early studies to serve as the solid electrolytes have enabled the Li-S cell with an excellent reversible capability of more than 1,000 mA h g^{-1}. Afterwards, nanosized porous β-Li_3PS_4 particles [113], which are prepared by the researchers through controlled wet chemical methods, have been used to act as the electrolytes. As is estimated, those particles show high ionic conductivity of about 1.6×10^{-4} S cm^{-1} at room temperature and overwhelming chemical stability in contrast to the traditional solid electrolytes. After that, all-solid-state Li-S batteries have been produced based on the nano-porous β-Li_3PS_4 particles and Li_3PS_{4+5} [115], which show great cycle stability and a surprising initial discharge capacity of 1,272 mA h g^{-1}.

8.7 SUMMARY AND PERSPECTIVE

According to the state-of-the art research, the status of the Li-S batteries is generally both promising and challenging, with considerably different operating principles and structure-property relationships compared to other batteries. For one thing, until now, magnificent progresses have been made in respect to the sulfur cathode, Li metal anode, the separator modification, and the design of the electrolytes, considerably promoting the performance of the Li-S batteries. However, for the other one, many of the underlying mechanisms and chemical reaction processes are still not explored and illustrated, fundamentally prohibiting the further development of the Li-S batteries. Therefore, in the future work, under the guidance of density functional theory and massive previous research achievements, more efforts should be made on the improvements of comprehensive properties of the Li-S batteries, including the excellent energy density, the great Coulombic efficiency, high reversible capacities, and long cycling life. In a bid to achieve this, it is of vital importance to concentrate on the design and synthesis of sulfur cathode, fabrication of the stable SEI between the anode and electrolytes, and exploration of novel electrolytes and modified separators to eliminate the shuttle effect. With these further advancements, the commercial production of Li-S batteries could be achieved in the near future.

REFERENCES

[1] Goodenough, J. B., How we made the Li-ion rechargeable battery. *Nature Electronics* 2018, *1* (3), 204.

[2] Manthiram, A.; Fu, Y.; Chung, S.-H.; Zu, C.; Su, Y.-S., Rechargeable lithium–sulfur batteries. *Chemical Reviews* 2014, *114* (23), 11751–11787.

[3] Bruce, P. G.; Freunberger, S. A.; Hardwick, L. J.; Tarascon, J. M., Li-O2 and Li-S batteries with high energy storage. *Nature Materials* 2011, *11* (1), 19–29.

[4] Liu, J.; Zhang, J.-G.; Yang, Z.; Lemmon, J. P.; Imhoff, C.; Graff, G. L.; Li, L.; Hu, J.; Wang, C.; Xiao, J.; Xia, G.; Viswanathan, V. V.; Baskaran, S.; Sprenkle, V.; Li, X.; Shao, Y.; Schwenzer, B., Materials science and materials chemistry for large scale electrochemical energy storage: From transportation to electrical grid. *Advanced Functional Materials* 2013, *23* (8), 929–946.

[5] Mikhaylik, Y. V.; Akridge, J. R., Polysulfide shuttle study in the Li/S battery system. *Journal of The Electrochemical Society* 2004, *151* (11).

[6] Zhang, S. S., Liquid electrolyte lithium/sulfur battery: Fundamental chemistry, problems, and solutions. *Journal of Power Sources* 2013, *231*, 153–162.

[7] Cheon, S.-E.; Ko, K.-S.; Cho, J.-H.; Kim, S.-W.; Chin, E.-Y.; Kim, H.-T., Rechargeable lithium sulfur battery. *Journal of the Electrochemical Society* 2003, *150* (6).

[8] Ryu, H.; Ahn, H.; Kim, K.; Ahn, J.; Lee, J.; Cairns, E., Self-discharge of lithium–sulfur cells using stainless-steel current-collectors. *Journal of Power Sources* 2005, *140* (2), 365–369.

[9] Ryu, H. S.; Ahn, H. J.; Kim, K. W.; Ahn, J. H.; Cho, K. K.; Nam, T. H., Self-discharge characteristics of lithium/sulfur batteries using TEGDME liquid electrolyte. *Electrochimica Acta* 2006, *52* (4), 1563–1566.

[10] Wang, L.; Hu, X., Recent advances in porous carbon materials for electrochemical energy storage. *Chemistry: An Asian Journal* 2018, *13* (12), 1518–1529.

[11] Wei Seh, Z.; Li, W.; Cha, J. J.; Zheng, G.; Yang, Y.; McDowell, M. T.; Hsu, P. C.; Cui, Y., Sulphur-TiO2 yolk-shell nanoarchitecture with internal void space for long-cycle lithium-sulphur batteries. *Nature Communications*2013, *4*, 1331.

[12] Evers, S.; Nazar, L. F., New approaches for high energy density lithium–sulfur battery cathodes. *Accounts of Chemical Research* 2013, *46* (5), 1135–1143.

[13] Wang, M.; Xia, X.; Zhong, Y.; Wu, J.; Xu, R.; Yao, Z.; Wang, D.; Tang, W.; Wang, X.; Tu, J., Porous carbon hosts for lithium–sulfur batteries. *Chemistry – A European Journal* 2019, *25* (15), 3710–3725.

[14] Zhou, J.; Guo, Y.; Liang, C.; Yang, J.; Wang, J.; Nuli, Y., Confining small sulfur molecules in peanut shell-derived microporous graphitic carbon for advanced lithium sulfur battery. *Electrochimica Acta* 2018, *273*, 127–135.

[15] Han, J.; Li, Y.; Li, S.; Long, P.; Cao, C.; Cao, Y.; Wang, W.; Feng, Y.; Feng, W., A low cost ultra-microporous carbon scaffold with confined chain-like sulfur molecules as a superior cathode for lithium–sulfur batteries. *Sustainable Energy & Fuels* 2018, *2* (10), 2187–2196.

[16] Wu, H. B.; Wei, S.; Zhang, L.; Xu, R.; Hng, H. H.; Lou, X. W., Embedding sulfur in MOF-derived microporous carbon polyhedrons for lithium-sulfur batteries. *Chemistry* 2013, *19* (33), 10804–10808.

[17] Wang, D. W.; Zhou, G.; Li, F.; Wu, K. H.; Lu, G. Q.; Cheng, H. M.; Gentle, I. R., A microporous-mesoporous carbon with graphitic structure for a high-rate stable sulfur cathode in carbonate solvent-based Li-S batteries. *Physical Chemistry Chemical Physics* 2012, *14* (24), 8703–8710.

[18] Wang, D.-W.; Zeng, Q.; Zhou, G.; Yin, L.; Li, F.; Cheng, H.-M.; Gentle, I. R.; Lu, G. Q. M., Carbon–sulfur composites for Li–S batteries: Status and prospects. *Journal of Materials Chemistry A* 2013, *1* (33).

[19] Xin, S.; Gu, L.; Zhao, N. H.; Yin, Y. X.; Zhou, L. J.; Guo, Y. G.; Wan, L. J., Smaller sulfur molecules promise better lithium-sulfur batteries. *Journal of the American Chemical Society* 2012, *134* (45), 18510–18513.

[20] Ji, X.; Lee, K. T.; Nazar, L. F., A highly ordered nanostructured carbon-sulphur cathode for lithium-sulphur batteries. *Nature Materials* 2009, *8* (6), 500–506.

[21] Rybarczyk, M. K.; Peng, H.-J.; Tang, C.; Lieder, M.; Zhang, Q.; Titirici, M.-M., Porous carbon derived from rice husks as sustainable bioresources: Insights into the role of micro-/mesoporous hierarchy in hosting active species for lithium–sulphur batteries. *Green Chemistry* 2016, *18* (19), 5169–5179.

[22] Chen, S.; Sun, B.; Xie, X.; Mondal, A. K.; Huang, X.; Wang, G., Multi-chambered micro/mesoporous carbon nanocubes as new polysulfides reserviors for lithium–sulfur batteries with long cycle life. *Nano Energy* 2015, *16*, 268–280.

[23] Zhang, Y.-B.; Zhao, Y.; Hao, X.-F.; Ma, Y.-c.; Wu, Y.; Li, G.-l.; Cao, J.-j.; Yan, Y.; Qiao, L.-z.; Hao, C., Sulfur encapsulated in a wafer-like carbon substrate with interconnected meso/micropores for high-performance lithium–sulfur batteries. *Inorganic Chemistry Frontiers* 2019, *6* (11), 3264–3269.

[24] Fu, A.; Wang, C.; Pei, F.; Cui, J.; Fang, X.; Zheng, N., Recent advances in hollow porous carbon materials for lithium-sulfur batteries. *Small* 2019, *15* (10), e1804786.

[25] Jayaprakash, N.; Shen, J.; Moganty, S. S.; Corona, A.; Archer, L. A., Porous hollow carbon@sulfur composites for high-power lithium-sulfur batteries. *Angewandte Chemie International Edition in English* 2011, *50* (26), 5904–5908.

[26] He, G.; Evers, S.; Liang, X.; Cuisinier, M.; Garsuch, A.; Nazar, L. F., Tailoring porosity in carbon nanospheres for lithium–sulfur battery cathodes. *ACS Nano* 2013, *7* (12), 10920–10930.

[27] Geim, A. K., Graphene: Status and prospects. *Science* 2009, *324* (5934), 1530–1534.

[28] Novoselov, K. S.; Fal'ko, V. I.; Colombo, L.; Gellert, P. R.; Schwab, M. G.; Kim, K., A roadmap for graphene. *Nature* 2012, *490* (7419), 192–200.

[29] Ren, W.; Cheng, H. M., The global growth of graphene. *Nature Nanotechnology* 2014, *9* (10), 726–730.

[30] Fang, R.; Zhao, S.; Sun, Z.; Wang, D. W.; Cheng, H. M.; Li, F., More reliable lithium-sulfur batteries: Status, solutions and prospects. *Advanced Materials* 2017, *29* (48).
[31] Wang, H.; Yang, Y.; Liang, Y.; Robinson, J. T.; Li, Y.; Jackson, A.; Cui, Y.; Dai, H., Graphene-wrapped sulfur particles as a rechargeable lithium-sulfur battery cathode material with high capacity and cycling stability. *Nano Letters* 2011, *11* (7), 2644–2647.
[32] Lin, T.; Tang, Y.; Wang, Y.; Bi, H.; Liu, Z.; Huang, F.; Xie, X.; Jiang, M., Scotch-tape-like exfoliation of graphite assisted with elemental sulfur and graphene–sulfur composites for high-performance lithium-sulfur batteries. *Energy & Environmental Science* 2013, *6* (4).
[33] Fang, R.; Zhao, S.; Hou, P.; Cheng, M.; Wang, S.; Cheng, H. M.; Liu, C.; Li, F., 3D interconnected electrode materials with ultrahigh areal sulfur loading for Li-S batteries. *Advanced Materials* 2016, *28* (17), 3374–3382.
[34] Ji, L.; Rao, M.; Zheng, H.; Zhang, L.; Li, Y.; Duan, W.; Guo, J.; Cairns, E. J.; Zhang, Y., Graphene oxide as a sulfur immobilizer in high performance lithium/sulfur cells. *Journal of the American Chemical Society* 2011, *133* (46), 18522–18525.
[35] Ding, B.; Yuan, C.; Shen, L.; Xu, G.; Nie, P.; Lai, Q.; Zhang, X., Chemically tailoring the nanostructure of graphenenanosheets to confine sulfur for high-performance lithium-sulfur batteries. *Journal of Materials Chemistry A* 2013, *1* (4), 1096–1101.
[36] Guo, J.; Xu, Y.; Wang, C., Sulfur-impregnated disordered carbon nanotubes cathode for lithium-sulfur batteries. *Nano Letters* 2011, *11* (10), 4288–4294.
[37] Fang, R.; Zhao, S.; Wang, D.-W.; Sun, Z.; Cheng, H.-M.; Li, F., Micro-macroscopic coupled electrode architecture for high-energy-density lithium–sulfur batteries. *ACS Applied Energy Materials* 2019, *2* (10), 7393–7402.
[38] Landi, B. J.; Ganter, M. J.; Cress, C. D.; DiLeo, R. A.; Raffaelle, R. P., Carbon nanotubes for lithium ion batteries. *Energy & Environmental Science* 2009, *2* (6).
[39] Jia, X.; Wei, F., Advances in production and applications of carbon nanotubes. *Topics in Current Chemistry* 2017, *375* (1), 18.
[40] Jin, F.; Xiao, S.; Lu, L.; Wang, Y., Efficient activation of high-loading sulfur by small CNTs confined inside a large CNT for high-capacity and high-rate lithium-sulfur batteries. *Nano Letters* 2016, *16* (1), 440–447.
[41] Yuan, L.; Yuan, H.; Qiu, X.; Chen, L.; Zhu, W., Improvement of cycle property of sulfur-coated multi-walled carbon nanotubes composite cathode for lithium/sulfur batteries. *Journal of Power Sources* 2009, *189* (2), 1141–1146.
[42] Zhang, S.-M.; Zhang, Q.; Huang, J.-Q.; Liu, X.-F.; Zhu, W.; Zhao, M.-Q.; Qian, W.-Z.; Wei, F., Composite cathodes containing SWCNT@S coaxial nanocables: Facile synthesis, surface modification, and enhanced performance for Li-ion storage. *Particle & Particle Systems Characterization* 2013, *30* (2), 158–165.
[43] Fang, R.; Li, G.; Zhao, S.; Yin, L.; Du, K.; Hou, P.; Wang, S.; Cheng, H.-M.; Liu, C.; Li, F., Single-wall carbon nanotube network enabled ultrahigh sulfur-content electrodes for high-performance lithium-sulfur batteries. *Nano Energy* 2017, *42*, 205–214.
[44] Huang, H.; Liu, C. H.; Wu, Y.; Fan, S., Aligned carbon nanotube composite films for thermal management. *Advanced Materials* 2005, *17* (13), 1652–1656.
[45] Moon, S.; Jung, Y. H.; Jung, W. K.; Jung, D. S.; Choi, J. W.; Kim, D. K., Encapsulated monoclinic sulfur for stable cycling of Li-S rechargeable batteries. *Advanced Materials* 2013, *25* (45), 6547–6553.
[46] Fu, C.; Oviedo, M. B.; Zhu, Y.; von Wald Cresce, A.; Xu, K.; Li, G.; Itkis, M. E.; Haddon, R. C.; Chi, M.; Han, Y.; Wong, B. M.; Guo, J., Confined lithium-sulfur

reactions in narrow-diameter carbon nanotubes reveal enhanced electrochemical reactivity. *ACS Nano* 2018, *12* (10), 9775–9784.
[47] Hu, G.; Sun, Z.; Shi, C.; Fang, R.; Chen, J.; Hou, P.; Liu, C.; Cheng, H. M.; Li, F., A sulfur-rich copolymer@CNT hybrid cathode with dual-confinement of polysulfides for high-performance lithium-sulfur batteries. *Advanced Materials* 2017, *29* (11).
[48] Zhao, Y.; Wu, W.; Li, J.; Xu, Z.; Guan, L., Encapsulating MWNTs into hollow porous carbon nanotubes: A tube-in-tube carbon nanostructure for high-performance lithium-sulfur batteries. *Advanced Materials* 2014, *26* (30), 5113–5118.
[49] Manthiram, A.; Chung, S. H.; Zu, C., Lithium-sulfur batteries: Progress and prospects. *Advanced Materials* 2015, *27* (12), 1980–2006.
[50] Fu, Y.; Su, Y.-S.; Manthiram, A., Sulfur-polypyrrole composite cathodes for lithium-sulfur batteries. *Journal of the Electrochemical Society* 2012, *159* (9), A1420–A1424.
[51] Wang, J.; Yang, J.; Wan, C.; Du, K.; Xie, J.; Xu, N., Sulfur composite cathode materials for rechargeable lithium batteries. *Advanced Functional Materials* 2003, *13* (6), 487–492.
[52] Wang, J.; Chen, J.; Konstantinov, K.; Zhao, L.; Ng, S. H.; Wang, G. X.; Guo, Z. P.; Liu, H. K., Sulphur-polypyrrole composite positive electrode materials for rechargeable lithium batteries. *Electrochimica Acta* 2006, *51* (22), 4634–4638.
[53] Yang, H.; Chen, J.; Yang, J.; Wang, J., Prospect of sulfurized pyrolyzed poly (acrylonitrile) (S@pPAN) cathode materials for rechargeable lithium batteries. *Angewandte Chemie International Edition in English* 2020, *59* (19), 7306–7318.
[54] Wu, F.; Chen, J.; Li, L.; Zhao, T.; Chen, R., Improvement of rate and cycle performence by rapid polyaniline coating of a MWCNT/sulfur cathode. *The Journal of Physical Chemistry C* 2011, *115* (49), 24411–24417.
[55] Wu, F.; Wu, S.; Chen, R.; Chen, J.; Chen, S., Sulfur–polythiophene composite cathode materials for rechargeable lithium batteries. *Electrochemical and Solid-State Letters* 2010, *13* (4).
[56] Chen, H.; Dong, W.; Ge, J.; Wang, C.; Wu, X.; Lu, W.; Chen, L., Ultrafine sulfur nanoparticles in conducting polymer shell as cathode materials for high performance lithium/sulfur batteries. *Scientific Reports* 2013, *3*, 1910.
[57] Manthiram, A.; Fu, Y.; Chung, S. H.; Zu, C.; Su, Y. S., Rechargeable lithium-sulfur batteries. *Chemical Reviews* 2014, *114* (23), 11751–11787.
[58] Yu, M.; Li, R.; Wu, M.; Shi, G., Graphene materials for lithium–sulfur batteries. *Energy Storage Materials* 2015, *1*, 51–73.
[59] Kim, J.-H.; Kim, H.-S.; Kang, Y.-M.; Song, M.-S.; Rajendran, S.; Han, S.-C.; Jung, D.-H.; Lee, J.-Y., Carbon-supported and unsupported Pt anodes for direct borohydride liquid fuel cells. *Journal of the Electrochemical Society* 2004, *151* (7).
[60] Whittingham, M. S., Lithium batteries and cathode materials. *Chemical Reviews* 2004, *104* (10), 4271–4302.
[61] Wang, D. Y.; Guo, W.; Fu, Y., Organosulfides: An emerging class of cathode materials for rechargeable lithium batteries. *Accounts of Chemical Research*2019, *52* (8), 2290–2300.
[62] Zhang, X.; Chen, K.; Sun, Z.; Hu, G.; Xiao, R.; Cheng, H.-M.; Li, F., Structure-related electrochemical performance of organosulfur compounds for lithium–sulfur batteries. *Energy & Environmental Science* 2020, *13* (4), 1076–1095.
[63] Zhou, J.; Qian, T.; Xu, N.; Wang, M.; Ni, X.; Liu, X.; Shen, X.; Yan, C., Selenium-doped cathodes for lithium-organosulfur batteries with greatly improved volumetric capacity and coulombic efficiency. *Advanced Materials* 2017, *29* (33).

[64] Bhargav, A.; Bell, M. E.; Karty, J.; Cui, Y.; Fu, Y., A class of organopolysulfides as liquid cathode materials for high-energy-density lithium batteries. *ACS Applied Materials & Interfaces* 2018, *10* (25), 21084–21090.

[65] Gomez, I.; Leonet, O.; Alberto Blazquez, J.; Grande, H.-J.; Mecerreyes, D., Poly (anthraquinonyl sulfides): High capacity redox polymers for energy storage. *ACS Macro Letters* 2018, *7* (4), 419–424.

[66] Preefer, M. B.; Oschmann, B.; Hawker, C. J.; Seshadri, R.; Wudl, F., High sulfur content material with stable cycling in lithium-sulfur batteries. *Angewandte Chemie International Edition* 2017, *56* (47), 15118–15122.

[67] Kang, H.; Kim, H.; Park, M. J., Sulfur-rich polymers with functional linkers for high-capacity and fast-charging lithium-sulfur batteries. *Advanced Energy Materials* 2018, *8* (32).

[68] Chung, W. J.; Griebel, J. J.; Kim, E. T.; Yoon, H.; Simmonds, A. G.; Ji, H. J.; Dirlam, P. T.; Glass, R. S.; Wie, J. J.; Nguyen, N. A.; Guralnick, B. W.; Park, J.; Somogyi, A.; Theato, P.; Mackay, M. E.; Sung, Y. E.; Char, K.; Pyun, J., The use of elemental sulfur as an alternative feedstock for polymeric materials. *Nature Chemistry* 2013, *5* (6), 518–524.

[69] Wu, M.; Cui, Y.; Bhargav, A.; Losovyj, Y.; Siegel, A.; Agarwal, M.; Ma, Y.; Fu, Y., Organotrisulfide: A high capacity cathode material for rechargeable lithium batteries. *Angewandte Chemie International Edition in English* 2016, *55* (34), 10027–10031.

[70] Wu, M.; Bhargav, A.; Cui, Y.; Siegel, A.; Agarwal, M.; Ma, Y.; Fu, Y., Highly reversible diphenyl trisulfide catholyte for rechargeable lithium batteries. *ACS Energy Letters* 2016, *1* (6), 1221–1226.

[71] Guo, W.; Wawrzyniakowski, Z. D.; Cerda, M. M.; Bhargav, A.; Pluth, M. D.; Ma, Y.; Fu, Y., Bis(aryl) tetrasulfides as cathode materials for rechargeable lithium batteries. *Chemistry* 2017, *23* (67), 16941–16947.

[72] Bhargav, A.; Patil, S. V.; Fu, Y., A phenyl disulfide@CNT composite cathode for rechargeable lithium batteries. *Sustainable Energy & Fuels* 2017, *1* (5), 1007–1012.

[73] Bhargav, A.; Ma, Y.; Shashikala, K.; Cui, Y.; Losovyj, Y.; Fu, Y., The unique chemistry of thiuram polysulfides enables energy dense lithium batteries. *Journal of Materials Chemistry A* 2017, *5* (47), 25005–25013.

[74] Paczesny, J.; Wolska-Pietkiewicz, M.; Binkiewicz, I.; Wróbel, Z.; Wadowska, M.; Matuła, K.; Dzięcielewski, I.; Pociecha, D.; Smalc-Koziorowska, J.; Lewiński, J.; Hołyst, R., Towards organized hybrid nanomaterials at the air/water interface based on liquid-crystal/ZnO nanocrystals. *Chemistry – A European Journal* 2015, *21* (47), 16941–16947.

[75] Cui, Y.; Ackerson, J. D.; Ma, Y.; Bhargav, A.; Karty, J. A.; Guo, W.; Zhu, L.; Fu, Y., Phenyl selenosulfides as cathode materials for rechargeable lithium batteries. *Advanced Functional Materials* 2018, *28* (31).

[76] Guo, W.; Bhargav, A.; Ackerson, J. D.; Cui, Y.; Ma, Y.; Fu, Y., Mixture is better: Enhanced electrochemical performance of phenyl selenosulfide in rechargeable lithium batteries. *Chemical Communications*2018, *54* (64), 8873–8876.

[77] Li, F.; Si, Y.; Liu, B.; Li, Z.; Fu, Y., Lithium benzenedithiolate catholytes for rechargeable lithium batteries. *Advanced Functional Materials* 2019, *29* (32), 1902223.

[78] Peng, H.-J.; Huang, J.-Q.; Cheng, X.-B.; Zhang, Q., Review on high-loading and high-energy lithium-sulfur batteries. *Advanced Energy Materials* 2017, *7* (24).

[79] Li, S.; Zhang, W.; Zheng, J.; Lv, M.; Song, H.; Du, L., Inhibition of polysulfide shuttles in Li–S batteries: Modified separators and solid-state electrolytes. *Advanced Energy Materials* 2020, *11* (2).

[80] Chung, S.-H.; Manthiram, A., Bifunctional separator with a light-weight carbon-coating for dynamically and statically stable lithium-sulfur batteries. *Advanced Functional Materials* 2014, *24* (33), 5299–5306.
[81] Chung, S. H.; Manthiram, A., A polyethylene glycol-supported microporous carbon coating as a polysulfide trap for utilizing pure sulfur cathodes in lithium-sulfur batteries. *Advanced Materials* 2014, *26* (43), 7352–7357.
[82] Chung, S. H.; Manthiram, A., High-performance Li-S batteries with an ultra-lightweight MWCNT-coated separator. *Journal of Physical Chemistry Letters* 2014, *5* (11), 1978–1983.
[83] Liao, H.; Zhang, H.; Hong, H.; Li, Z.; Lin, Y., Novel flower-like hierarchical carbon sphere with multi-scale pores coated on PP separator for high-performance lithium-sulfur batteries. *Electrochimica Acta* 2017, *257*, 210–216.
[84] Zeng, P.; Huang, L.; Zhang, X.; Zhang, R.; Wu, L.; Chen, Y., Long-life and high-areal-capacity lithium-sulfur batteries realized by a honeycomb-like N, P dual-doped carbon modified separator. *Chemical Engineering Journal* 2018, *349*, 327–337.
[85] Cheng, P.; Guo, P.; Liu, D.; Wang, Y.; Sun, K.; Zhao, Y.; He, D., Fe3O4/RGO modified separators to suppress the shuttle effect for advanced lithium-sulfur batteries. *Journal of Alloys and Compounds* 2019, *784*, 149–156.
[86] Zheng, J.; Tian, J.; Wu, D.; Gu, M.; Xu, W.; Wang, C.; Gao, F.; Engelhard, M. H.; Zhang, J. G.; Liu, J.; Xiao, J., Lewis acid-base interactions between polysulfides and metal organic framework in lithium sulfur batteries. *Nano Letters* 2014, *14* (5), 2345–2352.
[87] Li, N.; Cao, W.; Liu, Y.; Ye, H.; Han, K., Impeding polysulfide shuttling with a three-dimensional conductive carbon nanotubes/MXene framework modified separator for highly efficient lithium-sulfur batteries. *Colloids and Surfaces A: Physicochemical and Engineering Aspects* 2019, *573*, 128–136.
[88] Lee, D. H.; Ahn, J. H.; Park, M.-S.; Eftekhari, A.; Kim, D.-W., Metal-organic framework/carbon nanotube-coated polyethylene separator for improving the cycling performance of lithium-sulfur cells. *Electrochimica Acta* 2018, *283*, 1291–1299.
[89] Kim, S. H.; Yeon, J. S.; Kim, R.; Choi, K. M.; Park, H. S., A functional separator coated with sulfonated metal–organic framework/Nafion hybrids for Li–S batteries. *Journal of Materials Chemistry A* 2018, *6* (48), 24971–24978.
[90] Xu, R.; Sun, Y.; Wang, Y.; Huang, J.; Zhang, Q., Two-dimensional vermiculite separator for lithium sulfur batteries. *Chinese Chemical Letters* 2017, *28* (12), 2235–2238.
[91] Sun, W.; Sun, X.; Akhtar, N.; Li, C.; Wang, W.; Wang, A.; Wang, K.; Huang, Y., Attapulgite nanorods assisted surface engineering for separator to achieve high-performance lithium–sulfur batteries. *Journal of Energy Chemistry* 2020, *48*, 364–374.
[92] Yun, Q.; He, Y. B.; Lv, W.; Zhao, Y.; Li, B.; Kang, F.; Yang, Q. H., Chemical dealloying derived 3D porous current collector for Li metal anodes. *Advanced Materials* 2016, *28* (32), 6932–6939.
[93] Yang, C. P.; Yin, Y. X.; Zhang, S. F.; Li, N. W.; Guo, Y. G., Accommodating lithium into 3D current collectors with a submicron skeleton towards long-life lithium metal anodes. *Nature Communications* 2015, *6*, 8058.
[94] Lu, L. L.; Ge, J.; Yang, J. N.; Chen, S. M.; Yao, H. B.; Zhou, F.; Yu, S. H., Free-standing copper nanowire network current collector for improving lithium anode performance. *Nano Letters* 2016, *16* (7), 4431–4437.

[95] Cheng, X. B.; Peng, H. J.; Huang, J. Q.; Wei, F.; Zhang, Q., Dendrite-free nanostructured anode: entrapment of lithium in a 3D fibrous matrix for ultra-stable lithium-sulfur batteries. *Small* 2014, *10* (21), 4257–4263.

[96] Aurbach, D.; Pollak, E.; Elazari, R.; Salitra, G.; Kelley, C. S.; Affinito, J., On the surface chemical aspects of very high energy density, rechargeable Li–sulfur batteries. *Journal of the Electrochemical Society* 2009, *156* (8).

[97] Lin, Z.; Liu, Z.; Fu, W.; Dudney, N. J.; Liang, C., Phosphorous pentasulfide as a novel additive for high-performance lithium-sulfur batteries. *Advanced Functional Materials* 2013, *23* (8), 1064–1069.

[98] Zu, C.; Manthiram, A., Stabilized lithium-metal surface in a polysulfide-rich environment of lithium-sulfur batteries. *Journal of Physical Chemistry Letters*2014, *5* (15), 2522–2527.

[99] Jing, H.-K.; Kong, L.-L.; Liu, S.; Li, G.-R.; Gao, X.-P., Protected lithium anode with porous Al2O3layer for lithium–sulfur battery. *Journal of Materials Chemistry A* 2015, *3* (23), 12213–12219.

[100] Huang, C.; Xiao, J.; Shao, Y.; Zheng, J.; Bennett, W. D.; Lu, D.; Saraf, L. V.; Engelhard, M.; Ji, L.; Zhang, J.; Li, X.; Graff, G. L.; Liu, J., Manipulating surface reactions in lithium-sulphur batteries using hybrid anode structures. *Nature Communications* 2014, *5*, 3015.

[101] Ma, G.; Wen, Z.; Wu, M.; Shen, C.; Wang, Q.; Jin, J.; Wu, X., A lithium anode protection guided highly-stable lithium-sulfur battery. *Chemical Communications* 2014, *50* (91), 14209–14212.

[102] Zhang, R.; Cheng, X.-B.; Zhao, C.-Z.; Peng, H.-J.; Shi, J.-L.; Huang, J.-Q.; Wang, J.; Wei, F.; Zhang, Q., Conductive nanostructured scaffolds render low local current density to inhibit lithium dendrite growth. *Advanced Materials* 2016, *28* (11), 2155–2162.

[103] Cheng, X.-B.; Zhao, C.-Z.; Yao, Y.-X.; Liu, H.; Zhang, Q., Recent advances in energy chemistry between solid-state electrolyte and safe lithium-metal anodes. *Chem* 2019, *5* (1), 74–96.

[104] Scheers, J.; Fantini, S.; Johansson, P., A review of electrolytes for lithium–sulphur batteries. *Journal of Power Sources* 2014, *255*, 204–218.

[105] Guo, B.; Ben, T.; Bi, Z.; Veith, G. M.; Sun, X. G.; Qiu, S.; Dai, S., Highly dispersed sulfur in a porous aromatic framework as a cathode for lithium-sulfur batteries. *Chemical Communications* 2013, *49* (43), 4905–4907.

[106] Kolosnitsyn, V. S.; Karaseva, E. V.; Syng, D. Y.; Cho, M. D., Effect of ether nature on cycling of a sulfur electrode in sulfolane-based mixed electrolytes. *Elektrokhimiya* 2002, *38* (12), 1452–1456.

[107] Park, J.-W.; Yamauchi, K.; Takashima, E.; Tachikawa, N.; Ueno, K.; Dokko, K.; Watanabe, M., Solvent effect of room temperature ionic liquids on electrochemical reactions in lithium–sulfur batteries. *The Journal of Physical Chemistry C* 2013, *117* (9), 4431–4440.

[108] Shin, E. S.; Kim, K.; Oh, S. H.; Cho, W. I., Polysulfide dissolution control: the common ion effect. *Chemical Communications* 2013, *49* (20), 2004–2006.

[109] Aurbach, D.; Pollak, E.; Elazari, R.; Salitra, G.; Kelley, C. S.; Affinito, J., On the surface chemical aspects of very high energy density, rechargeable Li–sulfur batteries. *Journal of the Electrochemical Society* 2009, *156* (8), A694.

[110] Li, N.; Weng, Z.; Wang, Y.; Li, F.; Cheng, H.-M.; Zhou, H., An aqueous dissolved polysulfide cathode for lithium–sulfur batteries. *Energy & Environmental Science* 2014, *7* (10), 3307–3312.

[111] Hassoun, J.; Scrosati, B., Moving to a solid-state configuration: A valid approach to making lithium-sulfur batteries viable for practical applications. *Advanced Materials* 2010, *22* (45), 5198–5201.

[112] Jin, J.; Wen, Z.; Liang, X.; Cui, Y.; Wu, X., Gel polymer electrolyte with ionic liquid for high performance lithium sulfur battery. *Solid State Ionics* 2012, *225*, 604–607.

[113] Liu, Z.; Fu, W.; Payzant, E. A.; Yu, X.; Wu, Z.; Dudney, N. J.; Kiggans, J.; Hong, K.; Rondinone, A. J.; Liang, C., Anomalous high ionic conductivity of nanoporous beta-Li3PS4. *Journal of the American Chemical Society* 2013, *135* (3), 975–978.

[114] Lin, Zhan, Liu, Zengcai, Dudney, Nancy J., & Liang, Chengdu,Lithium Superionic Sulfide Cathode for All-Solid Lithium–Sulfur Batteries. *ACS Nano* 2013, *7*, 2829–2833.

[115] Kozen, Alexander C., Lin, Chuan-Fu, Pearse, Alexander J., Schroeder, Marshall A., Han, Xiaogang, Hu, Liangbing, Lee, Sang-Bok, Rubloff, Gary W., & Noked, Malachi,Next-Generation Lithium Metal Anode Engineering via Atomic Layer Deposition. *ACS Nano*2015, *9*, 5884–5892.

9 Solid-State Batteries and Interface Issues

Nan Sun, Fang Zhang, Xufeng Wang, Wei Zhao, Hanwen An, Han Wang, and Jiajun Wang
MIIT Key Laboratory of Critical Materials Technology for New Energy Conversion and Storage, School of Chemistry and Chemical Engineering, Harbin Institute of Technology, Harbin, People's Republic of China

CONTENTS

9.1 INTRODUCTION

Lithium-ion batteries (LIBs), as one of our life's necessities, play essential roles and have been widely used in consumer electronics, electric automobiles, and aerospace products. The rapidly growing demand for Li-ion batteries in daily life has raised

DOI: 10.1201/9781003133971-9

concerns regarding both safety and durability issues. Current Li-ion batteries containing liquid electrolytes have a significant risk of catching fire. The replacement of liquid electrolytes with solid electrolytes (SE) has attracted enormous attention, due to high thermal durable, no electrolyte leakage, and compatibility of lithium metal anode [1]. Lithium metal anode can provide the highest theoretical capacity and lowest electrochemical potential, which would greatly increase the energy density of batteries. Therefore, all-solid-state batteries (ASSBs) are regarded as one of the critical future energy storage and conversion systems.

Seeking suitable materials with high ionic conductivity and lithium-ion transference numbers that can serve as solid electrolytes is the basis of all-solid-state battery research. Currently, solid electrolytes include polymer solid electrolytes, oxide solid electrolytes, sulfide solid electrolytes, and their composite electrolytes. All solid polymer electrolytes are composed of lithium salt and polymer. Polymer solid electrolytes commonly include polyethylene oxide (PEO) [2], polyacrylonitrile (PAN) [3], polymethyl methacrylate (PMMA) [4], and polyvinylidene fluoride (PVDF) [5], which have good flexibility, tensile and shear properties, and are easy to assemble to flexible and bendable batteries. In the polymer solid electrolyte, lithium ions follow the peristaltic movement of polymer molecular chains to realize the directional movement. Thus, inhibiting polymer crystallization, which improves the peristaltic of polymer molecular chains, is often used to improve the ionic conductivity of polymer solid electrolytes. Cross-linking, copolymerization, and inorganic additives (SiO_2 [6], Al_2O_3 [7], ZrO_2 [8], $Li_7La_3Zr_2O_{12}$ [9], and Li-alloy [10], etc.) can effectively inhibit polymer crystallization and improve ionic conductivity. However, compared with inorganic solid electrolytes, polymer solid electrolytes still face low ion conductivity and ion transference numbers at room temperature. Therefore, improving the ionic conductivity of polymer solid electrolytes is the hotspot and important goal. Oxide solid electrolytes can be classified as crystalline and glassy from the structural point of view. Perovskite ($Li_{3+x}La_{2/3-x}TiO_3$ [11]), NASICON ($Li_{1+x}Al_xTi_{2-x}(PO_4)_3$ [12]), anti-perovskite (Li_3OCl [13]), and garnet ($Li_7La_3Zr_2O_{12}$ [14]) are all crystalline. The glassy oxide solid electrolyte is the LiPON-type electrolyte used in thin-film batteries. An oxide solid electrolyte has a dense morphology compared with sulfide, it has higher mechanical strength and excellent stability in the air environment. However, it is precise because of its higher mechanical strength, poor deformability, and softness, combined with the problem of interface contact that is difficult to improve, the problem of oxide electrolyte is also more prominent. Sulfide solid electrolytes are attracting many concentrations, due to their high ionic conductivity, which is comparable to that of liquid electrolytes. The oxygen ions of oxide solid electrolytes are replaced by sulfur ions to construct sulfide solid electrolytes. The lower electro-negativity and larger radius of sulfur ions than oxygen ions conduct more free-moving lithium ions and larger migration tunnels for lithium ions. Therefore, the sulfide solid electrolytes are beneficial for the migration of lithium ions. As a result, sulfide solid electrolytes exhibit higher ionic conductivities approximately 10^{-3}–10^{-4} S cm^{-1} at room temperature. Thio-lithium superionic conductor (thio-LISICON) $Li_{10}GeP_2S_{12}$ possesses a high ionic conductivity of 1.2×10^{-2} S cm^{-1} [15]. The $Li_{9.54}Si_{1.74}P_{1.44}S_{11.7}Cl_{0.3}$ presents the highest ionic conductivity (2.5×10^{-2} S cm^{-1}) in the reported papers [16]. Argyrodite-type crystals

Li_6PS_5X (X = Cl, Br, or I) also exhibit high ionic conductivities (>10^{-3} S cm^{-1}) [17]. However, most sulfide solid electrolytes are unstable in conventional environments. Therefore, the atmosphere of operating the sulfide solid electrolytes requires special consideration because most sulfide solid electrolytes can react with H_2O, releasing a highly toxic gas, H_2S [18]. Though sulfide solid electrolytes are softer than oxide solid electrolytes, constructing suitable solid-solid interfaces between electrolyte and electrode materials is another important criterion for all-solid-state lithium battery applications. Thus, the interface design of electrode materials should be taken into consideration deeply. In general, the key characters of solid-state electrolytes used in all-solid-state batteries should include: (1) High ionic conductivity (body phase and grain boundary) in a wide temperature range, (2) wide chemical window for lithium metal anode coupling and matching with a high-voltage cathode, (3) chemically and mechanically compatible interface with anode and cathode, (4) chemical stability in the environment, and (5) low interface resistance with electrodes.

For all solid-state battery designs, it is necessary to fully understand the electrode materials and interfaces in the solid-state battery system, including the composition and distribution of active materials, electrode/electrolyte interfaces, and their evolution during cycling. Therefore, it is necessary to develop and customize new advanced technologies based on existing characterization tools to discover the internal phenomena and problems of solid-state batteries and guide the design of solid-state batteries. Synchrotron X-ray three-dimensional imaging [19], solid-state nuclear magnetic resonance [20], and *operando* testing technologies [21] provide great support for solid-state battery research.

In this chapter, the developments of polymer, oxide, and sulfide solid-state batteries and the application of advanced characterization methods in solid-state batteries are reviewed.

9.2 OXIDE SOLID ELECTROLYTES

The oxide solid-state of electrolytes (O-SSEs) that have been studied in all-solid-state batteries mainly includes LISICON [22], NASICON [23], Garnet [24], Perovskite [25], and others. Among them, the cubic garnet-type $Li_7La_3Zr_2O_{12}$ (LLZO) draws more attention due to its high ionic conductivity (up to 2×10^{-4} S cm^{-1}) and wide voltage window at room temperature. Generally speaking, a higher sintering temperature (>1000°C) is required to obtain dense ceramics with high ionic conductivity of 10^{-4}–10^{-3} S cm^{-1} at room temperature [26]. The high temperature will decompose the electrode and electrolyte materials, and there will be a serious O-SSEs/electrode interface element diffusion phenomenon. Although the rigidity of the ceramic electrolyte can buffer the volume change of the electrode material during the cycle, it also causes the oxide electrolyte to lose close contact with the electrode. In addition, oxide solid electrolytes have a large number of grain boundaries and crystal defects during the preparation process. The low-cost preparation process of perfect and high-performance O-SSEs is still challenging, which further limits the application process of oxide-based solid-state batteries.

9.2.1 Interfaces on the Anode Side

The interface problems with the negative electrode are mainly in the following three aspects: 1) The LLZO material itself is very hard, and lithium carbonate will form on the surface after exposure to air, which leads to point contact with the electrode material at the interface and poor wettability [27]; 2) the uneven distribution of current density at the location promotes uneven decomposition and deposition of lithium, and dendrites will still grow along the grain boundaries, pierce the solid electrolyte, and short-circuit the battery [28]; 3) LLZO is not stable for metal lithium. Direct contact will still cause side reactions under electrochemical conditions, and the accumulation of by-products at the interface will destroy the cycle performance of the battery [29]. Therefore, an in-depth study of the interface reaction and optimization mechanism is essential to further improve the performance of all-solid-state batteries. At first, researchers believed that oxide solid electrolytes had good stability in the air. However, Cheng et al. [30]. found that the interface impedance of LLZO was significantly increased after being stored in the air for two months. Studies have shown that after LLZO is exposed to air, a contaminated Li_2CO_3 layer will be formed on the surface and at the grain boundary, and the ion conductivity of Li_2CO_3 itself is low (10^{-8} S cm^{-1}) and has poor wettability to lithium metal, which is the root cause of the increase in interface impedance. Simple dry polishing or wet polishing cannot significantly improve the wettability of the interface. Research has found that the use of additives, the interface structure, and rapid acid treatment can inhibit or remove Li_2CO_3.

To improve the interface contact and inhibit the growth of lithium dendrites, Sharafi [31] and Tsai et al. [32] first tried to use high pressure and heating methods to improve the interface contact and increase the Li^+ transmission capacity of the interface, but the harsh experimental conditions are difficult to apply to real batteries. Since 2016, researchers began to work on adding a buffer layer to reduce the interface resistance and improve the wettability of the lithium metal electrode. Guo's [33] and Sun's [34] group used magnetron-sputtering to prepare a 20 nm thick SnO_2 buffer layer and Cu_3N thin film on LLZO wafers (Figure 9.1). The interface conversion reaction was used to significantly improve the interface while inhibiting the growth of lithium dendrites. However, when the current density is higher than the critical current density, the dendrites will still pierce the solid electrolyte and cause a short circuit. At present, the critical current density of garnet-type solid electrolytes is far lower than the practically applicable current density of 10 mA cm^{-2}, and further improvement is needed.

9.2.2 Interfaces on the Cathode Side

Compared with the anode interfaces, the issues of the interface of O-SSEs/cathode is more challenging. The interface problem can be summarized as the following two: 1) Its high hardness and shear modulus result in a small electrolyte/positive electrode contact area, and certain structural defects are formed at the interface, which restricts the transmission of lithium ions [38]; 2) the interface is unstable. During high-temperature sintering or electrochemical cycling, inter-diffusion of

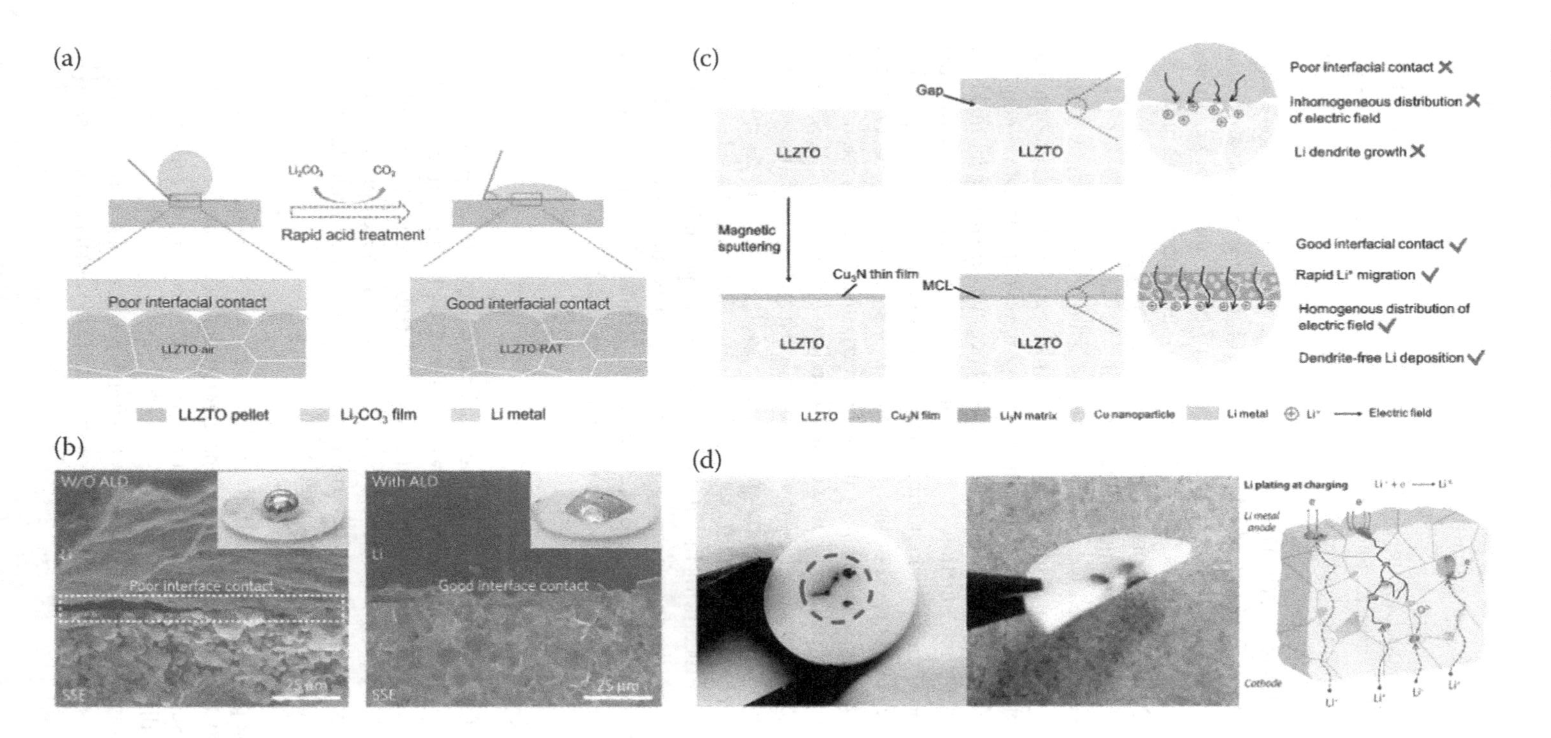

FIGURE 9.1 (a) Schematic illustration of LLZTO/Li interfaces before and after the rapid acid treatment [35]. (b) ALD Al_2O_3 layer to improve the lithiophilicity of O-SSEs identified by the decreased contact angle and close contact [36]. (c) The schematic of the mixed conductive intermediate layer (MCL) protected LLZTO/Li interface for dendrite-free Li metal solid batteries [34]. (d) Evidence and schematic of lithium dendrites growth across the LLZO inorganic electrolytes [37].

elements easily occurs, destroying the skeleton structure of the material and forming the third phase, hindering the transmission of Li^+, increasing the interface resistance, and with the progress of electrochemical cycles, the interfacial stress is introduced, and the gradual accumulation of stress will cause the solid-solid interface to fracture and separate, and ultimately lead to the failure of the all-solid battery [2,39]. Therefore, researchers have carried out a lot of research on how to improve the interface problem.

The pulsed laser deposition (PLD) technique was first used to try to improve the interface problem of the O-SSEs/cathode. Kato et al. [40] first deposited a layer of Nb with a thickness of about 10 nm on LLZO, and a Li-Nb-O amorphous interface layer is formed by heat-treating it at 600°C under O_2 atmosphere for 2 hours, as shown in Figure 9.2. The amorphous Li-Nb-O layer not only acts as a buffer layer to smooth the interface but also significantly inhibits the inter-diffusion of the interface element's behavior. However, the PLD has a high cost and low load of electrode materials, which is not suitable for large-scale preparation. In addition, the co-sintering method is also an important method to improve the interface contact, but the interface diffusion-reaction under a high-temperature environment makes it more complicated. The introduction of buffer materials essentially improves the compatibility of the interface and inhibits the interaction between the positive electrode and the electrolyte. The inter-diffusion of the elements between them reduces the interface impedance. Han et al. [43] cleverly used the characteristics of lithium carbonate on the surface of LLZO and introduced $Li_{2.3-x}C_{0.7+x}B_{0.3-x}O_3$ to react with lithium carbonate during high-temperature sintering to form an LCBO ion conductive matrix, which significantly improved the interface contact and avoided the phenomenon of inter-diffusion of elements. However, the selection of the composition and structure of the introduced modified materials, the direction, and the mechanism of the interface reaction all need to be understood and systematically studied. In summary, there is currently no particularly complete method to improve the interface between traditional cathode and O-SSEs.

9.2.3 Outlook

O-SSEs are one of the solid electrolyte materials most likely to be applied to all-solid-state lithium batteries due to their advantages such as high room temperature ion conductivity, wide electrochemical windows, and stable contact with lithium negative electrodes. Due to insufficient O-SSEs/electrode interface contact, it leads to a great interface impedance, which affects battery performance and hinders its application in all-solid-state lithium batteries.

1. To solve the problem of air stability and storage of solid electrolytes, it is necessary to use advanced in-situ characterization technology to essentially understand the reaction mechanism of solid electrolytes with water and carbon dioxide in the air.
2. For the Li/LLZO interface, many methods have been able to reduce the interface resistance to the point where the interface resistance is close to

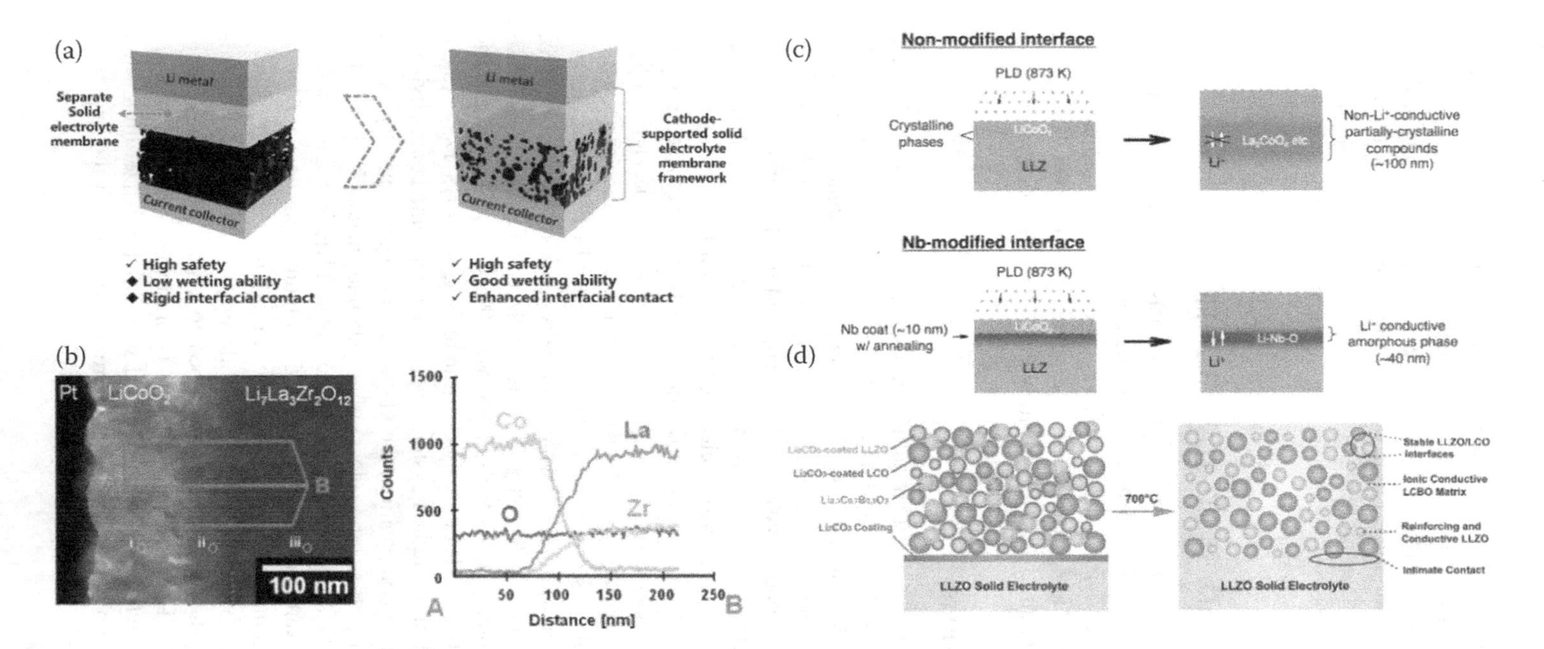

FIGURE 9.2 (a) Schematic of the novel cathodes with homogeneous active materials and electrolytes distribution in comparison with a conventional rigid cathode [38]. (b) Cross-sectional TEM image and the EDS line profile of an LLZ/$LiCoO_2$ thin-film interface [41]. (c) Using Nb as a buffer layer can modify the stability of the $LiCoO_2$/LLZO interface during the PLD process [42]. (d) Schematics of the interphase-engineered all-ceramic cathode/electrolyte by using the reaction of $Li_{2.3\text{-}x}C_{0.7+x}B_{0.3\text{-}x}O_3$ with Li_2CO_3 [43].

that of a liquid battery. Next, we should study how to inhibit the growth of lithium dendrites and increase the critical current density.

3. After the interface modification, the impedance on the interface side of the anode is greatly reduced, but there is still a certain interface impedance at the cathode. The interface problem of the O-SSEs/cathode still needs to invest more effort to realize the application of cobalt oxide all-solid-state batteries as soon as possible.

9.3 SULFIDE-BASED ALL-SOLID-STATE BATTERIES

Sulfide-based all-solid-state batteries have gained increasing attention because of their ultrahigh ionic conductivity. In addition, sulfide electrolytes exhibit the attractive mechanical feature of low rigidity, which is conducive to the preparation of densely packed interfaces. Here, we start with introducing electrolytes Thereafter, cathode/sulfide electrolyte and Li/sulfide electrolyte interfaces are discussed.

9.3.1 Sulfide Electrolytes

Over the past decades, as the heart of sulfide-based solid-state batteries, sulfide electrolytes have made a series of exciting achievements in ionic conductivity. In 2011, a superionic solid electrolyte $Li_{10}GeP_2S_{12}$ displays a high ionic conductivity of 12 mS cm^{-1} [44]. Subsequently, $Li_{9.54}Si_{1.74}P_{1.44}S_{11.7}Cl_{0.3}$ has been demonstrated to possess a higher ionic conductivity of 25 mS cm^{-1} [45]. The ultra-high ionic conductivity of sulfide electrolytes, which is equal to or greater than liquid electrolytes, makes them the most promising electrolyte for ASSLBs.

In general, sulfide electrolytes can be divided into three types according to crystal structure: Thio-LISICON, $Li_{11-x}M_{2-x}P_{1+x}S_{12}$ (M = Ge, Sn, Si), and Li_6PS_5X (X = Cl, Br, I). The crystal structures of several typical electrolytes are shown in Figure 9.3. The thio-LISICON come from LISICON-type γ-Li_3PO_4 solid electrolytes by replacing oxygen with sulfur, which normally possesses three orders of magnitude increase in ionic conductivity than the oxide counterparts [48]. Thio-LISICON includes the binary Li_2S–P_2S_5, Li_2S–MS_2 (M = Ge, Sn, Si) and the ternary $Li_{4-x}Ge_{1-x}P_xS_4$ ($0 < x < 1$) electrolytes. Among these, the Li_2S–P_2S_5 system has been the earliest and most studied system due to superior ionic conductivities. For electrolytes at the same composition, the glass-ceramic types generally exhibit higher ionic conductivity than the glass or crystalline materials [49]. The $Li_{11-x}M_{2-x}P_{1+x}S_{12}$ (M = Ge, Sn, Si) is another typical electrolyte with a unique 3-D framework structure. R. Kanno et al. first disclosed a lithium superionic conductor $Li_{10}GeP_2S_{12}$ (LGPS) with unprecedented conductivity of 12 mS cm^{-1} in 2011 [44]. Since then, there is a new wave of enthusiasm in developing LGPS-type electrolytes. $Li_{9.54}Si_{1.74}P_{1.44}S_{11.7}Cl_{0.3}$ was later developed to obtain the highest ionic conductivity to date [45]. Li_6PS_5X (X = Cl, Br, I) originated from the mineral argyrodite (Ag_8GeS_6), in which Ag^+ was replaced with Li^+ to realize lithium-ion mobility [47]. Li_6PS_5X with the argyrodite structure has been one of the important solid electrolyte types.

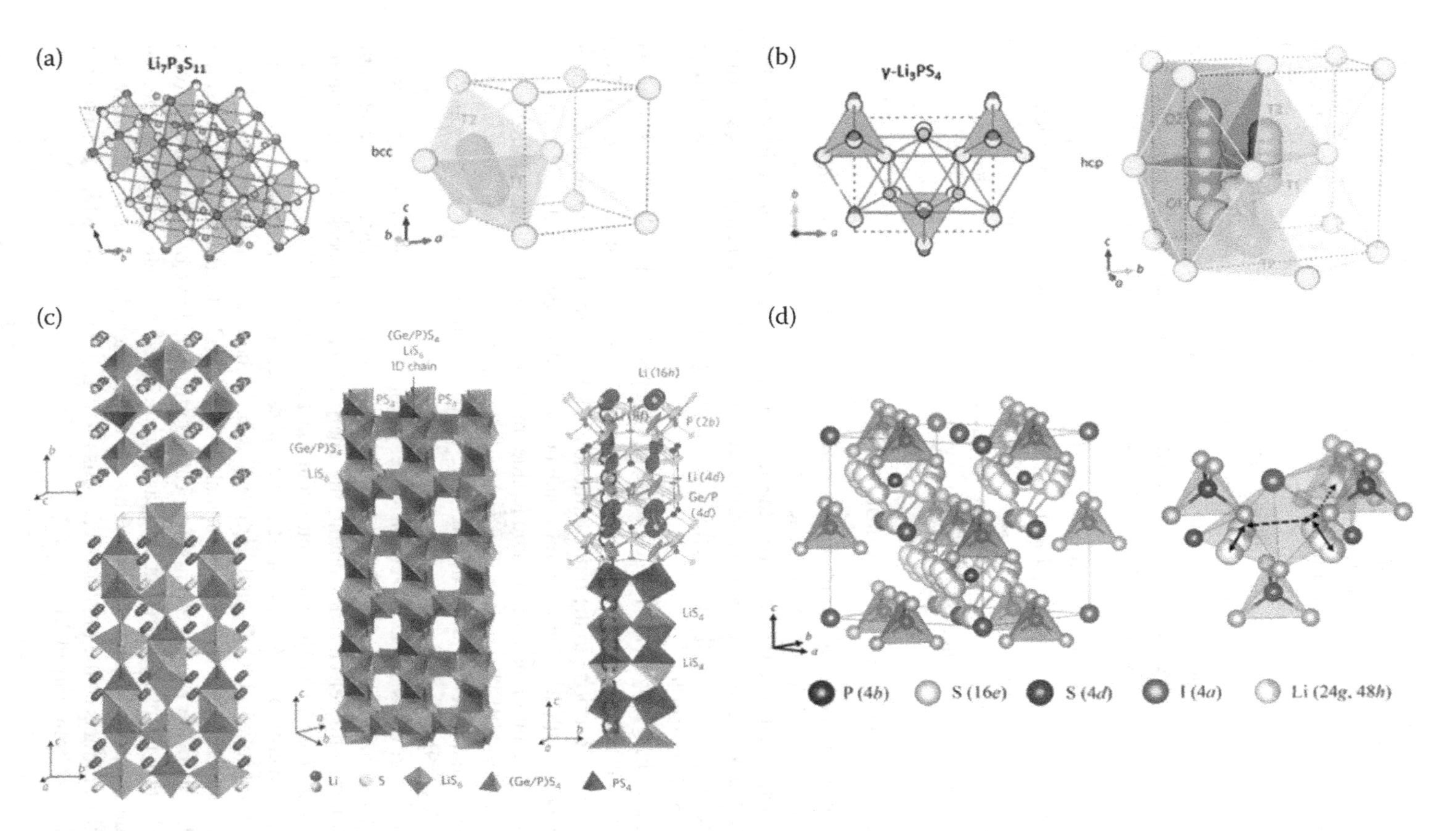

FIGURE 9.3 Crystal structure of the Li-ion conductors: (a) $Li_7P_3S_{11}$ [46], (b) γ-Li_3PS_4 [46], (c) $Li_{10}GeP_2S_{12}$ [44], (d) Li_6PS_5I [47].

Several methods have been used to synthesize sulfide electrolytes, including solid-state method [50], mechanical ball-milling [44,45,49,49,49,51], and liquid phase synthesis [52]. The solid-state method is also the melting-quenching method, which can obtain higher ion conductivity of the electrolyte by melting the precursor material mixture in stoichiometric ratios and then quenching the sample to room temperature. Mechanical ball-milling processes have relatively lower cost and simple operation, with the entire process being able to be completed at room temperature. The liquid phase synthesis is more desirable for the uniformity of resulting electrolytes. In addition, the controllable liquid phase synthesis takes into account the interfacial compatibility between the solid electrolyte and solid electrode.

The main drawbacks of sulfides are their narrow electrochemical stability window and sensitivity to H_2O and O_2. In addition, the widespread application of sulfide solid electrolytes are limited by their high cost. At present, substitution has been demonstrated to be an effective strategy for improving electrochemical properties of sulfide electrolytes [53].

9.3.2 Cathode/Sulfide Electrolyte Interface

The performance of all-solid-state batteries strongly depends on the solid-solid interface of solid-state batteries. The cathode/sulfide electrolyte interface where the electrochemical reactions occur is the most critical part of the all-solid-state batteries. However, the interface generally suffers from great challenges. As shown in Figure 9.4, major challenges include space charge layer effects, interfacial reactions, and contact loss.

Space charge layers are formed at the oxide cathodes/sulfide electrolyte interface due to their large chemical potential differences, causing lithium ions to move from the sulfide electrolyte to the oxide cathode [54]. Furthermore, because of the mixed conductors with ion and electron conductivities of oxide cathodes, electrons can eliminate lithium ions on the cathode side, resulting in the continuous diffusion of lithium ions [58]. The space charge layer effects can induce high interfacial resistances that severely affect the high-rate charge/discharge ability of all-solid-state batteries [59].

Many previous works based on either theoretical calculation or experimental data have unveiled undesirable interfacial reactions at the interface between cathodes and sulfide electrolytes during the charge/discharge process [55,60]. On the one hand, element diffusion often occurs at the interface. By using transmission electron microscopy, A. Sakuda et al. first observed a 10 nm interfacial layer at the interface between $LiCoO_2$ and $Li_2S–P_2S_5$ which showed the coexistence of Co, S, and P elements, indicating the elemental diffusion of Co, P, and S [55]. On the other hand, due to the narrow electrochemical stability windows of sulfide electrolytes, the electrolytes usually present a metastable intermediate state at high voltage, and sulfide electrolytes can be oxidized by oxide cathodes [56]. It is broadly accepted that the products of the interface reactions are highly insulating, which dramatically hinders the transport of Li^+.

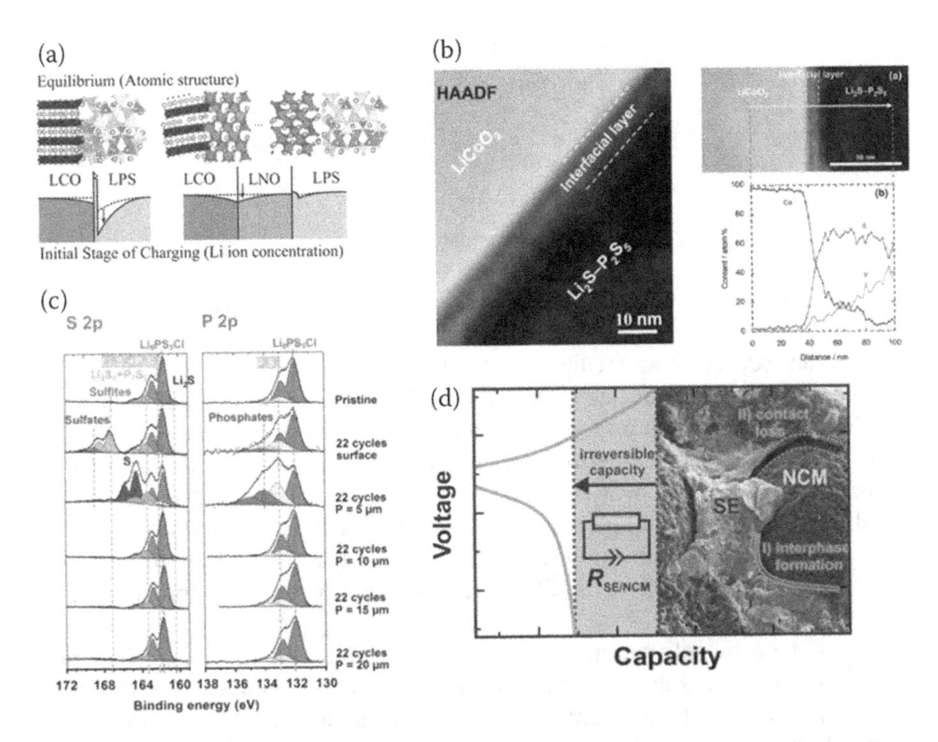

FIGURE 9.4 Cathode/Sulfide electrolyte interface issues: (a) space charge layer effects [54], (b) element diffusion [55], (c) electrolyte decomposition [56], and (d) contact loss [57].

To stabilize the cathode/sulfide electrolyte interface, a buffer layer is generally introduced between cathodes and sulfide electrolytes, aiming at preventing interfacial reactions and simultaneously suppressing the space charge layer. So far, typical $LiNbO_3$ [61], $Li_4Ti_5O_{12}$ [62], Li_2ZrO_3 [63], Li_3PO_4 [64], and Al_2O_3 [65] have been proven to effectively reduce the cathodic interface resistance.

The physical contact is prone to a "point to point" model because of the rigidity of cathodes and solid-state electrolytes. Physical contact loss is a common phenomenon that contributes to solid-state battery performance [57,66]. Due to the lack of ion transport path, sluggish transport kinetics in solid-state batteries cause the coulomb efficiency to be low and the capacity to decay rapidly. Especially for the materials with obvious volume change during Li^+ intercalation/de-intercalation, the interface contact failure is more severe [67]. With the proposal of the liquid phase synthesis of sulfide electrolytes, sulfide electrolytes can be used to coat or in-situ grow on the surfaces of cathode materials, thereby increasing the solid–solid ionic contact and preventing contact loss [68,69].

9.3.3 Li/Sulfide Electrolyte Interface

The application of lithium metal anodes in sulfide all-solid-state batteries is limited by the lithium dendrite growth and SE reduction by lithium metal.

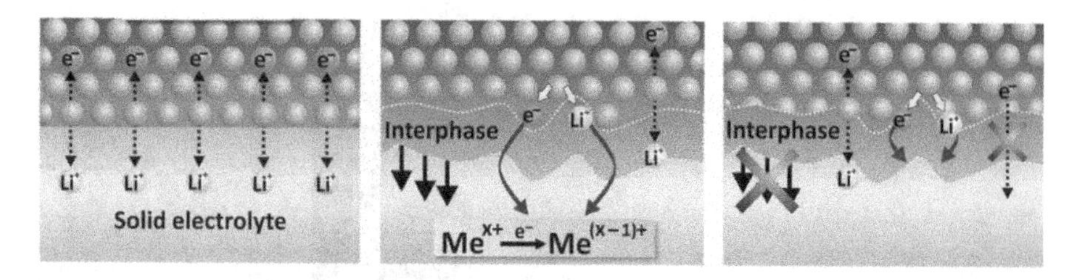

FIGURE 9.5 Types of interphases between Li metal and the solid electrolyte [74].

Dendrite-induced short circuits are one of the bottlenecks in the development of high-energy-density solid-state batteries. The dendrite growth mechanism through sulfide electrolytes is controversial. It is reported that the grain boundary, voids of solid electrolytes, and pre-existing defects are the intrinsic reasons for dendrite formation [70,71]. At a high current density, inhomogeneous Li deposition triggered large cracks that will provide the active sites for the lithium dendrite growth [72]. Recently, it has been proposed and discussed that the high electronic conductivity of the electrolyte is likely responsible for dendrite formation and growth [73].

Due to the extremely strong reducing ability of lithium metal, almost all sulfide electrolytes are thermodynamically unstable in contact with lithium. As shown in Figure 9.5, there are three different types according to the formation features of the Li/sulfide electrolyte interfaces: (1) In a perfect thermodynamically stable interface, no reactions occur at the interface; (2) a mixed conducting interface with electronic and ionic conductive results in a continuous decrease of the internal fresh solid electrolyte; and (3) a metastable interface with poor electronic conductivity for the sluggish reaction kinetics [74]. A completely stable interface like tape1 has not been reported yet, so the ideal Li/sulfide electrolyte interface should be ionically conductive but electronically insulating.

For the Li/sulfide electrolyte interface, we usually optimize it by adjusting the electrolyte composition and constructing an artificial interface layer. In addition, a lithium alloy anode with much lower density and good chemical stability against sulfide electrolytes also can stabilize the interface [48].

9.3.4 Outlook

Sulfide-based all-solid-state batteries are the most promising candidates for next-generation rechargeable batteries with high safety and high energy. Although there are many impressive achievements in recent years, the commercialization of sulfide-based all-solid-state batteries still face many challenges. In fact, it is insufficient to adopt a straightforward strategy alone, such as only improving electrolyte ion conductivity or interface compatibility. High-performance sulfide-based all-solid-state batteries need combined approaches to achieve a joint modification. Here, potential directions and perspectives regarding sulfide-based all-solid-state batteries are as follows:

1. Improvements of sulfide electrolytes. An ideal electrolyte possesses unparalleled ionic conductivities, sufficiently low electronic conductivity, good electrochemical/chemical stability, and stability to the ambient atmosphere. In the case of practical large-scale applications, it is necessary to consider machinability, flexibility, and cost.
2. Optimization of cathode/sulfide electrolyte interfaces. These interfaces should possess high ionic conductivity to ensure smooth lithium-ion migration as well as good compatibility with electrodes and stability to shield sulfide electrolytes from high electrode potentials.
3. Dendrite-free Li anode. Modification strategies include the construction of artificial interfaces, the design of 3-D lithium matrixes, and the adoption of lithium alloys. It can even be a combination of the above methods, a dendrite-free Li anode will be realized by uniform deposition of lithium. In addition, the lower electronic conductivity of sulfide electrolytes can also inhibit dendrite formation.
4. Sulfide-based all-solid-state batteries with high energy density. The following aspects deserve consideration: prepare ultra-thin electrolyte under the premise of ensuring ionic conductivity; reduce the amount of Li in an anode or even choose a Li-free anode; prepare a thick composite cathode. At the same time, optimize the battery assembly.

9.4 POLYMER AND COMPOSITE SOLID ELECTROLYTES

Since the discovery of poly(ethylene oxide) (PEO) with ionic transport in 1973 [[75], the solid-state polymer electrolytes (SPEs) without any liquid, are widely considered as an intriguing candidate to substitute for the liquid electrolytes currently for high-security and high-gravimetric-energy-density all-solid-state lithium battery [76–78]. Figure 9.6(a) shows a schematic illustration of an integrated all-solid-state lithium battery. SPEs contain polymers as a backbone matrix and lithium salts to supply Li-ion, which are compatible with large-scale manufacturing processes. The polymer hosts of SPEs are commonly PEO, poly(acrylonitrile) (PAN), poly(methyl methacrylate) (PMMA), poly(vinyl chloride) (PVC), poly(vinylidene fluoride) (PVDF), and so on [79–84]. Figure 9.6(b) shows the most commonly used

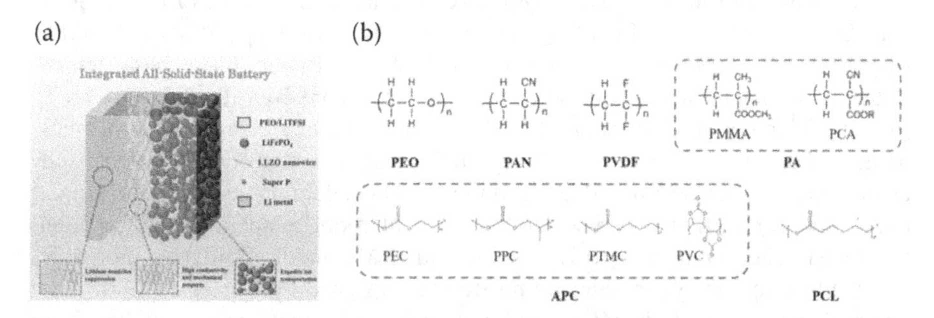

FIGURE 9.6 (a) Schematic illustration of an integrated all-solid-state polymer lithium battery [78]. (b) Chemical structures of common polymer matrices for PSEs [83].

polymer hosts and their structures. Among them, PEO-based SPE is the most widely studied. Lithium salts are commonly $LiClO_4$, LiTFSI LiFSI, LiBOB, LiDFOB, $LiAsF_6$, and $LiPF_6$.

SPEs confront several severe challenges: the trade-off between high ionic conductivity and good mechanical properties, interfacial instability with electrodes, the low Li^+ transference number, as well as electrochemical stability window. For oligoether-based SPEs, $LiCoO_2$ and $Li(Ni_xCo_yMn_z)O_2$ (NCM) cathode often present an insurmountable barrier as the breaking down the potential for ether linkage lies below 4.0 V [85]. To date, tremendous efforts have been made to overcome the above problems, which include: i) forming composite polymer electrolytes with inorganic or ceramic fillers; ii) modifying polymer structure via copolymerization and grafting; and iii) modification of interface layer with cathode/SPEs. In the following sections, we focus on strategies based on these challenges.

9.4.1 Composite Polymer Electrolytes (CPEs)

The room temperature ionic conductivities of PEO-based SPEs are below 10^{-4} S cm^{-1}, partially due to the crystalline phase of oligoether linkages that only melts at ~60 °C, and mainly due to the tight coupling of Li^+-mobility with the sluggish segmental motion of the PEO backbone [86]. Generally, ceramic fillers are classified as either active or passive [87].

Passive fillers (or inert fillers, such as Al_2O_3, SiO_2, TiO_2, and ZrO_2) decrease the glass transition temperatures to enhance the ionic conductivity (but do not participate in ionic transport) at room temperature from 10^{-6} to 10^{-5} S cm^{-1} and mechanical strengths [88–90]. Some inactive fillers play a role as anion receptors to immobilize anions from Li-salt and increase the Li^+ transference number. A novel cationic metal-organic framework (CMOF) is proposed to immobilize anions and guide Li^+ uniform distribution by Sun et al. [91], which has a high ionic conductivity of 3.1×10^{-5} S cm^{-1} at 25°C and a high Li^+ transference number of 0.72. The authors attributed this increase to the effect of anion adsorption on the surface of CMOF (Figure 9.7(a)). The CMOF grafted with a $-NH_2$ group protects the ether oxygen of polymer chains by hydrogen bonds, extending the electrochemical window to 4.97 V.

The addition of highly conductive active ceramic fillers, such as $Li_7La_3Zr_2O_{12}$ (LLZO), $Li_{6.75}La_3Zr_{1.75}Ta_{0.25}O_{12}$ (LLZTO), $Li_{1.5}Al_{0.5}Ge_{1.5}(PO_4)_3$ (LAGP), $Li_{10}GeP_2S_{12}$ (LGPS), and $Li_{1.3}Al_{0.3}Ti_{1.7}(PO_4)_3$ (LATP) are Li-ion conductors, further enhancing the ionic conductivity at room temperature from 10^{-6} to 10^{-4} S cm^{-1} [92–95]. Wang et al. [96] reported a CPE with PEO and a ceramic fiber or mats based on $La_{0.55}Li_{0.35}TiO_3$. The high conductivity of the active ceramic fillers itself helped the fast transport of Li^+ by forming a 3-D conduction network, showing a high ionic conductivity of 5×10^{-4} S cm^{-1} at room temperature and a high Li^+ transference number of 0.7. A CPE with PAN and well-aligned inorganic Li^+-conductive nanowires nanocomposite was designed by Cui et al. [97], who reported improvements in ionic conductivity and Li^+ transference number. The former is one order of magnitude higher than previous polymer electrolytes with randomly aligned nanowires; the latter increases from 0.27 to 0.42. The large

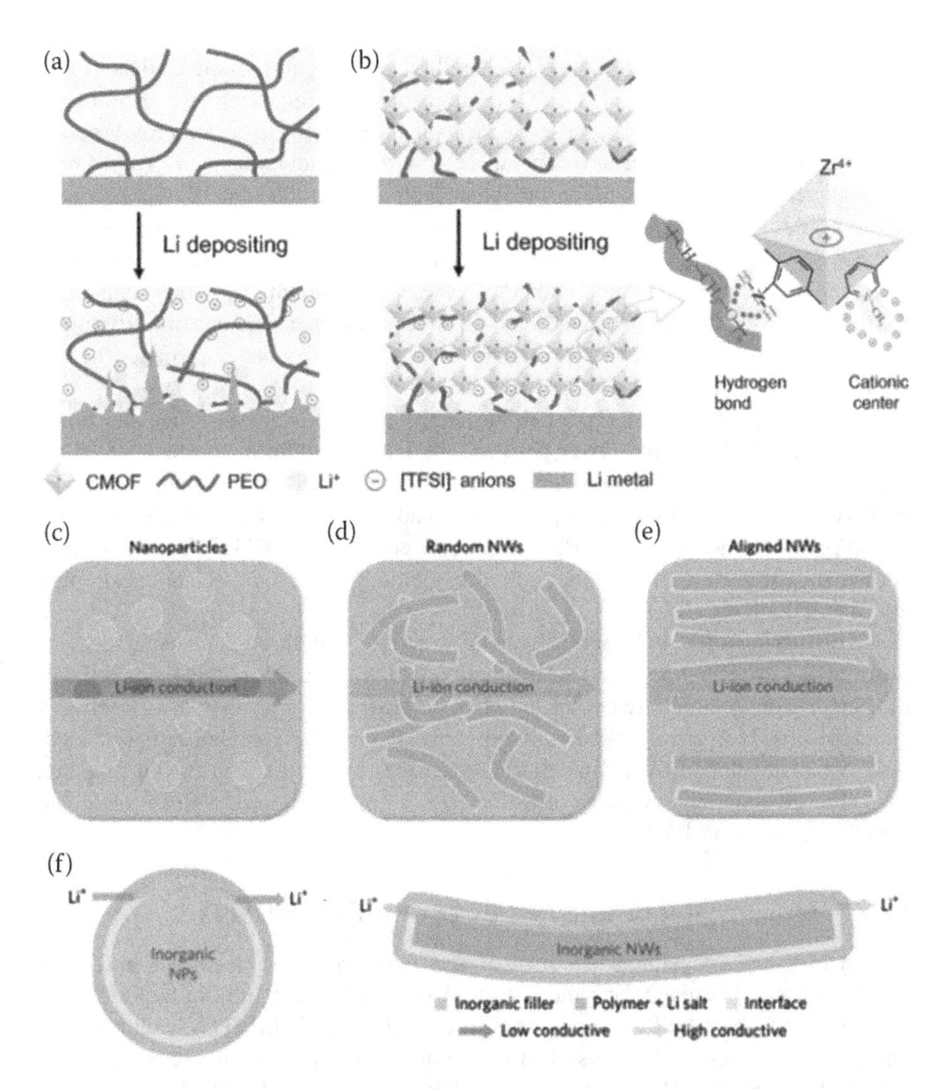

FIGURE 9.7 Schematic of the Li deposition behavior with (a) PEO(LiTFSI) electrolyte and (b) anion-immobilized P@CMOF electrolyte [91]. Li-ion conduction pathways in composite polymer electrolytes with (c) nanoparticles, (d) random nanowires, and (e) aligned nanowires. (f) The surface region of inorganic nanoparticles (NPs) and nanowires (NWs) acts as an expressway for Li-ion conduction [97].

enhancement is ascribed to a fast ion-conducting pathway without crossing junctions on the surfaces of the aligned nanowires (Figure 9.7(b)).

9.4.2 Copolymerization and Grafting

The main dilemma for SPE improvement is the unfortunate contradiction between the dual functions expected from its polymer backbone, which acts as both structural unit to provide dimensional stability and functional unit to dissolve lithium salt and conduct Li^+ [11]. More often than not, improvements in ion conduction were made at the expense of mechanical strength to a degree where the advantages of a SPE either severely suffered or completely vanished. To promote ion conduction while maintaining mechanical strength at the same time, one needs to decouple Li^+-movement from the segmental movement of a polymer. One strategy to achieve this decoupling is to employ copolymers, in which two or more different structural units would shoulder these two conflicting missions separately. As the almost exclusive functional units in SPE, oligoether linkages are difficult to replace; however, a rich chemical database of structural units is provided by the widely available structural polymer materials.

Poly(tetrahydrofuran) (PTHF) exhibits high lithium ion conductivity and transference number due to the fewer oxygen heteroatoms in the main chain of PTHF [98]. However, PTHF-based PSEs present the poor thermal stability and mechanical properties and low melting point [99,100]. To address these issues, Bao et al. [101] developed a thermally stable PSE based on cross-linked PTHF by introducing carbamate groups into the polymer network, yielding improved mechanical strength and thermal stability.

Cross-linking of a polymer is generally carried out through thermal decomposition of a cross-linking agent or irradiation by UV light, where the generated free radicals trigger chemical reactions between polymer chains to form a 3-D network structure [100]. A cross-linked poly(ethylene glycol) diglycidylether-based electrolyte was developed by a situ self-catalyzed reaction [102]. The in-situ prepared PSE showed ionic conductivity of 8.9×10^{-5} S cm^{-1} and a wide electrochemical stability window of 4.5 V.

9.4.3 Modification of Interface Layer with Cathode/SPEs

Satisfactory electrochemical stability of interfaces between electrodes and electrolyte is the prerequisite for safe and durable all-solid-state lithium batteries. For example, PEO-based SPEs show excellent stability with $LiFePO_4$ cathode with charge voltage below 4 V (vs. Li/Li^+), while they match with $LiCoO_2$ and Li $(Ni_xCo_yMn_z)O_2$ (NCM) cathode of above 4 V are instability [103]. Xia et al. [104] first reported that the oxidative degradation of PEO-based SPEs with a carbon composite electrode appeared at about 3.8 V (Figure 9.8(a)). Sun et al. [105] found that the conductive carbon accelerates the decomposition of PEO-based SPEs (Figure 9.8(b)). The high-nickel cathode materials may catalyze some chemical reactions of the solid electrolyte, leading to oxidative degradation of SPEs [106,107]. Apart from that, irreversible oxidative decomposition of free anions from Li-salt appears with the increasing voltage (Figure 9.8(c)), leading to further deterioration of the interfacial structure stability [108–111]. Degradation mechanisms of the electrode/SPEs interfaces include chemical, electrochemical, and mechanical degradations [103–112].

Passivating the catalytically active sites by inert materials is able to suppress PEO oxidation. Different protective layers have been developed with various technologies by atomic layer deposition (ALD). Among them, protective layers

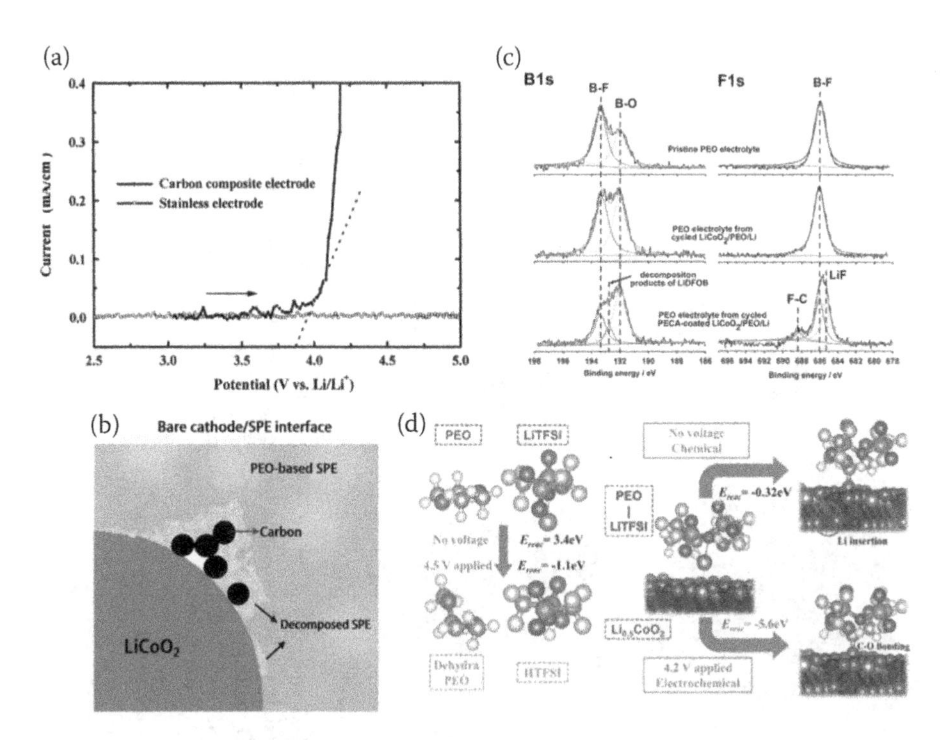

FIGURE 9.8 (a) Linear sweep voltammogram of a $PEO_{20}LiN(CF_3SO_2)_2$ electrolyte sandwiched between a lithium disc and a stainless disc or a carbon composite electrode. The scan rate was 0.05 mV/s, and $T = 80°C$ [104]. (b) Schematic diagram of the decomposition of SPEs on unprotected $LiCoO_2$ electrode [105]. (c) B1s and F1s XPS results of PEO electrolyte before and after charging/discharging 20 cycles [109]. (d) Reaction energies of PEO electrochemical oxidation over a carbon composite electrode (4.5 V) and a $LiCoO_2$ composite electrode (4.2 V) and chemical oxidation with carbon and $LiCoO_2$ composite electrodes determined by theoretical calculations [112].

materials have metal oxides (Al_2O_3, ZnO), phosphates ($FePO_4$, $AlPO_4$), and fluoride (AlF_3) [113–117]. For example, Sun et al. [35] reported that a thin ALD-derived $LiTaO_3$ coating on the $LiCoO_2$ electrode demonstrated good compatibility with PEO-based SPEs (Figure 9.9(b)), significantly enhancing the cycling performance of the ASSLBs. This is attributed to the fact that the protection of the conductive carbon/SPE interface helps reduce the electrochemical oxidation of PEO-based SPEs.

Although these methods can mitigate the interfacial problems to some extent, these inorganic modification layers make less contribution to mitigating the poor physical contact problem. Architecting an ionically conductive elastic interlayer between the electrode materials and SPEs would be an effective approach. Guo et al. [118] reported a poly(acrylonitrile-co-butadiene) polymer as a coating layer for $LiNi_{0.6}Mn_{0.2}Co_{0.2}O_2$ cathode material (Figure 9.9(d)), exhibit a favorable electrochemical performance. This is attributed to the modification of polymer

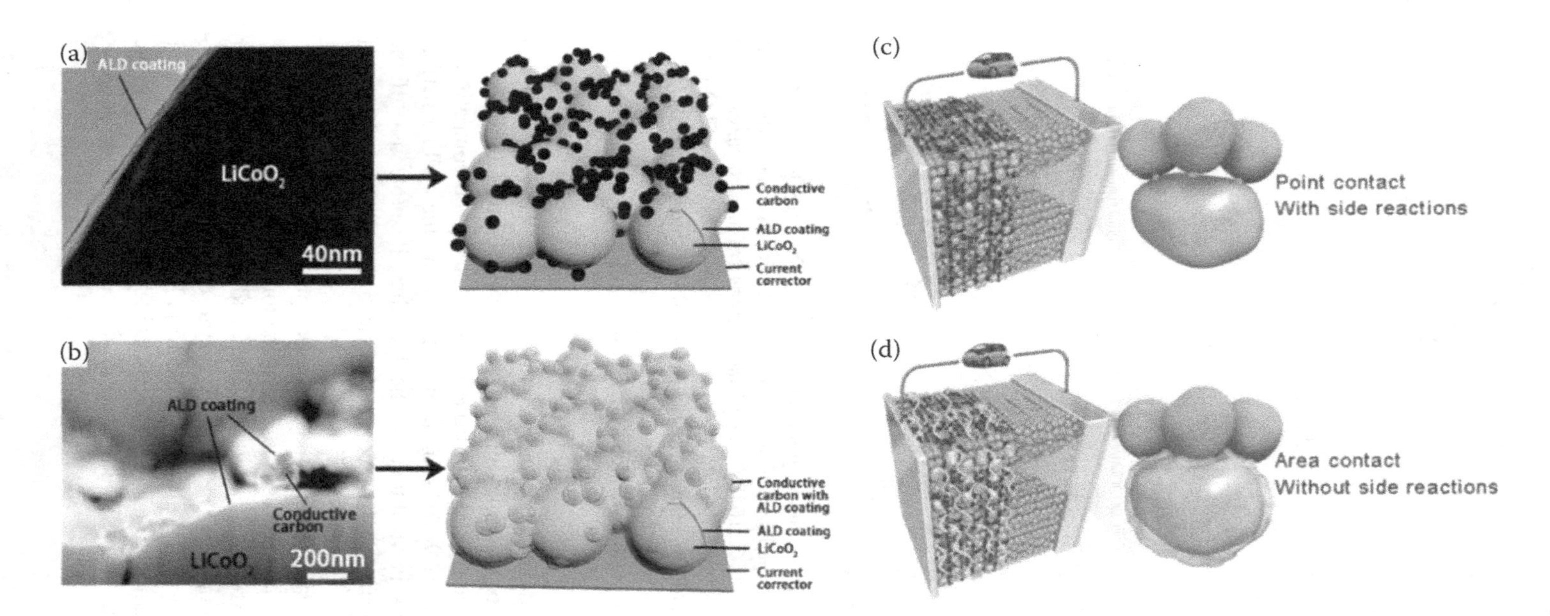

FIGURE 9.9 (a) A TEM image of the 10 cycles of ALD LTO (thickness is ~5 nm) coating on $LiCoO_2$ particles and its schematic diagram of the $LiCoO_2$ electrode with the LCO-coating where conductive carbon is not protected. (b) A SEM image in backscattered electron mode of the 20 cycles of ALD LTO (thickness is ~10 nm) coating on both conductive carbon and $LiCoO_2$ particles from the LCO + C3-coating sample after focused ion beam (FIB) cutting, and its schematic diagram showing the $LiCoO_2$ electrode where both $LiCoO_2$ and conductive carbon are protected [105]. (c) and (d) Schematic illustration of existing interfacial problems in all-solid-state battery [118].

layer, meliorating the physical contact between cathode and solid electrolyte during the cycling process and suppressing the side reactions.

9.4.4 Outlook

The current development of SPEs needs to adopt multiple strategies to overcome these challenges. Designing a SPE with high ionic conductivity, good electrochemical stability with electrodes, acceptable mechanical properties, high Li^+ transference number, and wide electrochemical stability window requires an integrated, in-depth approach between experiments and computational modeling, along with advanced state-of-the-art characterization techniques to understand the modification mechanisms. By the addition of a functional filler encapsulated in the polymer matrix, copolymerization and grafting can improve the room temperature ionic conductivity, mechanical strength, the Li^+ transference number, and electrochemical stability window of SPEs simultaneously. The electrode/SPE interfacial stability is crucial to the excellent rate performance and cycle stability of all-solid-state lithium batteries. Herein, a deeper understanding of the ion-transport mechanisms and the strain/stress behavior of the electrode/SPE interface would be instructive for the management of the electrode/electrolyte interface. Architecting an ionically conductive elastic interlayer between the electrode materials and SPEs would be an effective approach. In addition, new materials are expected to emerge in the coming years.

9.5 CHARACTERIZATION METHODS IN ALL-SOLID-STATE BATTERIES

9.5.1 Lithium Dendrites Characterization Method

The solid-state electrolytes (SSEs) with high mechanical strength are expected to block the lithium dendrite penetration. The unit transference number of Li-ions in SSEs should prevent concentration gradient-induced Li dendrite growth in SSEs. However, extensive investigations demonstrated that Li dendrites still easily grow in SSEs. The SSEs with a much higher mechanical strength show even lower dendrite suppression capability than that in conventional organic electrolytes [119].

In 1998, a combination of optical microscope and CCD camera was used to in-situ observe the lithium metal in the charging and discharging process of Li/PEO-LiTFSI/Li half-cell [120], which was designed to a hexagonal quasi-two-dimensional battery with a special structure. Based on the deposition behavior, Brissot proposed that the formation of lithium dendrites is inseparable from the current density, they believed that the short circuit of the battery was caused by the vertical lithium dendrites piercing the polymer electrolyte (Figure 9.10(a)). Although the lithium dendrite that parallels the electrolyte cannot cause the short circuit, it also greatly reduces the performance and cycle life. Electron microscopes exhibit higher resolution and better depth of field than optical microscopes and can observe lithium dendrites more intuitively. Researchers [121] used scanning electron microscopes to directly observe the deposition behavior of lithium dendrites in

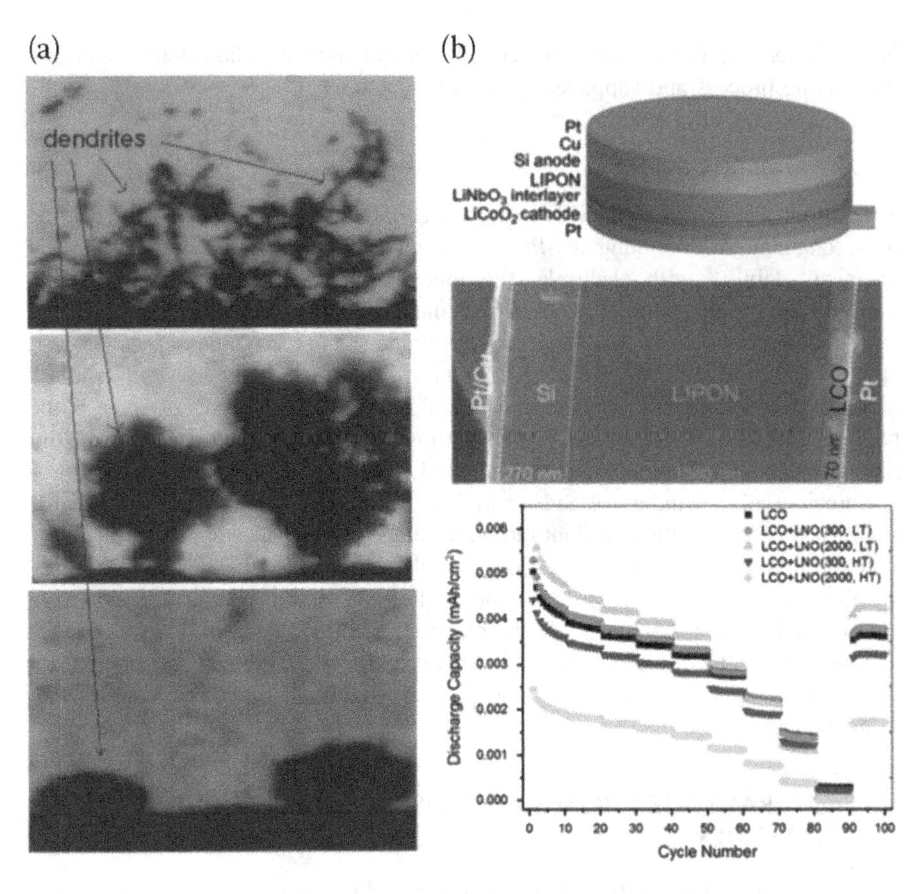

FIGURE 9.10 Schematic diagram of the characterization of lithium dendrites and space charge layers. (a) SEM OF typical lithium dendrites [120]; (b) example and cross-sectional SEM micrograph of full cell architecture, and discharge capacity of cells [123].

Li/LLZTO/Li half-cells. Lithium metal grows along the grain boundaries and pores. It is found that the battery will be short-circuited after a period of time at a high current density. The short-circuit time is positively correlated with the current density. To prevent the formation of lithium dendrites, the pores of SSEs must be reduced.

Han [73] used time-resolved in-situ neutron depth analysis (NDP) to monitor the dynamic evolution of the Li concentration distribution in three representative solid electrolytes during lithium electroplating: lithium phosphorus oxynitride (LiPON), LLZO, and amorphous Li_3PS_4. Although no significant changes in the lithium concentration in LiPON were observed, the direct deposition of Li inside the bulk LLZO and Li_3PS_4 can be seen. Their findings indicate that the high electronic conductivity of LLZO and Li_3PS_4 leads to the formation of lithium dendrites. Therefore, reducing the electronic conductivity of the SSEs is critical to all-solid-state lithium batteries. For the characterization of lithium dendrites, most researchers use the combination of imaging and electrochemical measurement. The in-situ observation method can be used to deeply understand the formation mechanism of lithium dendrites to effectively suppress the growth of lithium dendrites.

9.5.2 Space Charge Layer Characterization

The current insights of the space charge layer are in terms of macroscopic performance, and characterizing the internal voltage distribution of the battery via the existing technology remains a challenge. However, with the development of solid-state batteries, many excellent works have been carried out. Researchers [122] employed TEM and TEM in-space electron energy loss spectroscopy (SR-TEM-EELS) to observe the $Li_2OAl_2O_3$-TiO_2-P_2O_5 glass-ceramic solid electrolyte, and understand that the reason for its low interface resistance is the insertion of Li and the amorphous phase of the interface. Gittleson [123] used XPS technology to characterize the chemical coordination at the $LiCoO_2$-LIPON interface (Figure 9.10b), and found that the Co element in the positive active material increased in valence, which inferred the loss of lithium in the lithium cobaltate and the lithium in the electrolyte. The enrichment of the elements forms a lithium-poor region and a lithium-rich region, resulting in a space charge layer. They believe that this spontaneously formed charge layer will aggravate the charge separation at the interface and reduce the capacity of the cathode material by at least 15%. It is obvious that XPS, a method of analyzing the surface chemical composition and element valence of materials, is a powerful tool for studying the interface problems of solid-state batteries.

9.5.3 Characterization of Element Diffusion

Generally, it is processed by pressure and heating to obtain a compact battery, when the inorganic electrolyte is used to prepare a solid-state battery. However, in the process of high-temperature treatment or charging and discharging, element diffusion and new phases generation usually occur at the electrolyte and electrode interface, which increases the interface impedance and greatly reduces the cycle performance of the battery. The high-energy electron beam of the transmission electron microscope can penetrate thin samples. The transmission electron microscope can not only directly observe the structural changes of the materials, but also use its additional functions to perform elemental analysis on the surface of the material. It is a powerful tool to characterize the diffusion of elements.

Sakuda et al. [124] used focused ion beam cutting (FIB) to prepare samples for TEM characterization, and study the interface between $LiCoO_2$ and sulfide electrolyte Li_2S-P_2S_5 through transmission electron microscopy with energy-dispersive X-ray spectroscopy (EDX). They directly observed the existence of an interface layer between the two and found that Co, P, and S were all diffused (Figure 9.11(a)).

Researchers have also combined electrochemical impedance spectroscopy (EIS) with X-ray photoelectron spectroscopy to understand the capacitance and impedance changes of the electrolyte and electrode interface [125] and have carried out an in-depth study of the area between $LiCoO_2$ and Li-La-Zr-O electrolyte. The

(a)

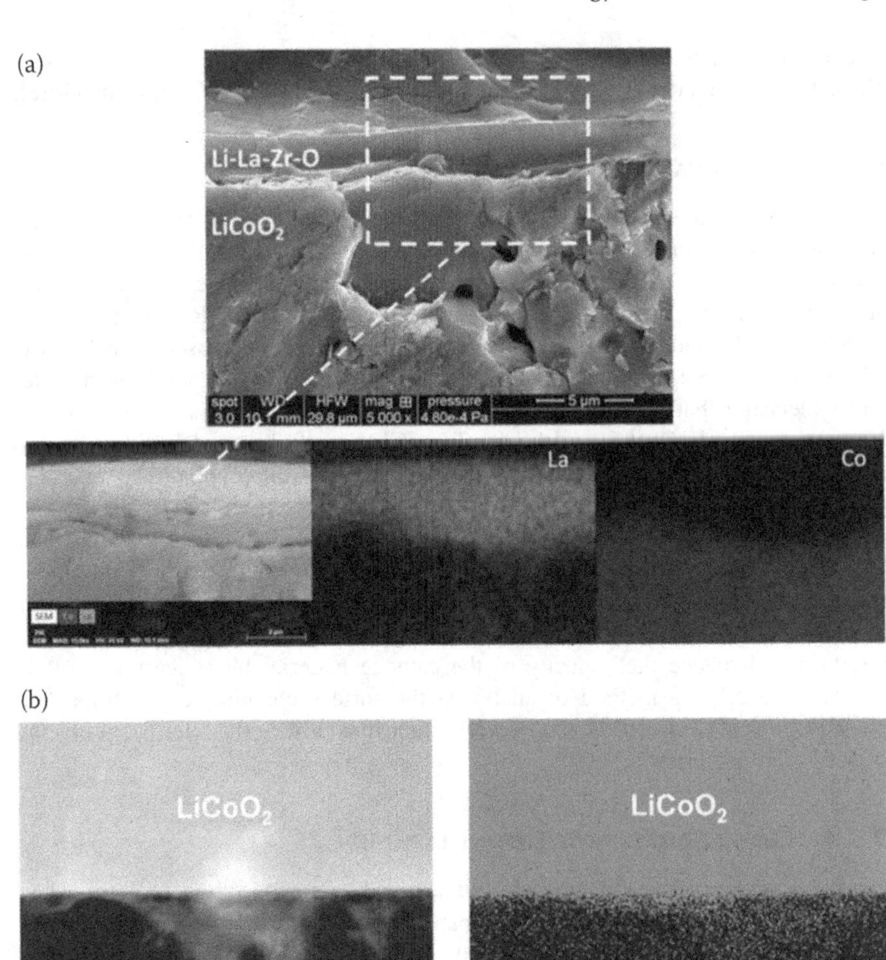

FIGURE 9.11 Schematic representation of element diffusion and interface reaction (a) SEM cross-section image of Li-La-Zr-O on $LiCoO_2$ substrate, along with mapping, showing dense electrolyte [125]; (b) cross-sectional HAADF-STEM image and the corresponding EDX mapping for the Co element near the $LiCoO_2$ electrode/Li_2S-P_2S_5 solid electrolyte interface after initial charging [124].

results showed that CO, La, and Zr have all diffused, and they believe that it is very important to form a protective layer between the positive electrode and the electrolyte (Figure 9.11(b)). For the influence of element diffusion, researchers do not yet have a deeper understanding. Whether element diffusion will bring more complex effects to the battery, and whether there is a more effective solution besides the introduction of a buffer layer are all the directions of future research.

9.5.4 Characterization of Interfacial Reaction

At present, the characterization of interface reactions and structural changes are generally direct observation methods or elemental analysis speculation. Hovington et al. used an in-situ scanning electron microscope to observe the electrolyte/lithium metal anode interface of polymer solid-state batteries and found that the thickness of the lithium metal anode changed from 35 μm to 42 μm after the first charge and discharge [126]. It shows that the lithium metal has expanded, the battery continued to cycle, and the thickness of the lithium was still increasing slowly, indicating that the situation was still deteriorating. The researchers used synchrotron X-ray nano-tomography technology combined with electron microscopy to characterize the FeS_2 electrode in the sulfide battery [67], providing direct evidence for the sulfide electrolyte/FeS_2 interface problem for the first time, demonstrating the heterogeneous phase transition and the internal electrode strain prevents further electrochemical reactions, thereby reducing the reversible capacity of the battery.

9.5.5 Characterization of the Internal Structure Evolution of the Positive Electrode

The researchers explained the capacity decay and structural instability of the nickel-rich single-crystal NCM622 material during the charging and discharging process [127]. They studied the evolution of the material by operating X-ray spectroscopy imaging, nano-tomography, and spectroscopy. It is found that the heterogeneity of the Ni oxidation state distribution, reaction heterogeneity, and high irreversibility in the single crystal particles can be attributed to the characteristics of the original NCM crystal surface chemistry (Li^+/Ni^{2+} disorder), which may cause internal heterogeneity in the particles' internal strain, and further lead to performance degradation. The mixed surface phase caused by the mixing of cations may inhibit the transportation of lithium ions, which triggers the transformation of the surface phase from a layered structure to a rock salt structure and induces uneven distribution of lithium ions (Figure 9.12).

Liu et al. used two-dimensional and three-dimensional full-field transmission X-ray microscopes for the first time to detect NCA secondary particles collected from circulating NCA-PEO polymer batteries [128]. This method can quantify the morphology that occurs in solid-state polymer-based lithium-ion batteries and electrochemical changes. By combining scanning electron microscopy and other methods, it is concluded that the lack of liquid electrolyte will seriously affect the transportation of ruptured particles and, due to the low uniformity of current density, it will accelerate the rupture. Intergranular cracks significantly increase the diffusion path length of charge carriers, electrons, and lithium ions. The intergranular cracks of the NCA particles cast in the polymer matrix seem to cause the

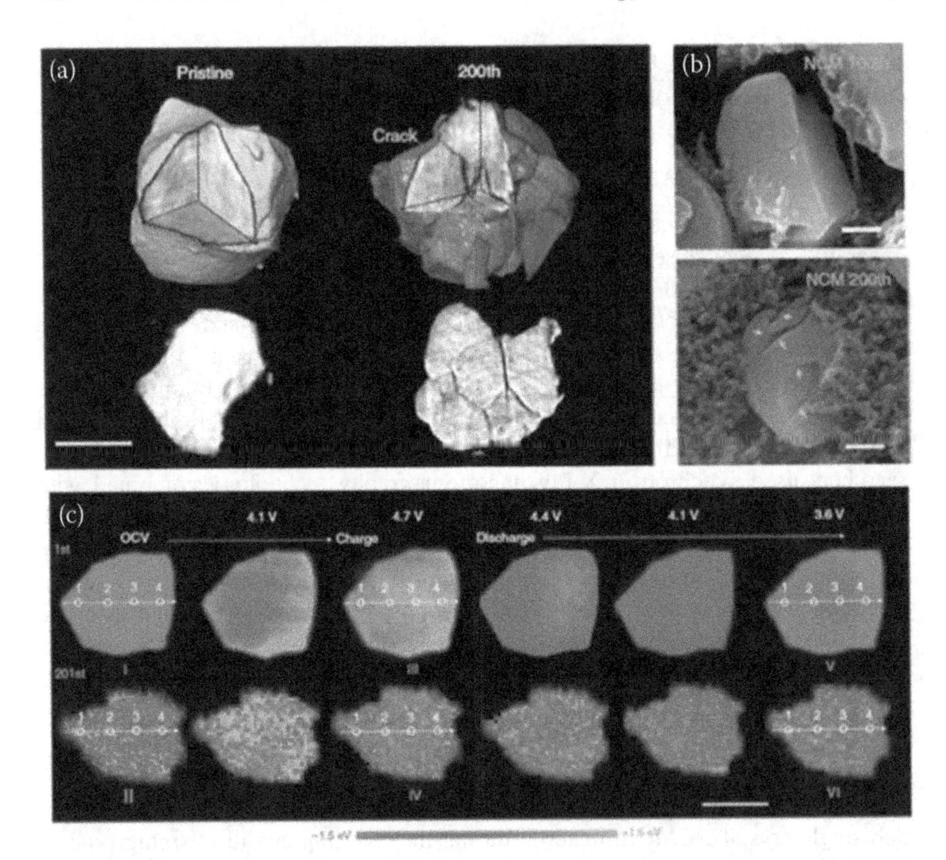

FIGURE 9.12 Schematic diagram of positive fragmentation: (a) X-ray nano-tomography reconstruction with volume rendering shows the morphological evolution of NCM after cycling; (b) SEM images of the NCM microsphere after 100 and 200 cycles; (c) Operando 2-D chemical phase mappings at the Ni K-edge of NCM particles during the first and 201st cycles [127].

isolation and deactivation of the primary particles in the core of the secondary particles, resulting in capacity decay and reduction in the oxidation state of nickel.

At present, for the study of solid-state battery interface problems, researchers mostly use different combinations of characterization methods, and for different aspects of problems, certain specific characterization methods can be used to study and to maximize the use of characterization methods to explore unknown problems. In further exploration, researchers not only need to continue to develop new and higher-level characterization techniques, such as a new generation of synchrotron radiation light sources, solid-state nuclear magnetism but more importantly, rationally develop and utilize existing characterization methods to solve current problems.

REFERENCES

[1] Liu, H., et al., Controlling Dendrite Growth in Solid-State Electrolytes. *ACS Energy Letters* (2020) 5 (3), 833.
[2] Fenton, D. E., *et al.*, Complexes of alkali metal ions with poly (ethylene oxide). *Polymer* (1973) 14 (11).
[3] Hu, P., et al., Progress in nitrile-based polymer electrolytes for high performance lithium batteries. *Journal of Materials Chemistry* (2016) 4 (26), 10070.
[4] Ghelichi, M., et al., Conformational, thermal, and ionic conductivity behavior of PEO in PEO/PMMA miscible blend: Investigating the effect of lithium salt. *Journal of Applied Polymer Science* (2013) 129 (4), 1868.
[5] Liu, F., et al., Progress in the production and modification of PVDF membranes. *Journal of Membrane Science* (2011) 375 (1), 1.
[6] Lin, D., et al., A Silica-Aerogel-Reinforced Composite Polymer Electrolyte with High Ionic Conductivity and High Modulus. *Advanced Materials* (2018) 30 (32).
[7] Croce, F., et al., Nanocomposite polymer electrolytes for lithium batteries. *Nature* (1998) 394 (6692), 456.
[8] Liu, W., et al., Improved Lithium Ionic Conductivity in Composite Polymer Electrolytes with Oxide-Ion Conducting Nanowires. *ACS Nano* (2016) 10 (12), 11407.
[9] Li, Z., et al., Three-Dimensional Garnet Framework-Reinforced Solid Composite Electrolytes with High Lithium-Ion Conductivity and Excellent Stability. *ACS Applied Materials & Interfaces* (2019) 11 (30), 26920.
[10] Liu, Y., et al., Constructing Li-Rich Artificial SEI Layer in Alloy-Polymer Composite Electrolyte to Achieve High Ionic Conductivity for All-Solid-State Lithium Metal Batteries. *Advanced Materials* (2021) 33 (11).
[11] Stramare, S., et al., Lithium Lanthanum Titanates: A Review. *Chemistry of Materials* (2003) 15 (21), 3974.
[12] Jian, Z., et al., NASICON-Structured Materials for Energy Storage. *Advanced Materials* (2017) 29 (20).
[13] Reckeweg, O., et al., Li5OCl3 and Li3OCl: Two Remarkably Different Lithium Oxide Chlorides. *Zeitschrift fur Anorganische und Allgemeine Chemie* (2012) 638, 2081.
[14] Thangadurai, V., et al., Garnet-type solid-state fast Li ion conductors for Li batteries: critical review. *Chemical Society Reviews* (2014) 43 (13), 4714.
[15] Kamaya, N., et al., A lithium superionic conductor. *Nature Materials* (2011) 10 (9), 682.
[16] Kato, Y., et al., High-power all-solid-state batteries using sulfide superionic conductors. *Nature Energy* (2016) 1 (4).
[17] Epp, V., et al., Highly Mobile Ions: Low-Temperature NMR Directly Probes Extremely Fast Li+ Hopping in Argyrodite-Type Li6PS5Br. *Journal of Physical Chemistry Letters* (2013) 4 (13), 2118.
[18] Kerman, K., et al., Review—Practical Challenges Hindering the Development of Solid State Li Ion Batteries. *Journal of The Electrochemical Society* (2017) 164, A1731.
[19] Sun, N., et al., Anisotropically Electrochemical-Mechanical Evolution in Solid-State Batteries and Interfacial Tailored Strategy. *Angewandte Chemie* (2019) 58 (51), 18647.
[20] Yu, C., et al., Accessing the bottleneck in all-solid state batteries, lithium-ion transport over the solid-electrolyte-electrode interface. *Nature Communications* (2017) 8 (1), 1086.

[21] Wood, K. N., et al., Operando X-ray photoelectron spectroscopy of solid electrolyte interphase formation and evolution in Li 2 S-P 2 S 5 solid-state electrolytes. *Nature Communications* (2018) 9 (1).

[22] Okumura, T., et al., LISICON-Based Amorphous Oxide for Bulk-Type All-Solid-State Lithium-Ion Battery. *ACS Applied Energy Materials* (2020) 3 (4), 3220.

[23] Jian, Z., et al., NASICON-Structured Materials for Energy Storage. *Advanced Materials* (2017), 1601925.

[24] Xiaogang Han, Y. G., et al., Negating interfacial impedance in garnet-based solid-state Li metal batteries. *Nature Materials* (2017) 16, 572-579.

[25] Xu, J., et al., Efficiently photo-charging lithium-ion battery by perovskite solar cell. *Nature Communications* (2015) 6, 8103.

[26] Zhu, J., et al., A Multilayer Ceramic Electrolyte for All-Solid-State Li Batteries. *Angewandte Chemie International Edition in English* (2021) 60 (7), 3781.

[27] Delluva, A. A., et al., Decomposition of Trace Li2CO3 During Charging Leads to Cathode Interface Degradation with the Solid Electrolyte LLZO. *Advanced Functional Materials* (2021) 31, 2103716.

[28] Gao, J., et al., Rational Design of Mixed Electronic-Ionic Conducting Ti-Doping Li7La3Zr2O12 for Lithium Dendrites Suppression. *Advanced Functional Materials* (2020) 31 (2).

[29] Luo, W., et al., Reducing Interfacial Resistance between Garnet-Structured Solid-State Electrolyte and Li-Metal Anode by a Germanium Layer. *Advanced Materials* (2017), 1606042.

[30] Lei Cheng, E. J. C., et al., The origin of high electrolyte-electrode interfacial resistances in lithium cells containing garnet type solid electrolytes. *Physical Chemistry Chemical Physics* (2014) 16 (34), 18294.

[31] Sharafi, A., et al., Characterizing the Li–Li7La3Zr2O12 interface stability and kinetics as a function of temperature and current density. *Journal of Power Sources* (2016) 302, 135.

[32] Tsai, C. L., et al., Li7La3Zr2O12 Interface Modification for Li Dendrite Prevention. *ACS Applied Materials & Interfaces* (2016) 8 (16), 10617.

[33] Chen, Y., et al., Nanocomposite intermediate layers formed by conversion reaction of SnO2 for Li/garnet/Li cycle stability. *Journal of Power Sources* (2019) 420, 15.

[34] Huo, H., et al., Design of a mixed conductive garnet/Li interface for dendrite-free solid lithium metal batteries. *Energy & Environmental Science* (2020) 13 (1), 127.

[35] Huo, H., et al., In-situ formed Li2CO3-free garnet/Li interface by rapid acid treatment for dendrite-free solid-state batteries. *Nano Energy* (2019) 61, 119.

[36] Han, X., et al., Li_7 Negating interfacial impedance in garnet-based solid-state Li metal batteries. *Nature Materials* (2016) 8, 10617–10626.

[37] Aguesse, F., et al., Nanocomposite intermediate layers formed by conversion reaction of SnO2 for Li/garnet/Li cycle stability. *ACS Applied Materials & Interfaces* (2017) 9 (4), 3808.

[38] Chen, X., et al., Enhancing interfacial contact in all solid state batteries with a cathode-supported solid electrolyte membrane framework. *Energy & Environmental Science* (2019) 12, 938–944.

[39] Kim, Y., et al., Thermally Driven Interfacial Degradation between Li7La3Zr2O12 Electrolyte and LiNi0.6Mn0.2Co0.2O2 Cathode. *Chemistry of Materials* (2020) 32 (22), 9531-9541.

[40] Sastre, J., et al., Fast Charge Transfer across the Li7La3Zr2O12 Solid Electrolyte/LiCoO2 Cathode Interface Enabled by an Interphase-Engineered All-Thin-Film Architecture. *ACS Applied Materials & Interfaces* (2020) 12 (32), 36196-36207.

[41] Kim, K. H., et al., Characterization of the interface between LiCoO2 and Li7La3Zr2O12 in an all-solid-state rechargeable lithium battery. *Journal of Power Sources* (2011) 196 (2), 764-767.

[42] Kato, T., et al., In-situ Li7La3Zr2O12/LiCoO2 interface modification for advanced all-solid-state battery. *Journal of Power Sources* (2014) 260, 292.

[43] Han, F., et al., Interphase Engineering Enabled All-Ceramic Lithium Battery. *Joule* (2018) 2 (3), 497.

[44] Kamaya, N., et al., A lithium superionic conductor. *Nature Materials* (2011) 10 (9), 682.

[45] Kato, Y., et al., High-power all-solid-state batteries using sulfide superionic conductors. *Nature Energy* (2016) 1 (4), 16030.

[46] Wang, Y., et al., Design principles for solid-state lithium superionic conductors. *Nature Materials* (2015) 14 (10), 1026.

[47] Kraft, M. A., et al., Inducing High Ionic Conductivity in the Lithium Superionic Argyrodites $Li_{6+x}P_{1-x}Ge_xS_5I$ for All-Solid-State Batteries. *Journal of the American Chemical Society* (2018) 140 (47), 16330.

[48] Wu, J., et al., Lithium/Sulfide All-Solid-State Batteries using Sulfide Electrolytes. *Advanced Materials* (2021) 33 (6), e2000751.

[49] Minami, K., et al., Crystallization Process for Superionic $Li_7P_3S_{11}$ Glass-Ceramic Electrolytes. Journal of the American Ceramic Society. *Journal of the American Ceramic Society* (2011) 94 (6), 1779.

[50] Seino, Y., et al., A sulphide lithium super ion conductor is superior to liquid ion conductors for use in rechargeable batteries. *Energy & Environmental Science* (2014) 7 (2), 627.

[51] Wu, F., et al., Advanced sulfide solid electrolyte by core-shell structural design. *Nature Communications* (2018) 9 (1), 4037.

[52] Oh, D. Y., et al., Wet-Chemical Tuning of $Li_{3-x}PS_4$(0</=x</=0.3) Enabled by Dual Solvents for All-Solid-State Lithium-Ion Batteries. *ChemSusChem* (2020) 13 (1), 146.

[53] Wu, J., et al., All-Solid-State Lithium Batteries with Sulfide Electrolytes and Oxide Cathodes. *Electrochemical Energy Reviews* (2020) 4 (1), 101.

[54] Haruyama, J., et al., Space–Charge Layer Effect at Interface between Oxide Cathode and Sulfide Electrolyte in All-Solid-State Lithium-Ion Battery. *Chemistry of Materials* (2014) 26 (14), 4248.

[55] Sakuda, A., et al., Interfacial Observation between $LiCoO_2$ Electrode and $Li_2S–P_2S_5$ Solid Electrolytes of All-Solid-State Lithium Secondary Batteries Using Transmission Electron Microscopy. *Chemistry of Materials* (2010) 22 (3), 949.

[56] Auvergniot, J., et al., Interface Stability of Argyrodite Li_6PS_5Cl toward $LiCoO_2$, $LiNi_{1/3}Co_{1/3}Mn_{1/3}O_2$, and $LiMn_2O_4$ in Bulk All-Solid-State Batteries. *Chemistry of Materials* (2017) 29 (9), 3883.

[57] Koerver, R., et al., Capacity Fade in Solid-State Batteries: Interphase Formation and Chemomechanical Processes in Nickel-Rich Layered Oxide Cathodes and Lithium Thiophosphate Solid Electrolytes. *Chemistry of Materials* (2017) 29 (13), 5574.

[58] Wang, L., et al., In-situ visualization of the space-charge-layer effect on interfacial lithium-ion transport in all-solid-state batteries. *Nature Communications* (2020) 11 (1), 5889.

[59] Zhang, J., et al., Unraveling the Intra and Intercycle Interfacial Evolution of Li_6PS_5Cl-Based All-Solid-State Lithium Batteries. *Advanced Energy Materials* (2019) 10 (4).

[60] Haruyama, J., et al., Cation Mixing Properties toward Co Diffusion at the $LiCoO_2$ Cathode/Sulfide Electrolyte Interface in a Solid-State Battery. *ACS Applied Materials & Interfaces* (2017) 9 (1), 286.

[61] Ohta, N., et al., $LiNbO_3$-coated $LiCoO_2$ as cathode material for all solid-state lithium secondary batteries. *Electrochemistry Communications* (2007) 9 (7), 1486.
[62] Ohta, N., et al., Enhancement of the High-Rate Capability of Solid-State Lithium Batteries by Nanoscale Interfacial Modification. *Advanced Materials* (2006) 18 (17), 2226.
[63] Zhao, F., et al., Tuning bifunctional interface for advanced sulfide-based all-solid-state batteries. *Energy Storage Materials* (2020) 33, 139.
[64] Deng, S., et al., Dual-functional interfaces for highly stable Ni-rich layered cathodes in sulfide all-solid-state batteries. *Energy Storage Materials* (2020) 27, 117.
[65] Zhou, A., et al., Al_2O_3 surface coating on $LiCoO_2$ through a facile and scalable wet-chemical method towards high-energy cathode materials withstanding high cutoff voltages. *Journal of Materials Chemistry A* (2017) 5 (46), 24361.
[66] Wu, X., et al., Operando Visualization of Morphological Dynamics in All-Solid-State Batteries. *Advanced Energy Materials* (2019) 9 (34), 1901547.
[67] Sun, N., et al., Anisotropically Electrochemical-Mechanical Evolution in Solid-State Batteries and Interfacial Tailored Strategy. *Angewandte Chemie International Edition in English* (2019) 58 (51), 18647.
[68] Yue, J., et al., Long Cycle Life All-Solid-State Sodium Ion Battery. *ACS Applied Materials & Interfaces* (2018) 10 (46), 39645.
[69] Zhang, Q., et al., Nickel sulfide anchored carbon nanotubes for all-solid-state lithium batteries with enhanced rate capability and cycling stability. *Journal of Materials Chemistry A* (2018) 6 (25), 12098.
[70] Lewis, J. A., et al., Linking void and interphase evolution to electrochemistry in solid-state batteries using operando X-ray tomography. *Nature Materials* (2021) 20 (4), 503.
[71] Kasemchainan, J., et al., Critical stripping current leads to dendrite formation on plating in lithium anode solid electrolyte cells. *Nature Materials* (2019) 18 (10), 1105.
[72] Ning, Z., et al., Visualizing plating-induced cracking in lithium-anode solid-electrolyte cells. *Nature Materials* (2021) 20 (8),1121.
[73] Han, F., et al., High electronic conductivity as the origin of lithium dendrite formation within solid electrolytes. *Nature Energy* (2019) 4 (3), 187.
[74] Wenzel, S., et al., Interphase formation on lithium solid electrolytes—An in situ approach to study interfacial reactions by photoelectron spectroscopy. *Solid State Ionics* (2015) 278, 98.
[75] Fenton, D.E., Parker, J.M. and Wright, P.V., Complexes of alkali metal ions with poly (ethylene oxide). Polymer (1973) 14, 589.
[76] Xu, B., et al., Interfacial Chemistry Enables Stable Cycling of All-Solid-State Li Metal Batteries at High Current Densities. *Journal of the American Chemical Society* (2021) 143 (17), 6542.
[77] Schwietert, T. K., et al., Clarifying the relationship between redox activity and electrochemical stability in solid electrolytes. *Nature Materials* (2020) 19 (4), 428.
[78] Wan, Z., et al., Low Resistance-Integrated All-Solid-State Battery Achieved by $Li_7La_3Zr_2O_{12}$ Nanowire Upgrading Polyethylene Oxide (PEO) Composite Electrolyte and PEO Cathode Binder. *Advanced Functional Materials* (2019) 29 (1).
[79] Chen-Yang, Y. W., Lin, H. C. C. F. J., Chen, C. C., Polyacrylonitrile electrolytes: A novel highconductivity composite polymer electrolyte based on PAN, $LiClO_4$ and α-Al_2O_3. *Solid State Ionics* (2002) 150, 327.
[80] Di Noto, V. Z., Inorganic-organic polymer electrolytes based on PEG400 and Al $[OCH(CH_3)_2]_3$ Synthesis and vibrational characterizations. *Journal of the Electrochemical Society* (2004) 151, A216.

[81] Liu, Y., et al., In situ preparation of poly(ethylene oxide)–SiO_2 composite polymer electrolytes. *Journal of Power Sources* (2004) 129 (2), 303.
[82] Magistris, A., Quartarone, P. M. E., Tomasi, C., Transport and thermal properties of (PEO)n–LiPF6 electrolytes for super-ambient applications. *Solid State Ionics* (2000) 136, 1241.
[83] Marcinek, M., et al., Ionic association in liquid (polyether–AlO–LiClO) composite electrolytes. *Solid State Ionics* (2005) 176 (3–4), 367.
[84] Xi, G., et al., Polymer-Based Solid Electrolytes: Material Selection, Design, and Application. *Advanced Functional Materials* (2020) 31 (9).
[85] Xu, K., Nonaqueous Liquid Electrolytes for Lithium-Based Rechargeable Batteries. *Chemical Reviews* (2004) 104 (10), 4303.
[86] Manthiram, A., et al., Lithium battery chemistries enabled by solid-state electrolytes. *Nature Reviews Materials* (2017) 2 (4).
[87] Xu, K., Electrolytes and interphases in Li-ion batteries and beyond. *Chemical Reviews* (2014) 114 (23), 11503.
[88] Croce, F., Persi, G. B. A. L., Scrosati, B., Nanocomposite polymer electrolytes for lithium batteries. *Nature* (1998) 394, 456.
[89] Lin, D., et al., High Ionic Conductivity of Composite Solid Polymer Electrolyte via In Situ Synthesis of Monodispersed SiO_2 Nanospheres in Poly(ethylene oxide). *Nano Letters* (2016) 16 (1), 459.
[90] Bae, J., et al., A 3D Nanostructured Hydrogel-Framework-Derived High-Performance Composite Polymer Lithium-Ion Electrolyte. *Angewandte Chemie International Edition in English* (2018) 57 (8), 2096.
[91] Huo, H., et al., Anion-immobilized polymer electrolyte achieved by cationic metal-organic framework filler for dendrite-free solid-state batteries. *Energy Storage Materials* (2019) 18, 59.
[92] Choi, J.-H., et al., Enhancement of ionic conductivity of composite membranes for all-solid-state lithium rechargeable batteries incorporating tetragonal $Li_7La_3Zr_2O_{12}$ into a polyethylene oxide matrix. *Journal of Power Sources* (2015) 274, 458.
[93] Chen, B., et al., A new composite solid electrolyte PEO/$Li_{10}GeP_2S_{12}$/SN for all-solid-state lithium battery. *Electrochimica Acta* (2016) 210, 905.
[94] Zhai, H., et al., A Flexible Solid Composite Electrolyte with Vertically Aligned and Connected Ion-Conducting Nanoparticles for Lithium Batteries. *Nano Letters* (2017) 17 (5), 3182.
[95] Zhang, X., et al., Synergistic Coupling between $Li_{6.75}La3Zr_{1.75}Ta_{0.25}O_{12}$ and Poly (vinylidene fluoride) Induces High Ionic Conductivity, Mechanical Strength, and Thermal Stability of Solid Composite Electrolytes. *Journal of the American Chemical Society* (2017) 139 (39), 13779.
[96] Wang, C., Appleby, X.-W. Z. A. J., Solvent-free composite PEO-ceramic fiber/mat electrolytes for lithium secondary cells. *Journal of the Electrochemical Society* (2005) 152, A205.
[97] Liu, W., et al., Enhancing ionic conductivity in composite polymer electrolytes with well-aligned ceramic nanowires. *Nature Energy* (2017) (17035).
[98] Alamgir, M., Abraham, R. M. K., Li+-conductive polymer electrolytes derived from poly (1, 3-dioxolane) and polytetrahydrofuran. *Electrochimica Acta* (1991) 36, 773.
[99] Akbulut, O., et al., Conductivity hysteresis in polymer electrolytes incorporating poly(tetrahydrofuran). *Electrochimica Acta* (2007) 52 (5), 1983.
[100] Liao, Y. P., et al., Replies to comments contained in "Conductivity hysteresis in polymer electrolytes incorporating poly(tetrahydrofuran)". *Electrochimica Acta* (2007) 52 (24), 7173.

[101] Mackanic, D. G., et al., Crosslinked Poly(tetrahydrofuran) as a Loosely Coordinating Polymer Electrolyte. *Advanced Energy Materials* (2018) 8 (25).

[102] Cui, Y., et al., High Performance Solid Polymer Electrolytes for Rechargeable Batteries: A Self-Catalyzed Strategy toward Facile Synthesis. *Advanced Science* (2017) 4 (11), 1700174.

[103] Li, Z., et al., Mitigating Interfacial Instability in Polymer Electrolyte-Based Solid-State Lithium Metal Batteries with 4 V Cathodes. *ACS Energy Letters* (2020) 5 (10), 3244.

[104] Xia, Y. Y., et al., Thermal and electrochemical stability of cathode materials in solid polymer electrolyte. *Journal of Power Sources* (2001) 92, 234.

[105] Liang, J., et al., Engineering the conductive carbon/PEO interface to stabilize solid polymer electrolytes for all-solid-state high voltage $LiCoO_2$ batteries. *Journal of Materials Chemistry A* (2020) 8 (5), 2769.

[106] Bae, S.-H., et al., Seamlessly Conductive 3D Nanoarchitecture of Core-Shell Ni-Co Nanowire Network for Highly Efficient Oxygen Evolution. *Advanced Energy Materials* (2017) 7 (1).

[107] Masa, J., et al., Ultrathin High Surface Area Nickel Boride (NixB) Nanosheets as Highly Efficient Electrocatalyst for Oxygen Evolution. *Advanced Energy Materials* (2017) 7 (17).

[108] Fu, C., et al., Capacity degradation mechanism and improvement actions for 4 V-class all-solid-state lithium-metal polymer batteries. *Chemical Engineering Journal* (2020) 392.

[109] Ma, J., et al., A Strategy to Make High Voltage $LiCoO_2$ Compatible with Polyethylene Oxide Electrolyte in All-Solid-State Lithium Ion Batteries. *Journal of the Electrochemical Society* (2017) 164 (14), A3454.

[110] Nakayama, M., et al., Factors affecting cyclic durability of all-solid-state lithium polymer batteries using poly(ethylene oxide)-based solid polymer electrolytes. *Energy & Environmental Science* (2010) 3 (12).

[111] Sheng, O., et al., In Situ Construction of a LiF-Enriched Interface for Stable All-Solid-State Batteries and its Origin Revealed by Cryo-TEM. *Advanced Materials* (2020) 32 (34), e2000223.

[112] Nie, K., et al., Increasing Poly(ethylene oxide) Stability to 4.5 V by Surface Coating of the Cathode. *ACS Energy Letters* (2020) 5 (3), 826.

[113] Liu, W., et al., Significantly improving cycling performance of cathodes in lithium ion batteries: The effect of Al_2O_3 and $LiAlO_2$ coatings on $LiNi_{0.6}Co_{0.2}Mn_{0.2}O_2$. *Nano Energy* (2018) 44, 111.

[114] Singh, G., et al., Electrochemical and Structural Investigations on ZnO Treated 0.5 Li_2MnO_3-0.5$LiMn_{0.5}Ni_{0.5}O_2$ Layered Composite Cathode Material for Lithium Ion Battery. *Journal of the Electrochemical Society* (2012) 159 (4), A470.

[115] Wang, Z., et al., Effect of amorphous $FePO_4$ coating on structure and electrochemical performance of $Li_{1.2}Ni_{0.13}Co_{0.13}Mn_{0.54}O_2$ as cathode material for Li-ion batteries. *Journal of Power Sources* (2013) 236, 25.

[116] Olivares-Marin, M., et al., Spatial Distributions of Discharged Products of Lithium-Oxygen Batteries Revealed by Synchrotron X-ray Transmission Microscopy. *Nano Letters* (2015) 15 (10), 6932.

[117] Zheng, J., et al., Functioning Mechanism of AlF_3 Coating on the Li- and Mn-Rich Cathode Materials. *Chemistry of Materials* (2014) 26 (22), 6320.

[118] Wang, L. P., et al., Ameliorating the Interfacial Problems of Cathode and Solid-State Electrolytes by Interface Modification of Functional Polymers. *Advanced Energy Materials* (2018) 8 (24).

[119] Ji, X., et al., Solid-State Electrolyte Design for Lithium Dendrite Suppression. *Advanced Materials* (2020) 32 (46), e2002741.

[120] Brissot, C., et al., In situ study of dendritic growth inlithium/PEO-salt/lithium cells. *Electrochimica Acta* (1998) 43 (10), 1569.
[121] Ren, Y., et al., Direct observation of lithium dendrites inside garnet-type lithium-ion solid electrolyte. *Electrochemistry Communications* (2015) 57, 27.
[122] Yamamoto, K., et al., Nano-scale simultaneous observation of Li-concentration profile and Ti-, O electronic structure changes in an all-solid-state Li-ion battery by spatially-resolved electron energy-loss spectroscopy. *Journal of Power Sources* (2014) 266, 414.
[123] Gittleson, F. S., El Gabaly, F., Non-Faradaic Li(+) Migration and Chemical Coordination across Solid-State Battery Interfaces. *Nano Letters*(2017) 17 (11), 6974.
[124] Sakuda, A., et al., Interfacial Observation between LiCoO2 Electrode and Li2S–P2S5 Solid Electrolytes of All-Solid-State Lithium Secondary Batteries Using Transmission Electron Microscopy. *Chemistry of Materials* (2009) 22 (3), 949.
[125] Zarabian, M., et al., X-ray Photoelectron Spectroscopy and AC Impedance Spectroscopy Studies of Li-La-Zr-O Solid Electrolyte Thin Film/LiCoO2Cathode Interface for All-Solid-State Li Batteries. *Journal of the Electrochemical Society* (2017) 164 (6), A1133.
[126] Hovington, P., et al., New lithium metal polymer solid state battery for an ultrahigh energy: nano C-LiFePO(4) versus nano Li1.2V(3)O(8). *Nano Letters* (2015) 15 (4), 2671.
[127] Zhang, F., et al., Surface regulation enables high stability of single-crystal lithium-ion cathodes at high voltage. *Nature Communications* (2020) 11 (1), 3050.
[128] Besli, M. M., et al., Mesoscale Chemomechanical Interplay of the LiNi0.8Co0.15Al0.05O2 Cathode in Solid-State Polymer Batteries. *Chemistry of Materials* (2018) 31 (2), 491.

10 Key Electrode Materials for Lithium-Ion Capacitor Batteries

Kedi Cai, Xiaoshi Lang, and Shuang Yan
College of Chemistry and Materials Engineering,
Bohai University, Jinzhou, People's Republic of China

CONTENTS

10.1 RESEARCH BACKGROUND

With rapid economic development and increasing public attention to environmental issues, the search for clean and renewable energy has become a subject of common concern for researchers worldwide [1–3]. At present, two complementary electrochemical energy storage systems represented by lithium-ion batteries (LIB_S) and supercapacitors (SCs) occupy an important position in the market for electric vehicles and portable electronic devices. Due to the repeated disintercalation of lithium during charge and discharge, LIBs exhibit high energy density, but low power density and a limited life cycle. On the other hand, on account of the highly reversible adsorption and desorption capacity of ions at the electrode/electrolyte interface, SCs show high power density, excellent cycling stability, and rapid

DOI: 10.1201/9781003133971-10

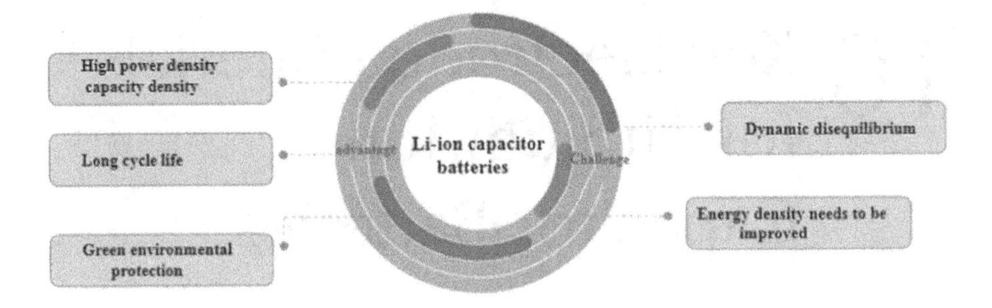

FIGURE 10.1 Diagram of advantages of Li-ion capacitor batteries.

charge discharge capacity, but problems remain such as unsatisfactory energy density [4–6]. To meet the requirements of high energy density and power density at the same time, a new type of special energy storage device called a lithium-ion capacitor (LICs) battery has gradually become the focus of research [7–9]. Figure 10.1 shows an advantage diagram of the LICs.

LICs, also known as lithium-ion capacitors or hybrid lithium-ion ultracapacitors, are usually composed of battery-type materials (negative electrode) and capacitive materials (positive electrode) in an electrolyte-containing lithium salt. As early as the 1990s, Evans first proposed the concept of the lithium-ion capacitor group. Then, at the beginning of the 21st century, Amatucci et al. [10] first reported the LIC system known as "asymmetric hybrid-energy storage battery", which was constructed in an organic electrolyte based on $LiPF_6$, with activated carbon (AC) as capacitive anode and nanostructured $Li_4Ti_5O_{12}$ (LTO) as battery anode. The device produced three to four times the energy density of a normal SC. It also presented high-power performance and excellent cycle stability, with capacity utilization reaching 90% at 10 times the current density, and capacity retention rate remains at approximately 80% of the initial capacity after 4,500 charge–discharge cycles. Professor Xia Yongyao's research group [11] at Fudan University constructed a "rocking chair" power storage device with capacitive activated carbon as negative electrode and battery-type lithium-embedded material as positive electrode, laying a good foundation for the field of LICs in China. At present, domestic and foreign companies have successively produced new-type LICs with good performance. For example, JM Energy has introduced a new type of LICs with a small unit. Compared with the original flat rectangular parallelepiped unit, this unit has been reduced to less than half of its volume. It has achieved miniaturization and can be used in construction machinery, industrial machinery, and areas with limited space for power-supply devices such as automobiles. Maxwell uses a combined system of LIB and super capacitor as an auxiliary technology to meet the needs of various types of power grids. Aowei Technology Development Co. Ltd. has developed LICs battery modules from 3.6 V to 720 V, which are used in starting various vehicles and internal combustion engines, as well as traction of light vehicles and electric buses and other fields, and has successfully established the first capacitor bus line in commercial operation in the world [12].

In recent years, the industrial development of LICs has been gradually promoted in China and abroad. More excellent electrode materials need to be developed to solve the problem of dynamic imbalance between positive and negative electrode materials. Many scientific research teams have combined the related principles of LIBs and SCs with LICs to achieve a breakthrough in the field of LICs. This study reviews the research progress in electrode materials of LICs in recent years, expounds on the advantages and disadvantages of various materials, and presents prospects for the development trend of LICs.

10.2 ANODE MATERIALS

10.2.1 Transition Metal Oxides

Transition metal oxides (mxoym = Fe, Co, Cu, Mo, and others) have become a hot spot in the field of LICs due to its rich source, low cost, stable performance, and higher specific capacitance than capacitor electrode materials [13]. However, this type of material has a large volume change during the long cycle, which leads to problems of low initial coulomb efficiency, poor rate capacity, and rapid capacity decay. The following solutions are proposed by researchers. On the one hand, the morphology of the materials is treated to form porous nanostructures, maintain structural integrity, and promote an electrochemical reaction. On the other hand, carbon materials are integrated into oxides to prepare nanocomposites. Under the protection of carbonaceous materials, the volume expansion of oxide in the cycle can be buffered, and the cycle reversibility and conductivity can be greatly improved.

As widely known, Fe_2O_3 is used in the field of LICs because of its low cost, nontoxicity, and high theoretical specific capacitance [14–17]. However, due to the large size of the original nanoparticles and insufficient porosity, the lithium storage performance is not ideal, which affects the conductivity and cycle stability. Dong Research Group [18] prepared fusiform a-Fe_2O_3 and reduced graphene (rGO) composites through hydrothermal method and applied them to capacitive batteries. The specific capacitance of the material is 455 F/g at 1 A/g and 87.5% of the initial capacitance after 10,000 cycles, showing high specific capacitance (SC) and good cycle stability. Subsequently, Li et al. [19] prepared 1-D mesoporous Fe_2O_3 nanowires growing on the matrix of strain steel (SS) by a two-step method and used them as cathode materials for lithium batteries. The synthesis process was shown in Figure 10.2. At 100 mA/g, the discharge capacity of the electrode after 200 cycles was 1,460 mAh/g, and at 800 mA/g, the reversible capacity can reach 700 mAh/g. Such excellent cycling performance is attributed to the unique mesoporous nanowire structure generated by the capacitance effect. These results indicated that the modified materials have potential applications in LICs.

Co_3O_4 is also a commonly used oxide anode material, but the electrode stability is poor and the specific capacity is low. To improve its performance, Yin et al. [20] prepared GO-MOFs coated rGO/Co_3O_4 composite materials by using temperate coprecipitation method, and innovatively applied MOFs materials to LICs. When zoomed in to 200 nm, the structure appears sparse and porous. This unique structure

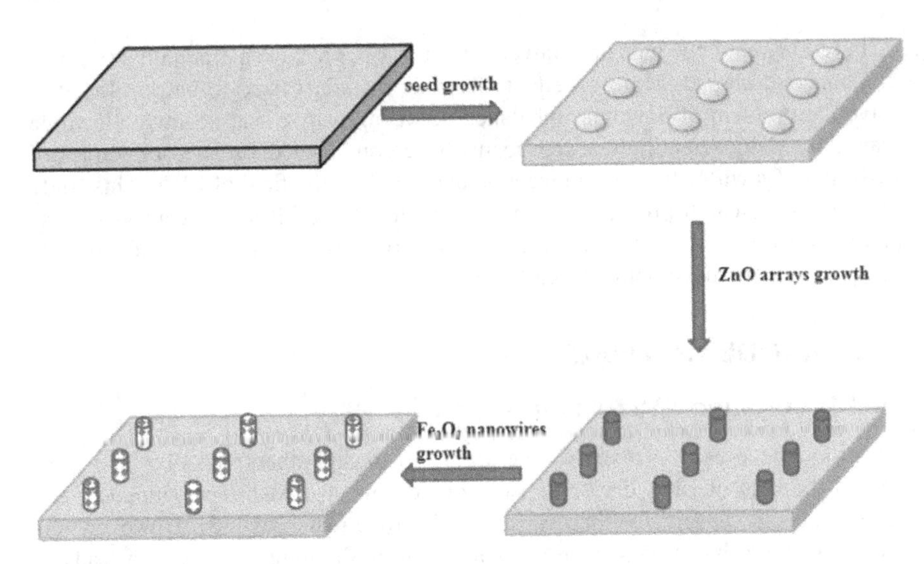

FIGURE 10.2 Synthesis of SS@Fe_2O_3 mesoporous nanowires [19].

not only made the electrode material exhibit excellent lithium storage performance but also had high initial discharge specific capacity (at current density of 100 mA/g and capacity of 1,451 mAh/g), good cycle stability (capacity retention rate of more than 96% after 100 cycles), and excellent pseudo-capacitance performance (546 F/g). In addition, this method provided a prospect for the synthesis of other advanced electrode materials through reasonable structural design, and also made the GO-MOFS-derived transition metal oxide composite a promising electrode material for the next generation of LICs.

Then, the researchers studied other metal oxides and found that the properties of the treated materials had been greatly improved. For example, Zhao et al. [21] synthesized carbon-free MoO_2 nanosheets with a crystalline hybrid structure through a one-step solvothermal reaction. The uneven crystalline structure formed a unique boundary that could effectively increase the diffusion rate of Li^+. LICs was assembled with it as negative electrode, the SC could reach 190 F/g at 1 A/g, and after 7,500 cycles at 20 A/g, the SC still had 12 F/g. Lee et al. [22] prepared graphite/CuO composites by air sintering, and developed a high-energy density LICs system with porous carbon as the positive electrode. The manufacturing process was simple and cost-effective. When the power density was 1.3 kW/kg, the energy density of the system could reach 212.3 Wh/kg. These research results provide new insights to assist researchers in related fields.

10.2.2 Carbon Materials

Carbon materials have attracted attention from various energy-storage fields due to their low cost, high abundance, and stable physical/chemical properties. These materials have been successfully applied to the negative electrode of LICs. Carbon materials are mainly divided into graphite, hard carbon, and soft carbon. Graphite is considered an ideal anode material for high-power capacitor batteries due to its low cost, high conductivity, and large theoretical capacity, and has been commercialized since 1990. However, due to the orientation of graphite, the graphite layer easily falls off, which affects the high rate charge and discharge of the battery. At the same time, the pseudo-capacitance effect reduces the capacitance characteristics. Therefore, the performance of the battery can be improved by using graphene-based composites, doping, and other strategies.

Graphene is the basic building block of graphite material. It has a large specific surface area, excellent electrical conductivity, and high mechanical flexibility, thereby showing excellent electrochemical performance in lithium memory. Wu et al. [23] proposed heteroatom-doped graphene composites for the first time, and the test results showed that the first reversible capacity of boron-doped graphene was as high as 1,549 mAh/g, much higher than that of pure graphene. When the current density was increased to 25 A/g, the reversible capacities of boron-doped and nitrogen-doped graphene were 235 mAh/g and 199 mAh/g, respectively. The power densities of 29.1 kW/kg and 34.9 kW/kg correspond to the energy densities of 226 Wh/kg and 320 Wh/kg. Recently, Arnaiz et al. [24] made a breakthrough in this field by designing an ultra-fast, ultra-stable all-graphene electrode with excellent electrochemical performance, as well as an unprecedented capacity retention of 98.9% after 50,000 cycles, while providing an ultra-high power of 53.5 kW/kg. These results demonstrated that a fast and stable LICs system can be achieved by combining the graphene structure with the matching method.

Soft carbon has high abundance, controllable crystallinity, and is compatible with electrolyte. Although soft carbon has good rate performance and electrochemical stability, it faces the problem of low diffusion rate of Li^+. The storage performance of lithium can be improved by modifying soft carbon through nitrogen doping and structural design. By combining hetero-atomic control and structural design, Zhang et al. [25] obtained ternary-layered porous soft carbon (NPSC) doped with nitrogen, phosphorus, and sulfur, which was used as the negative electrode and $LiVPO_4F$ as the positive electrode to assemble LICs. Figure 10.3 was the schematic diagram of layered porous soft carbon-lithium intercalation. At 100 mA/g, the discharge capacity was 770 mAh/g, and after 500 cycles at 500 mA/g, the capacity reaches 500 mAh/g. This excellent performance showed that the synergistic doping effect could be applied to improve the electrochemical performance of traditional carbon materials.

Hard carbon is synthesized from single graphene and has an sp^3-induced turbine structure, which has great advantages in chemical stability. However, the development of hard carbon is still limited by poor rate performance and large irreversible capacity. Therefore, doping and surface modification of hard carbon can improve its performance. Applying the carbonization process of low-cost biological soap residue, Jiang et al. [26] prepared high-defect, N-doped hard carbon as an anode without adding a template and a catalyst. As a result of N doping and defect engineering, the electrode had high specific capacity, excellent rate capability, and long cycle stability. The discharge capacity was 580.3 mAh/g at 0.05 A/g, and could be stably circulated more than 1,000 times. Electrochemical kinetic analysis and density functional theory calculation also confirmed its outstanding pseudo-

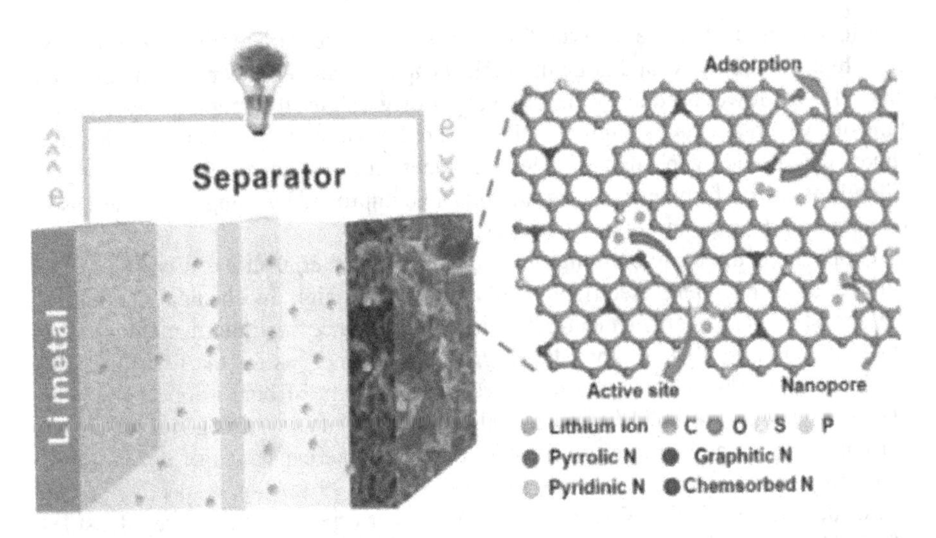

FIGURE 10.3 Schematic illustration of lithium intercalation in hierarchical porous soft carbon with homogeneous doping [25].

capacitance performance and excellent Li^+ storage capacity. As expected, the LICs have excellent energy density.

10.2.3 Lithium-Embedded Compounds

Titanium oxide (TiO_2) is a type of common lithium-embedded compound. Compared with graphite, TiO has higher electrochemical working potential and better safety. Lithium titanate $Li_4Ti_5O_{12}$ and TiO_2 have been used as high-power-density cathode materials in LICs.

$Li_4Ti_5O_{12}$ has unique "zero strain" characteristics and excellent cycle stability. However, extremely low electronic conductivity and Li^+ diffusion coefficient seriously limit its performance at high rates. The conductivity of $Li_4Ti_5O_{12}$ can be significantly improved by composite or nanostructure treatment. Lee et al. [27] prepared a $Li_4Ti_5O_{12}$ active carbon composite anode and selected an active carbon cathode to assemble LICs. The results showed that the battery had high specific capacity, superior speed capability, and cycle stability. When the current increased from 0.5 A/g to 5 A/g, the discharge capacity was 81.9% of the initial capacity. After 7,000 cycles at 3 A/g, the capacity retention rate was still 92.8%. Liu et al. [28] constructed porous $Li_4Ti_5O_{12}$ nanoparticles on self-supporting carbon nanotube films (CNT) by using hard template and soft template methods of F127, and prepared a new self-supporting anode of lithium titanate CNT@pLTO. The synthesis process was shown in Figure 10.4. At the current density of 0.175 A/g and 8.75 A/g, the corresponding discharge capacities were 148 mAh/g and 85 mAh/g, respectively, which were obviously better than that of pure LTO. The improvement of battery performance could be attributed to the increase of Li^+ diffusion coefficient mainly due to the particle size of the nano-structure and the transmission

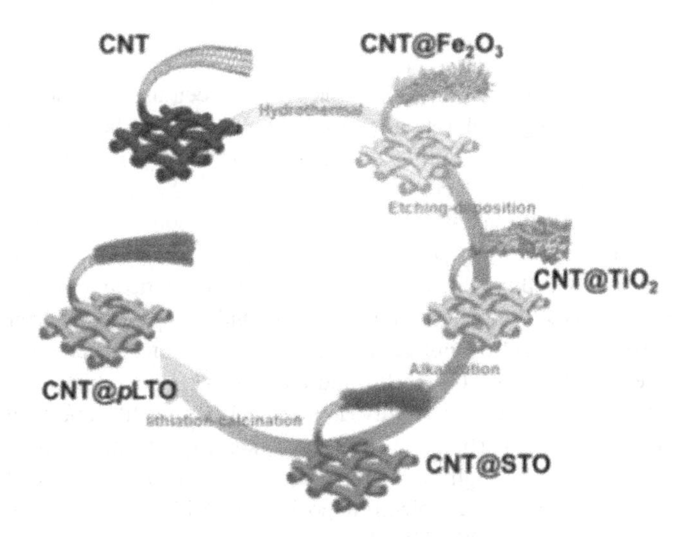

FIGURE 10.4 Schematic diagram of the synthesis of CNT@pLTO [28].

channel through lithium ion. At the same time, when the power density was 436.1 W/kg, the capacitor battery system achieved maximum energy density of 101.8 Wh/kg. This study provided a new idea for the realization of high-performance LICs.

Compared with $Li_4Ti_5O_{12}$, TiO_2 is a more reasonable anode material to ensure that high-power lithium batteries have stable chemical properties, simple synthesis, low cost, and environmentally friendly features [29]. Its theoretical specific capacity is at least twice that of $Li_4Ti_5O_{12}$, and the anatase and bronze phases in TiO_2 have also attracted much attention in recent years. However, in the process of lithium intercalation and lithium removal, the lattice change of TiO_2 is slightly more obvious than that of lithium titanate, which leads to the decrease of cycling ability during long-term charge–discharge process. Therefore, various methods have been proposed to improve the electrochemical performance of TiO_2, including carbon coating, design of hybrid materials, and doping of metals and nonmetals. Zhou et al. [30] synthesized a banded TiO_2-Co_9S_8 nanocomposite by hydrothermal method, and both bronze and anatase phases were detected in the product after calcination. When the current density was 100 mA/g, the first reversible capacity could reach 714 mAh/g, and the coulombic efficiency of the cell was 74%. During further cycles, the lithium storage capacity of the material increases, and the capacity stabilizes at 886 mAh/g after 100 cycles. More importantly, as the skeleton of a Co_9S_8 nanoparticle layer, anatase/TiO_2 nanoribbon can still achieve a charging capacity of 260 mAh/g at a high rate of 5 A/g. This result indicates that the material has a potential application in energy storage.

In addition to $Li_4Ti_5O_{12}$ and TiO_2, other Ti-containing lithium-embedded compounds, such as $LiCrTiO_4$ and $LiTi_2(PO_4)_3$, have been gradually applied to the anode materials of LICs in recent years. Carbon coating, doping, and material composite treatment methods are also applicable to them.

10.2.4 Sulfide

Sulfides (such as NiS, CoS_2, MoS_2, and others) have been used as anode materials for LICs due to their high theoretical capacity, low cost, and better electrode dynamic performance. However, sulfides have many limitations in practical application. First of all, the poor conductivity of sulfide leads to the poor rate capability of the battery and slow electrochemical reaction kinetics. Second, in the process of Li^+ embedding/exiting, serious agglomeration, repeated absorption of lithium, and volume expansion caused by particle fracture lead to a rapid decline in capacity. To address this issue, nanotreatment of metal sulfide and composite materials can effectively improve the electrochemical reaction efficiency and utilization of active electrode materials, and can also buffer and adapt to volume expansion. The NiS electrode material has been successfully applied in LICs mainly due to its unique chemical and physical properties, which can provide a good transfer path for ions on the electrode surface through the Faraday process. Tran et al. [31] synthesized a NiS@NF composite electrode by electrodeposition of NiS on nickel foam at room temperature. After testing, NiS@NF had extremely high SC and good cycle stability, SC reached 1,553 F/g at 2.35 A/g, and retained 95.7% of its initial SC after cycling 2,000 times. In 2019, Chen et al. [32] developed a novel 2D/3D NiS/Ni_3S_4 composite using a one-pot hydrothermal method. At 1 A/g, the SC of the composite material was as high as 1,796 F/g, and the capacitance retention rate could reach 80.5% of the initial capacitance. The results show that the 2-D/3-D NiS/Ni_3S_4 composite has the potential to be used as electrode material for capacitor batteries.

CoS_2 is a promising energy storage material. In recent years, CoS_2 and CoS_2 matrix composites with different nanostructures have been developed as electrode materials for high-performance capacitor batteries. For example, Jia et al. [33] prepared porous four-shell hollow COS_2 by template-guided synthesis. At 1 A/g, the SC value was 375.2 F/g, and the capacitance retention was 92.1% after 10,000 cycles. Then, Govindasamy et al. [34] first used the two-step hydrothermal method to grow the $NiCo_2S_4$@CoS_2 nanostructure on carbon fabric (CC). The synthesis process was shown in Figure 10.5. In addition, when the power density was 242.8 W/kg, the energy density of the composite material was 17 Wh/kg. All these results indicate that $NiCo_2S_4$@CoS_2@CC is a good candidate electrode for high-performance LICs.

MoS_2 is an affordable and easy-to-synthesize material, which has the morphology of nanoscale sheet, large surface area, and excellent electrical conductivity. Due to its unique properties, MoS_2 has attracted attention from scientists in the field of energy storage. Gao et al. [35] took SiO_2 nanospheres as templates, prepared MoS_2 nanosheets by simple hydrothermal process, and assembled them into LICs. At 1 A/g, the maximum SC of the MoS_2 nanospheres was 683 F/g, and the capacitance retention of the MoS_2 nanospheres was 85.1% after 10,000 cycles. When the energy density was 20.42 Wh/kg, the power was 750.31 kW/kg, showing excellent electrochemical performance. Wang Research Group [36] synthesized a continuous ladder-like flower ridge similar to MoS_2/N-doped carbon nanocomposites (MoS_2/NC) by single-pot hydrothermal and sintering methods induced by chitosan. The synthesis diagram was shown in Figure 10.6. SEM results show that the

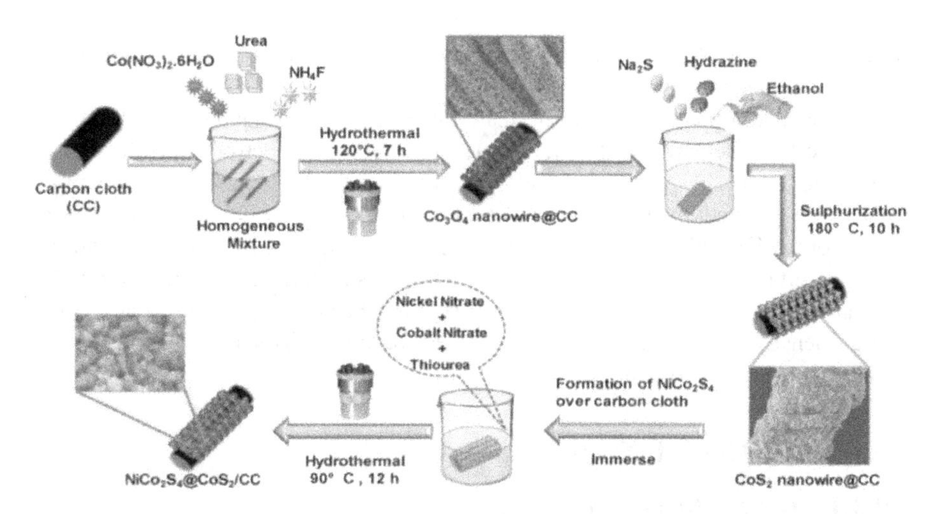

FIGURE 10.5 Schematic diagram of $NiCo_2S_4@CoS_2$ growing on CC [34].

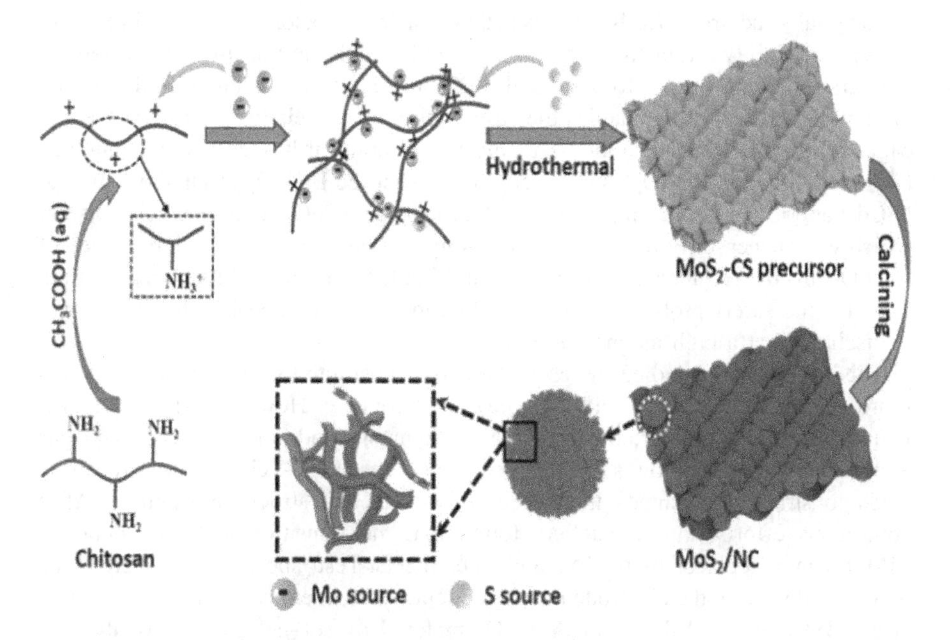

FIGURE 10.6 Synthesis of MoS_2/NC composite material [36].

MoS_2nanosheets with less than three layers form the interconnection nanosheets that grow vertically on the 2-D N-doped carbon nanosheets to form a MoS_2/NC composite electrode. The unique layered nanostructure caused the electrode material to have excellent ratio performance and cyclic stability. Under the current density of 500, 1,000, and 2,000 mA/g, 742, 686, and 534 mAh/g can be released after 400 cycles. The synthesis strategy is a good choice for the construction of other transition metal disulfide and carbon nanostructures to improve their electrochemical performance.

FeS_2 is also a potential electrode material for energy storage devices due to its low price, excellent conductivity, and large number of electrochemically active sites. Sun et al. [37] prepared FeS_2 nano-ellipsoids for LICs using a simple microwave-assisted method. When the current density was 1 A/g and 20 A/g, the SC of the FeS_2 electrode could reach 515 F/g and 355 F/g, respectively, and the initial specific capacitance of the FeS_2 electrode was maintained at 90% after 5,000 cycles. At the power density of 271.2 W/kg, the FeS_2 nano-ellipsoid had a high energy density of 64 Wh/kg, which indicated that the FeS_2 nano-ellipsoid was an ideal high-power energy storage device.

10.3 CATHODE MATERIALS

10.3.1 LAYERED METAL OXIDES

10.3.1.1 Layered $LiMO_2$ Compound (M Is a Transition Metal)

The crystal structure of $LiMO_2$ series cathode materials belongs to layered rock salt phase. The tightly arranged oxygen ions combine with the transition metal ions in the octahedral position to form a stable MO_2 layer, and Li^+ is embedded between the layers. This structure enables the structure to have a relatively large theoretical capacity of mass and volume. The most common $LiMO_2$ compounds include $LiCoO_2$(LCO) and $LiNiO_2$(LNO). Ferg et al. proposed $LiCoO_2$ as an alternative to NiCd batteries in 1994, and researchers turned to $LiCoO_2$ as a research hotspot, which was successfully implemented by Sony in 1990. Although the emergence of $LiCoO_2$ fills the technical space, it is less used in LICs because of its high cost, high toxicity, and safety problems caused by the rapid temperature rise when the power is discharged at a high magnification rate.

$LiNiO_2$ is widely studied by researchers as a substitute for LCO, where nickel is more abundant than cobalt and is a cheaper alternative. However, trivalent nickel is unstable at high temperature, which leads to harsh conditions for LNO synthesis and strongly depends on specific conditions. The Jahn–Teller phase transition leads to structural changes that affect life cycle and structural stability. After continuous efforts, the researchers found that the transition metal elements in $LiMO_2$ can be replaced by Ni, Mn, and Al elements. Tests showed that the electrode materials doped by the electrode have excellent electrochemical properties. Ohzuku et al. [38] prepared $LiNi_{0.5}Co_{0.5}Mn_{0.5}O_2$ material by sol–gel method for the first time. The charging and discharging voltage platform of $LiNi_{0.5}Co_{0.5}Mn_{0.5}O_2$ was between 3.6 V and 4.3 V, and its thermal stability was good. The maximum reversible capacity was up to 200 mAh/g, and the capacity does not decay after 100 cycles. The energy density was greater than $LiCoO_2$, but the mobility of Li^+ was one order of magnitude lower than that of $LiCoO_2$, which led to its relatively low multiplication performance. To further improve its overall performance, researchers studied the element doping on this basis, resulting in a new generation of ternary NCM/NCA cathode materials, which was the most important achievement in the research on layered $LiMO_2$ composite doping.

10.3.1.2 Spinel $LiMn_2O_4$ Compounds

$LiMn_2O_4$ (LMO) with spinel structure has attracted much attention in recent years due to its low price and environmental friendliness compared with LCO and LNO. $LiMn_2O_4$ is classified as part of the fd3m space group, and its crystal structure is cubic compact packing, which has good rate performance. However, the Jahn–Teller effect introduced by manganese ions leads to the deterioration of the cycle performance. In addition, $LiMn_2O_4$ has obvious capacity degradation at high temperature. It is found that doping and coating can improve the cycling performance of the materials. Li et al. [39] used the co-precipitation method to coat a layer of porous Al_2O_3 nanoparticles on $LiMn_2O_4$. Compared with the original LMO, the Al_2O_3@LMO sample had a unique structure. With regard to electrochemical performance, the capacity gap between the two was not large at 25°C, and the cycle performance of the Al_2O_3@LMO sample was significantly improved at 55°C. The capacity retention rate after 100 cycles was 91.9% of the initial value. Other metal ions such as Cu^{2+} and Mo^{2+} al could also be doped into $LiMn_2O_4$. Thus, the overall performance of the battery has been improved to varying degrees.

10.3.1.3 Olivine $LiFePO_4$ Material

$LiFePO_4$ has become one of the most mature cathode materials among olivine phosphates due to its environmental friendliness, low price, non-toxicity, abundance in nature, and good stability. Early studies on $LiFePO_4$ can be traced back to the 1980s, but the material gradually attracted research attention only when it was discovered in 1997. So far, $LiFePO_4$ has been used in LICs.

The crystal structure of $LiFePO_4$ can be considered as a hexagonal compact accumulation with slight distortion. The compact structure leads to the low diffusion coefficient of Li^+, resulting in poor conductivity, which is not conducive to charge and discharge at large magnification. Therefore, carbon coating or other conductive materials can be used to improve the conductivity of $LiFePO_4$ to meet the requirements of high current charge and discharge [40]. Wang et al. [41] conducted an in-depth study on the effect of fluorine-doped carbon (FC) modification on the electrochemical properties of $LiFePO_4$ cathode materials. The results show that the optimal content of FC is approximately 2.8 wt%, and the electrode exhibited remarkable electrochemical properties, including high capacity close to the theoretical value, good cycle stability, and high rate performance. The discharge capacity was still maintained at 100.2 mAh/g after 1,000 stable cycles under high current of 20 C. This work proved that the FC coating strategy can open up a new way to maximize the performance of $LiFePO_4$ cathode materials.

$LiMPO_4F$ is a derivative series of olivine structure. This type of cathode material combines the advantages of PO_4^{3-} (inductive effect) and F^- (high electronegativity) to provide high working potential [42]. Zhang et al. [43] successfully prepared $LiFePO_4$ nanospheres for the first time through a solid-state route combined with chemical-induced precipitation. The initial discharge capacity of the prepared sample is 110.2 mAh/g at 0.5 C. The stable cycling performance (94.4% capacity retention after 200 cycles) can be attributed to the uniform nanospherical morphology, which was beneficial to increase the contact area between electrode and electrolyte, shorten the ion transport path, and improve the kinetics of lithium ion.

10.3.1.4 Ternary Cathode Materials

In recent years, the ternary material $LiNi_{1-x-y}Co_xMn_yO_2$ has attracted increasing attention due to its comprehensive advantages in specific capacity, operation potential, safety, cost, and environmental protection. The material has been successfully applied in LICs. The common acronym for this material is NCM or NAM, which numerically denotes a fractional fraction of the element (e.g., NCM523 = $LiNi_{0.5}Co_{0.2}Mn_{0.3}O_2$) and is also known as Co or Al-doped $LiNiO_2$ and $LiMnO_2$. Among these materials, +IV has an oxidation state that does not change during the charge discharge cycle, and can be used as a structural stabilizer. Ni^{2+}/Ni^{3+} and Ni^{3+}/Ni^{4+} have strong redox activity, which contributes most of the specific capacity of the material. In addition, the presence of cobalt is conducive to the synthesis of low cationic stoichiometric compounds. Thus, the properties of the materials can be adjusted by adjusting the ratio of Ni, CO, and Mn. $LiNi_{0.5}Co_{0.5}Mn_{0.5}O_2$ is the first ternary cathode material synthesized by Ohzuku et al. [38]. In the following years, other ternary positive electrode materials were gradually developed, such as $LiNi_{0.5}Co_{0.2}Mn_{0.3}O_2$, $LiNi_{0.6}Co_{0.2}Mn_{0.2}O_2$, nickel-rich, and lithium-rich ternary materials.

Despite the rapid development of ternary cathode materials, some challenges remain in application, such as high first-period irreversibility, rapid capacity decay, and others. Therefore, in addition to the strict requirements on the synthesis conditions, some necessary steps are further improving the structural stability and improving the performance through doping, surface coating, and recombination at the level of composition/structure/morphology. For example, Shi et al. [44] synthesized spherical lithium-rich $Li_{1.2}Mn_{0.56}Ni_{0.16}Co_{0.08}O_2$ material through microwave hydrothermal method. The synthesis schematic diagram was shown in Figure 10.7. The prepared material was composed of spherical particles (2–3 μm) and porous particles (150–250 nm). X-ray diffraction and specific surface area (Brunauer, Emmett, and Teller) that this material had a good layered structure and large specific surface area. The porous structure was conducive to the contact between the active material and the electrolyte. At a current density of 0.2 A/g, the discharge specific capacity is 235.6 mAh/g, and at a current density of 1 A/g and 2 A/g, the discharge specific capacity still reached 168.6 mAh/g and 131.2 mAh/g (Figure 10.8).

Yang et al. [45] successfully synthesized a layered $LiNi_{0.88}Co_{0.095}Mn_{0.025}O_2$ cathode with the primary grain size aligned by the precursor optimization design. During the sintering process, the introduction of nano-Al_2O_3 caused Al^{3+} to follow the radial primary grains. The particles enter the block to realize the uniform distribution of Al^{3+} in the entire material. The electrochemical results were shown in Figure 10.9. When the molar mass of Al^{3+} was 2%, the electrochemical performance was the best. After 150 cycles at 1 C, the capacity retention rate of the electrode reached 91.57%. Even at high current densities of 5 C and 10 C, the cathode material can obtain high reversible capacities of 172.3 mAh/g and

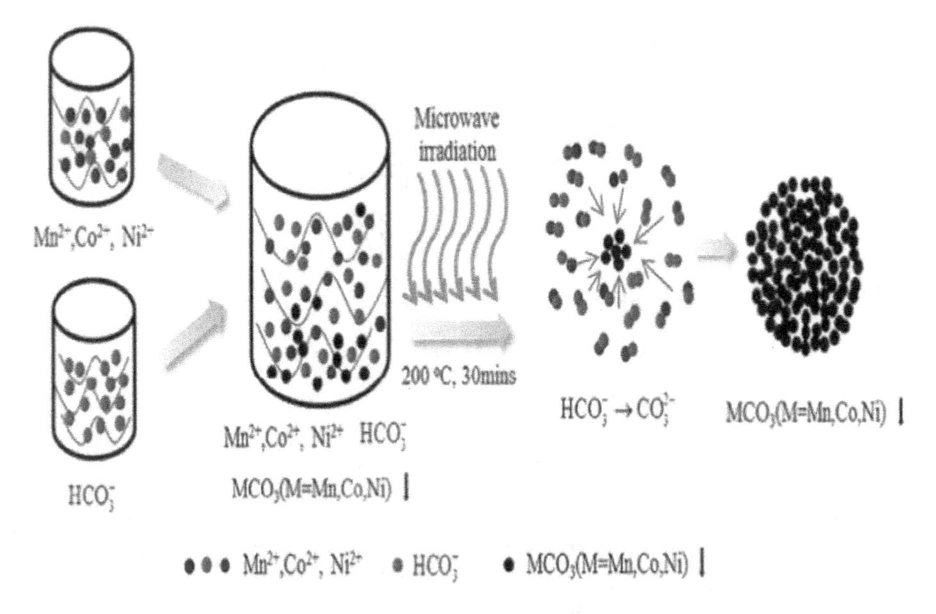

FIGURE 10.7 Preparation process of $Li_{1.2}Mn_{0.56}Ni_{0.16}Co_{0.08}O_2$ [44].

165.7 mAh/g. In addition, the results of median voltage and DQ/DV curve show that uniform Al^{3+} doping can significantly inhibit the voltage deterioration, which indicated that Al^{3+} doping provided an effective and practical method to improve the rate performance and voltage stability of nickel-rich ternary materials.

10.3.1.5 Vanadium Oxide V_2O_5

V_2O_5, as a layered compound for electrochemical energy storage, has become one of the cathode materials with promising development prospects in recent years due to its high energy density, low cost, abundant sources, and good safety performance. V_2O_5 is a typical interlayer compound [46]. Its interlayer structure is composed of van der Waals forces to combine the layers. This layered structure makes the theoretical capacity of V_2O_5 (294 mAh/g) much higher than that of other cathode materials. However, it has poor conductivity, and is easy to agglomerate and dissolve in liquid electrolytes, thereby resulting in unsatisfactory rate and cycle performance [47].

V_2O_5 can be transformed into nanostructures or coupled with carbon matrix composites to improve its conductivity [48–50]. Using a thick film CNT/V_2O_5, Qu et al. [51] designed and fabricated a composite cathode that was used in combination with organic electrolyte, and activated carbon was used as anode to prepare a capacitor cell. The results show that the system had excellent electrical conductivity, high specific capacitance, and large voltage window, enabling it to achieve an energy density of up to 40 Wh/kg at a power density of 210 W/kg. Even at a high power density of 6,300 W/kg, the energy density of the device was close to 7.0 Wh/kg. Zhao Research Group [52] synthesized V_2O_5 nanostructures composed of ultra-thin nanosheets as electrode materials by hydrothermal method. The

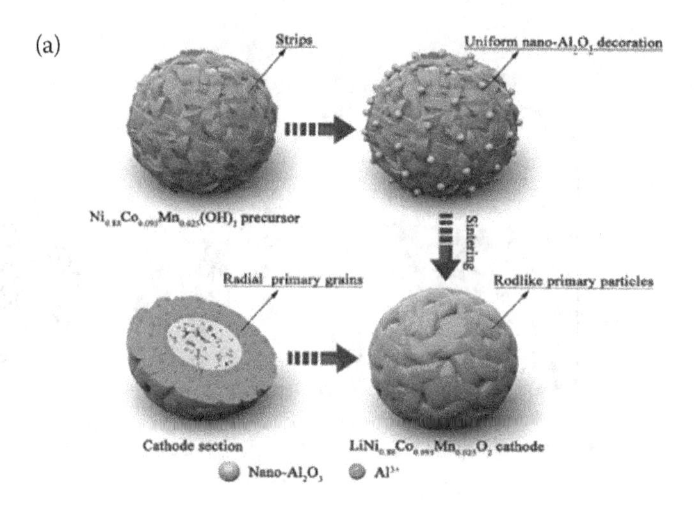

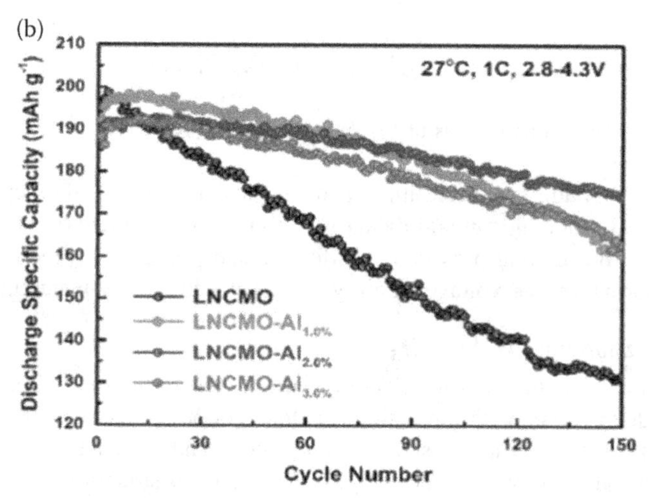

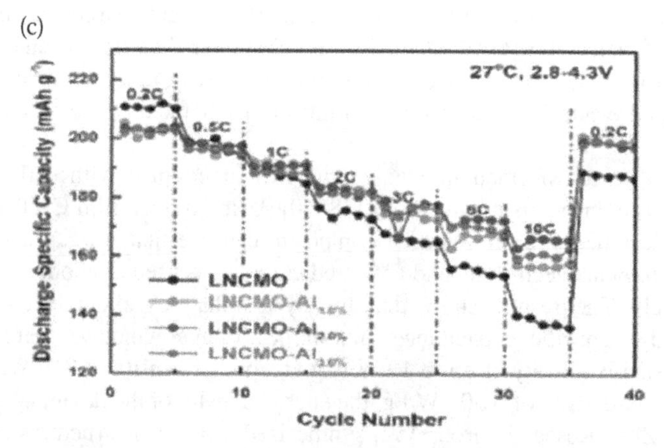

FIGURE 10.8 (a) Synthesis of $LiNi_{0.88}Co_{0.095}Mn_{0.025}O_2$@Al composite material (b) cycle capability at 1C rate (c) rate capability under different C-rates [45].

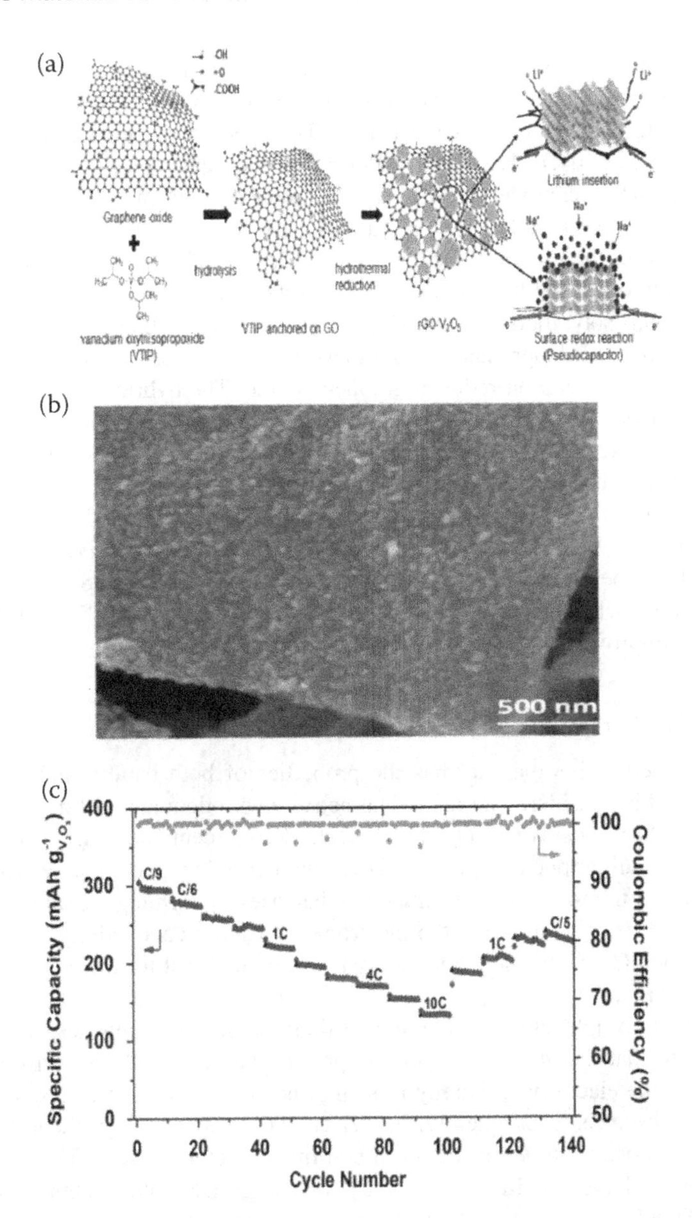

FIGURE 10.9 (a) Schematic diagram of the synthesis of rGO/V_2O_5; (b) SEM images of V_2O_5 nanoparticles anchored on rGO nanosheets after annealing in vacuum at 300°C for 2 hours; (c) the rate performance at various rates from C/9 to 10 C [54].

specific capacitance of the electrode at 1 A/g was 704.17 F/g. After 4,000 cycles, the capacitance retention rate was 89%. At this time, the power density was 800 W/kg and the energy density was 29.49 Wh/kg, which greatly improved the electrochemical performance. These results show that the preparation of nanocomposites is an effective way to improve the energy density and power density of LICs.

Constructing a 3-D structure is important in solving the agglomeration problem of vanadium-based materials. Yao Research Group [53] prepared a rGO/V_2O_5 nanocomposite cathode with 3-D structure. The specific capacitance of this electrode at 1 A/g was 704.17 F/g. After 4,000 cycles, the capacitance retention rate was 89%. At this time, the power density was 800 W/kg and the energy density reached 29.49 Wh/kg, which greatly improved the electrochemical performance. Due to its unique structure, the electrode had a capacitance of 225.6 F/g at 0.5 A/g and an energy density of 31.3 Wh/kg at a power density of 249.7 W/kg. Pandey et al. [54] used a two-step solvothermal method and a vacuum annealing method to slowly hydrolyze vanadyl triisopropanol to synthesize a mesoporous hybrid of V_2O_5 nanoparticles immobilized on reduced graphene oxide. The hybrid material had high surface area and mesoscale porosity, significantly improved electronic conductivity, accelerated electrolyte ion diffusion, and improved electrical energy storage performance. When used as the cathode of a LICs, rGO-V_2O_5 exhibited good cycle stability and discharge capacity. At current densities of C/9, 1 C, and 10 C, its discharge capacity can reach 295, 220, and 132 mAh/g, respectively. After 150 cycles at 1 C, the electrode retained approximately 83% discharge capacity, and only 0.12% of each cycle capacity was reduced. The construction of a 3-D structure can greatly improve the energy density and cycle stability.

10.3.2 Graphene Composite Cathode

Graphene-based active material has the properties of both battery and capacitive materials, and is considered an ideal dual-active electrode material for LICs. It can be used as either a positive or a negative electrode. As mentioned, Arnaiz et al. [31] synthesized an all-graphene capacitor battery with excellent performance. Graphene composites can improve the performance of batteries by forming a dual continuous conductive network to improve the electronic and ionic conductivity of the electrode, of which GO is the most commonly used raw material for the construction of porous networks.

Wang et al. [55] developed 3-D porous $LiFePO_4$/N/rGO composites, which have demonstrated that the graphene layer can protect the active substance from direct contact with the electrolyte, thereby reducing the side reactions and dissolution of the active substance. Subsequently, Xie et al. [56] prepared a graphene-modified $LiFePO_4$ composite cathode using an in-situ microreactor strategy. The composite material had a large specific surface area for charge adsorption, abundant active sites for Faraday reaction, and ideal kinetic characteristics of electron and ion transport. Therefore, it showed extremely fast surface dynamic-controlled lithium storage behavior. In terms of rate performance, the discharge capacity at 200 C was 78 mAh/g, and the capacity retention rate at 20 C was 90% after 1,000 cycles, showing a quasi-capacitive mechanism and excellent cycle stability.

10.3.3 Other New Cathode Materials

Other new cathode materials include $LiMBO_3$ borate and Li_2MSO_4 silicate. Borate $LiMBO_3$ has a high theoretical capacity of 220 mAh/g due to its low molecular weight of Bo_3^{3-}. In 2001, Legagneur et al. [57] studied $LiMBO_3$ for the first time. During the initial charging process, the amount of lithium removed per unit was only 0.02, and the discharge capacity was only 9 mAh/g, which limited the application of $LiMBO_3$ in the field of LICs. Thereafter, Yamada et al. [58] proposed that the reason for the low specific capacity of $LiMBO_3$ was not the material itself but the hydrolysis of the positive electrode surface in contact with water in the air. Therefore, on this basis, the $LiFeBO_3$ was coated with carbon to avoid contact with air, so that the specific capacity of the modified material can be as high as 190 mAh/g, which had attracted widespread attention from scholars.

Silicate Li_2MSO_4 also attracted much attention due to its high theoretical specific capacity. The first report of using Li_2MSO_4 as cathode can be traced back to 2005, when Nyten et al. [59] pointed out that Li_2FeSiO_4 had double electron capacity, that was, two redox peaks: 2.8 V (Fe^{2+}/Fe^{3+}) and 4.0–4.8 V (Fe^{3+}/Fe^{4+}). Subsequently, they explored the transition mechanism between cation and anion redox activities in this material and found that it has a significantly high energy density [60]. However, the transport mechanism of ions and electrons in the lattice remains a key problem that can be better understood through further experiments.

10.4 SUMMARY AND PROSPECTS

As a new type of energy-storage device, LICs concentrates high energy density, high power density, and long-term cycle stability in one system, filling the gap between the LIB and the SCs. The choice of electrode material directly affects the overall performance of the capacitor battery. The anode materials of LICs mainly include transition metal oxides, carbon materials, lithium-embedded compounds and sulfides, among which carbon materials are the most widely used. At present, the negative electrode materials have the main problems of unstable structure, poor cycling stability, and poor plasticity in the process of charge and discharge. In general, the combination of carbon coating and carbon nanofibers as well as the addition of conductive additives (such as graphene and carbon nanotubes) are used to improve the performance of negative electrode materials. Cathode materials mainly include layered metal oxides and graphene composites. Most of the layered metal oxides are traditional cathode materials that have a better specific capacity and a wider range of working potential. However, the structure is prone to phase transition in the process of charge and discharge, thereby resulting in poor cyclic performance. Meanwhile, the electrochemical performance is not so stable at a high rate, which is the main problem of all cathode materials.

Although progress has been made in the research on electrode materials for LICs, practical commercial application will take time. As low energy density is still the main problem of hybrid capacitors, future research can be conducted from the following aspects: (1) Combined with high conductivity materials, this can effectively improve electron transfer dynamics. (2) To greatly improve the voltage window of LICs, developing new electrode materials with a unique structure is necessary to obtain a large specific surface area and accelerate the electron conduction rate. Based on the synergistic effect, the combination of different electrode materials should be considered as an effective method to enhance the specific surface area and large potential window. (3) Further optimization of anode and cathode matching is necessary to improve the overall performance of the device. Ensuring high power density and high energy density of LICs is inevitable in future development. Therefore, in the future design and preparation of LIC electrodes, the dynamic balance between positive and negative materials should be fully considered to obtain materials with low toxicity, low cost, and good electrical conductivity, so as to realize the sustainable development and utilization of LICs.

REFERENCES

[1] Wang Y, Song Y, Xia Y. Electrochemical capacitors: mechanism, materials, systems, characterization and applications[J]. *Chemical Society Reviews*, 2016, 45(21): 5925–5950.

[2] Hou H, Banks C E, Jing M, Zhang Y, Ji X. Carbon quantum dots and their derivative 3D porous carbon frameworks for sodium-ion batteries with ultralong cycle life[J]. *Advanced Materails*, 2015, 27: 7861–7866.

[3] Shi D, Zheng R, Liu C S, Chen D M, Zhao J, Du M. Dual-functionalized mixed keggin-and lindqvist-type Cu_{24}-based POM@ MOF for visible-light-driven H_2 and O_2 Evolution[J]. *Inorganic Chemistry*, 2019, 58(11): 7229–7235.

[4] Li Z X, Zou K Y, Zhang X, Han T, Yang Y. Hierarchically flower-like N-doped porous carbon materials derived from an explosive 3-fld interpenetrating diamondoid copper mtal-organic framework for a supercapacitor[J]. *Inorganic Chemistry*, 2016, 55: 6552–6562.

[5] Huang J, Zhao B, Liu T, Mou J, Jiang J, Li H, Liu M. Wood-derived materials for advanced electrochemical energy storage devices[J]. *Advanced Function Materails*, 2019, 29: 1902255.

[6] Wang F, Wu X, Yuan X, Liu Z, Zhang Y, Fu L, Zhu Y, Zhou Q, Wu Y, Huang W. Latest advances in supercapacitors: From new electrode materials to novel device designs[J]. *Chemical Society Reviews*, 2017, 46: 6816–6854.

[7] Jagadale A, Zhou X, Xiong R, Dubal D P, Xu J, Yang S. Lithium ion capacitors (LICs): Development of the materials[J]. *Energy Storage Mater*, 2019, 19: 314–329.

[8] Han P, Xu G, Han X, Zhao J, Zhou X, Cui G. Lithium ion cpacitors in organic electrolyte system: Scientific problems, mterial development, and key technologies [J]. *Advanced Energy Materails*, 2018, 8: 1801243.

[9] Ding J, Hu W, Paek E, Mitlin D. Review of hybrid ion capacitors: from aqueous to lithium to sodium[J]. *Chemical Reviews*, 2018, 118: 6457–6498.

[10] Amatucci G G, Badway F, Du P, Pasquier T, Zheng J. An asymmetric hybrid nonaqueous energy storage cell[J]. *Journal of the Electrochemical Society*, 2001, 148(8), A930–A939.

[11] Liu Q L, Lan G C, Xia Y Y. A hybrid electrochemical supercapacitor based on a 5 V Li-ion battery cathode and active carbon[J]. *Electrochemical and Solid State Letters*, 2005, 8: 433–436.

[12] Yang Y. Research progress and demonstration application of lithium ion capacitors [J]. *Foreign Rolling Stock*, 2019, 56(06): 25–28.

[13] Liu J, Zheng M, Shi X, Zeng H, Xia H. Amorphous FeOOH quantum dots assembled mesoporous film anchored on graphene nanosheets with superior electrochemical performance for supercapacitors[J]. *Advanced Functional Materials*, 2016, (26): 919–930.

[14] Jiao Y, Liu Y, Yin B, et al. Hybrid α-Fe_2O_3@ NiO heterostructures for flexible and high performance supercapacitor electrodes and visible light driven photocatalysts [J]. *Nano Energy*, 2014, 10: 90–98.

[15] Zhong Y, Ma Y F, Guo Q B. Controllable synthesis of TiO_2@Fe_2O_3 core-shell nanotube arrays with double-wall coating as superb lithium-ion battery anodes. *Scientific Reports*, 2017, (7): 40927.

[16] Zheng X, Jiao Y, Chai F.Template-free growth of well-crystalline alpha-Fe_2O_3 nanopeanuts with enhanced visible-light driven photocatalytic properties[J]. *Journal of Colloid and Interface Science*, 2015, 457: 345–352.

[17] Wang P, Zheng Z, Cheng X, et al. Ionic liquid-assisted synthesis of α-Fe_2O_3 mesoporous nanorod arrays and their excellent trimethylamine gas-sensing properties for monitoring fish freshness. *Journal of Materials Chemistry A*, 2017, 5: 19846–19856.

[18] Dong Y, Xing L, Hu F, Umar A, Wu X. α-Fe_2O_3 /rGO nanospindles as electrode materials for supercapacitors with long cycle life. *Materials Research Bulletin*, 2018, 107: 391–396.

[19] Li H, Wu L J, Zhang S G, et al. Facile synthesis of mesoporous one-dimensional Fe_2O_3 nanowires as anode for lithium ion batteries[J]. *Journal of Alloys and Compounds*, 2020, 832: 155008.

[20] Yin D, Huang G, Sun Q. RGO/Co_3O_4 composites prepared using GO-MOFs as precursor for advanced lithium-ion batteries and supercapacitors electrodes. *Electrochimica Acta*, 2016, 215: 410–419.

[21] Zhao X, Wang H E, Cao J, et al. Amorphous/crystalline hybrid MoO_2 nanosheets for high-energy lithium-ion capacitors[J]. *Chemical Communications*, 2017, 53(77): 10723–10726.

[22] Lee S H, Yoo G, Cho J, Seokgyu R,Youn S K, Jeeyoung Y. Expanded graphite/ copper oxide composite electrodes for cell kinetic balancing of lithium-ion capacitor. *Journal of Alloys and Compounds*, 2020, 154566.

[23] Wu Z S, Ren W C, Xu L, Li F, Cheng H M. Doped graphene sheets as anode materials with superhigh rate and large capacity for lithium ion batteries. *ACS Nano*, 2011, 5: 5463–5471.

[24] Gomez-Urbano J L, Moreno-Fernandez G, Arnaiz M. Graphene-coffee waste derived carbon composites as electrodes for optimized lithium ion capacitors. *Carbon*, 2020, 162: 273–282.

[25] Sun B, Zhang Q, Xiang H, et al. Enhanced active sulfur in soft carbon via synergistic doping effect for ultra–stable lithium–ion batteries[J]. Energy Storage *Materials*, 2020, 24: 450–457.

[26] Jiang J, Zhang Y, Li Z, et al. Defect-rich and N-doped hard carbon as a sustainable anode for high-energy lithium-ion capacitors[J]. *Journal of Colloid and Interface Science*, 2020, 567: 75–83.

[27] Lee S H, Kim J M. Improved performances of hybrid supercapacitors using granule $Li_4Ti_5O_{12}$/activated carbon composite anode[J]. *Materials Letters*, 2018, 228: 220–223.

[28] Liu Y, Wang W, Chen J, Li X W, Cheng Q L. Fabrication of porous lithium titanate self-supporting anode for high performance lithium-ion capacitor. *Journal of Energy Chemistry*, 2020, 50.

[29] Lang X, Zhao Y, Cai K, Li L, Zhang Q A. Facile synthesis of stable TiO_2/TiC composite material as sulfur immobilizers for cathode of lithium–sulfur batteries with excellent electrochemical performance. *Energy Technology*, 2019, 7(12): 1900543.

[30] Zhou Y, Zhu Q, Tian J, Jiang F. TiO_2 nanobelt@Co_9S_8 composites as promising anode materials for lithium and sodium ion batteries. *Nanomaterials*, 2017, 7(9): 252.

[31] Tran V C, Sahoo S, Shim J J. Room-temperature synthesis of NiS hollow spheres on nickel foam for high-performance supercapacitor electrodes. *Materials Letters*, 2018, 210: 105–108.

[32] Chen T H, Liu Z L, Liu Z S. Fabrication of interconnected 2D/3D NiS/Ni_3S_4 composites for high performance supercapacitor. *Materials Letters*, 2019, 248: 1–4.

[33] Jia H, Wang Z, Zheng X. Controlled synthesis of MOF-derived quadruple-shelled CoS_2 hollow dodecahedrons as enhanced electrodes for supercapacitors. *Electrochimica Acta*, 2019, 312: 54–61.

[34] Govindasamy M, Shanthi S, Elaiyappillai E, Wang S F, Johnson P M. Fabrication of hierarchical $NiCo_2S_4@CoS_2$ nanostructures on highly conductive flexible carbon cloth substrate as a hybrid electrode material for supercapacitors with enhanced electrochemical performance. *Electrochimica Acta*, 2019, 293: 328–337.

[35] Gao Y, Huang K J, Wu X, Hou Z, Liu Y. MoS_2 nanosheets assembling three-dimensional nanospheres for enhanced-performance supercapacitor. *Journal of Alloys and Compounds*, 2018, 741: 174–181.

[36] Wang X, Tian J, Cheng X, Na, Ren, Shan Z. Chitosan-induced synthesis of hierarchical flower ridge-like MoS_2/N-doped carbon composites with enhanced lithium storage. *ACS Applied Materials & Interfaces*, 2018, 10.

[37] Sun Z, Yang X, Lin H, Zhang F, Wang Q, Qu F. Bifunctional iron disulfide nanoellipsoids for high energy density supercapacitor and electrocatalytic oxygen evolution applicationsElectronic supplementary information (ESI). *Inorganic Chemistry Frontiers*, 2019, 6(3): 659–670.

[38] Ohzuku T, Makimura Y. Layered lithium insertion material of $LiNi_{1/2}Mn_{1/2}O_2$: A possible alternative to $LiCoO_2$ for advanced lithium-ion batteries[J]. *Chemistry Letters*, 2001, 30(8): 744–745.

[39] Li S, Zhu K, Zhao D, Zhao Q, Zhang N. Porous $LiMn_2O_4$ with Al_2O_3 coating as high-performance positive materials. *Ionics* , 2019, 25: 1991-1998

[40] Li L, Wu L, Wu F, Song S P, Zhang X Q, Fu C, Yuan D D, Xiang Y. Recent research progress in surface modification of $LiFePO_4$ cathode materials[J]. *Journal of the Electrochemical Society*, 2017, 164(9): A2138–A2150.

[41] Wang X, Feng Z, Hou X, Liu L, He M, He H. Fluorine doped carbon coating of $LiFePO_4$ as a cathode material for lithium-ion batteries[J]. *Chemical Engineering Journal*, 2020, 379: 122371.

[42] Julien C, Mauger A, Vijh A, Zaghib K. *Lithium batteries: Science and technology*. Springer International Publishing, Cham, 2016, pp. 201–268.

[43] Zhang Y, Liang Q, Huang C, Gao P, Shu H, Zhang X, Yang X, Liu L, Wang X. Nearly monodispersed $LiFePO_4F$ nanospheres as cathode material for lithium ion batteries[J]. *Solid State Electrochem*, 2018, 22 (7): 1995–2002.

[44] Shi S, Wang T, Cao M, et al. Rapid self-assembly spherical $Li_{1.2}Mn_{0.56}Ni_{0.16}Co_{0.08}O_2$ with improved performances by microwave hydrothermal method as cathode for lithium-ion batteries[J]. *ACS Applied Materials & Interfaces*, 2016, 8(18): 11476–11487.

[45] Yang X, Tang Y, Shang G, et al. Enhanced cyclability and high-rate capability of $LiNi_{0.88}Co_{0.095}Mn_{0.025}O_2$ cathodes by homogeneous Al^{3+} doping[J]. *ACS Applied Materials & Interfaces*, 2019, 11(35): 32015–32024.

[46] Cai K, Li Y Y, Lang X, et al. Synergistic effect of sulfur on electrochemical performances of carbon-coated vanadium pentoxide cathode materials with polyvinyl alcohol as carbon source for lithium-ion batteries[J]. *International Journal of Energy Research*, 2019, 43(13): 7664–7671.

[47] Liu H Q, Tang Y P, Wang C. A lyotropic liquid-crystal-based assembly avenue toward highly oriented vanadium pentoxide/graphene films for flexible energy storage[J]. *Advanced Functional Materials*, 2017, 27(12): 1606269.

[48] Zhang Y, Lai J, Gong Y, Hu Y, Liu J, Sun C. A safe high-performance all-solid-state lithium-vanadium battery with a freestanding V_2O_5 nanowire composite paper cathode[J]. *ACS Applied Materials & Interfaces*, 2016, (8): 34309–34316.

[49] Ya Y, Li B, Guo W, Pang H, Xue H G. Vanadium based materials as electrode materials for high performance supercapacitors[J]. *Journal of Power Sources*, 2016, (329): 148–169.

[50] Qu Q T, Zhu Y S, Gao X W, Wu Y P. Core–shell structure of polypyrrole grown on V_2O_5 nanoribbon as high performance anode material for supercapacitors[J]. *Advanced Energy Materials*, 2012, (2): 950–955.

[51] Chen Z, Augustyn V, Wen J, et al. High-performance supercapacitors based on intertwined CNT/V_2O_5 nanowire nanocomposites[J]. *Advanced Materials*, 2011, 23(6): 791–795.

[52] Xing L L, Zhao G G, Huang K J, Wu X. A yolk–shell V_2O_5 structure assembled from ultrathin nanosheets and coralline-shaped carbon as advanced electrodes for a high-performance asymmetric supercapacitor[J]. *Dalton Trans*, 2018, 47(7): 2256–2265.

[53] Yao L, Zhang C R, Huang N T. Three-dimensional skeleton networks of reduced graphene oxide nanosheets/vanadium pentoxide nanobelts hybrid for high-performance supercapacitors[J]. *Electrochimica Acta*, 2019, 295: 14–21.

[54] Pandey G P, Liu T, Brown J E, Yang Y, Li Y, Sun X. Mesoporous hybrids of reduced graphene oxide and vanadium pentoxide for enhanced performance in lithium-ion batteries and electrochemical capacitors. *ACS Applied Materials & Interfaces*, 2016: 9200.

[55] Wang B, Abdulla A L, Wang D, Zhao X S. A three-dimensional porous $LiFePO_4$ cathode material modified with a nitrogen-doped graphene aerogel for high-power lithium ion batteries[J]. *Energy & Environmental Science*, 2015, 8(3): 869–875.

[56] Wang B, Xie Y, Liu T, Zhao X S. $LiFePO_4$ quantum-dots composite synthesized by a general microreactor strategy for ultra-high-rate lithium ion batteries[J]. *Nano Energy*, 2017, 42: 363–372.

[57] Legagneur V, An Y, Mosbah A, Portal R, La S A, Verbaere D, Guyomard Y. $LiMBO_3$ (M=Mn, Fe, co): Synthesis, crystal structure and lithium deinsertion/insertion properties. *Solid State Ionics*, 2001, 139(1/2): 37–46.

[58] Yamada A, Iwane N, Harada Y, et al. Lithium iron borates as high-capacity battery electrodes[J]. *Advanced Materials*, 2010, 22(32): 3583–3587.

[59] Sivaraj P, Nalini B, Abhilash K P, et al. Study on the influences of calcination temperature on structure and its electrochemical performance of Li_2FeSiO_4/C nano cathode for lithium ion batteries[J]. *Journal of Alloys and Compounds*, 2018, 740: 1116–1124.

[60] Zheng J, Teng G, Yang J, et al. Mechanism of exact transition between cationic and anionic redox activities in cathode material Li_2FeSiO_4[J]. *The Journal of Physical Chemistry Letters*, 2018, 9(21): 6262–6268.

11 Solar-Induced CO_2 Electro-Thermochemical Conversion and Emission Reduction Principles

Bachirou Guene Lougou
School of Energy Science and Engineering, Harbin Institute of Technology, Harbin, People's Republic of China

MIIT Key Laboratory of Critical Materials Technology for New Energy Conversion and Storage, Harbin Institute of Technology, Harbin, People's Republic of China

Azeem Mustapha, Enkhbayar Shagdar, and Yong Shuai
School of Energy Science and Engineering, Harbin Institute of Technology, Harbin, People's Republic of China

Zhijiang Wang
MIIT Key Laboratory of Critical Materials Technology for New Energy Conversion and Storage, Harbin Institute of Technology, Harbin, People's Republic of China

CONTENTS

DOI: 10.1201/9781003133971-11

11.1 INTRODUCTION

The unprecedented consumption of natural resources—petroleum, coal, and natural gas—to fulfill the worldwide energy requirements of the continuously growing population. Presently, more than 80% of the energy demand is addressed by the daily utilization of 10 Mt of petroleum, 20 Mt of coal, and 10 B cubic meters of natural gas. Their combined combustion generates 30 Bt of carbon dioxide into our environment and 2% of this CO_2 is absorbed by the ocean water and utilized by the plants, resulting in the acidification of ocean water. The CO_2 concentration in the atmosphere was around 275 ppm at the start of the industrial revolution. It became nearly 400 ppm before the start of this decade and is expected to be 575 ppm by 2100. Undoubtedly, these large anthropogenic CO_2 emissions have caused the existing deleterious climate change, resulting in detrimental effects on the ecosystem and are also a severe threat to the present living community. Therefore, the conversion of this greenhouse gas into valuable chemicals is an imperative issue for researchers and scientists. The aim is to provide the solution for climate change in addition to decrease the uninterrupted burning of natural resources. The utilization of CO_2 by converting it to industrially important commodities can efficiently close the carbon loop and also a sustainable path for the development of green fuels. The novel technologies such as photochemical, biochemical, electrochemical, plasma-chemical, and solar thermochemical are being extensively practiced due to the impediments in traditional technologies. Also, the development of solar-assisted power generation or the hybridization of conventional coal power systems is pursuing to reduce fossil fuel consumption and meantime avoiding enormous CO_2 emission. Ultimately, this chapter provides comprehensible insights into the principles underpinning the theory and practice of the solar electro-thermochemical CO_2 conversion strategies and solar-aided CO_2 emission reduction in the energy section that are considered difficult to decarbonize.

11.2 THERMOCHEMICAL-BASED CO_2 CONVERSION MECHANISM

Thermal effect on the raw physical component induced rapid changes in the physicochemical properties of the activated materials resulting in the formation of new phases including solid active catalysts and mixed gases such O_2, CO, H_2, CH_4, etc. depending on the composition of the working fluid. Thermochemical catalysis driven by heat generated by sunlight is an emerging sub-discipline of CO_2 gas-solid heterogeneous catalytic conversion into fuels and chemicals. This technology has been proven as an effective and promising strategy for converting CO_2 to synthetic fuels and chemicals. Providing a comprehensible insight into the principles underpinning the theory and practice of thermochemical CO_2 catalysis systems would ultimately boost the general research trend in such a field. The redox material preparation method-based thermodynamic criteria, reactor engineered design, and CO_2 cyclic conversion mechanisms and kinetics are the most significant challenges as for scientific, techno-economic, and environment often encounter in pursuing solar-driven thermochemical CO_2 reduction.

11.2.1 CO_2 Conversion Thermodynamics and Materials

11.2.1.1 Non-stoichiometric Perovskites and Ceria

Perovskites-based thermochemical redox materials have been widely used because of their tunable chemical properties allowing large oxygen storage capacities and exchange capabilities that suit well the high-temperature thermochemical CO_2 conversion applications. As shown in Figure 11.1(a), the perovskites are generally characterized by A- and B-site easing the substitution of active oxide/catalyst. In the development of Mn-containing perovskite oxides $La_{1-x}Sr_xMnO_{3-\delta}$ for CO_2-splitting, Demont et al. [1] have incorporated Ba^{2+}, Ca^{2+}, or Y^{3+} on the A-site while the B-site is substituted by Al^{3+} and Mg^{2+} single reducible cations. The newly synthesized material thermochemical redox performances are analyzed based on the redox thermodynamic principles typically referring to the high-temperature thermal reduction and relatively low-temperature CO_2-splitting via oxidation of the thermally activated catalysts. The thermal redox process induced the active material weight changes and microstructure modification due to the O_2 release/uptake phenomenon that can be quantified by thermogravimetric analysis, differential scanning calorimetry, scanning electron microscopy, and energy-dispersive X-ray [2] as shown in Figure 11.1(b). The following characteristics and thermal chemical mechanistic conversion underpinning the theory of the thermochemical thermodynamics of active material are analyzed based on the simplified standard change in Gibbs energies for the reactions as follows:

$$\Delta g^o = \Delta h^o - T\Delta s^o \tag{11.1}$$

The thermal-activated material's chemical-physical properties modification and meantime derived chemical product formations is a complex process, thanks to the two-step thermochemical redox cycles providing a comprehensible understanding of such a technology. The two-step thermochemical redox cycles typically referring

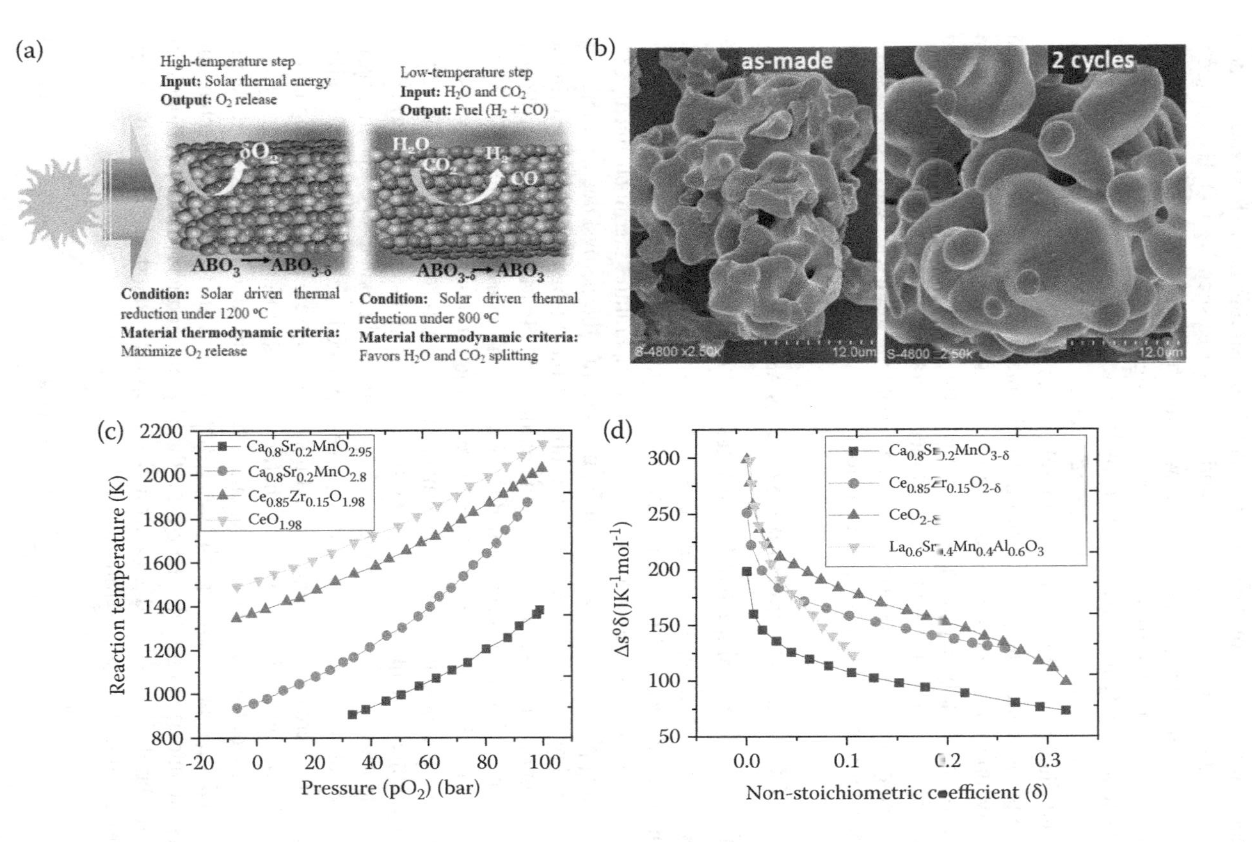

FIGURE 11.1 (a) Tunable perovskite material design; (b) high-temperature induced material chemical properties modification [1]; (c–d) molar entropy variation as a function of non-stoichiometric coefficient and temperature changes as a function of the pressure of a number of perovskite and ceria material [6].

to the high-temperature thermal energy charging via solar IR absorption inducing the raw material reduction following by the chemical energy release via oxidation at relatively low temperature. In the case of perovskites and ceria, the thermodynamic fundamentals of the material are analyzed by considering the partial reduction of the oxide creating O-defects at the surface of the active material. The redox material may exhibit high thermal stability when it could maintain its underlying micro-structural and crystal motif. The redox thermodynamics are analyzed by quantifying equilibrium between O_2 release and uptake and remaining material weight and weight gain during reduction and oxidation, respectively. The redox reaction mechanism of perovskites (ABO_3) and ceria (CeO_2) can be described by the following equation considering the non-stoichiometric coefficient δ, reaction temperature T, and O_2 partial pressure p_{O2}.

$$ABO_3 \xleftrightarrow{\delta(T,p_{O2})} ABO_{3-\delta} + \frac{\delta}{2}O_2 \tag{11.2}$$

$$CeO_2 \xleftrightarrow{\delta(T,p_{O2})} CeO_{2-\delta} + \frac{\delta}{2}O_2 \tag{11.3}$$

Perovskites are novel chemical energy storage materials widely used in thermochemical CO_2 and H_2O-splitting. The possibility of tuning A- and B-sites provide a variety of raw oxide material such $CaMnO_3$ and $SrFeO_3$ by substituting A-site into Ca or Sr and B-site into Mn or Fe, respectively. A-site is mainly characterized by alkali earth metals (A^{2+}) including Mg, Sr, Ba, Ca, etc., while B-site is substituted by transition metals ($B^{3+\text{ or }4+}$) including Fe, Co, Cu, Mn, etc. Other metals can be added to the main material to improve the composite oxide catalyst thermochemical reactivity. For instance, $CaMnO_3$ can be modified into $Ca_{0.8}Sr_{0.2}MnO_3$ by adding a small quantity of Sr nanoparticles. By the similar strategy, a wide variety of redox materials including $La_2O_3CaCO_3MnO_2$ [1], $Y_2O_3SrCO_3MnO_2$ [1], $La_{0.5}Sr_{0.5}Mn_{0.9}Mg_{0.1}O_3$-$CeO_2$ [3], $LaCo_{0.7}Zr_{0.3}O_3$ [4], $Pr_{0.18}Sr_{0.80}Mn_{0.99}O_{2.951}$ [5] can be synthesized resulting in 210 $\mu mol{\cdot}g^{-1}$, 269 $\mu mol{\cdot}g^{-1}$, 387 $\mu mol{\cdot}g^{-1}$, 1066.6 μmol/g, and 637.6 μmol/g of CO yield from CO_2 conversion, respectively. Since the thermal reduction of the non-stoichiometric oxide materials is generally characterized by partial reduction, the usage of partial molar quantities better describes the equilibrium thermodynamic calculations. Considering the reaction mechanism described in Equations (11.2)–(11.3), the equilibrium thermodynamics can be described by the following equations:

$$\Delta g_\delta = \Delta h_\delta^o - T\Delta s_\delta^o + RT\ln\left(\frac{pO_2{}^{0.5}}{p^o}\right) = 0 \tag{11.4}$$

$$\Delta s_\delta^o(\delta) = 0.5sO_2 + \Delta s_{ph} + \frac{\partial \Delta s_{con}(\delta)}{\partial \delta} \tag{11.5}$$

$$\Delta h^o = 100 - 500 kJmol^{-1} \tag{11.6}$$

where sO_2 is the entropy of O_2 gained, Δs_{ph} is the vibrational state changes, and $\Delta s_{con}(\delta)$ is the configuration entropy. The non-stoichiometric coefficient dependence of the partial entropy changes and temperature versus pressure of some materials is shown in Figure 11.1(c,d).

11.2.1.2 Stoichiometric Ferrite-Based Oxides

The analysis shows that ferrite-based oxides are typically subjected to a stoichiometric phase change during high-temperature thermal reduction. Fundamental understanding of the redox materials and their implications and limitations concerning thermodynamic constraints revealed the stoichiometric materials may have much larger specific energy storage compared to the non-stoichiometric materials undergoing partial reduction [6]. However, the faster kinetics and better activity exhibiting ceria and perovskite material class at low temperature induce a trade-off between the material selectivity. The thermal reduction of the stoichiometric materials is characterized by phase change reduction, decomposition of the raw oxide into multiple phases following by stoichiometric O_2 release as described by the following equations:

$$M_yO_x \xleftrightarrow{T,pO_2} M_yO_{x-1} + 0.5O_2 \tag{11.7}$$

The stoichiometric oxide materials typically consist of a wide range of metal oxide material including single oxides (Fe_3O_4, Co_3O_4, ZnO, CuO, Mn_3O_4, etc.) and ferrite oxides ($NiFe_2O_4$, $CuFe_2O_4$, $CoFe_2O_4$, $ZnFe_2O_4$, etc.). Ferrites and Ceria-supported ferrite oxide are the oxide materials design strategies for improving the raw material thermochemical activity and major product selectivity. For example, $Fe_{0.35}Ni_{0.65}O_x$ [7], 30 wt% $MgFe_2O_4$ [8], CeO_2-Fe_2O_3 [9], 70CeO_2-$NiFe_2O_4$ [10] resulted in higher CO selectivity in the process of CO_2 thermochemical conversion. The thermodynamic calculation of such a process described in Equation (11.7) refers to the difference in enthalpy and entropy of the reactants and products as follows:

$$\Delta h^o = h^o_{MyOx} + 0.5h^o_{O2} - h^o_{MyOx-1} \tag{11.8}$$

$$\Delta s^o = s^o_{MyOx} + 0.5s^o_{O2} - s^o_{MyOx-1} \tag{11.9}$$

To date, a chemical energy storage density of 1,070 kJ/kg was experimentally reached using $(Mn_{0.33}Mg_{0.66})_3O_4/3(Mn_{0.33}Mg_{0.66})O$ cycle under 1500°C that could be outperformed up to 2356.46 and 2897.13 kJ/kg under ~1143.26°C by sulfate modified $MnSO_4$/MnO and $MgSO_4$/MgO cycles, respectively [11]. The thermochemical energy conversion/storage potential was mainly attributed to the materials' oxygen exchange kinetics in terms of O_2 extraction/uptake efficiency across the reaction temperature changes. The material presenting a higher reduction extent could result in a higher ratio of chemical energy released per mole of the oxygen carriers. To this end, an engineered design of composite redox material easing higher possible O_2 exchanges at relatively low thermal reduction temperatures in the chemically reacting media could overcome the constraints related to thermodynamic. Perovskites are emerging as an alternative to metal oxide materials due to their flexibility of tuning metal oxide materials in a broad range of chemical compositions and allowing them to operate under 1000°C, relevant to large-scale facilities (Figure 11.2).

11.2.2 Thermochemical CO_2 Conversion Reactor Engineering

Start-ups developing alternatives fuels from CO_2 were found increasing thanks to great pieces of knowledge in advanced thermochemical reactor design able to effectively convert CO_2, CH_4, H_2O, and other flue gases into value-added products, ultimately net-zero energy use. Harnessing solar energy to convert CO_2 into net-zero fuels typically consist of the development of a thermal catalysis system utilizing a mirror field to concentrate solar radiation onto the receiver, converting it into high-temperature process heat that activates the redox oxide material utilized as a burner and catalysts. Then, the thermochemical reactor services as a machine for recycling CO_2 into fuels and chemicals. The oxide catalyst absorber is designed in the catalyst bed by considering heat and mass transfer fundamentals. The most significant in the material selectivity and composition is based on the material physicochemical properties able to expose to extreme heat, give up O_2 during reduction, and retrieves O_2 when cooling, which follow up the concept of a two-step cyclic process based on the reduction and oxidation reactions as shown in Figure 11.3(a). O_2 exchange in the chemical reacting media is highly favored by temperature and pressure swings. From the mechanistic aspect, the active material can be heated to high-temperature ~1500°C under reduced pressure to generate O_2 in the first step, then, oxidize with CO_2 at a relatively low temperature ~900°C to generate CO in the second step. The same mechanism can be reputed into one to several cycles depending on the material thermal and chemical stability.

As shown in Figure 11.3(b,d), two kinds of solar reactor design including fluidized bed reactor (Figure 11.3(b)) and membrane reactor (Figure 11.3(d)) models are significantly impacting the progress of solar fuel and chemical synthesis. The reactor design strongly refers to the reaction mechanisms extracted from the thermodynamic analysis. The difference between the reaction mechanisms described in Figure 11.3(a,c) can be directly identified in Figure 11.3(b,d). In terms of innovation in the reactor design, the membrane reactor model overcomes the issues associated with material stresses and energy irreversibility and meantime favors the conduction of oxygen ions, electrons, and vacancies driven by the O_2 chemical potential gradient employing the membrane consisting of the burner/absorber [13]. The solar-to-fuel energy conversion efficiency is the key metric for solar reactor performance. Still, the membrane reactor exhibits a solar-to-fuel efficiency of 1% much lower than 5.25% of that reported with the fluidized bed reactor model showed in Figure 11.3(b).

The cavity geometry is designed to enable efficient absorption of transmitted radiative flux across the heat exchanger serving as the chemical reaction interfaces.

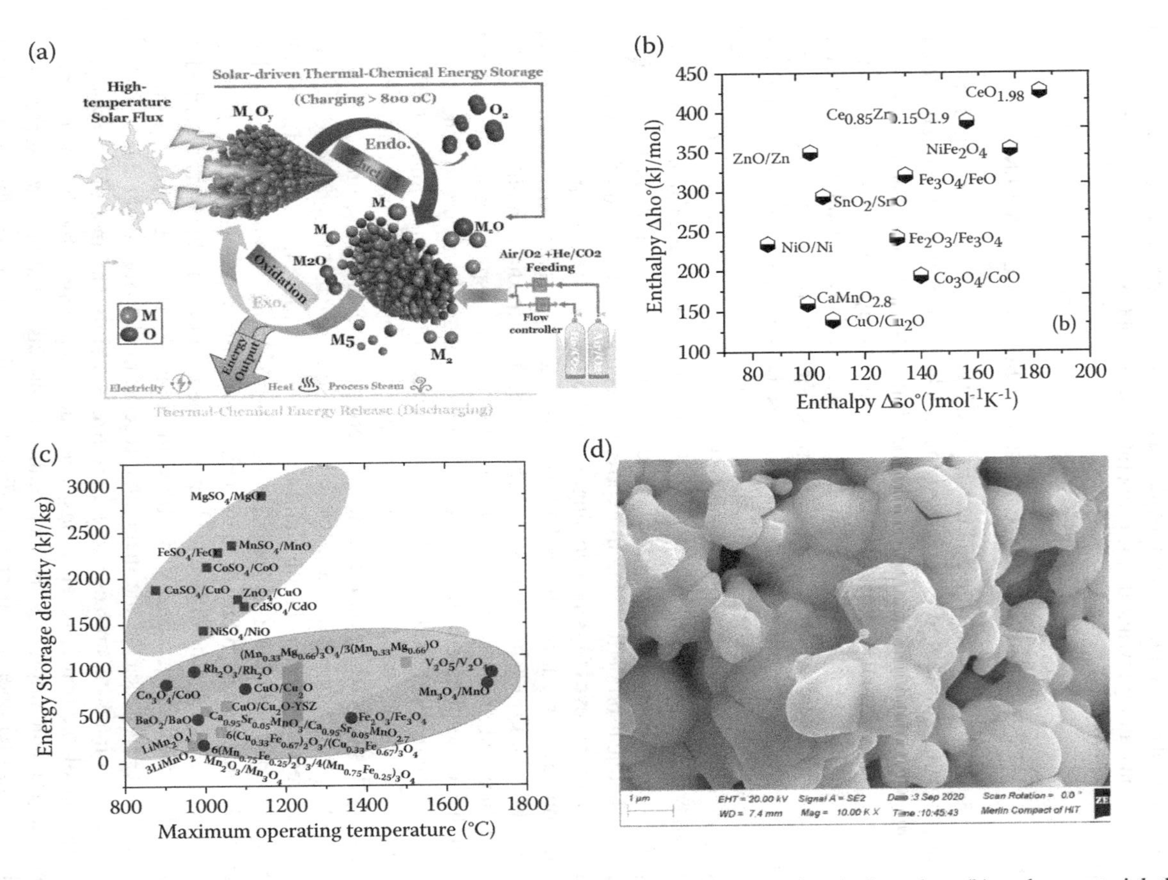

FIGURE 11.2 (a) Reaction mechanism of stoichiometric material based on two-step thermochemical cycles; (b) redox material thermodynamic properties in term of reaction enthalpy Δh^o and entropy Δs^o; (c) chemical energy released as a function of reaction temperature; (d) $NiFe_2O_4@ZrO_2$ microstructural behavior after two thermochemical cycles of CO_2-splitting [12].

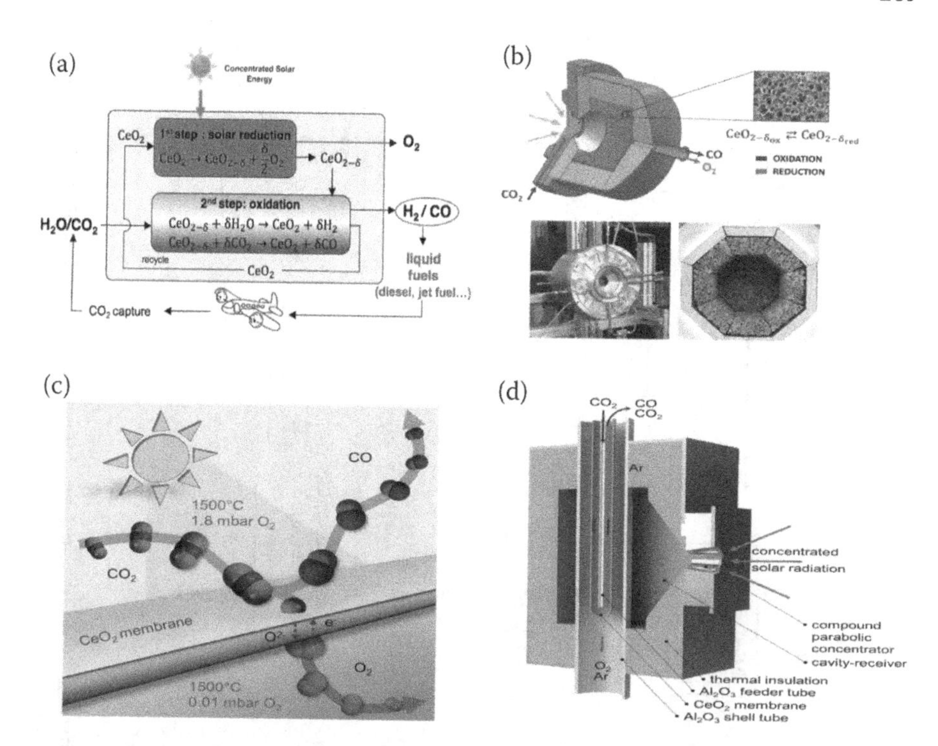

FIGURE 11.3 (a) CO_2-splitting reaction mechanism in a fluidized bed reactor; (b) solar reactor splitting CO_2 into streams of CO and O_2; (c) CO_2-splitting reaction mechanism across a redox membrane [13]; (d) Ceria redox membrane reactor configuration for CO_2-splitting [13].

The initial materials selectivity, design strategy, and thermal stability are the most important factors considered in the design of a thermally insulated cavity receiver. In most cases, ceramic porous media is preferably selected for improving heat and mass transfer, especially transport phenomena. The numerical model consists of a 2- or 3-D porous-medium-filled domain used as a computational domain and the physic interface of heat transfer in a porous medium using the physical models including radiation heat transfer models. Commercial computational modeling stools including Fluent, COMSOL Multiphysics, CHEMKIN, OpenFOAM, and other relevant software are often used for the theoretical thermal analysis. A basic schematic diagram describing the numerical solution method is shown in Figure 11.4. A coupled heat transfer model including conduction, convection, and radiation with different transfer fluids and porous matrix properties are modeled for providing a novel solar reactor configuration able to process into industrial manufacture for further experimental performances.

11.2.3 Kinetics Modeling of CO_2 Thermochemical Conversion

The thermochemical reaction kinetic is measured by oxygen exchange dynamics and fuel release rate in the reactor. It is a complex coupling finite-volume

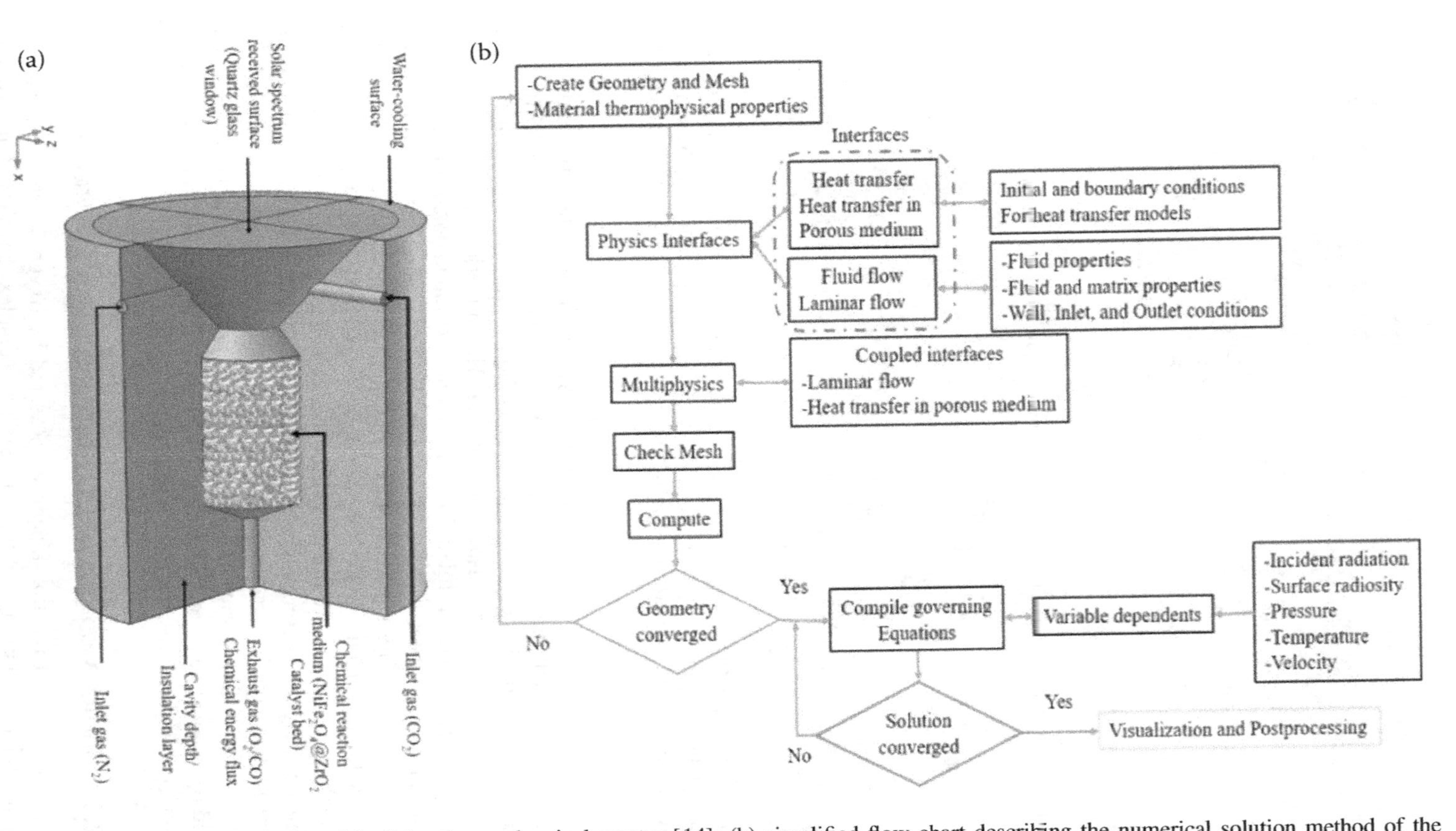

FIGURE 11.4 (a) Numerical model of the thermochemical reactor [14]; (b) simplified flow chart describing the numerical solution method of the geometry model and heat transfer section of a solar thermochemical reactor [15].

computational fluid dynamic (heat transfer and fluid flow model of the solar reactor) and chemical reaction mechanisms. The heat and mass transfer problems and chemical conversion rates in the reactor are solved by developing fluid phase and solid phase numerical models accounting for mass, momentum, energy conservation equations, and the chemical species mass balance equation. A simplified model of high-temperature heat transfer coupled to the chemical reaction kinetics in the porous media-filled solar reactor can be described by the following expressions:

$$\frac{\partial \varepsilon_p \rho_f}{\partial t} + \nabla \cdot (\rho_f U) = Q_m \tag{11.10}$$

$$\rho_f \frac{1}{\varepsilon_p} \rho_f \frac{\partial \mathbf{u}}{\partial t} + \frac{1}{\varepsilon_p} \rho_f (U \cdot \nabla) U \frac{1}{\varepsilon_p} = -\nabla P + \nabla \cdot \left[\mu_f \frac{1}{\varepsilon_p} (\nabla U + (\nabla U)^T) - \frac{2}{3} \mu_f \frac{1}{\varepsilon_p} \nabla U \right] - \left(\frac{\mu}{\kappa_p} + \frac{Q_m}{\varepsilon_p{}^2} \right) U + F \tag{11.11}$$

$$(\rho C_p)_{eff} \frac{\partial T}{\partial t} + \rho_f C_{p,f} U \cdot \nabla T = \nabla \cdot (k_{eff} \nabla T) + \nabla \cdot \dot{q}_{rad}(r) + Q_{conv} + S_{chem} \tag{11.12}$$

where ρ_f is the fluid density, U is the velocity vector, ε_p is the porosity of porous medium, $Q_m = R_{chem,i} + R_{chem,p}$ is the mass source of the porous medium, μ_f is the dynamic viscosity of the fluid, κ_p is the permeability of the porous medium, F is the Forchheimer drag, $(\rho C_p)_{eff}$ is the effective volumetric heat capacity, k_{eff} is the effe ctive thermal conductivity, k_R is the irradiance conductivity, S_{chem} is the enthalpy source term due to the chemical reaction, Q_{conv} is the convective chemical energy flux, and Q_{rad} is the radiative heat flux.

The species mass balance can be calculated by the following equation:

$$\varepsilon_p \rho_f \frac{\partial (\rho Y_i)}{\partial t} = -\nabla \cdot (\rho_f U Y_i) + \nabla (\rho_f \mu_{eff}(T) \nabla Y_i) + R_{chem} \tag{11.13}$$

where Y_i is the mass fraction of chemical species $i = CO_2$, H_2O, CH_4, Ar, etc., μ_{eff} is the effective dynamic viscosity, R_{chem} is the chemical species volumetric production and consumption rate term calculated by considering different kinetic models such as irreversible Arrhenius kinetic rate expressions and Langmuir-Hinshelwood kinetic rate expressions.

The enthalpy changes due to the chemical reaction S_{chem} function of R_{chem}, convective Q_{conv}, and radiative heat flux $Q_{rad} = \nabla \cdot \dot{q}_{rad}(r)$ can be calculated by the following equations. For example, R_{chem} is calculated considering the modified irreversible Arrhenius kinetic rate expressions:

$$S_{chem} = \Delta h_i^0 R_{chem} \tag{11.14}$$

$$R_{chem} = \rho\left(M_{w,i}\sum_{r=1}^{N} R_{i,r} + S_r\right) \tag{11.15}$$

$$R_{i,r} = k_{i,r}\prod_{r=1}^{N_{species}} C_r^{\nu_{i,r}} \tag{11.16}$$

$$k_i = A_i T^{\beta_i} e^{-\frac{T_{a,i}}{T}} \tag{11.17}$$

$$Q_{conv} = Nu\frac{6fvk_f}{d_p^2}(T_f - T_s) \tag{11.18}$$

$$Q_{rad} = \nabla\cdot\dot{q}_{rad}(r) = 4\pi\int_{\lambda=0}^{\infty}\kappa_\lambda(r)\left(i_{\lambda,b}(r) - \frac{1}{4\pi}\int_0^{4\pi} i_\lambda(r,\Omega)d\Omega\right)d\lambda \tag{11.19}$$

where Δh_i^0 is the chemical enthalpy of formation of the species i at $T_0 = 298.15$ K and $P_0 = 1$ atm, $M_{w,i}$ is the molecular weight of ith species, S_r is the rate of other chemical reaction sources term in reaction r, $R_{i,r}$ is the rate of chemical change of ith species in reaction r, k_i is the rate constant of species i, A_i is the pre-exponential factor, β_i is the temperature exponent, $T_{a,i}$ is the ratio of $\frac{E_{a,i}}{T}$, Nu is the Nussle number, $\nabla\cdot\dot{q}_{rad}(r)$ is the divergence of radiative heat flux across r direction, κ_λ is the spectral absorption coefficient for wavelength λ and Ω is the solid angle, and $i_{\lambda,b}$ is the spectral radiative intensity.

The thermodynamics and kinetics modeling of solar thermochemical CO_2 conversion is performed considering the set of the chemical reaction mechanisms described in Equations (11.2), (11.3), and (11.7). CO_2 thermal catalytic conversion and fuel yield are becoming more effective by mixing CO_2 feedstock with other reactive gases such as CH_4, H_2O, and H_2 thereby resulting in dry methane reforming, chemical looping redox, and CO_2 hydrogenation reaction mechanisms that are highly investigated to date. The operational parameters of the solar thermal catalysis system include reactive gas (CO_2, H_2O, CH_4, etc.), feeding rate, inert gas (Ar, N_2, He) flow rate, solar concentration ratio, or solar radiant energy flow or temperature, and total pressure. These parameters are the determining factors for calculating reactor performance and solar-to-fuel efficiency. The thermodynamics is combined with the kinetics analysis to provide a comprehensible insight into the effects of reaction temperature and transient and steady-state product formation rates. The thermal modified active materials physicochemical properties, temperature distribution, and product formation rates are shown in Figure 11.5(a–d). Figure 11.5(d) indicated pure splitting of CO_2 into CO and O_2 is likely impossible due to the high-energy penalty. Besides, spin-polarized density functional theory (DFT) calculations can also be developed to investigate the active

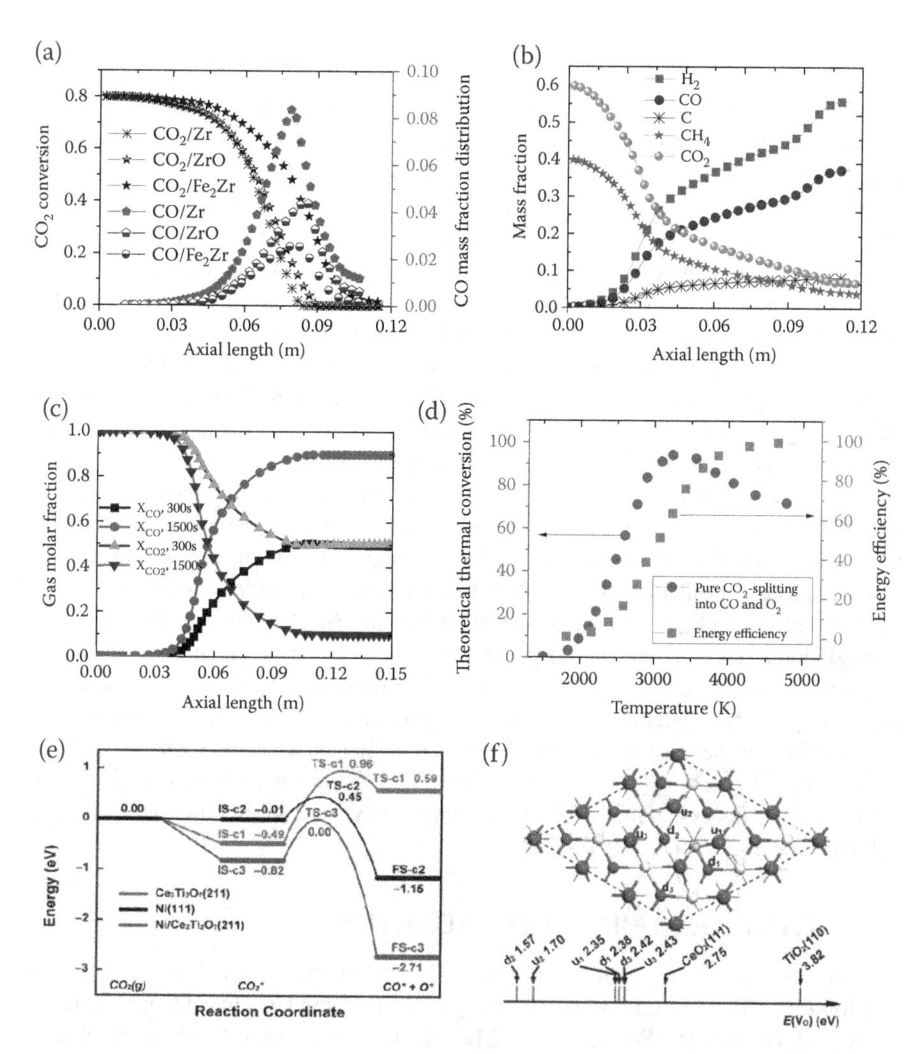

FIGURE 11.5 (a) CO_2 conversion and CO yield as a function of Zr-Fe_x catalyst composition at 1400 K; (b) CO_2 conversion and product gases formation at 1300 K across $NiFe_2O_4$-Al_2O_3; (c) CO_2 conversion in the Al_2O_3-TiC catalyst bed at 400 kW/m^2; (d) pure splitting of CO_2 into CO and O_2 and energy conversion efficiency [16]; (e-f) DFT calculation results of CO_2-splitting and O^{2-} vacancies formation energies at different sites of Ti-doped $CeO_2(111)$, $CeO_2(111)$, and $TiO_2(110)$ [17].

catalyst surface structures with oxygen vacancies, electron density difference diagrams, energy change during the reaction, and their catalytic performance in CO_2 adsorption and dissociation. The DFT calculation performed on CO_2 conversion can be seen in Figure 11.5(e,f).

11.2.4 Thermochemical Redox Cycling CO_2-Splitting Performance

Because of the complexity of physical phenomena occurring in the thermochemical reactor, the numerical models are sometimes limited to accurate prediction of the thermal-chemical conversion mechanisms, product selectivity, and fuel formation. Considering the two-step thermochemical cycle in which the redox metal oxides enable the conversion of CO_2, the bulk composition, short-range structural and defect chemical changes presenting the active material during the oxidation/reduction cycle are crucial for understanding the science behind the redox catalyst's O_2 exchange capacity, CO_2 conversion, and product formation kinetics. As shown in Figure 11.6, the lab-scale demonstration of the CO_2-based renewable fuel processing through a fixed-bed reactor coupled to a solar simulator has proved the effectiveness and feasibility of such a technology. The physical perception of the intrinsic lattice oxygen (O^{2-}) exchange capacity and conversion mechanism is based on the active material surface characteristics in terms of cation expansion and vacancy formation behavior. In-situ methodologies presenting detail in the material micro physicochemical properties are highly developing to gain more insights into the mechanisms and principles of the CO_2 thermochemical conversion. Sediva et al. [11] have recently developed the Time-Resolved Raman Spectroscopy method to show beyond doubt the fundamental insights into vacancy formation at the surface of $Ce_{0.9}La_{0.1}O_2$ during a CO_2-splitting reaction. The dynamic of O^{2-} vacancy exchange induced by the raw material oxygen chemical potential gradient is further investigated via other in-situ measurements technology such as thermogravimetric analysis, X-ray diffraction and energy dispersive spectroscopy, and X-ray photoelectron spectroscopy for the further understanding of the reaction pathways, design of new materials, and validation of fuel production stability.

11.3 ELECTROCHEMICAL CO_2 REDUCTION

Concerning the challenge of worldwide CO_2 emissions, electrochemical CO_2 reduction (eCO_2R) is an emerging technology to transmute CO_2 into valuable energy fuels and chemicals (Figure 11.7). The eCO_2R progresses thermodynamically through the transfer of different numbers of electrons and coupling of protons in an electrolyte placed in an electrocatalytic cell to produce C_1 and C_2 products through different reaction pathways (Figure 11.8). The electrocatalytic cell consists of a catalyst-loaded cathode, an anode-counter electrode to keep the balance of the charges, and a reference electrode. The cathodic and anodic sections are separated using a membrane which is also further utilized for charge balancing and to avoid re-oxidation of the eCO_2R products. The electrolyte functions as a charge transporting medium and adsorb CO_2. eCO_2R takes place on the surface of the catalyst and typically involves three critical stages: (1) CO_2 adsorption over the catalyst surface; (2) transfer of charge to break the C=O bond to generate the C–O or C–H bond; and (3) end product obtaining from the catalyst surface.

The eCO_2R approach is greatly advantageous for other CO_2 conversion technologies in terms of working conditions (temperature and pressure) and applications, for instance, the carbon-neutralization of industries, decreasing the large CO_2 emissions,

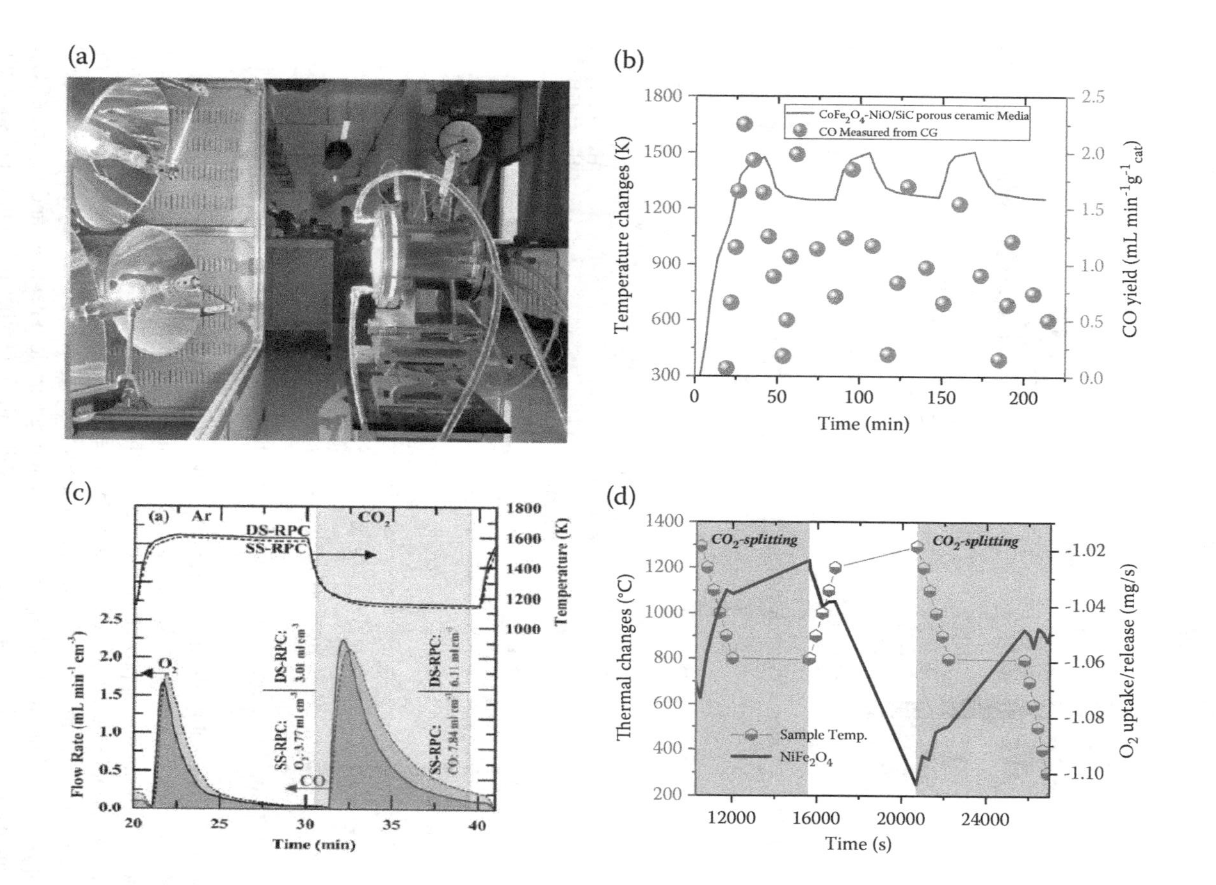

FIGURE 11.6 (a, b) Lab-scale performance of CO_2 thermochemical conversion at Harbin Institute of Technology's thermochemical lab; (c) volume-specific flow rate of O_2 and CO measured in a single-scale porosity (SS-RPC) (dashed lines) and dual-scale porosity (DS-RPC) (solid lines) [18]; (d) $NiFe_2O_4$ thermal changes and thermochemical reactivity with CO_2 [14].

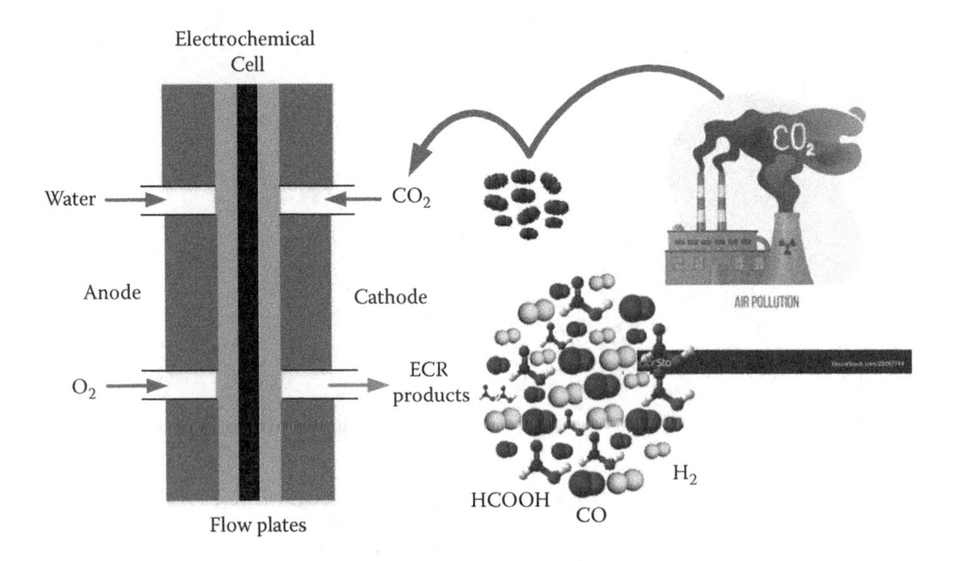

FIGURE 11.7 Operation of electrochemical CO_2 reduction (eCO_2R).

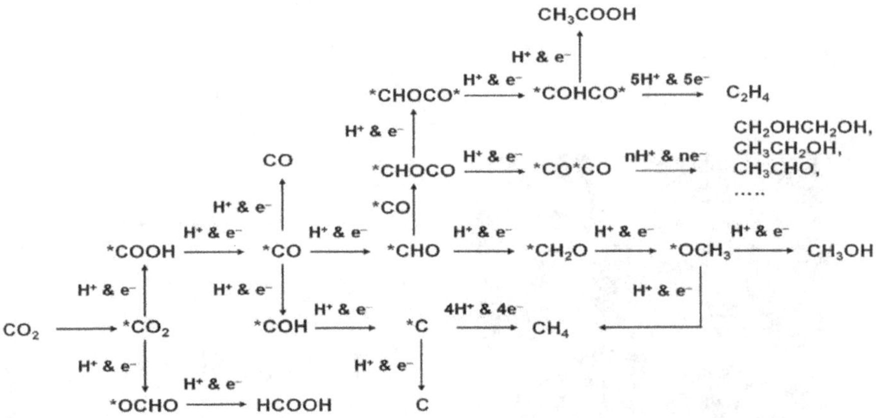

CH_3COOH
H^+ & e^-
$*CHOCO*$ —(H^+ & e^-)→ $*COHCO*$ —($5H^+$ & $5e^-$)→ C_2H_4
H^+ & e^-
$*CHOCO$ —(H^+ & e^-)→ $*CO*CO$ —(nH^+ & ne^-)→ CH_2OHCH_2OH, CH_3CH_2OH, CH_3CHO, ...
CO
H^+ & e^-
$*CO$
$*COOH$ —(H^+ & e^-)→ $*CO$ —(H^+ & e^-)→ $*CHO$ —(H^+ & e^-)→ $*CH_2O$ —(H^+ & e^-)→ $*OCH_3$ —(H^+ & e^-)→ CH_3OH
H^+ & e^-
H^+ & e^-
H^+ & e^-
CO_2 → $*CO_2$
$*COH$ —(H^+ & e^-)→ $*C$ —($4H^+$ & $4e^-$)→ CH_4
H^+ & e^-
H^+ & e^-
$*OCHO$ —(H^+ & e^-)→ HCOOH
C

FIGURE 11.8 Different reaction routes to produce useful chemicals through eCO_2R [19].

storage of renewable electric energy into chemicals and fuels, and development of the carbon-neutral economy. Furthermore, eCO_2R has several plus points compared to other CO_2 reduction approaches, such as: (1) the hydrocarbons can be facilely produced using renewable electricity, CO_2/H_2O; (2) the eCO_2R process is compact, effective, on-demand, and scalable; (3) the overall process is convenient and easy to administer by simply controlling the working conditions and electrode potentials; and (4) the eCO_2R mechanism can utilize clean energy sources like tidal, geothermal, wind, and solar plus additional electricity obtained from hydroelectric and nuclear sources.

The achievements to date in eCO_2R have been promising, and there are many potential benefits. Nonetheless, this approach still experiences several challenges

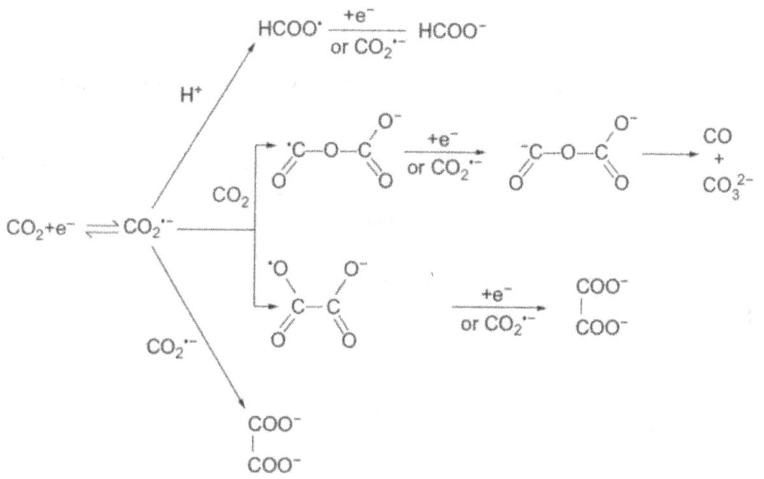

FIGURE 11.9 Mechanism of eCO_2R while using aqueous and non-aqueous electrolytes [20].

such as: (1) the passive reaction kinetics of CO_2 reduction, despite using higher electrode potentials and finely tuned catalysts. (2) The poor energy efficiency which attributed to the parasitic or decomposition or parasitic solvent reaction even at higher reduction potentials. (3) The higher utilization of energy. It is extensively recognized that the great challenge in eCO_2R is the poor performance of catalysts exhibiting insufficient catalytic activity, stability, and durability.

The selection of electrolytes can greatly influence the performance of eCO_2R (Figure 11.9). Several important factors such as the cationic and anionic composition and concentration having intricate relations with local reaction conditions can generate alterations in proton donor's availability, pH, buffer capacity, and electrostatic interactions. These alterations aren't always convenient to detect because of several intertwining impacts.

11.3.1 Electrocatalyst

The imperative role of electrocatalysts in eCO_2R is widely accepted for the selective formation of C_1 and C_2 products. There are four categories of metal electrodes in aqueous electrolytes and three main types in non-aqueous electrolytes. Metallic (1) indium, tin, mercury, and lead are selectively used to produce formic acid; (2) zinc, silver, and gold yield carbon monoxide; (3) copper demonstrates superior catalytic activity to form alcohols, aldehydes and hydrocarbons; and (4) aluminum, gallium, and group VIII elements, other than palladium exhibit low catalytic activity in eCO_2R while using aqueous electrolytes. In non-aqueous media, (1) lead, thallium, and mercury produce oxalic acid; (2) copper, gold, indium, zinc, and tin form carbonates and carbon monoxide, whereas nickel, palladium, and platinum specifically yield carbon monoxide; and (3) aluminum, gallium, and VIII elements (excluding Nickel, palladium, and platinum) generate oxalic acid and carbon monoxide. Additional categorization criteria for the phenomenon taking

TABLE 11.1
Different products produced by metal electrodes belonging to different groups

Metals	Aqueous solution	Non-aqueous solution
	sp group	
Cu, Zn, Sn	HCOOH	–
In, C, Si, Sn, Pb, Bi, Cu, Zn, Cd, Hg	Hydrocarbons, HCOOH, CO	–
In, Sn, Pb, Cu, Au, Zn, Cd	–	CO, hydrocarbons, CO_3^{2-}
In, Sn, Au, Hg	–	CO
Ln, Tl, Sn, Pd, Zn, Hg	–	Oxalic acid
	d group metals	
Ni, Pt	–	CO, CO_3^{2-}
Ni, Pd, Rh, Ir	CO, HCOOH	–
Fe, Ru, Ni, Pd, Pt	Hydrocarbon	–
Ti, Nb, Cr, Mo, Fe, Pd	–	Oxalic acid
Mo, W, Ru, Os, Pd, Pt	MeOH	–
Zr, Cr, Mn, Fe, Co, Rh, Ir	CO	–

place in eCO_2R depends on the electrode catalytic properties and electrolyte, which will provide better insights about reactions and better systematization into the processes administering the phenomena. The variation in the electrocatalytic properties of electrode materials belonging to the d and sp group metals was determined in the 1970s, and it is a critical parameter in estimating the selectivity of the electrode materials [21,22]. A summary of electrode materials belonging to different groups and producing different products in the aqueous and non-aqueous electrolyte is provided in Table 11.1.

The preparation of nanocatalysts with acceptable shapes and sizes has significantly advanced nanoscience research and such kind of nanomaterials greatly assist in improving the electrocatalytic activity, selectivity, and stability. Many research studies have reported that nanomaterials-based electrocatalysts demonstrate a disparate behavior in eCO_2R than their bulk materials [23,24]. Recently, this field has received great attention and showed the successful synthesis of nanomaterials and their improved catalytic activities. The nanomaterials possess additional benefits of having more active sites over the elevated surfaces than their bulk counterparts. The increased number of active sites can greatly enhance the electrocatalytic activity, representing a proportional association between active sites and catalytic activity [25]. Furthermore, the morphology of nanomaterials has many under-coordinated sites such as corner sits and edges, facilitating the eCO_2R performance compared to the planar and smooth surfaces [24]. Electrolytes having dense metal impurities improve the electrocatalytic stability, which in turn increases the eCO_2R performance on nanomaterial-based catalysts.

11.3.2 Electrochemical Cell

An electrochemical cell is a device whereby reactions are influenced by the difference between cell open circuit potential and external voltage. Two independent half chemical reactions: (1) reduction and (2) oxidation half-reaction, constitute the complete electrochemical reaction in an electrochemical cell. For eCO_2R, the commonly adopted catalytic system usually contains anode, cathode, and catalyst for CO_2RR and OER, electrolyte permitting fast mass transport of the reacting species and obtained products, and a membrane separating the anodic and cathodic sections. During eCO_2R, ions crossover the membrane from a section to another employing an external circuit. A reference electrode is also employed for potential reduction calculations. Other than the finding suitable electrode and catalysts, the kind of membrane together with aqueous electrolyte greatly impact the reduction reactions, i.e., the ions crossing over the membrane and transmission direction will be completely different. Several studies have reported that product distribution and current density for eCO_2R doesn't only affect by the catalysts or other working conditions, but are also greatly impacted by the configuration of the electrochemical cell [26].

Presently, an H-type cell is still extensively used on a lab scale for CO_2 reduction and a commercially available reactor. Its cathodic section contains working and reference electrodes and the anodic section has a counter electrode (Figure 11.10(a)). A membrane is used to separate these sections and further utilized to avoid the oxidation of produced chemicals. To address the issue of limited mass transfer in H-type cells occurred due to the low CO_2 solubility in aqueous electrolytes, the flow cell with improved mass transfer characteristics is developed for commercial-scale applications of eCO_2R. The flow cells can be classified into three types: (1) solid-oxide electrolysis cell (SOEC), (2) microfluidic-flow cell (MFC), and (3) polymer electrolyte membrane (PEM) flow cell (Figure 11.10(b–d)). The feedstock properties like flow rate, electrolyte supply, GDEs, and membrane types of the flow cells need careful monitoring for efficient cell designing because all these parameters strongly affect the performance of eCO_2R. Nevertheless, different kinds of flow cells have demonstrated great potential for the development of electrochemical cells, it is important to mention that the present structure of flow cells can't be used for industrial-scale eCO_2R, whereby the electrode sizes are less than 10-meter squares.

Another important type of electrochemical cell is differential electrochemical mass spectrometry (DEMS) reactor (Figure 11.10(e)) for efficiently detecting produced chemicals in actual time and can complete detection for different values of applied potential sharply in linear-sweep voltammetry, quicker compared to other electrochemical cells employing conventional detection techniques. Moreover, products obtained through eCO_2R can be conveniently detected at the interface of electrode and electrolyte in actual time using DEMS cell, assisting the researchers to analyze the local reaction conditions during the reduction reactions, providing useful insights for eCO_2R mechanism for the selective production using different catalysts. Table 11.2 provides a summary of the configurations, compositions, and objectives of each kind of cell along with their advantages and disadvantages.

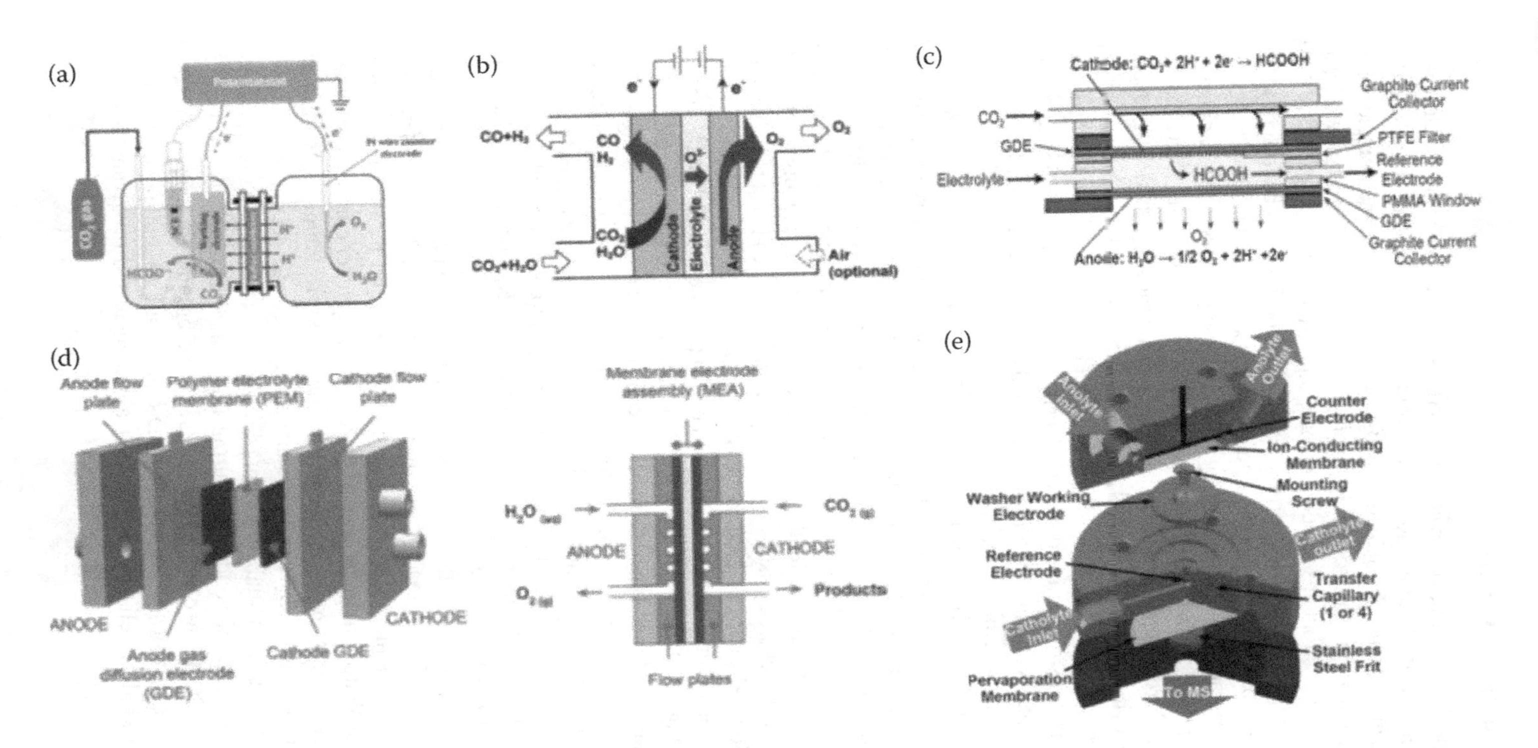

FIGURE 11.10 Schematic diagram of electrochemical cells: (a) H-type cell, (b) SOEC, (c) microfluidic reactor, (d) PEM cell, (e) DEMS cell [26,27].

TABLE 11.2
Different types of electrochemical cells, their properties, merits, and demerits [26,27]

Cell type		Properties	Merits	Demerits
H-type	Sandwich-type	High S/V ratio	Efficient determination of liquid products, low solution resistance	Limited mass transfer, strong generation of bubbles.
	Traditional H-type	Batch reactor	Convenient and facile operation, many kinds of electrodes can be used, excellent for electing and quantifying catalysts, available on a commercial scale.	Limitation of trace liquid products detection; Limitation of mass transport and large-scale application
Flow cell	SOEC	Solid electrolyte	It can operate at high-temperature values, Efficient mass transport, high current densities.	Facilitate production of a limited number of products.
	Microfluidic flow cell	Flow channel	Permitting use of alkaline electrolyte, fewer water management challenges, no specific issue of pH decrease, the elevated mass transport mechanism	The sensitivity of pressure generating from pressure variations along the membrane, a major issue for commercial-scale utilization.
	PEM flow cell	MEA configuration	Elevated mass transport, low cell resistance, inexpensive structural material, GDLs, and the layer of electrocatalysts	The high overpotentials cause corrosion, absence of reference electrode intricates the potential calculations
DEMS cell		Pervaporation membrane	Efficient product detection in actual time.	Restriction of electrode choice, mass transport, and commercial-scale utilization.

11.3.3 Commercialization of eCO2R

The commercial-scale utilization of any novel processor technology is not simple and an easy stage, it requires continuous development and research on small-scale study. Up to now, we have presented important factors influencing the selectivity and activity of important product generation in the eCO_2R process. Nevertheless, while moving from lab-scale to commercial scale, many issues like catalyst durability, strong reaction environment, reaction kinetics, required product selectivity need to be carefully monitored. The novel approach like eCO_2R or any other technology doesn't linearly extend, it is often very problematic to estimate how commercial scale path will conveniently and safely perform.

Many research studies have reported the technical possibility of commercial-scale eCO_2R, but research investigations concerning economic feasibility depending upon the recent approaches indicate a negative trend. Nevertheless, these studies relating to commercial-scale economic feasibility don't sufficiently incorporate the techno-economic analysis of several key points, for instance, electric power cost, functioning cost, capital cost (machinery for CO_2 capture, gas separation, and electrolyzer), and supply and demand of different products in the market. As a result, it is not facile and convenient to estimate the actual financial viability. Although few reports have analyzed the economic feasibility of eCO_2R, and different models were employed for the formation of several important products, commercialization may be required from more actual performance indicators. Moving from lab-scale to industrial-scale eCO_2R, several important observations are included in Table 11.3.

Employing different experimental and theoretical techniques, many types of research are conducted in the field of eCO_2R. The literature represents that the accomplishments to date in eCO_2R are very encouraging, and the potential advantages are high in numbers. Albeit many fixation mechanisms like electrocatalytic conversions, hydrogenation, and high temperature-based homogeneous electrocatalysis have been used, the superiorly crucial content in CO_2 reduction is to deal it as high low energy input to avoid the further formation of CO_2. Consequently, eCO_2R conducted at lower temperatures founds to be very promising. Higher negative potentials (−2.2 V vs. SCE) are needed for direct eCO_2R on different electrodes to yield a variety of useful products, and the reaction conditions greatly decide the product distribution. This has motivated the scientific

TABLE 11.3
Improvements needed in possible parameters while moving from lab-scale to commercial-scale [28,29]

Sr. No.	Lab-Scale production	Commercial-scale production
1	The electrode employed is usually of traditional H-type	GDE possessing microfluidic flow cell will be utilized, this has commercial-scale utilization.
2	Required selectivity of several products (more than 99).	Required selectivity of several products is not the only big point for commercialization.
3	Requires to conduct eCO_2R in gram scale (less than 10 gram).	Requires to conduct eCO_2R in kg scale (more than 100 kg).
4	Operational expenses are not performance specific.	Operational expenses are quite important and specific for eCO_2R performance.
5	The need for operational electric power is smaller than the industry.	Needs a continuous and cheap supply of electric power, which plays an important role.
6	Product separation in lab-scale is entirely facile.	It is very complicated and requires additional energy, this component also needs modifications.

community to explore and develop suitable electrocatalytic systems to shape eCO_2R into a more productive and efficient approach. Homogenous/heterogeneous catalysis is also an attractive path utilizing metal electrodes and solution modifiers. Concerning the improvement of electrocatalysts to optimize the catalytic activity, product selectivity, and durability, the underlying principles of eCO_2R through theoretical and experimental research investigations should be comprehended in detail. We should understand the eCO_2R system and its relation with the active sites and configurations of catalysts through detailed experimental and simulation (DFT, COMSOL) techniques to synthesize novel electrocatalysts. Likewise, to prevent the catalyst early degradation, many approaches can be utilized to analyze the degradation and failure systems. For instance, a variety of probing techniques (RRDE, CV, RDE, and EIS) and structure studying spectroscopies (XRD, HPLC, GC, NMR, SEM, and TEM) can be employed to recognize the characteristics of the catalyst in experimental analysis.

A more detailed understanding will help to develop novel mitigation pathways. The future of eCO_2R primarily bases on finding novel configurations of catalysts besides improving the electrochemical technologies, and yes, it is hard to design novel catalytic systems since the development of newly constructed catalysts must comply with particular electrochemical parameters such as faster electron transfer, appropriate intermediate stability, and potentials. Besides improving the catalyst stability, electrode and reactor designing also play a major role in the selective formation of valuable chemicals and fuels. The well-structured electrodes contributing to eCO_2R at suitable current density should be developed. For example, the recently designed GDEs and electrocatalytic cells having membrane-based electrolytes are the promising choice that reduces the internal resistance and facilitate mass transfer. Finally, efforts should be initiated to design stable and durable catalysts and improve the system design. The future research analysis may include discussions on improving the catalytic activity and stability by designing novel electrocatalysts and optimizing the systems well-suited for commercial eCO_2R. Other than electrocatalysts, the current density and product distribution in eCO_2R are greatly impacted by the structure and configuration of the electrochemical cell. The role of flow cells will be very crucial in analyzing the performance of electrocatalysts in multi-process mechanisms attributed to their high mass transfer efficiency. Albeit several electrocatalytic cells have demonstrated excellent properties for eCO_2R, it is worth mentioning that their present status is not up to the mark for commercial-scale eCO_2R. Recently, the design of electrolytic cells has attracted much attention. However, their applications for eCO_2R are still at the lab scale, and more practical research studies are a need of time to optimize its stability, efficiency, and volume so that large CO_2 amounts could be converted into energy-dense chemicals. For eCO_2R commercialization, several challenges should be addressed. The issue of limited mass transport efficiency in the aqueous media due to low CO_2 solubility can be solved by designing such flow cells that can enable the movement of reactants and products incessantly between the electrodes and the feedstock characteristics must be included while cell designing. The choice of suitable electrolyte membranes and the CO_2 stream phase also influence the eCO_2R. Furthermore, the CO_2 stream effect concerning the CO_2 concentration on the electrode surface, product separation, purity of inlet gas, anode improvement, and channel length requires more research studies to optimize the cell design for industrial-scale applications.

11.4 SOLAR-ASSISTED POWER GENERATION TECHNOLOGY FOR CO_2 EMISSION REDUCTION

11.4.1 Solar-Assisted Power Generation Concept

Nowadays, the increasing worldwide energy demands associated with the rising greenhouse gas effects and environmental pollution hazards are the main challenges and problems of power generation technology development. Besides, fossil fuel prices are increasing unceasingly while its reserves are decreasing. According to the statistical review of the world energy 2018, the reserves of oil and natural gas are a limited range of 40–60 years whereas the reserves of coal can last for more than 150 years [30]. The issues related to environmental pollution, limitless sources of fossil fuels, and continuously growing fuel costs and energy demands have recently led to the development of next-generation sustainable and clean technology for fossil fuels-based power generation. Therefore, finding innovative approaches for solving these issues, such as reducing O&M and production costs, and environmental impacts could be necessary for guaranteeing a reliable and clean energy future for sustainable development. Solar thermal power plants have been developed in different scales using CSP technologies. However, solar thermal power plants could not be competing against conventional thermal power plants, due to the higher production costs, lower operational efficiency, and limited capacity. On the other hand, to solve the solar energy the stand-alone CSP plants need to be integrated with the thermal energy storage (TES) system, as a result in a further increase in the investment cost [31]. To avoid the disadvantages associated with CSP plants and conventional thermal power plants, these two power plants need to be integrated [32,33]. However, this hybrid system would further increase the techno-economic indices and eliminate their disadvantages.

Many scientists and engineers related to power engineering have to find innovative approaches to increase the operating performance and ecological indices of the power plants by improving operational flexibility and reducing environmental impact. Solar-assisted power generation (SAPG) system is one of the potential solutions for improving operating performance and ecological indices in the short and midterm. As depicted in Figure 11.11, regarding power generation systems and their associated issues, the coal-fired thermal power plant alone can satisfy the increasing energy demand with limited fuel sources, high pollutant emissions, and average operation efficiency. However, clean energy generation with limited capacity, larger required land, and high investment cost accompanied by solar radiation fluctuation and low operation efficiency can be obtained by replacing the conventional system consisting of the coal-fired power plant with the solar thermal power plant. Concerning today's energy demand and associated impacts on the environment, the major scientific and engineering challenges related to the thermal power plant is designing appropriately a solar assisted power generation system that

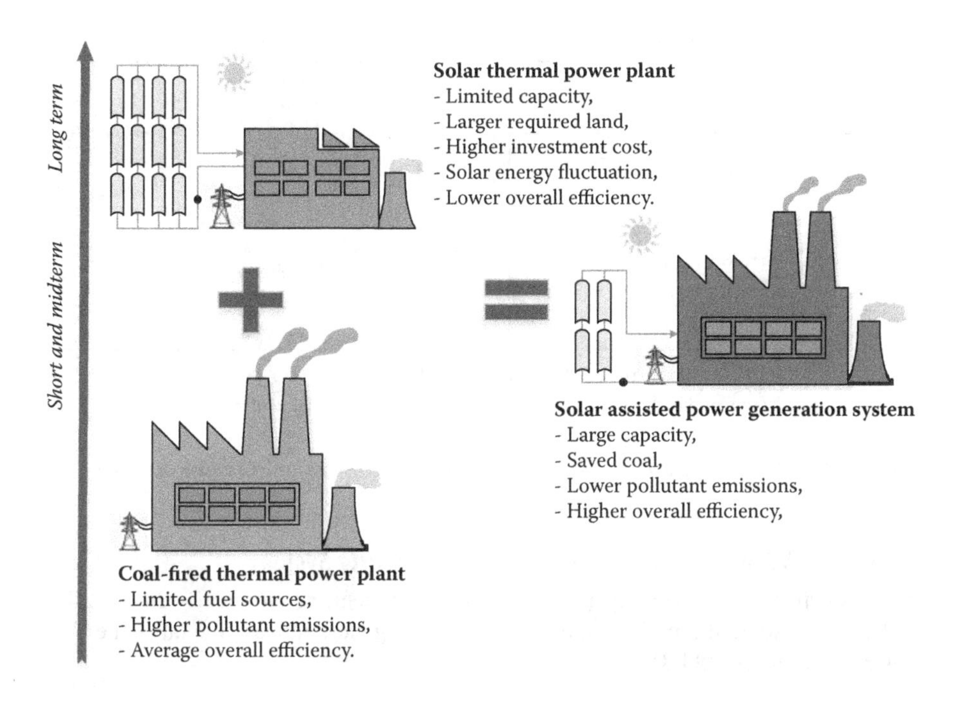

FIGURE 11.11 The concept of solar-assisted power generation system [34].

can provide short and midterm power generation with a large capacity, saving coal, lowering pollutant emission, and higher operational efficiency. Another advantage related to the SAPG system is that the operating performance of the SAPG system would be less affected by the solar radiation fluctuation compared to the solar thermal power plant. The SAPG system is suitable for countries with large solar and coal resources, such as Mongolia, the USA, Australia, China, and India.

In the SAPG system, there are two operation modes, such as fuel-saving (FS) and power-boosting (PB) at base and peak loads, respectively. Figure 11.12 clearly describes the FS and PB operation modes of SAPG. In FS operation mode, the SAPG system consumes less fuel and generates the same power output as the reference power plant. The benefit of the FS operation mode is better fuel consumption, which could indirectly reduce power generation costs and pollutant emissions. However, the FS mode could reduce the operating efficiency of a steam boiler as well. In PB operation mode, the SAPG system generates additional power output with the same fuel consumption as the reference power plant. The maximum permissible power output is 5–10% higher than the nominal load in a short time operation according to manufacturer standards when the electricity demand is at peak load. However, the maximum permissible power output is limited by the maximum capacity of the electric generator. Both operation modes show technical potential.

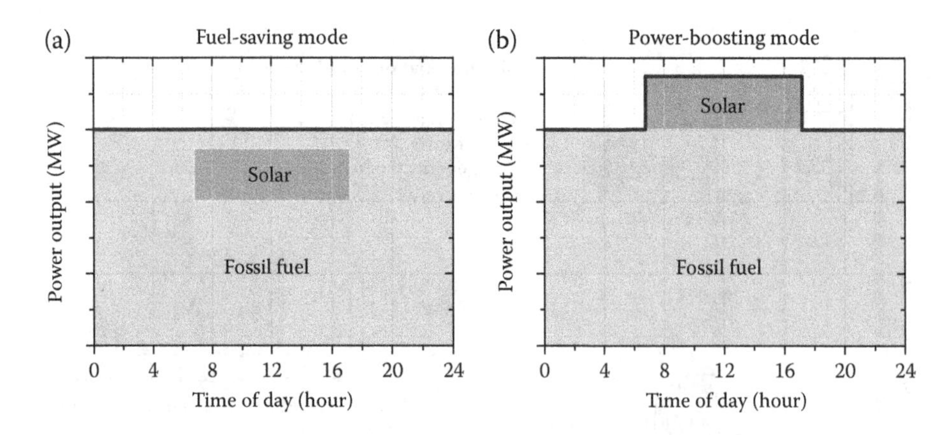

FIGURE 11.12 The operation modes of SAPG system: (a) fuel-saving, (b) power-boosting [34].

11.4.1.1 Advantages and Disadvantages of SAPG System

The concept of the SAPG system is the most efficient, economical, eco-friendly, and reliable solar thermal technology for power generation as it retains the following advantages [34,35]:

- The thermal efficiency of the SAPG system is higher than the conventional thermal power plant and CSP plant with the same capacity;
- The operating performance and ecological indices of the SAPG system are higher than the conventional thermal power plant with the same capacity;
- The SAPG system is regarded as the green power generation with minimum pollutants and higher efficiency achieved through solar thermal energy;
- The SAPG system could be applied to new power plants, as well as using to expand the existing thermal power plant with lower investment cost and higher economic and environmental benefits;
- The investment cost of a solar field system is balanced by the saved fuel, reduced pollutant emissions, and the increased output power;
- Various types of heat transfer fluids (HTF) can be used in solar field system;
- Different types of CSP technologies can be used in solar field system as a heat source;
- A wide range of temperature scale applications can be used in solar field systems.
- The SAPG system operation is more flexible as it could operate in two operation modes, i.e., fuel-saving and power-boosting at peak and base loads.

In addition, the SAPG system has several limitations and disadvantages such as reducing the efficiency of the steam boiler in the FS mode, increasing heat loads of non-replaced feedwater preheaters, steam reheater, and condenser in the PB mode. To avoid such limitations, the power output of the turbo-generator should be reduced for increasing the vacuum of the condenser.

11.4.1.2 Potential Selection of CSP Technology and Integration Mechanism in SAPG System

In this section, a brief introduction to CSP technologies will be discussed. As known, CSP technologies can be classified into four types, such as parabolic trough collector (PTC), linear Fresnel collector (LFC), parabolic dish (PD) system, and solar power tower (SPT) [36,37]. The PTC and LFC technologies are the most mature and field-proven of the CSP technologies for low and medium-temperature scale applications (below 400°C) in power plants. Moreover, the SPT technology is almost fully developed, which is more suitable for high-temperature scale applications [38,39]. Therefore, it is beneficial to select the CSP technology depending on the utilization purpose of solar energy in the thermal power plant.

The integration of solar energy into a coal-fired thermal power plant is an interesting topic of solar power generation, which has been studied in the last few decades. Depending on the utilization purpose, there are many possible integration mechanisms (methods) for combining solar energy and coal-fired thermal power plants, such as feedwater preheating, air preheating, steam superheating, steam reheating, lignite drying, CO_2 capturing, and flue gas cleaning, etc. According to the literature review [32,40], all integration mechanisms show technically and economically potential.

11.4.2 Computational Approaches for the Numerical Modeling

In recent years, numerous studies have conducted detailed investigations on the performance analysis and techno-economic evaluation of the SAPG system with various capacities under different operation conditions. Moreover, the numerical simulations of various types of SAPG systems have been carried out by the process simulation tools including IPSEpro, Cycle Tempo, Epsilon, Transys, Thermosolv, Thermoflex, SAM, Gate Cycle, Apros, Matlab, and Aspen Plus using the material and energy balance approach based on the thermodynamics' first and second laws [32,34].

The hierarchical structure of modeling for the SAPG system using PTC is illustrated as an example in Figure 11.13. The simulation model of the SAPG system consists of two main subsystems, including the model of the solar field and the model of the power unit. The thermodynamic equations of the model of the SAPG system are not presented in detail since there were clearly described in the previous studies [34,41].

The logic flow diagram of the numerical modeling of the SAPG system for analyzing the performance and techno-economic feasibility is depicted in Figure 11.14. The input parameters of the solar field model were the geographical location (latitude, longitude, and time zone), solar time (hour of the day, day of the year, DNI), collector characteristics (dimensions, length, and mirror properties), and climate data (ambient temperature, pressure, wind speed, and relative humidity).

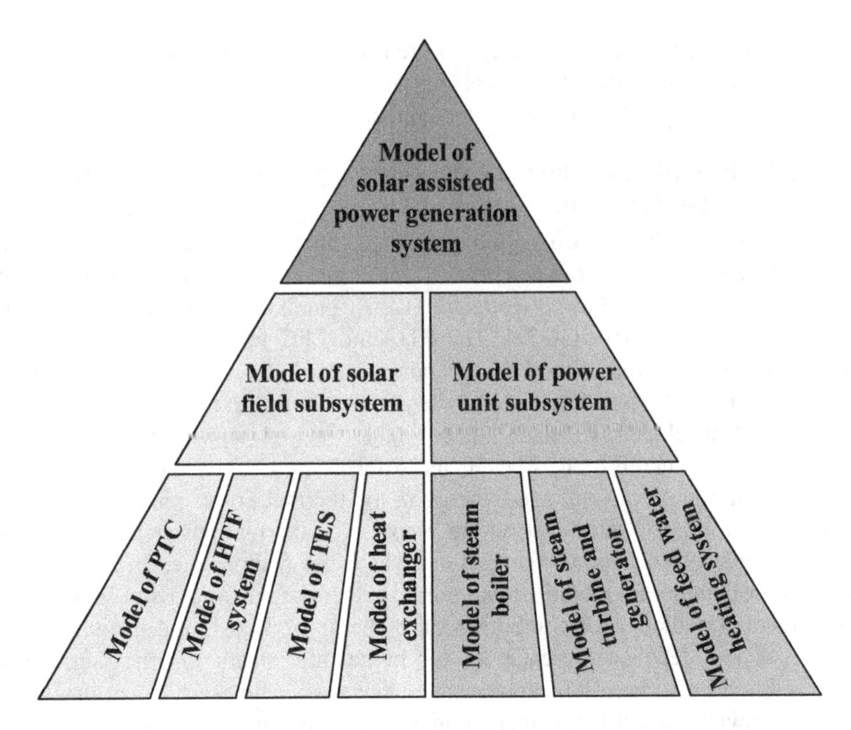

FIGURE 11.13 The hierarchical structure of modeling for the SAPG system.

These parameters allowed estimating the operating performance of the solar field. For the SCHPG system, the performance and techno-economic feasibility are influenced by the variations of the power output and solar radiation.

11.4.3 Coal Savings and CO_2 Emission Reduction in the SAPG System

Pollutant emissions, especially CO_2 generated from fossil fuel combustion is the key factor causing global problems such as greenhouse gas emissions, global warming, and air pollution. The CO_2 emissions derived from industrial processes and fossil fuel combustion have contributed to the largest position of total greenhouse gas emissions from 1970 to 2010, with a share of about 78% [42]. The SAPG system can significantly contribute to the reduction of pollutant emissions from power plants without an additional flue gas treatment system. Several potential examples of coal savings and CO_2 emission reduction related to the SAPG system are also introduced in the following section. Many studies have been conducted on the hybridization of solar energy and coal-fired power plant for improving the operating performance of the proposed system by reducing coal consumption and environmental impacts.

Xu et al. [43] have examined energy, exergy, and economic analyses of the 600 MW SAGP based on solar-assisted lignite drying. The results show that the proposed system can produce additional 29.6 MW electricity with an overall energy and exergy efficiencies improvement of 0.5 and 1.6 percentage points, respectively,

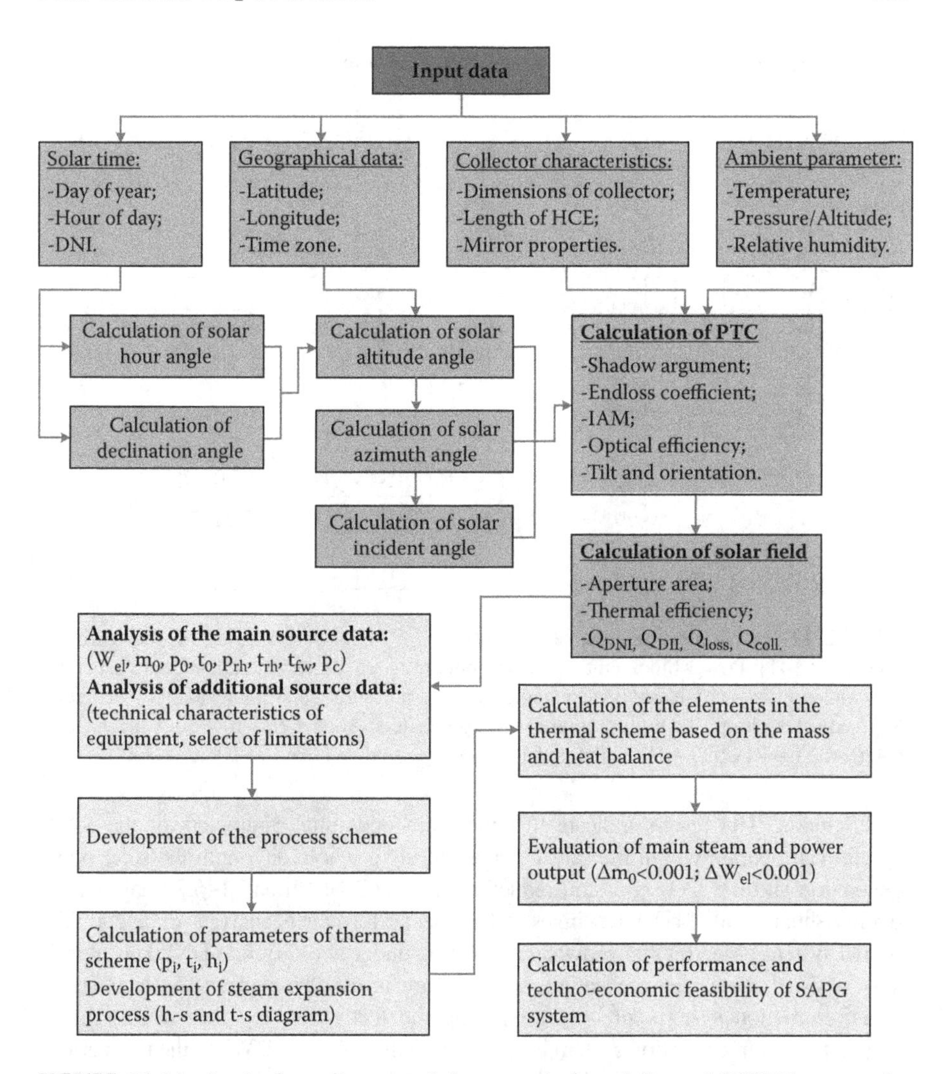

FIGURE 11.14 Logic flow diagram of the numerical modeling of SCHPG system for analyzing the performance and techno-economic feasibility [41].

as 57.1 MW solar thermal energy is introduced. Besides, after the drying process, the moisture content of the lignite reduces from 32.2% to 14.0%, with the lower heating value of the lignite increasing from 16.1 to 21.1 MJ/kg. Li et al. [44] have performed the thermodynamic analysis of the 300 MW SAPG system based on solar power-assisted steam reheating. Compared with the reference power plant, the specific equivalent coal consumption of the STACP system in PB and FS modes was reduced by 35.98 and 34.99 g/kWh, respectively, whereas the CO_2 emission rate was reduced by 101.79 and 98.99 g/kWh, respectively. Zhang et al. [45] have performed full-day dynamic characteristics analysis of the 330 MW SAPG system

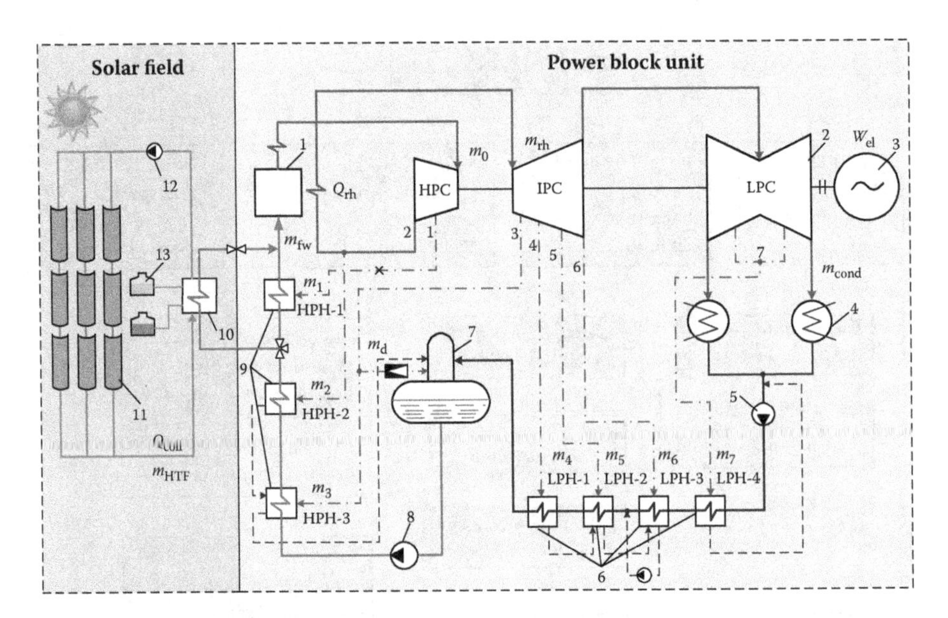

FIGURE 11.15 Process scheme of the 200 MW SAPG system based on the feedwater preheating [41]: Power block unit: 1—steam boiler; 2—steam turbine; 3—electric generator; 4—condenser; 5—condensate water pump; 6—low-pressure water preheaters; 7—deaerator; 8—feedwater pump; 9—high-pressure water preheaters; Solar field: 10—oil-water heat exchanger; 11—PTC; 12—HTF pump; 13—TES.

in FS mode. The results indicate that the main operating parameters of the SAPG system can operate within the safety ranges during a whole day, both the solar power generation and the CO_2 emission reduction are 207.7 MWh and 186.7 t/day, respectively. Shuai et al. [41] have investigated the performance analysis of a 200 MW SAPG system based on the feedwater preheating under two different DNI days. For the solar-assisted feedwater preheating, solar energy is used to replace the regenerative steam extraction system for only preheating the feedwater. In other words, the regenerative steam extraction system is replaced by the solar field. When the regenerative steam extraction system is replaced by the solar field, the major percent of replaced extraction steam is saved. Therefore, coal consumption and pollutant emissions are directly reduced. The process scheme of the 200 MW SAPG system is depicted in Figure 11.15. In the design condition, a decline in fuel consumption of 4.15 tons per hour and an increase in power output of 7.4 MW per hour are caused by FS and PB operation modes. Moreover, specific equivalent coal consumption was reduced by 2.91% compared to the reference power plant.

Figure 11.16 illustrates the power output and coal consumption in two operating modes under two different DNI days. The power output in PB mode increases above the nominal load whereas, in FS mode, the power output is constant when the solar field is operating. The coal consumption in PB mode is slightly higher compared to the reference power plant. On the 10th of January and July, power production is increased by 12.6 MW and 75.9 MW during the operation period of the solar field, respectively. Coal consumption in FS mode decreases from the

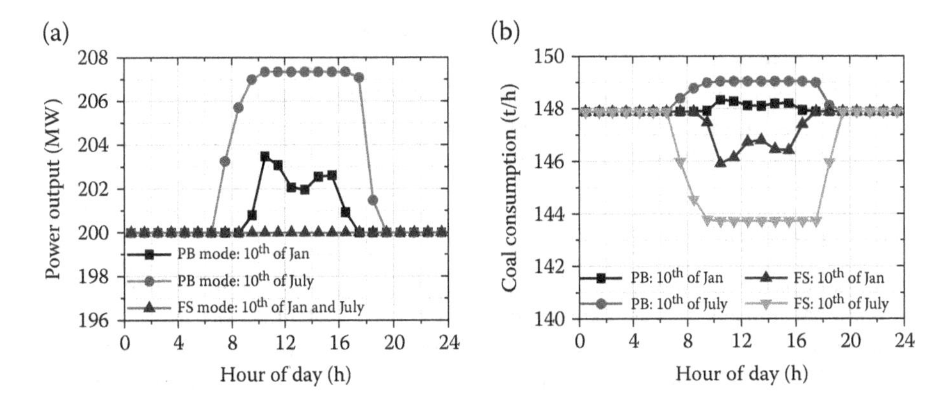

FIGURE 11.16 Effect of DNI on the (a) power output and (b) coal consumption of the SAPG system [41].

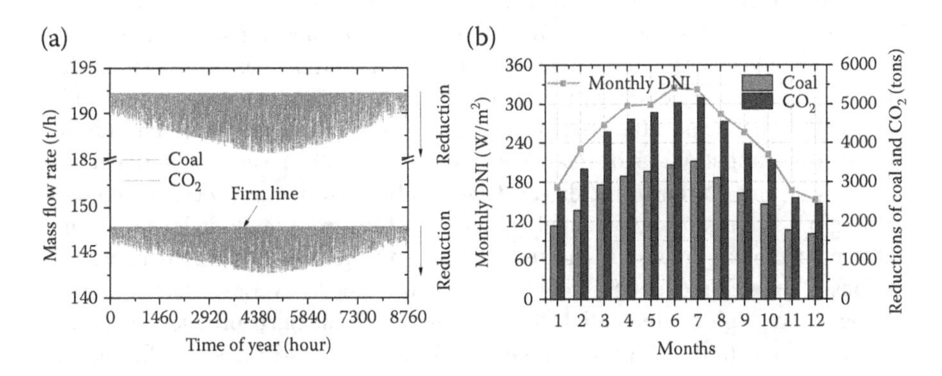

FIGURE 11.17 (a) The annual reductions of coal consumption and CO_2 emission [41]; (b) the annual reductions of coal and CO_2 emission of the 300 MW SAPG system.

nominal load when the solar field is operating whereas, in PB mode, coal consumption is slightly higher than the design value. On the 10th of January and July, coal consumption is saved by 7.5 tons and 44.5 tons during the working period of the solar field, respectively.

The main benefit of the SAPG system is the CO_2 emission reduction without any flue gas treatment system. The annual reductions of coal consumption and CO_2 emission in a typical year are illustrated in Figure 11.17(a). In the FS mode, the annual reductions of coal consumption and CO_2 emission were estimated at 9,910 and 11,000 tons, respectively, which is a greater contribution to environmental protection. Shagdar et al. [34] have performed the performance analysis and techno-economic evaluation of the 300 MW SAPG system based on the solar-assisted feedwater preheating in the various operating conditions. Figure 11.17(b) describes the annual reductions of coal and CO_2 emission of the 300 MW SAPG system in FS operation mode by monthly summary. The annual reductions of coal and CO_2 emission are calculated based on the nominal load. When the power output is at part-load, the annual reductions of coal and CO_2 emission require higher than the nominal load. SAPG system with FS mode can largely contribute to high ecological performance for power generation. The annual reductions of coal and CO_2 emissions are approximately 32,150 tons and 47,000 tons when the SAPG system operates at nominal load.

11.5 SUMMARY

From the sustainable development perspective, the integration of solar energy into conventional thermal power plants is one of the most potential solutions to clean power generation due to the higher techno-economic performance and lower pollutant emissions. Considering the hourly, daily, monthly, and yearly amount of saved coals and reduced pollutant emissions in the SAPG system, FS mode can significantly contribute to environmental protection. In addition to the CO_2 emission reduction, more sophisticated methods and technics continued to predict the effectiveness of CO_2 conversion into energized products H_2, C1 fuels such as CO, CH_4, HCOOH, MeOH, and syngas that can be converted into C2+ fuels and chemicals. Although fuels-based CO_2 presenting an exciting and attractive perspective for a sustainable energy future and in light of the current performances of the thermochemical and electrochemical mechanistic pathways, this major field is affected by the final achievable efficiencies remaining insufficient. The scientific and technological limitations and challenges associated with a single coordinate of sunlight heat and current force-driven CO_2 catalytic conversion were addressed, revealing the success in such technology require effort from multiple levels. While the thermochemical process is mostly affecting by the redox thermodynamic restrictions, the electrochemical process is influenced by various technological challenges like high overpotential, the generation of entangled products, material instability of electrodes, high consumption of energy, small Faradic efficiencies, kinetic barriers, and low CO_2 solubility in aqueous solution. Generally, the biggest challenge in both technologies is the poor performance of thermal and electrocatalysts because of insufficient stability of catalysts, poor catalyst activity, and product selectivity.

ACKNOWLEDGMENTS

This work was supported by the China National Key Research and Development Plan Project (2018YFA0702300), National Natural Science Foundation of China (No. 51950410590 and No. 51876049).

REFERENCES

[1] A. Demont and S. Abanades, Solar thermochemical conversion of CO2 into fuel via two-step redox cycling of non-stoichiometric Mn-containing perovskite oxides, *Journal of Materials Chemistry A*, 2015, 3, 3536–3546.

[2] Y. G. Dessie, Q. Hong, B. G. Lougou, J. Zhang, B. Jiang, J. Anees and E. B. Tegegne, Thermochemical Energy Storage Performance Analysis of (Fe, Co, Mn) Ox Mixed Metal Oxides, *Catalysts*, 2021, 11.

[3] A. Haeussler, S. Abanades, A. Julbe, J. Jouannaux and B. Cartoixa, Two-step CO2 and H2O splitting using perovskite-coated ceria foam for enhanced green fuel production in a porous volumetric solar reactor, *Journal of CO_2 Utilization*, 2020, 41.

[4] L. Wang, T. Ma, S. Dai, T. Ren, Z. Chang, L. Dou, M. Fu and X. Li, Experimental study on the high performance of Zr doped LaCoO3 for solar thermochemical CO production, *Chemical Engineering Journal*, 2020, 389.

[5] G. Takalkar and R. R. Bhosale, Solar thermocatalytic conversion of CO2 using PrxSr (1 − x) MnO3− δ perovskites, *Fuel*, 2019, 254.

[6] B. Bulfin, J. Vieten, C. Agrafiotis, M. Roeb and C. Sattler, Applications and limitations of two step metal oxide thermochemical redox cycles; a review, *Journal of Materials Chemistry A*, 2017, 5, 18951–18966.

[7] S. Zhai, J. Rojas, N. Ahlborg, K. Lim, C. H. M. Cheng, C. Xie, M. F. Toney, I.-H. Jung, W. C. Chueh and A. Majumdar, High-capacity thermochemical CO2 dissociation using iron-poor ferrites, *Energy & Environmental Science*, 2020, 13, 592–600.

[8] Y. Fu, J. Zhang, S. Li, J. Huang and Y. Sun, Self-regeneration of ferrites incorporated into matched matrices for thermochemical CO2 splitting, *Journal of Materials Chemistry A*, 2016, 4, 5026–5031.

[9] R. R. Bhosale, A. Kumar, F. AlMomani and I. Alxneit, Sol–gel derived CeO2–Fe2O3 nanoparticles: Synthesis, characterization and solar thermochemical application, *Ceramics International*, 2016, 42, 6728–6737.

[10] Z. Ma, S. Zhang, R. Xiao and J. Wang, Inhibited Phase Segregation to Enhance the Redox Performance of NiFe2O4 via CeO2 Modification in the Chemical Looping Process, *Energy & Fuels*, 2020, 34, 6178–6185.

[11] E. Sediva, A. J. Carrillo, C. E. Halloran and J. L. M. Rupp, Evaluating the Redox Behavior of Doped Ceria for Thermochemical CO2 Splitting Using Time-Resolved Raman Spectroscopy, *ACS Applied Energy Materials*, 2021, 4, 1474–1483.

[12] B. Guene Lougou, Y. Shuai, H. Zhang, C. Ahouannou, J. Zhao, B. B. Kounouhewa and H. Tan, Thermochemical CO2 reduction over NiFe2O4@ alumina filled reactor heated by high-flux solar simulator, *Energy*, 2020, 197.

[13] M. Tou, R. Michalsky and A. Steinfeld, Solar-driven thermochemical splitting of CO2 and in situ separation of CO and O2 across a ceria redox membrane reactor, *Joule*, 2017, 1, 146–154.

[14] Y. Shuai, H. Zhang, B. Guene Lougou, B. Jiang, A. Mustafa, C.-H. Wang, F. Wang and J. Zhao, Solar-driven thermochemical redox cycles of ZrO2 supported NiFe2O4 for CO2 reduction into chemical energy, *Energy*, 2021, 223.

[15] Y. Shuai, B. Guene Lougou, H. Zhang, J. Zhao, C. Ahouannou and H. Tan, Heat transfer analysis of solar-driven high-temperature thermochemical reactor using NiFe-Aluminate RPCs, *International Journal of Hydrogen Energy*, 2021, 46, 10104–10118.

[16] R. Snoeckx and A. Bogaerts, Plasma technology–a novel solution for CO2 conversion?, *Chemical Society Reviews*, 2017, 46, 5805–5863.

[17] C. Ruan, Z.-Q. Huang, J. Lin, L. Li, X. Liu, M. Tian, C. Huang, C.-R. Chang, J. Li and X. Wang, Synergy of the catalytic activation on Ni and the CeO2–TiO2/Ce2Ti2O7 stoichiometric redox cycle for dramatically enhanced solar fuel production, *Energy & Environmental Science*, 2019, 12, 767–779.

[18] M. Takacs, S. Ackermann, A. Bonk, M. Neises-Von Puttkamer, P. Haueter, J. R. Scheffe, U. F. Vogt and A. Steinfeld, Splitting CO2 with a ceria-based redox cycle in a solar-driven thermogravimetric analyzer, *AIChE J*, 2017, 63, 1263–1271.

[19] J. Xie, Y. Huang, M. Wu and Y. Wang, Cover feature: electrochemical carbon dioxide splitting, *ChemElectroChem*, 2019, 6, 1585.

[20] M. Aulice Scibioh and B. Viswanathan, Carbon dioxide to chemicals and fuels, *Elsevier*, 2018, 307–371.

[21] W. Zhu, R. Michalsky, Ö. Metin, H. Lv, S. Guo, C. J. Wright, X. Sun, A. A. Peterson and S. Sun, Monodisperse Au nanoparticles for selective electrocatalytic reduction of CO2 to CO, *Journal of the American Chemical Society*, 2013, 135, 16833–16836.

[22] B. Innocent, D. Pasquier, F. Ropital, F. Hahn, J.-M. Léger and K. B. Kokoh, FTIR spectroscopy study of the reduction of carbon dioxide on lead electrode in aqueous medium, *Applied Catalysis B: Environmental*, 2010, 94, 219–224.

[23] V. C. Hoang, V. G. Gomes and N. Kornienko, Metal-based nanomaterials for efficient CO2 electroreduction: Recent advances in mechanism, material design and selectivity, *Nano Energy*, 2020, 78, 105311.

[24] J. Rosen, G. S. Hutchings, Q. Lu, S. Rivera, Y. Zhou, D. G. Vlachos and F. Jiao, Mechanistic insights into the electrochemical reduction of CO2 to CO on nanostructured Ag surfaces, *ACS Catalysis*, 2015, 5, 4293–4299.

[25] Q. Lu, J. Rosen and F. Jiao, Nanostructured metallic electrocatalysts for carbon dioxide reduction, *ChemCatChem*, 2015, 7, 38–47.

[26] S. Liang, N. Altaf, L. Huang, Y. Gao and Q. Wang, Electrolytic cell design for electrochemical CO2 reduction, *Journal of CO2 Utilization*, 2019, 0–1.

[27] A. Mustafa, B. G. Lougou, Y. Shuai, Z. Wang, S. Razzaq, J. Zhao and H. Tan, Theoretical insights into the factors affecting the electrochemical reduction of CO2, *Sustainable Energy & Fuels*, 2020, 4, 4352–4369.

[28] J. M. Spurgeon and B. Kumar, A comparative technoeconomic analysis of pathways for commercial electrochemical CO2 reduction to liquid products, *Energy & Environmental Science*, 2018, 11, 1536–1551.

[29] M. Rumayor, A. Dominguez-Ramos, P. Perez and A. Irabien, A techno-economic evaluation approach to the electrochemical reduction of CO2 for formic acid manufacture, *Journal of CO2 Utilization*, 2019, 34, 490–499.

[30] BP, Statistical review of world energy, *Sustainability*, 2018, 10(3195), 17, https://www. bp. com/en/global/corporate/energy-economics/statistical-review-of-world-energy. html. (accessed on 4 September 2018).

[31] M. J. Blanco and L. R. Santigosa, Advances in concentrating solar thermal research and technology, *Joe Hayton*, 2017.

[32] J. Qin, E. Hu and X. Li, Solar aided power generation: A review, *Energy and Built Environment*, 2020, 1, 11–26.

[33] S. Mills, Combining solar power with coal-fired power plants, or cofiring natural gas, *Clean Energy* 2018, 2, 1–9.

[34] E. Shagdar, B. Guene Lougou, Y. Shuai, J. Anees, C. Damdinsuren and H. Tan, Performance analysis and techno-economic evaluation of 300 MW solar-assisted power generation system in the whole operation conditions, *Applied Energy,* 2020, 264.

[35] E. Hu, Y. Yang, A. Nishimura, F. Yilmaz and A. Kouzani, Solar thermal aided power generation, *Applied Energy* 2010, 87, 2881–2885.

[36] E. Shagdar, Y. Shuai, B. Guene Lougou, E. Ganbold, O. P. Chinonso and H. Tan, Analysis of heat flow diagram of small-scale power generation system: Innovative approaches for improving techno-economic and ecological indices, *Science China Technological Sciences,* 2020, 63, 1–19.

[37] A. Bilal Awan, M. N. Khan, M. Zubair and E. Bellos. Commercial parabolic trough CSP plants: Research trends and technological advancements, *Solar Energy,* 2020, 211, 1422–1458.
[38] E. Shagdar, Y. Shuai, B. Guene Lougou, T. O. Jagvanjav, F. Wang and H. Tan. Comparative performance assessment of 300 MW solar-coal hybrid power generation system under different integration mechanisms, *Energy Technology*, 2021, 9, 1–15.
[39] E. Shagdar, B. Guene Lougou, Y. Shuai, E. Ganbold, O. P. Chinonso and H. Tan. Process analysis of solar steam reforming of methane for producing low-carbon hydrogen, *RSC Advances*, 2020, 10, 12582–12597.
[40] M. Chitakure, W. R. Ruziwa and D. Musademba. Optimization of hybridization configurations for concentrating solar power systems and coal-fired power plants: A review, *Renewable Energy Focus*, 2020, 35, 41–55.
[41] Y. Shuai, E. Shagdar, B. Guene Lougou, A. Mustafa, B. Doljinsuren and D. Han. Performance analysis of 200 MW solar coal hybrid power generation system for transitioning to a low carbon energy future, *Applied Thermal Engineering*, 2020, 183, 116140.
[42] R. K. Pachauri and L. Meyer. *Climate Change 2014: synthesis report. Contribution of Working Groups I, II and III to the fifth assessment report of the Intergovernmental Panel on Climate Change* (p. 151). Ipcc. Geneva, Switzerland. 2014.
[43] C. Xu, X. Li, G. Xu, T. Xin, Y. Yang and W. Liu, Energy, exergy and economic analyses of a novel solar-lignite hybrid power generation process using lignite pre-drying, *Energy Conversion and Management,* 2018, 170, 19–33.
[44] C. Li, R. Zhai, B. Zhang and W. Chen, Thermodynamic performance of a novel solar tower aided coal-fired power system, *Applied Thermal Engineering*, 2020, 171, 115127.
[45] N. Zhang, G. Yu, C. Huang, L. Duan, H. Hou and E. Hu, Full-day dynamic characteristics analysis of a solar aided coal-fired power plant in fuel saving mode, *Energy*, 2020, 208, 118424.

12 CO_2 Electrochemical Reduction to CO: From Catalysts, Electrodes to Electrolytic Cells and Effect of Operating Conditions

Qiqi Wan and Changchun Ke
Institute of Fuel Cells, School of Mechanical Engineering, Shanghai Jiao Tong University, Shanghai, People's Republic of China

CONTENTS

DOI: 10.1201/9781003133971-12

12.1 INTRODUCTION

Since the Industrial Revolution, the excessive utilization and continuous consumption of fossil fuels, these limited unrenewable resources cannot keep up with the increasing demand [1], leading to the energy crisis and a series of environmental problems, such as air pollution, water pollution and land pollution [2,3]. Besides, climate change is considered to be one of the most serious environmental problems due to substantial CO_2 emissions from fossil fuels [4]. As reported, the CO_2 level in the atmosphere has reached 415 ppm in 2019, well above the safe limit of 350 ppm, and is still rising rapidly [5]. To mitigate the effects of global climate change, more than 170 nations signed the Paris Agreement, committed to significant reductions in greenhouse gas (GHG) emissions over the next several decades [6]. Consequently, there is an urgent need to attenuate CO_2 emissions in the atmosphere [7].

Presently, the technologies that utilize CO_2 mainly include thermochemical, biological, electrochemical, photoelectric and photochemical conversion. Among them, CO_2 electrochemical reduction is one of the most promising strategies due to its mild operation conditions, high conversion efficiency and pollution-free. The reaction path of CO_2 electrochemical reduction reaction (CO_2ERR) can be controlled by adjusting the reaction conditions (e.g., applied potential, temperature, catalysts, pressure, electrolyte), converting CO_2 into value-added chemicals selectivity, including carbon monoxide [8–13], formic acid [14], methane [15], ethylene [16,17], methanol [17,18], and multi-carbon alcohols [19]. The process has attracted much attention for its several advantages: (1) it allows CO_2 to be recycled to reduce CO_2 emission in the atmosphere; (2) CO_2ERR provides a possible way to level the output from strong but intermittent renewable electricity sources; (3) it supplies carbon-neutral fuels acted as storable energy carriers.

However, the electrolysis of CO_2 remains obvious challenges, such as unwanted competing reactions (commonly the hydrogen evolution), causing the extra consumption of energy; the slow CO_2 reduction kinetics, as CO_2 is a completely oxidized molecule and thermodynamically stable; the low solubility of CO_2 in water, which limits the mass transfer and results in a low current density; insufficient stability of electro-catalysts, leading to a short lifetime.

On this account, we introduce the general principles of CO_2ERR, and CO as a versatile feedstock for various fuels and chemicals and fuels which has a promising value and market size, is selected to be the key object, discussing how the electrocatalysts, electrodes, electrolytic cells, operating conditions affect the performance of CO_2ERR, as well as various degradation mechanisms.

12.2 PRINCIPLE OF CO_2ERR

CO_2 is inactive under atmospheric temperature and pressure conditions, which requires relatively high extra energy to drive a reduction reaction to get carbon-based fuels and chemicals. As shown in Figure 12.1, CO_2 electrolytic cell is a device that utilizes electricity to convert CO_2 into carbon-based products, where two half-cell reactions happen simultaneously. CO_2 molecules on the cathode combine with the electrons provided by an external power supply to form various possible products under the action of electrocatalysts, meanwhile, the water molecules (or OH^-) are converted to oxygen molecules and electrons occurring on the anode. In this way, charge balance is achieved by the transfer of protons and electrons. Generally, the process can be divided into three basic steps: (1) chemical adsorption of CO_2 molecules to the surface of the catalyst, (2) multiple proton-coupled electron transfer reactions, and (3) the desorption of products from electrocatalyst surface. The potentials corresponding to possible cathodic half-cell

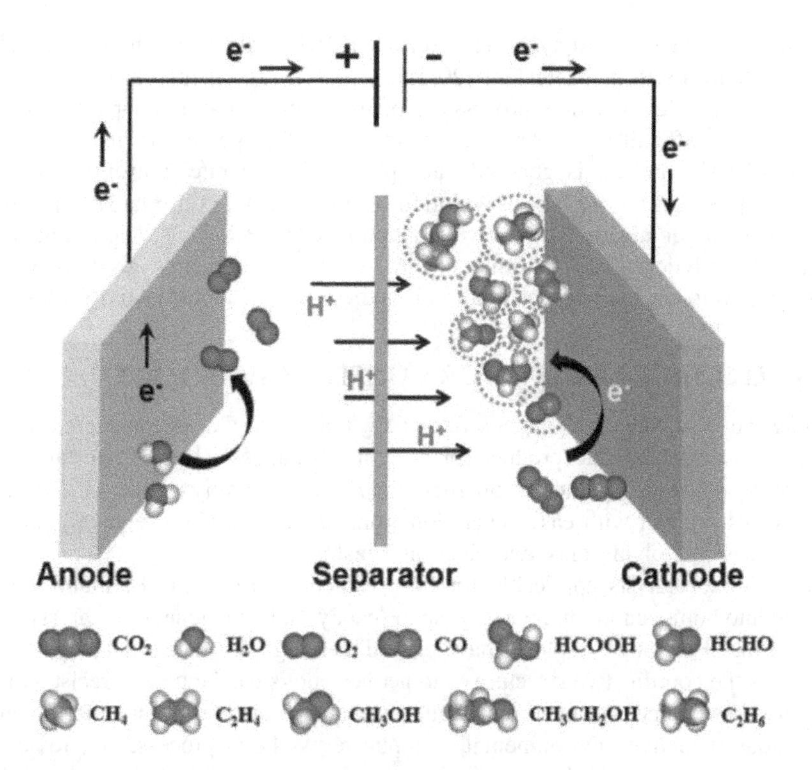

FIGURE 12.1 An illustration of electrocatalysts for CO_2 reduction reaction and possible products created in an electrochemical cell from [22].

TABLE 12.1
Electrochemical potentials of possible cathodic half-cell reactions in aqueous solutions

Possible cathodic half-cell reaction	Electrode potentials (V vs SHE)
$2H^+ + 2e^- \rightarrow H_2(g)$	−0.41
$CO_2(g) + e^- \rightarrow {}_*COO^-$	−1.9
$CO_2(g) + 2H^+ + 2e^- \rightarrow CO(g) + H_2O(l)$	−0.53
$CO_2(g) + 2H_2O(l) + 2e^- \rightarrow CO(g) + 2OH^-$	−0.52
$CO_2(g) + 2H^+ + 2e^- \rightarrow HCOOH(l)$	−0.61
$CO_2(g) + H_2O(l) + 2e^- \rightarrow HCOO^-(aq) + OH^-$	−0.43
$CO_2(g) + 4H^+ + 4e^- \rightarrow HCHO(l) + H_2O(l)$	−0.48
$CO_2(g) + 3H_2O(l) + 4e^- \rightarrow HCHO(l) + 4OH^-$	−0.89
$CO_2(g) + 6H^+ + 6e^- \rightarrow CH_3OH(l) + H_2O(l)$	−0.38
$CO_2(g) + 5H_2O(l) + 6e^- \rightarrow CH_3OH(l) + 6OH^-$	−0.81
$CO_2(g) + 8H^+ + 8e^- \rightarrow CH_4(g) + 2H_2O(l)$	−0.24
$CO_2(g) + 6H_2O(l) + 8e^- \rightarrow CH_4(g) + 8OH^-$	−0.25

reactions versus a standard hydrogen electrode (SHE) in aqueous solutions (at pH 7 and 25 °C) are listed in Table 12.1 [20,21].

As the complex reaction process involves multiple proton-coupled electron transfer, it is difficult to improve the selectivity of the specific product. The formation of $CO_2^{\cdot-}$ radical is generally accepted to be the rate-controlling step of CO_2ERR [23] and the hydrogen evolution reaction (HER) at a relatively positive potential is the dominant competitive reaction of CO_2ERR. Thus, appropriate catalysts, electrolytic cells, and other electro-reduction parameters need to be investigated to lower the overpotential and increase the product selectivity [24].

12.3 ELECTROCATALYSTS FOR CO_2 REDUCTION TO CO

Among the value-added chemicals from CO_2 reduction, CO is considered as the most promising candidate product for several advantages: (1) an important constituent of syngas to synthesize oil fuels via a Fischer-Tropsch reaction [25–27]; (2) gaseous product with easy separation from liquid; and (3) a relatively simple reaction path, involving only two-electron transfer.

The electrocatalysts applicable for CO_2 reduction to CO can be mainly categorized into homogenous and heterogeneous catalysts. Homogeneous catalysts such as some metal-organic complexes have proved their high Faradaic efficiency for CO (FEco) as the coordinative structures and active centers can be tuned precisely. But they also have some unwanted disadvantages, such as insufficient stability under a continuous reductive environmental, complicate synthesis process, and toxic effects. Heterogeneous metal electrocatalysts, such as metals, metal alloys, have also shown similar high FEco and reactive current density to homogenous catalysts with low toxicity [28]. Additionally, metal-free catalysts, such as some carbon materials, as another emergent alternative for CO_2ERR, have been developed in recent years. The development of these representative electrocatalysts for CO_2 reduction to CO will be introduced in the following sections.

12.3.1 Metal–Organic Complexes

The major classes of metal-organic complexes explored for CO_2 to CO conversion to date include metal centers with macrocyclic ligands, bipyridine ligands, and phosphine ligands.

In the 1970s, Meshitsuka et al. first applied metal phthalocyanines (M-Pc) in CO_2ERR and compared the activity of transition metal atoms (Co, Ni, Mn, and Fe) [29]. Since then, numerous related researches using metal–macrocyclic complexes have come forth. Shaoxuan et al. achieved CO_2ERR using cobalt phthalocyanine at current densities $\geq$ 150 mA cm^{-2} with FEco > 95% in a zero-gap membrane flow reactor [30]. Jaecheol et al. immobilized cobalt phthalocyanine on chemically converted graphene via π–π stacking to enhance the catalytic activity [31]. Zheng et al. elucidated the reaction mechanisms of metal–N_4 sites using metal phthalocyanines, combining density-functional theory calculations and experimental studies [32]. They revealed that the Co center with moderate *CO binding energy has the optimum activity for CO_2 reduction to CO and delivers a maximal FEco reaching 99%. Except for phthalocyanine ligands, metal–macrocyclic complexes composed of transition metal atom and porphyrin ligands are also a kind of typical electrocatalysts exhibiting high electrocatalytic activity. Maryam Abdinejad et al. designed synthesis of Co- and Fe-porphyrins with variable amino groups and monosubstituted amino-porphyrins, which was observed to have great efficiency to improve turnover number (TON) and turnover frequency (TOF) [33].

Bipyridine complex based on non-noble abundant and inexpensive metals were also considered as promising molecular electrocatalysts for electrochemical conversion of CO_2 to CO. Jiaju et al. reported a composite catalyst synthesized by anchoring Au nanoparticles on Cu nanowires via 4,4'-bipyridine (bipy) to achieve highly efficient electrochemical CO_2 reduction reaction, originating from the synergistic effects among Au (for CO_2 to CO) [34]. Marc Bourrez et al. first demonstrated that [Mn(bipyridyl)$(CO)_3$Br] complexes exhibit an excellent selectivity, stability, and efficiency for reducing CO_2 to CO, and the process occurs at lower overpotential [35].

Additionally, the introduction of a phosphine donor to a labile ligand is another powerful strategy to obtain active catalysts for CO_2 reduction. Donovan et al. developed two new Zn(II) complexes with phosphine groups and facilitated the electrochemical conversion of CO_2 to CO [36]. Lee et al. reported a Ru complex with a mixed phosphine–pyridine ligand, which provides a novel strategy to reduce the overpotential for reducing CO_2 to CO [37].

As metal-organic complexes with clear structure–property relationships have shown their unique activities towards CO_2 to CO conversion, more exploration of metal complex catalysts is deserved to expect.

12.3.2 Metals

Metal electrocatalysts, such as Ag, Au, and Zn, are typically electrocatalysts used for reducing CO_2 to CO, as their weak proton and CO binding strengths on the surface prevent further CO reduction, leading to a high selectivity for CO [38,39]. Microstructure of Ag catalysts can have a significant impact on CO2ERR, such as enlarged electrochemical surface area, and more exposed active facets and defect-rich surface. Liu et al. [40] prepared triangular silver nanoplates (Tri-Ag-NPs) in 0.1 M $KHCO_3$, exhibiting an increasing current density and higher Faradaic efficiency for CO (96.8%) and energy efficiency (61.7%). Li et al. prepared Ag nanowires anchored on carbon support to expose more active sites on the high specific surface area and showed a remarkably high selectivity (approximately 100%) for CO [41]. Zhijie et al. employed hydrophobically modified Ag_3 nanoclusters in aqueous solutions, facilitating the CO_2 electroreduction and suppressing H_2O reduction [42].

Except for Ag, Au is another typical catalyst used for the conversion of CO_2 to CO. Verma et al. [12] reported a supported gold catalyst in an alkaline flow electrolyzer and obtained an onset cell potential of just −1.50 V, a low-onset cathode potential of just −0.04 V *vs.* RHE, and the partial current density of CO as high as 99 and 158 mA cm^{-2} at corresponding energetic efficiencies for CO of 63.8 and 49.4%, close to the economic viability criteria. Shi et al. systematically modified the wettability of Au/C gas diffusion electrode to adjust the structure of gas-liquid-solid interfaces and revealed that the Cassie-Wenzel coexistence state is the ideal three-phase structure that favors the transfer of CO_2 from the gas phase to Au active sites at high current densities [43].

In addition, Binhai et al. [44] prepared a series of Zn catalysts with different crystal facet ratios of Zn(002) to Zn(101) to investigate the relationship between crystal facet and CO selectivity and revealed the superior catalysis capability of Zn (101) facets in CO_2ERR to CO. And earth-abundant Ni single-atom catalysts have also been found highly selective to CO, yielding a current density above 100 mA cm^{-2}, with nearly 100% selectivity for CO and around 1% toward the hydrogen evolution side reaction [13].

12.3.3 Metal Alloys

Metal alloys can enhance the electrocatalytic reactivity and selectivity of CO_2 to CO over each single metal catalyst due to the significant cooperative effect of both metal centers.

Ara Jo et al. reported bimetallic Ag-Zn catalysts formed on polypyrrole-coated electrodes and enhanced the CO selectivity and production rate over each single metal catalyst [45]. Kun et al. gave insights into the intermediate binding and geometric effect of Cu/Au core-shell nanoparticles in CO_2ERR [46]. The Au-coated Cu nanowires achieved a larger current density and produced syngas ($CO+H_2$) with high efficiency and stability. They ascribed the higher current density to the unique surface morphology with a larger surface area of the catalysts.

Also, non-noble bimetallic catalysts are used to efficiently and selectively reduce CO_2 to CO over a wide potential range. Cu–Sn bimetallic catalyst has shown a

FEco greater than 90% and a current density of -1.0 mA cm^{-2} at -0.6 V vs RHE in aqueous solutions [47], while the pristine monometallic surfaces (both Cu and Sn) fail to selectively convert CO_2 into CO. Because the bimetallic electrocatalysts generate a surface that inhibits adsorbed *H and results in a high FEco. Weiwei et al. reported Cu–Co bimetallic nanoparticles dispersed on porous carbon and reached FEco of 97.4% with a current density of 62.1 mA cm^{-2}. The incorporation of Co into Cu led to a higher electrochemical surface area with more space for mass diffusion and stronger CO_2 adsorption [48]. Besides, Pavel et al. developed an optimized $Zn_{94}Cu_6$ foam alloy and achieved FEco of 90% at -0.95 V vs RHE [49]. The related researches provide the potential to utilize non-noble metals to efficiently reduce CO_2 to CO.

12.3.4 Inorganic Metal Compounds

Metal oxides, chalcogenides, and carbides are another class of promising catalysts applied in CO2ERR. Constructing the metal-oxide interface is a method to enhance the conversion to CO and the existing paper showed a CO geometric current density 1.6 times of that over Au/C at -0.89 V vs. RHE, resulting in a CO Faradaic efficiency of 89.1%, indicating that Au-CeOx had much higher activity and Faradaic efficiency than Au or CeOx alone for CO_2ERR [50]. The introduction of oxygen vacancies into catalysts is also a method to enhance the activation of CO_2, which increases the charge density around the valence band maximum. ZnO nanosheets with oxygen vacancies has exhibited a current density of -16.1 mA cm^{-2} with FEco of 83 % at -1.1 V versus RHE [51]. Also, Ag_2O/layered zeolitic imidazolate framework (ZIF) composite structure has been constructed and reached 26.2 mA cm^{-2} with FEco of 80.5% at -1.2 V vs RHE [52], which is probably attributed to the synergistic effect between Ag2O NPs and the layered ZIF, as well as facilitated mass transport due to the high specific surface area of the Ag_2O/layered ZIF.

As reported, heterogeneous Ag_2S/Ag stabilized the *COOH intermediate during the CO_2ERR process and achieved a large current density of 421.7 ± 14.4 mA cm^{-2} at -0.70 V vs RHE and maintained steadily at a current density of 244.5 ± 31.8 mA cm^{-2} and FEco of 99.1 ± 0.8% at -0.49 V vs. RHE for 50 hours in a flow cell reactor [53]. Nickel and nitrogen-doped porous carbon catalyst (Ni–N–C) without precious group metal applied in CO_2ERR has shown a CO partial current density above 200 mA cm^{-2} and provide stable Faradaic CO efficiencies around 85% for up to 20 hours [54]. Besides, the Ni–N–C powder electrocatalysts allow pH flexibility in operating conditions. In addition, Fe-N-C materials have been tuned to control CO/H_2 ratios and the materials containing only FeN_4 sites are able to reduce CO_2 in aqueous solution with FEco of over 90% at low overpotentials [55].

12.3.5 Metal-Free Electrocatalysts

Except for catalysts mentioned previously, carbon materials, molecular catalysts, and other types of new catalysts have been developed to improve selectivity and activity in recent years.

Duan et al. [56] pointed out that various carbon materials with heteroatom doping (e.g., N, S, and B) can be used as metal-free catalysts for the CO_2ERR, and ref. [9] reported graphitic carbon nitride (g-C_3N_4) embedded with transition metals and the Ni-C_3N_4-CNT catalyst presented a high FECO (~90%) under a wide potential range (-~0.6 V -~0.9 V vs. RHE) with a low overpotential of 0.39 V. Pranav et al. [57] investigate the role of defects and defect density on the selectivity of nitrogen-doped carbon nanotubes with DFT calculations. The presence of graphitic and pyridinic nitrogen was found to significantly decrease the overpotential (ca. −0.18 V) and increase the selectivity (ca. 80%) towards the formation of CO. A facile route to synthesize N and S co-doped porous carbon nanofiber (NSHCF) membranes on scalable production have also been reported. It has readily achieved CO with 94% Faradaic efficiency and −103 mA cm^{-2} current density about 1.2 mg catalyst loading, which is comparable with metal-based catalysts.

Compared to metal-based catalysts, the relevant investigations of metal-free electrocatalysts for CO_2 reduction are still quite a few. Emerging metal-free electrocatalysts for CO_2ERR with low cost, resistance to acids and bases and environmental friendliness need to be explored with more extended researches.

12.4 ELECTROLYTIC CELLS AND GAS DIFFUSION ELECTRODES (GDEs)

Besides electrocatalysts, electrocatalytic cells including related components, where two independent half-cell reactions as reduction half-reaction (CO_2ERR) and oxidation half-reaction (OER) occur, are another essential issue in electrochemical CO_2 reduction. Different reactor types are introduced here and their structural features, advantages, and disadvantages in electrocatalytic conversion of CO_2 are listed.

12.4.1 H-Type Cell

As Figure 12.2 shows, an H-type cell commonly consists of a cathodic compartment, an anodic compartment, and a polymer electrolyte membrane between them to separate two electrodes, which presents H-shape. During the electrocatalytic process, gas-phase CO_2 is bubbled into cathodic electrolyte continuously via a glass frit, and usually the flow rate is controlled by a mass flow controller at the inlet. The generated gaseous products are delivered into a gas chromatograph (GC) directly to be detected and analyzed, thus, the reactor is designed to be sealed to get an accurate calculation. Furthermore, there is a magnetic stirring equipped to eliminate the impact of concentration polarization through a uniform concentration distribution of the reactant and product.

As the working electrode is placed in the cathodic compartment, the counter electrode is placed in the anodic compartment, and the reference electrode is placed close to the working electrode to monitor the potential, the H-type cell forms a three-electrode system naturally, facilitating the study of kinetics and the reaction mechanism of CO_2ERR. Therefore, to date, the H-type cell is still the most commonly used reactor for CO_2 electroreduction in the laboratory to work on the screening of cathodic material.

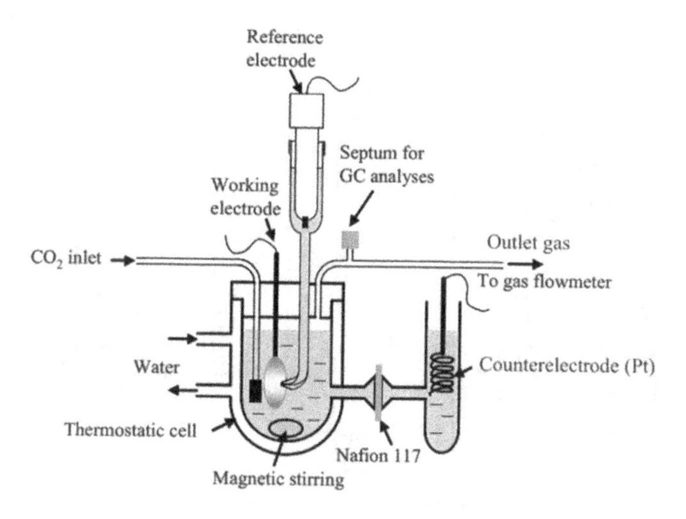

FIGURE 12.2 A schematic diagram of the conventional H-type cell used by [58].

However, the notably low current density and excessive reactor volume limit its industrial applications. The solubility of CO_2 in aqueous is low and large electrolyte volume makes CO_2 even more difficult to diffuse to the catalyst surface with a longer distance, leading to a higher ohmic resistance, resulting in mass transfer limitation and higher working cell voltages , reducing the energy efficiency.

In general, the H-type cell is a typical lab-scale reactor for quantifying and selecting suitable catalysts for CO_2 reduction to various value-added chemicals. But for further development, more efficient cells with compact structures, higher current densities, and product yield are required.

12.4.2 DEMS Cell

Since conventional H-type cell detects samples from bulk electrolyte, hard to analyze the reaction-relevant species near the cathodic surface and quickly reach a detectable level for liquid-phase products, differential electrochemical mass spectrometry (DEMS) as an analytical technique provides a new promising approach to observe transient intermediate reaction products close proximity with the electrode surface on the order of a second [59]. DEMS cell utilizes a pervaporation membrane coated the electrocatalyst directly to separate electrochemical reaction, as well as to transfer volatile species into the mass spectrometer (MS) immediately. In this way, only the DEMS cell is allowed to observe the composition of the local reaction environment and discuss its effects on intrinsic kinetics of the electrochemical reduction of CO_2.

Clark et al. [60] have applied a DEMS cell to measure the reactant concentration of carbon dioxide and analyze products rapidly near the cathode surface in real-time. In the DEMS setup (Figure 12.3), the structure included a core electrochemical cell, a set of electrolyte reservoirs, and a peristaltic pump. The working and counter electrodes were set parallel to have a uniform potential distribution

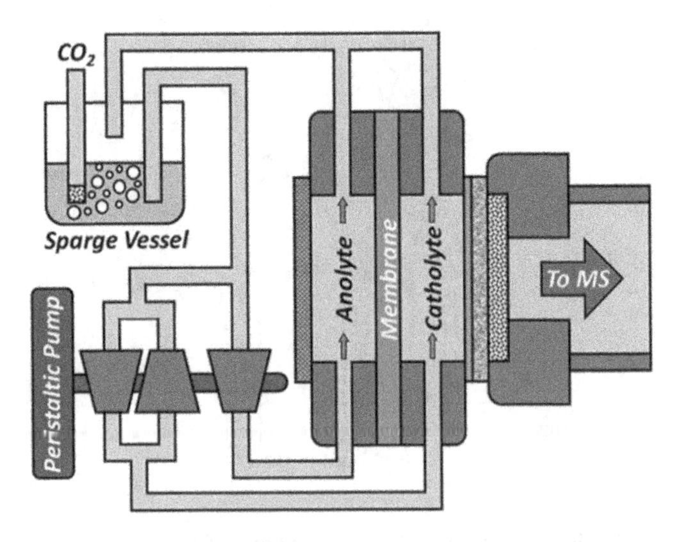

FIGURE 12.3 A schematic diagram of the DEMS setup used by [60].

across the electrode surface and separated by an ion-conducting membrane to avoid products reoxidation at the anode through the crossing. The pump created enough pressure to recirculate the electrolyte from the reservoir to the cell, preventing CO_2 depletion and providing better mass transfer to and away from the cathode. The electrolyte volume between the cathodic electrode and the pervaporation membrane should be minimized to achieve acceptable delay time between reduction products and MS, as well as not dilute the liquid-phase products to accumulate more quickly to reach the detectable level. Furthermore, the working electrode with a larger exposed surface area should be applied so that the concentration of liquid-phase reduction products can be maximized [59].

In this way, it can measure the CO_2 concentration and volatile reaction products in the immediate vicinity of the cathode directly to study the mechanisms of CO_2 reduction to multicarbon products over catalysts. Furthermore, the electrochemical reduction of CO_2 is found sensitive to electrolyte polarization, causing gradients in local pH and the concentration of CO_2 near the cathode surface, which leads to a significant difference from the bulk of the electrolyte. Thus, the DEMS cell provides a feasible approach to probe the effects of electrolyte hydrodynamics on the electrocatalytic activity and relate the electrocatalytic activity variety to the composition of the local reaction environment. Ref. [60] has shown that the amounts of CO_2 reduction (H_2 and CO) over Ag agreed with that in a traditional H-type cell measured by gas chromatography with the same electrolyte and catalyst. Nevertheless, the DEMS cell obtained the result nearly two orders of magnitude faster than the H-type cell. Notably, depletion of CO_2 in the vicinity of the cathode surface attributed to reaction with hydroxyl anions can be observed using the DEMS, where the process cannot be detected through other conventional analytical techniques. In addition, the findings over the Cu electrode suggested that acetaldehyde is a precursor to ethanol and propionaldehyde, which provides insight into the selectivity-determining steps for multicarbon compounds.

Therefore, the DEMS cell is also a lab-scale reactor with an analytical technique to observe the composition of the local reaction environment in the immediate vicinity of the cathode surface and discuss the possible reaction mechanism.

12.4.3 Solid Oxide Electrolysis Cell

Co-electrolysis of CO_2 and H_2O under high temperature using solid oxide electrolysis cell (SOEC) has presented a promising alternative to achieve high current and efficiency at the same time, attracting great attention in the last few years. As illustrated in Figure 12.4, a typical single SOEC includes three main parts: a solid ion-conducting electrolyte, an anode for OER, and a cathode where the electrolysis reaction takes place, showing a "sandwich" form. Moreover, ion-conducting electrolytes can be mainly divided into oxygen ionic conductors and protonic conductors with inappreciable electronic conductivity. As yttria-stabilized zirconia (YSZ) is the most common solid electrolyte material used in SOEC, showing high ionic conductivity as well as stable thermal and chemical performance under the high operation temperatures (800–1000°C) [61], the reaction paths usually follow Figure 12.4(a). CO_2/H_2O mixture flows to the cathode and further ionizes into CO,

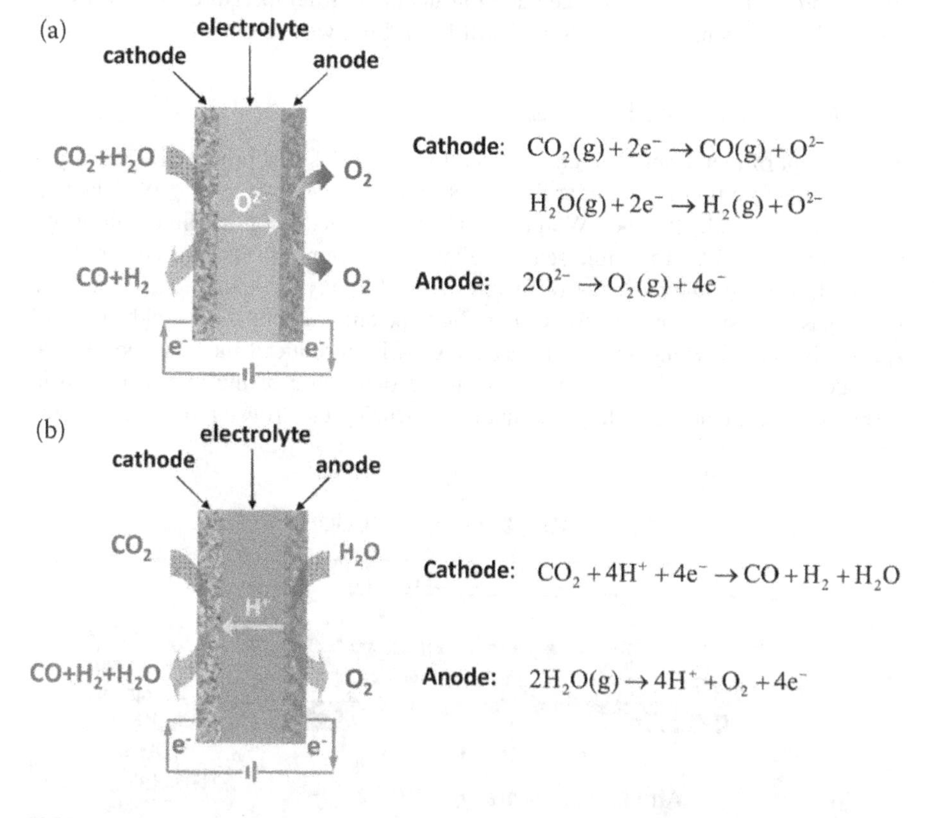

FIGURE 12.4 A schematic diagram for CO_2/H_2O co-electrolysis based on (a) oxygen ionic conductors and (b) protonic conductors used by [64].

H_2, and O_2^- ions with electrons applied from an external circuit. Then O_2^- ions pass through the electrolyte and are oxidized to oxygen gas on the anode.

As SOEC uses gaseous inputs for both electrodes fed by porous and diffusive ceramic materials, it easily breaks the limit of mass transport and ignores water management which is a grand challenge in the cells operated under ambient conditions. Besides, it avoids the risk of leakage and corrosion compared to low-temperature liquid electrolysis, and the sandwich-shape structure also facilitates the fabrication of cell assembles to be commercialized. In addition, the high operating temperature makes it easier to activate carbon-oxygen double-bond of CO_2, beneficial to fast kinetics and low internal resistance, which favors increasing energy efficiency and reducing low cost. Thus, SOEC has achieved considerable high current density over 1 A cm^{-2} and high thermal efficiency [62], approaching a commercial reality.

However, there are still some challenges remaining to be solved. The cathode materials such as Ni-YSZ, are prone to occur nickel aggregation and carbon deposition on the cathodic surface [63], as well as sensitive to gas purity, putting forward an even more strict requirement for active and stable cathodes. Besides, in practical applications of SOEC, it will encounter some problems such as sealing, limited products, and material degradation under high temperature. So, as a promising development direction, SOEC still has a long way to go.

12.4.4 Microfluidic Flow Cell

The concept of membrane-less continuous-flow cells was first proposed by Ferrigno et al. [65], exploiting the laminar flow to eliminate convective mixing of reduction and oxidation products. Then Whipple et al. [66] applied the microfluidic flow cell (MFC) to reduce CO_2 to formic acid. As Figure 12.5 demonstrates, the cell contains an anode flow field plate, gas diffusion anode for oxygen evolution, electrolyte layer, gas diffusion cathode for CO_2 reduction, and cathode flow field plate respectively. The flowing stream of gaseous CO_2 is introduced into one side of the cathode while the electrolyte stream is on the other side of the exposed surface, forming an efficient multiple-phase interface. During electrolysis, the two streams

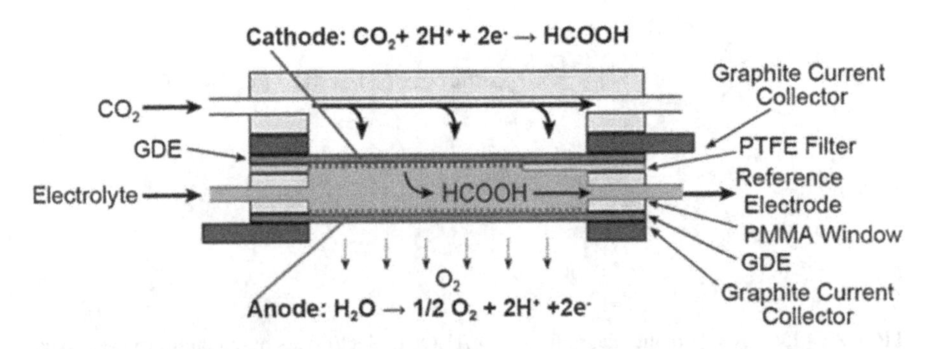

FIGURE 12.5 A schematic of the microfluidic reactor for CO_2 conversion used by [66].

flow together, maintaining a parallel co-laminar flow in the microchannel and the electrolyte with products finally flows out of the channel [1].

Most differently, MFCs utilize a laminar flow electrolyte instead of the conventional solid membrane to separate the streams of redox couple. Compared to the H-type cell,, it minimizes ohmic losses by combing membrane-free operation and thin electrolyte flow channels. Besides, MFC eliminates the membrane-related issues: water management, high manufacturing cost, cathode flooding, membrane degradation, and so on [67]. Moreover, the flow electrolyte contributes to regulating and maintaining the reaction environment. In this way, MFCs can achieve a higher current density under a strong alkaline environment to improve the kinetics of CO_2 conversion without extra depletion, as it starts the reduction reaction immediately once CO_2 diffuses to the triple boundary phase. Edwards et al. [68] reported a MFC using a silver catalyst for CO production with the combination of three points: minimal electrode spacing with 0.25 mm flow field, high pressurization of 50 bar, and 5 M KOH as a high-alkaline electrolyte. The strategy enables the improved system to operate at current densities as high as 941 mA/cm^2 with a respectable energy efficiency of 47%, applicable to large-scale electrolyzers. Moreover, effects including microchannel shape, thickness, flow rate, electrolyte pH, and CO_2 concentration are studied to optimize cell performances [1,67,69].

However, MFC is particularly sensitive to the pressure due to the membraneless structure, and reduction products existing in electrolyte inevitably transfer to the anode and be re-oxidized, reducing the total efficiency of the cell. Therefore, there is a need to introduce relatively dense solid electrolyte membranes to inhibit the migration of products between the electrodes.

12.4.5 Polymer Electrolyte Membrane-Based Flow Cell

To make the electrocatalytic conversion of CO_2 efficient and practically viable, researches are likely driving the shift in current technologies from H-cell experiments involving bulk liquid electrolyte to a continuous-flow reactor containing solid polymer electrolyte membrane (PEM).

12.4.5.1 Flow Cell With Electrolyte Chamber

As Figure 12.6 illustrates, in a typical membrane-based flow cell with electrolyte chamber, which consists of a cathode plate with the flow channel, anode plate, anode and cathode flow chamber, and a polymer electrolyte membrane (PEM) to separate the electrodes, gaseous CO_2 is delivered to the cathode, while catholyte and anolyte are continuously circulated through flow chambers. Such structure provides a stable reaction interface and the membrane contributes to avoiding the migration of products [70]. Besides, mass-transport limitations can be overcome by continuously circulating the reactants and products to and away from the electrodes.

However, the introduction of catholyte also leads to a high resistance and reaction overpotential, causing to a decreased overall energy efficiency. Additionally, the catholyte tends to penetrate the electrode and block the gas transfer channels [71].

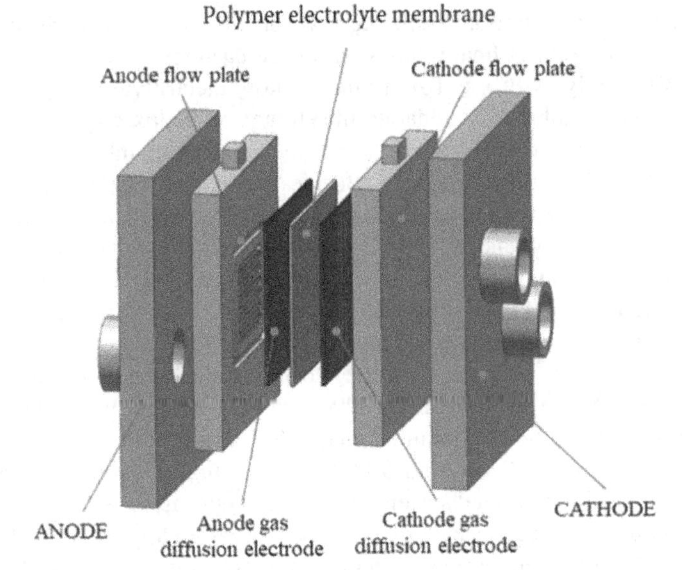

FIGURE 12.6 Exploded diagrams of polymer electrolyte membrane-based flow cell.

When the electrolyte finally saturates the electrode, CO_2 can no longer react at the catalyst surface and the competition reaction (hydrogen evolution reaction, HER) becomes dominant. Besides, the scale-up of the flow cell with an electrolyte chamber is complicated and challenging due to the increased cross-talk of the two half reactions as the size of the flow cell increases.

12.4.5.2 Zero-Gap Flow Cell

To minimize ohmic resistances and scale up, a zero-gap flow cell, which removes the cathode electrolyte, has been developed. As Figure 12.7 illustrates, the zero-gap flow cell contains a core membrane electrode assembly (MEA) consisting of the anode and cathode gas diffusion electrodes (GDEs) on either side of the polymer electrolyte membrane, forming a zero-gap cell configuration with decreased cell resistance.

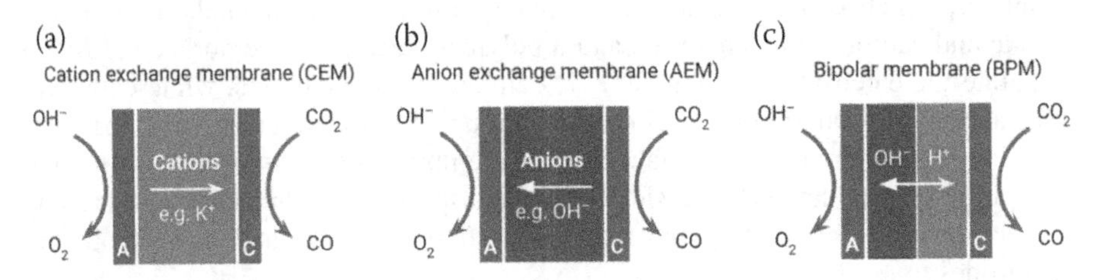

FIGURE 12.7 Exploded diagrams of the zero-gap flow cell.

The application of zero-gap cells enables the operation in multi-cell stacks due to the modular characteristic of the technology, which has already been proven to be commercially available in fuel cell [72]. Lin Zhuang and coworker have used the zero-gap cell to attain a current density of 0.5 A cm^{-2} at 3 V for CO production, with a CO faradaic efficiency over 90% [73]. A. Danyi and coworker developed a direct CO_2 gas-fed, zero-gap electrolyzer cell, achieving CO formation partial current densities above 250 mA cm^{-2}. Besides, they showed the assembly and operation of a multilayer electrolyzer stack consisting of three zero-gap cells for the first time, bringing CO_2ERR closer to its industrial implementation [74].

12.4.5.3 Gas Diffusion Electrode

Cathode gas diffusion layer (GDE, shown in Figure 12.8) is prepared by immobilizing electrocatalyst on a gas diffusion layer (GDL), a porous material consisting of a dense array of carbon fibers (typically carbon paper or carbon cloth) to support catalyst layer (CL) and a denser microporous layer (typically carbon nanofibers or compressed carbon particles) with appropriate pore distribution, which is conducive to mass transfer of CO_2 [75]. Such structure enables a prolonged contact between the reactants and catalytic sites, sustaining a high current density. For an anode, not only GDL-supported electrocatalyst but also noncarbon gas diffusion materials (e.g., metal foam, mesh, etc.) are applied [76].

Dinh et al. [77] reported a modified composite multilayered porous electrode containing a polytetrafluoroethylene membrane to stabilize the gas diffusion electrode in basic media, Ag catalysts for electro-catalysis, and a carbon current distributor on the surface of Ag catalyst. The system separating the gas and current distributor provided endurance and achieved a CO production Faradaic efficiency of over 90% at current densities above 150 mA/cm^2 under both neutral and alkaline conditions in a flow cell for over 100 hours of operation. To obtain more catalytic active sites at the three-phase interface, Li et al. [78] adopted a highly flexible, nanoporous, hydrophobic polyethylene membrane sputtered with gold nanocatalyst on one side to construct a bilayer pouch-type artificial alveolus structure with high gas permeability but low water diffusibility, beneficial to achieve high local pH tuning in the pouch. The innovative design generated a high Faradaic efficiency of 92% for CO with a thin catalyst thickness of 20–80 nm in an aqueous system.

Except for the GDL, the CL is also crucial in CO_2ERR, where the reaction occurs. An efficient reaction region, which is advantageous for mass transfer and electrons conduction, is the origin of the favorable CO_2-to-CO current density and selectivity. Q. Wan et al. reported CO_2-to-CO partial current densities of up to 110 mA cm^{-2} in a flow cell, and Faradaic efficiency for CO (FE_{CO}) clearly above 99%, by changing the composition of the catalyst ink [79]. EW Lees and coworker modulate the Faradaic efficiency for formate by 20%, which is considered to be inconsequential relative to CO when using Ag electrocatalysts in a flow cell, by a mere 5 wt% change in perfluorinated sulfonic acid content in GDE. Also, they provided a method to precisely control the relative amount of ionomer to electrocatalyst for each GDE [80].

Moreover, as flow cell where gas-phase CO_2 reaches the interior of cathode GDE and contact with the electrolyte outside the GDE simultaneously forms an important

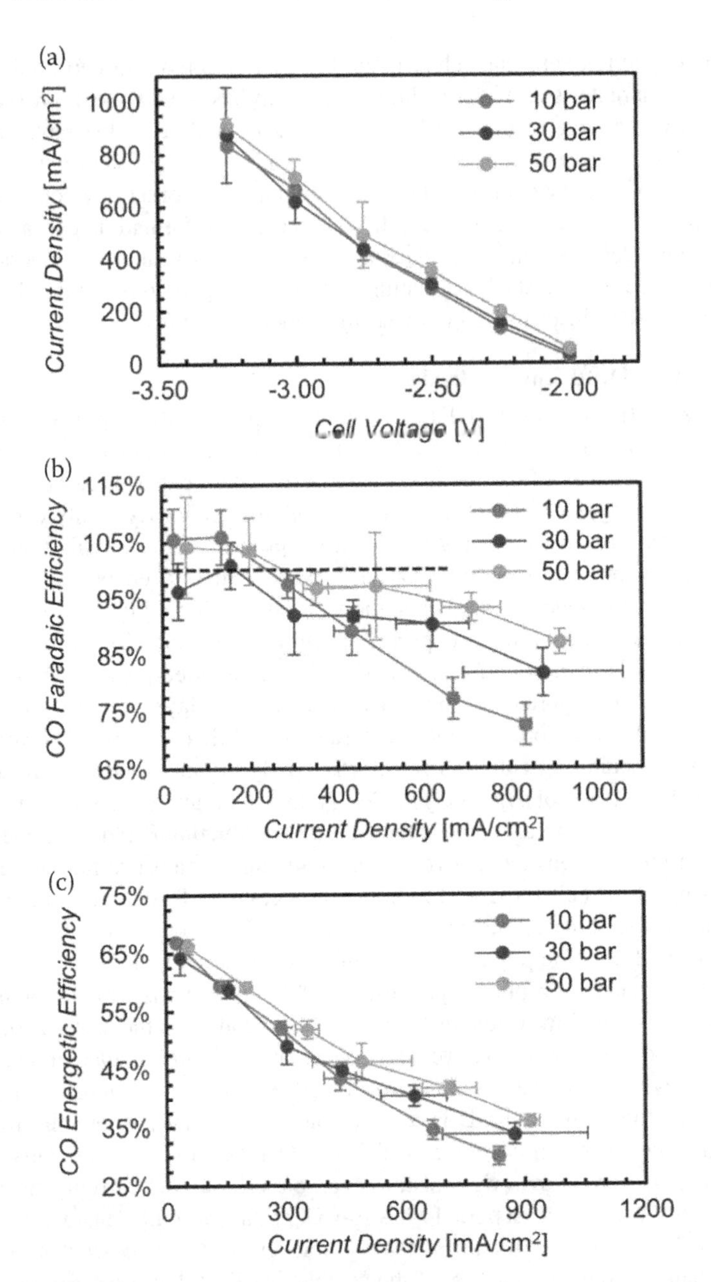

FIGURE 12.8 GDE with a catalyst layer and a gas diffusion layer (consisting of a microporous layer and a carbon paper as substrate) used by [79].

three-phase interface, the effect of GDL hydrophobic/hydrophilic properties and structure parameters including thickness, density, and porosity on cell performance should draw more attention [81]. Electrode optimization may allow achieving further improvements in electrocatalytic conversion devices.

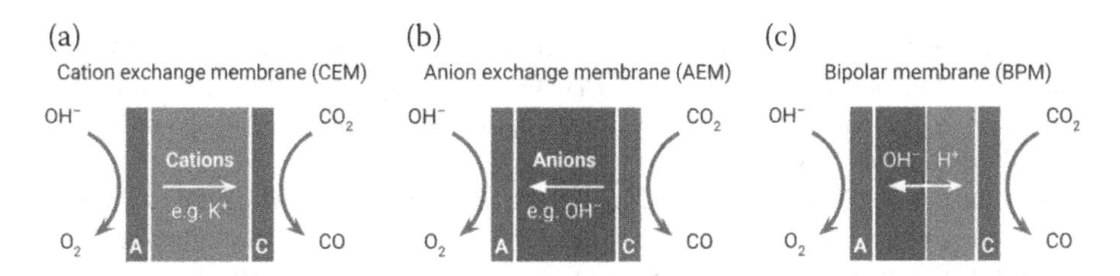

FIGURE 12.9 Overview of the different ion transport pathways between the anode (labeled "A") and the cathode (labeled "C") used by [75].

12.4.5.4 Polymer Electrolyte Membrane

Currently, there are three classes of ion-exchange membranes which have been deployed in the configuration of gaseous CO_2 flow cells: cation exchange membranes (CEMs), anion exchange membranes (AEMs), and bipolar membranes (BPMs). As Figure 12.9 shows, each type of PEM has its own unique ion transport pathway between the anode and cathode, accordant with the selection of electrolyte conditions to ensure efficient water and ion transport.

CEMs (e.g., Nafion) selectively transfer cations (such as H^+, K^+, etc.) from the anode to the cathode, which have been widely applied in proton-exchange-membrane fuel cells (PEMFCs). Delacourt et al. [82] firstly designed an electrochemical CO_2 cell similar to PEMFC using Nafion as the membrane, Ag as cathode, and Pt/Ir as the anode to form the MEA. But the result was shown to be unfavorable for CO_2 reduction as almost 100% of the current efficiency for H_2 at a current density of 20 mA/cm^2. The unprosperous outcome is mainly because protons diffuse from the anode to the catholyte continuously, contributing to acidification in the cathodic reducing environment and shifting the selectivity toward hydrogen evolution reaction. Besides, CEMs matched to acid anolyte require for noble metal catalysts to active oxygen evolution reaction at the anode, resulting in higher costs. Some improved configurations based on Nafion membranes have been proposed to suppress OER. Ma et al. [83] adopted a buffer layer ($KHCO_3$) between GDL and CEM to main the cathodic reaction environment, exhibiting a good ability to reduce the current efficiency of H_2. However, the electrolyte in the buffer layer needs a circulation flow to be renewed for steady operation, making the system much more complex.

AEMs inversely facilitate the flow of negative ions (e.g., OH-, HCO_3-, etc.) from the cathode to the anode. It seems more suitable for CO_2 conversion because the delivery of OH^- rather than H^+ is conducive to a relatively stable cathode reaction environment and high ion transfer conductivity. In addition, AEMs allow for the operation under alkaline conditions where non-precious metals can be applied at the anode for OER, and cathode potential can be reduced. Gabardo et al. [84] once used an AEM combined with pressurization to achieve a highly alkaline reaction environment to improve reaction kinetics and obtained a high selectivity for CO under high current densities. However, in such an alkaline environment, CO_2 could react with OH^- immediately during electrolysis to form bicarbonate (HCO_3^-) and carbonate (CO_3^{2-}) anions, not only decreasing the CO_2 concentration but also increasing ohmic polarization losses as a relatively lower ion mobility of HCO_3- and CO_3^{2-}. Furthermore, the transfer of HCO_3^- and CO_3^{2-} from the cathode to the anode is equivalent to transfer reactant CO_2 to the anode, observably reducing overall conversion efficiency [75].

The third type of PEM is the BPM, which consists of anion and cation exchange membranes, dissociating water into H^+ and OH^- under reverse bias at the interface and then H^+ and OH^- are transferred to the cathode and anode separately under the electric field [24]. In this way, BPM can keep a constant pH on both sides of the cell, which is a distinct advantage over AEM and CEM. Salvatore et al. [76] proposed a modified BPM-containing configuration with a solid-supported aqueous layer set between the Ag-based catalyst layer and the BPM to enhance the selectivity for CO. It obtained a higher current density (200 mA/cm^2) and demonstrated that adequate hydration of the CO_2 inlet stream is beneficial to the long-term stability. Nevertheless, it is required to consume partial input of energy to drive bipolar membrane hydrolysis during the electrolysis process, reducing the energy conversion efficiency of a single cell.

In conclusion, structural features and applications of each type of cells mentioned above are expounded here and their advantages and disadvantages are summarized as shown in Table 12.2.

TABLE 12.2
Summarize of advantages and disadvantages of different CO2RR reactors

Reactor type	Advantages	Disadvantages
H-type cell	Simple structure; convenient for catalyst selection with; wide selectivity of reduction products; convenient operation	Large volume; limited mass transfer with high electrolyte resistance; limited liquid products detection
DEMS cell	Detect local reaction environment in real time; observe transient intermediate reaction products	Limitation of large-scale application
SOEC	Efficient mass transport; high current density; avoids the risk of electrolyte leakage	Extreme conditions; carbon deposition on the cathodic surface; limited reduction species
MFC	Stable cathode reaction environment; adjustable electrolyte; high mass transport	Sensitive to pressure; crossover of reduction products
PEM flow cell	High mass transport; zero-gap cell configuration beneficial to commercial application; low cell resistance; avoid crossover	Corrosion under the high overpotentials; stability under the high current densities

12.5 OPERATING CONDITIONS

During electrochemical reduction of CO_2, the outcome is determined by the integrated effect of electrocatalysts and operating conditions, including the electrolyte, pressure, and temperature. Small changes in the reaction environment significantly changes the reaction kinetics and reaction selectivity of CO_2ERR, so it is necessary to optimize the operating conditions.

12.5.1 ELECTROLYTE SOLUTION

As the supporting electrolyte is still required in most of the electrolytic cells at present, the electrolyte solution is one of the main factors affecting the reaction process, roughly classified into two types: aqueous solutions and nonaqueous solvents.

Aqueous electrolytes as an excellent conductor for protons are commonly used to provide abundant H^+ for further proton-coupled steps. However, due to the low solubility of CO_2 (1.45 g/L), limiting the mass transport to the active site where the chemistry occurs, aqueous-fed systems are constrained to ~35 mA cm^{-2} when the commercial goal is to achieve >200 mA cm^{-2} to meet practical applications [72]. Therefore, CO_2ERR in an aqueous electrolyte has mainly focused on the mechanisms of CO_2 reduction to multicarbon products and catalyst screening.

Some studies have confirmed the effect of pH on both of the half reactions: OER and dioxide reduction reaction (CO_2ERR). Compared to acidic conditions, it is conducive to operate OER in an alkaline environment as OER kinetics have been improved and higher efficiencies have been realized [85–87]. In addition, OER under alkaline conditions can employ non-noble metal catalyst such as Ni-based materials [88–90], reducing cost. Also using alkaline electrolytes at the cathode can help to reduce ohmic losses via improved conductivity and stabilize CO_2 reaction intermediates, thus leading to higher electrolyte concentrations and providing additional benefits to cell operation [68], although there are some problems to be solved of CO_2ERR operating in high-alkaline electrolytes: electrolytes need to be constantly updated as it reacts with CO_2 to form carbonate. The pH operating conditions alter the selectivity of CO_2ERR at the same time. Gabardo et al. [90] demonstrated operating CO_2ERR in alkaline environments can suppress the competing HER and high bulk pH conditions leads to a decrease in the Faradaic Efficiency (FE) for CO linked to an increased FE contribution from formate on a silver (Ag)-based system. Dinh et al. [91] reported a copper (Cu) electrocatalyst in highly alkaline media reducing CO_2 to ethylene with 70% faradaic efficiency at a potential of –0.55 volts versus a reversible hydrogen electrode (RHE). Previous reports have also demonstrated the effects of alkali metal cations in CO_2ERR. Resasco et al. [98] led to the conclusion that the increased size of the alkali metal cation in the electrolyte increased the partial current densities for $HCOO^-$ and CO on polycrystalline Ag, Sn, and Cu. Besides, electrolyte cation size affects the rate of C-C bond formation, the rate increasing with increasing cation size [93].

Aprotic solvents such as organic solvent, ionic liquid and molten salt are drawing more attention to bypass the issues existing in aqueous electrolyte. Shaughnessy et al. [94] presented organic solvent-based CO_2 Expanded Electrolytes (CXEs) which solubilize multi-molar amounts of CO_2 at moderate pressures, improving the solubility of CO_2 and achieving a high FE_{CO} approaching 80% on polycrystalline Au electrodes. The ionic liquid is an attractive solvent currently due to the high solubility of CO_2 [95,96] and intrinsic ionic conductivity. Sun et al. [97] showed the ionic liquid 1-ethyl-3-methylimidazoliumbis(trifluoromethylsulfonyl)imide offers new ways to modulate CO_2ERR and shifts CO_2ERR course by promoting the formation of carbon monoxide instead of oxalate anion at Pb electrode. Molten salts perform as the reaction medium for high-flux absorption of CO_2 and as the electrolyte for electrochemical reduction of CO_2 simultaneously is also a good alternative [98]. Kaur et al. [99] reached almost 95% Faradaic efficiency of the CO_2 to CO conversion using molten salts in the solid oxide electrolysis cell.

12.5.2 Pressure

Pressure is another factor that plays a key role in CO_2ERR performance. A higher pressure with a higher solubility of CO_2 is supposed to mitigate the mass transport limitation, increasing current density.

Edwards et al. [100] pointed out that the use of high pressure suppressed the decay in FEco at high current densities, suppressing $HCOO^-$ generation in favor of CO. Pressurization overcame the mass transport limitation and enabled an uncorrected full cell energy efficiency of 67% at an industrially viable current density over 200 mA/cm^2 at 50 bar (Figure 12.10). Hara et al. [101] also obtained a high FEco of 86% at a very large current density of 200 mA/cm^2 under 20 atm using a Ag-based gas diffusion electrode. And a very large CO partial current density 3.05 A cm^{-2} was achieved under 30 atm on the Ag-GDE. Pressurization combined with high alkalinity is an effective strategy to improve voltage efficiency and unity selectivity. Gabardo et al. reported a new record for the highest half-cell EE (>80%) for CO production at 300 mA cm^{-2} via operation in a highly alkaline reaction environment and system pressurization [102].

As shown, pressure can improve reaction kinetics and tune product selectivity: (1) facilitating CO_2 transfer; (2) altering the relative surface coverage and binding energies of adsorbed CO_2 and reaction intermediates. Notably, operating at high current densities faces greater challenges for the stability of CO_2 electrolyzer systems with run times on the order of hours [103], requiring a further comprehensive understanding of pressurized electrolyzer systems.

12.5.3 Temperature

CO_2 electroreduction generally operates in ambient temperature and pressure, but recent researches have proved temperature has an effect on product distribution and current density.

Armin et al. suggested an optimal temperature range between 0 °C and 35 °C with HER becoming increasingly predominant since CO_2 solubility drops with increasing temperature. In contrast to decreased CO_2 solubility, reaction kinetics and diffusion coefficients increase, which compensates for the lowered CO_2 solubility partly [104]. Dufek et al. increased the temperature with Ag base GDE from room temperature to

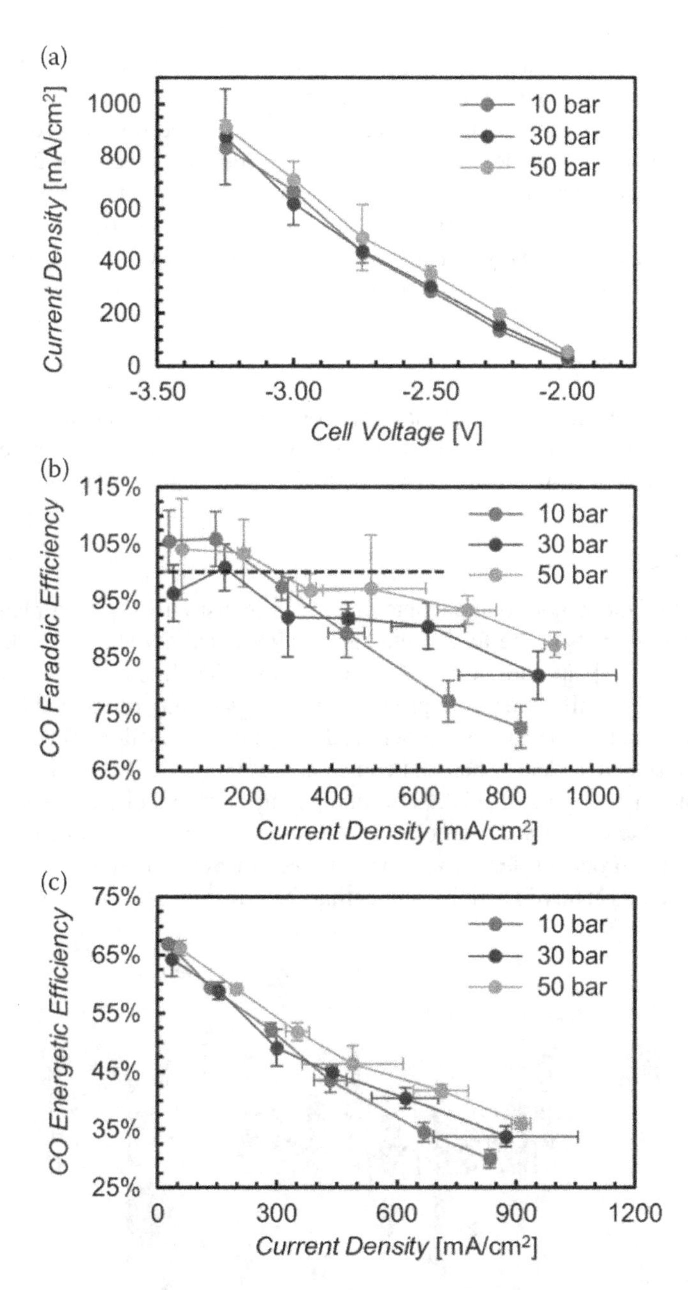

FIGURE 12.10 Effect of pressure on (a) Current density; (b) FE_{CO}; and (c) full cell EE_{CO} (electrolyte: 1 M KOH, 0.25 mm electrode spacing). Dashed line indicates 100% FE. Values above 100% are attributed to experimental fluctuations and measurement delay from [100].

70 °C, resulting in a decreased overall cell potential of 1.57 V at 70 mA cm^{-2}. The decrease can be attributed to both the anode and cathode potentials. Firstly, the main effect of the temperature increase is on the kinetics at the catalyst surfaces, which has been enhanced to lower the reduction potential for both CO_2ERR and HER.

Secondly, the decreased anode potential and ohmic resistances of the electrolyte and membrane can also account for the drop in cell potential, as ionic conductivity and contact resistances of the electrolyte and membrane have been improved [105,106].

When operating in an electrolyte-supporting system, the temperature demonstrates contradictory effects on the CO_2ERR performance. With the rising temperature, the CO_2 solubility drops and the reaction path changes with complex selectivity while reaction kinetics and ionic conductivity increase, resulting in larger higher currents. Therefore, it is important to tailor the operating temperature to reach maximum performance.

12.6 IMPURITIES

Because industrial carbon dioxide sources from fossil fuels often contain numerous contaminants, including SO_2 and NO_x, it is critical to understand the potential interactions between contaminants and CO_2-to-CO for practical applications.

12.6.1 SO_2

SO_2 is one of the major contaminants present in industrial CO_2 from plants and a trace amount of SO_2 in the feed is enough to alter reactivity and product.

Luc et al. [107] has chosen 1% SO_2 for initial studies. It is found that compared to CO_2, electrocatalysts tend to preferentially reduce SO_2 of favorable thermodynamical, reducing the energy efficiency of CO_2 electroreduction. The influence of SO_2 on the selective electrochemical conversion of CO_2 to CO was first studied on Ag. As shown in Figure 12.11, due to competing reactions of SO_2 over CO_2, the total CO_2 reduction Faradaic efficiency decreased but there was no change in the CO_2 reduction product selectivity. When the feed of SO_2 was stopped, Ag catalyst has recovered its catalytic activity, indicating the effect of SO_2 on Ag is reversible.

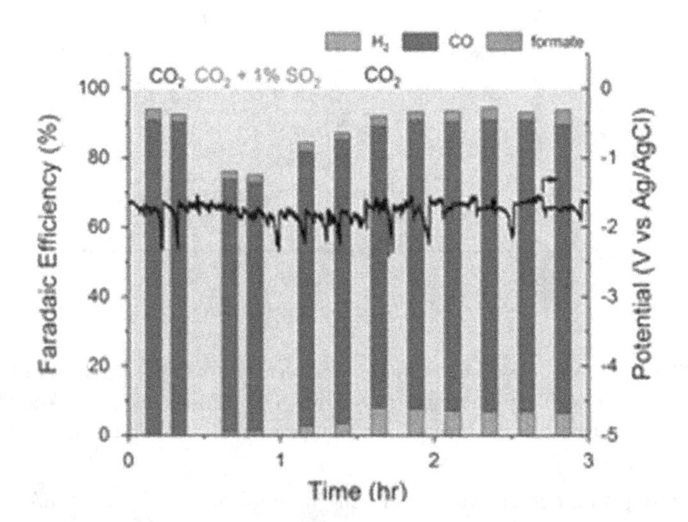

FIGURE 12.11 Performance of CO_2 + SO_2 electrolysis over Ag catalyst at a constant current density of 100 mA cm^{-2} in 1 M $KHCO_3$ over the span of 3 hours from [107].

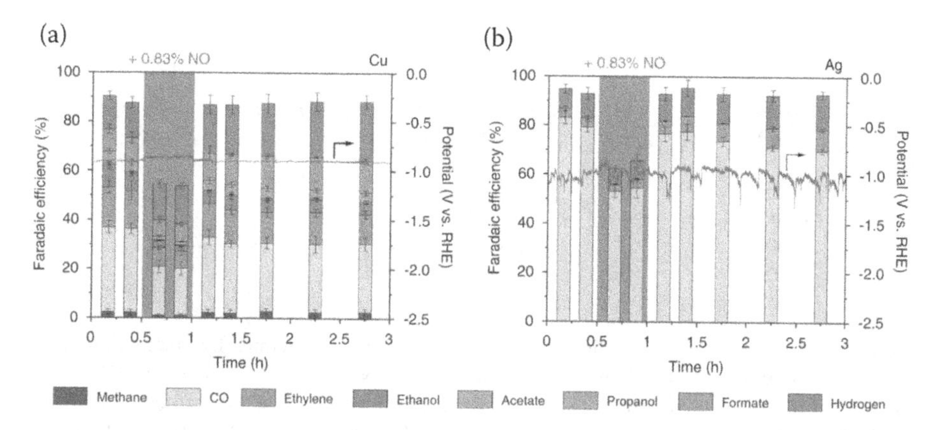

FIGURE 12.12 Faradaic efficiency and applied potential vs. time on (a) Cu, (b) Ag catalysts at a constant current density of 100 mA cm^{-2} in 1 M $KHCO_3$ for 3 hours from [108]. Gas feeds were 83.3% CO_2 and 16.7% Ar, and 83.3% CO_2, 15.87% Ar, and 0.83% NO (green). 0.83% NO was introduced at 0.5 hour for 0.5 hour.

12.6.2 NO_x

NO_x (i.e., NO, NO_2, and N_2O) is another common component present in the flue gas. Ag as one of the most studied catalysts is selected to explore the effects of NO_x on the conversion of CO_2 to CO [108].

As shown in Figure 12.12, when 0.83% NO was introduced into the flow, the total Faraday efficiency of CO_2ERR decreased apparently, which indicates the preferential reduction of NO over CO_2. When the feed of NO was stopped, the CO_2ERR performance and the total Faradaic efficiency quickly recovered. Also, there is no obvious change in selectivity, suggesting that the exposure of Ag surface to NO did not alter the catalytic property. Besides, a lower concentration of 0.0083% NO was evaluated and the results show negligible effects [108].

The influences of NO_2 and N_2O were further investigated by Hee Ko et al. as shown in Figure 12.13 [108]. Similarly, when NO_2 or N_2O was introduced, the total Faraday efficiency of $CO2_ERR$ decreased. The loss of the Faraday efficiency ofCO_2ERR was ascribed to the preferential reduction of NO_2 and N_2O over CO_2. When a pure CO_2 feed was fed again, the total Faraday efficiency was recovered, suggesting that the effect of NO_2 and N_2O on Ag is reversible.

Since a trace amount of impurities has already shown their non-negligible effects on the performance of CO_2ERR, which has rarely been explored in the literature, additional work should focus on it.

12.7 DURABILITY MECHANISM

Besides high current density and Faradaic efficiency, the long-term durability of the CO_2ERR system is also an important technical index for attention. We highlight the most common mechanisms that cause degradation in cathodes including applied electrocatalysts and GDEs.

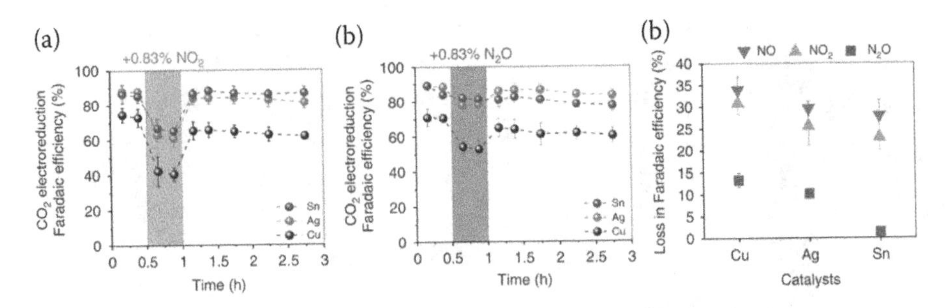

FIGURE 12.13 CO_2 electroreduction Faradaic efficiency, excluding hydrogen Faradaic efficiency, vs. time with the introduction of (a) 0.83% NO_2 (yellow) and (b) 0.83% N2O (blue) on Cu, Ag, and Sn catalysts at a constant current density of 100 mA cm^{-2} in 1 M KHCO3 for 3 hours. Gas feeds were 83.3% CO_2 and 16.7% Ar, and 83.3% CO_2 and 15.87% Ar with 0.83% NO_2 or 0.83% N_2O. NO_2 and N_2O were introduced at t = 0.5 hour for 0.5 hour. (c) Loss in Faradaic efficiency during CO_2 electroreduction from the introduction of 0.83% NO, 0.83% NO_2, and 0.83% N_2O on Cu, Ag, and Sn catalysts.

12.7.1 Catalysts

The degradation of catalysts usually originates from the adsorption of unwanted species, changes in the surface morphology, chemical composition, and crystalline structure.

The Sn nanoparticle was found that the SnO_x on the cathode surface was reduced to Sn during electrolysis and led to a decrease in FE [109]. Smith and co-workers [110] tested Ag (Ag_2CO_3)-foil catalyst at an applied potential of –0.55 V vs. RHE in $KHCO_3$ electrolyte for 100 hours. They found a decline in FEco after 37 hours and ascribed it to the adsorption of impurities on the surface of the catalyst. Catalyst morphology changes in Au nanoparticles have also been observed by Rogers et al. [111]. During long-term electrolysis, Au nanoparticles presented aggregation, leading to decreased active surface area and a sharp loss of catalytic selectivity (≈50% drop in FECO in 3 hours). In addition, the insufficient stability of electrocatalyst under a continuous reductive environment is also an important issue. Some related porphyrin complexes have been shown to undergo hydrogenation on their pyrrole rings at a negative potential and the undesirable reduction of ligands can deteriorate the selectivity [112].

12.7.2 Gas Diffusion Electrodes

GDEs (Figures 12.8) are used to allow a continuous supply of gaseous CO_2 into the catalyst surface and provide a larger triple-phase interface of gas/electrolyte/catalyst. Various mechanisms of degradation can affect each of these layers during long-term operation.

Loss of hydrophobicity of the GDE is a major issue for CO_2ERR. The microstructure of GDE and the binder that provide the necessary adhesion of catalysts to GDL suffer from various forms of physical and chemical degradation during prolonged exposure to the electrolyte and CO_2. The loss of binder functionality and the destruction of microstructure gradually leads to the loss of hydrophobicity. The GDL without sufficient hydrophobicity/wet-proofing cannot effectively separate the liquid electrolyte from the gaseous CO_2. Finally, the electrolyte will flow into the GDL and prevent CO_2 from reaching the catalyst surface. Accordingly, the triple-phase interface has moved into the microporous layer and carbon fiber layer from the catalyst layer and HER becomes the dominant reduction product.

Carbonate formation in the GDE is another serious issue. When the electrolyte saturates the GDL and HER dominates, residual salt from the electrolyte gradually accumulates on the inside of and at the back of the GDE, which blocks the inner pores of CO_2 transfer. The formation of carbonates thereby halts the function of the electrolyzer [113]. B. Endrődi and co-workers [114] have developed an operando activation and regeneration process to overcome the formation of precipitates in GDE by infusing alkali cation-containing solutions in a zero-gap electrolyzer cell.

In conclusion, new and more stable catalysts and electrodes need to be developed for the long-term operation of electrolytic cells.

12.8 CHALLENGES AND PERSPECTIVES OF CO_2 ELECTROREDUCTION

The CO_2 electrochemical reduction o (CO_2ERR) has been regarded as a promising strategy to alleviate excess CO_2 emissions by using produced CO_2 as a feedstock for value-added chemical production and storing intermittent renewable energies (such as solar and wind energy) in carbon-neutral chemicals. Many groups have completed significant research in the CO_2ERR field in recent years and new catalyst materials with high activity and selectivity.

To realize the commercial application of $CO2_E$RR, the catalyst at the present stage still faces some challenges: (1) the catalyst should gradually get rid of the reliance on precious metals, or reduce the loading of precious; (2) the stability of catalysts should be enhanced (beyond 2,000 hours, supposing a replacing every three months); (3) the catalyst should have specific selectivity towards CO_2 and avoid the effects of possible contaminants in flue gas.

Besides, an integrated system on a large scale should be developed and optimized. Beyond the challenges of a larger electrolytic cell, the process of scaling up can result in unexpected practical hurdles: altering product selectivity, uneven mass transport, difficult separation, and purification. Also, the applied electrodes should be optimized to better improve the mass transfer process during long-term electrolysis.

REFERENCES

[1] T. Ouyang, J. Chen, G. Huang, J. Lu, C. Mo, N. Chen, A novel two-phase model for predicting the bubble formation and performance in microfluidic fuel cells, *Journal of Power Sources*, 457 (2020).

[2] M.A. Qyyum, Y.D. Chaniago, W. Ali, K. Qadeer, M. Lee, Coal to clean energy: Energy-efficient single-loop mixed-refrigerant-based schemes for the liquefaction of synthetic natural gas, *Journal of Cleaner Production*, 211 (2019) 574–589.

[3] J. Liu, Y. Yin, S. Yan, Research on clean energy power generation-energy storage-energy using virtual enterprise risk assessment based on fuzzy analytic hierarchy process in China, *Journal of Cleaner Production*, 236 (2019) 117471.
[4] M. Mikkelsen, M. Jørgensen, F.C. Krebs, The teraton challenge. A review of fixation and transformation of carbon dioxide, *Energy & Environmental Science*, 3 (2010) 43–81.
[5] Y. Zhang, S.-X. Guo, X. Zhang, A.M. Bond, J. Zhang, Mechanistic understanding of the electrocatalytic CO_2 reduction reaction – New developments based on advanced instrumental techniques, *Nano Today*, 31 (2020) 100835.
[6] J.B. Greenblatt, D.J. Miller, J.W. Ager, F.A. Houle, I.D. Sharp, The technical and energetic challenges of separating (photo)electrochemical carbon dioxide reduction products, *Joule*, 2 (2018) 381–420.
[7] N.S. Lewis, D.G. Nocera, Powering the planet: Chemical challenges in solar energy utilization, *Proceedings of the National Academy of Sciences of the United States of America*, 104 (2007) 20142–20142.
[8] C. Cometto, L. Chen, P.-K. Lo, Z. Guo, K.-C. Lau, E. Anxolabéhère-Mallart, C. Fave, T.-C. Lau, M. Robert, Highly selective molecular catalysts for the CO_2-to-CO electrochemical conversion at very low overpotential. Contrasting Fe vs Co quaterpyridine complexes upon mechanistic studies, *ACS Catalysis*, 8 (2018) 3411–3417.
[9] C. Ding, C. Feng, Y. Mei, F. Liu, H. Wang, M. Dupuis, C. Li, Carbon nitride embedded with transition metals for selective electrocatalytic CO_2 reduction, *Applied Catalysis B: Environmental*, 268 (2020).
[10] Q. Li, J. Fu, W. Zhu, Z. Chen, B. Shen, L. Wu, Z. Xi, T. Wang, G. Lu, J.J. Zhu, S. Sun, Tuning Sn-catalysis for electrochemical reduction of CO_2 to CO via the core/shell Cu/SnO2 structure, *Journal of the American Chemical Society*, 139 (2017) 4290–4293.
[11] C. Long, X. Li, J. Guo, Y. Shi, S. Liu, Z. Tang, Electrochemical reduction of CO2 over heterogeneous catalysts in aqueous solution: Recent progress and perspectives, *Small Methods*, 3 (2019) 1800369.
[12] S. Verma, Y. Hamasaki, C. Kim, W. Huang, S. Lu, H.-R.M. Jhong, A.A. Gewirth, T. Fujigaya, N. Nakashima, P.J.A. Kenis, Insights into the low overpotential electroreduction of CO_2 to CO on a supported gold catalyst in an alkaline flow electrolyzer, *ACS Energy Letters*, 3 (2017) 193–198.
[13] T. Zheng, K. Jiang, N. Ta, Y. Hu, J. Zeng, J. Liu, H. Wang, Large-scale and highly selective CO_2 electrocatalytic reduction on nickel single-atom catalyst, *Joule*, 3 (2019) 265–278.
[14] W. Lee, Y.E. Kim, M.H. Youn, S.K. Jeong, K.T. Park, Catholyte-free electrocatalytic CO_2 reduction to formate, *Angewandte Chemie International Edition in English*, 57 (2018) 6883–6887.
[15] Z.Q. Liang, T.T. Zhuang, A. Seifitokaldani, J. Li, C.W. Huang, C.S. Tan, Y. Li, P. De Luna, C.T. Dinh, Y. Hu, Q. Xiao, P.L. Hsieh, Y. Wang, F. Li, R. Quintero-Bermudez, Y. Zhou, P. Chen, Y. Pang, S.C. Lo, L.J. Chen, H. Tan, Z. Xu, S. Zhao, D. Sinton, E.H. Sargent, Copper-on-nitride enhances the stable electrosynthesis of multi-carbon products from CO_2, *Nature Communications*, 9 (2018) 3828.
[16] C.-T. Dinh, T. Burdyny, M.G. Kibria, A. Seifitokaldani, C.M. Gabardo, F.P. García de Arquer, A. Kiani, J.P. Edwards, P. De Luna, O.S. Bushuyev, C. Zou, R. Quintero-Bermudez, Y. Pang, D. Sinton, E.H. Sargent, CO2 electroreduction to ethylene via hydroxide-mediated copper catalysis at an abrupt interface, *Science*, 360 (2018) 783–787.
[17] S. Ma, M. Sadakiyo, R. Luo, M. Heima, M. Yamauchi, P.J.A. Kenis, One-step electrosynthesis of ethylene and ethanol from CO_2 in an alkaline electrolyzer, *Journal of Power Sources*, 301 (2016) 219–228.

[18] M. Luo, Z. Wang, Y.C. Li, J. Li, F. Li, Y. Lum, D.H. Nam, B. Chen, J. Wicks, A. Xu, T. Zhuang, W.R. Leow, X. Wang, C.T. Dinh, Y. Wang, Y. Wang, D. Sinton, E.H. Sargent, Hydroxide promotes carbon dioxide electroreduction to ethanol on copper via tuning of adsorbed hydrogen, *Nature Communications*, 10 (2019) 5814.
[19] T.-T. Zhuang, Z.-Q. Liang, A. Seifitokaldani, Y. Li, P. De Luna, T. Burdyny, F. Che, F. Meng, Y. Min, R. Quintero-Bermudez, C.T. Dinh, Y. Pang, M. Zhong, B. Zhang, J. Li, P.-N. Chen, X.-L. Zheng, H. Liang, W.-N. Ge, B.-J. Ye, D. Sinton, S.-H. Yu, E.H. Sargent, Steering post-C–C coupling selectivity enables high efficiency electroreduction of carbon dioxide to multi-carbon alcohols, *Nature Catalysis*, 1 (2018) 421–428.
[20] J. Schneider, H. Jia, J.T. Muckerman, E. Fujita, Thermodynamics and kinetics of CO_2, CO, and H+ binding to the metal centre of CO_2 reduction catalysts, *Chemical Society Reviews*, 41 (2012) 2036–2051.
[21] W. Zhang, Y. Hu, L. Ma, G. Zhu, Y. Wang, X. Xue, R. Chen, S. Yang, Z. Jin, Progress and perspective of electrocatalytic CO_2 reduction for renewable carbonaceous fuels and chemicals, *Advanced Science*, 5 (2018) 1700275.
[22] M. Tekalgne, H. Do, A. Hasani, Q. Le, H. Jang, S. Ahn, S.Y. Kim, Two-dimensional materials and metal-organic frameworks for the CO_2 reduction reaction, *Materials Today Advances*, 5 (2019).
[23] G. Centi, S. Perathoner, G. Win, M. Gangeri, Electrocatalytic conversion of CO_2 to long carbon-chain hydrocarbons, *Green Chemistry*, 9 (2007).
[24] S. Liang, N. Altaf, L. Huang, Y. Gao, Q. Wang, Electrolytic cell design for electrochemical CO_2 reduction, *Journal of CO_2 Utilization*, 35 (2020) 90–105.
[25] T. Yamamoto, D.A. Tryk, A. Fujishima, H. Ohata, Production of syngas plus oxygen from CO_2 in a gas-diffusion electrode-based electrolytic cell, *Electrochimica Acta*, 47 (2002) 3327–3334.
[26] L. Mascaretti, A. Niorettini, B.R. Bricchi, M. Ghidelli, A. Naldoni, S. Caramori, A. Li Bassi, S. Berardi, Syngas evolution from CO_2 electroreduction by porous Au nanostructures, *ACS Applied Energy Materials*, 3 (2020) 4658–4668.
[27] C. Delacourt, P.L. Ridgway, J.B. Kerr, J. Newman, Design of an electrochemical cell making syngas ($CO+H_2$) from CO_2 and H_2O reduction at room temperature, *Journal of the Electrochemical Society*, 155 (2008) B42.
[28] M. Li, S. Garg, X. Chang, L. Ge, L. Li, M. Konarova, T.E. Rufford, V. Rudolph, G. Wang, Toward excellence of transition metal-based catalysts for CO_2 electrochemical reduction: An overview of strategies and rationales, *Small Methods*, 4 (2020) 2000033.
[29] S. Meshitsuka, M. Ichikawa, K. Tamaru, Electrocatalysis by metal phthalocyanines in the reduction of carbon dioxide, *Journal of the Chemical Society, Chemical Communications*, (1974) 158–159.
[30] S. Ren, D. Joulié, D. Salvatore, K. Torbensen, M. Wang, M. Robert, C.P. Berlinguette, Molecular electrocatalysts can mediate fast, selective CO_2 reduction in a flow cell, *Science*, 365 (2019) 367–369.
[31] J. Choi, P. Wagner, S. Gambhir, R. Jalili, D.R. MacFarlane, G.G. Wallace, D.L. Officer, Steric modification of a cobalt phthalocyanine/graphene catalyst to give enhanced and stable electrochemical CO_2 reduction to CO, *ACS Energy Letters*, 4 (2019) 666–672.
[32] Z. Zhang, J. Xiao, X.-J. Chen, S. Yu, L. Yu, R. Si, Y. Wang, S. Wang, X. Meng, Y. Wang, Z.-Q. Tian, D. Deng, Reaction mechanisms of well-defined metal-N_4 sites in electrocatalytic CO2 reduction, *Angewandte Chemie International Edition*, 57 (2018) 16339–16342.
[33] M. Abdinejad, A. Seifitokaldani, C. Dao, E.H. Sargent, X.-a. Zhang, H.B. Kraatz, Enhanced electrochemical reduction of CO_2 catalyzed by cobalt and iron amino porphyrin complexes, *ACS Applied Energy Materials*, 2 (2019) 1330–1335.

[34] J. Fu, W. Zhu, Y. Chen, Z. Yin, Y. Li, J. Liu, H. Zhang, J.J. Zhu, S. Sun, Bipyridine-assisted assembly of Au nanoparticles on Cu nanowires to enhance the electrochemical reduction of CO_2, *Angewandte Chemie*, 131 (2019) 14238–14241.
[35] M. Bourrez, F. Molton, S. Chardon-Noblat, A. Deronzier, [Mn(bipyridyl)(CO)3Br]: An abundant metal carbonyl complex as efficient electrocatalyst for CO_2 reduction, *Angewandte Chemie International Edition*, 50 (2011) 9903–9906.
[36] E.S. Donovan, B.M. Barry, C.A. Larsen, M.N. Wirtz, W.E. Geiger, R.A. Kemp, Facilitated carbon dioxide reduction using a Zn (II) complex, *Chemical Communications*, 52 (2016) 1685–1688.
[37] S.K. Lee, M. Kondo, G. Nakamura, M. Okamura, S. Masaoka, Low-overpotential CO_2 reduction by a phosphine-substituted Ru (ii) polypyridyl complex, *Chemical Communications*, 54 (2018) 6915–6918.
[38] J. Rosen, G.S. Hutchings, Q. Lu, S. Rivera, Y. Zhou, D.G. Vlachos, F. Jiao, Mechanistic insights into the electrochemical reduction of CO_2 to CO on nanostructured Ag surfaces, *ACS Catalysis*, 5 (2015) 4293–4299.
[39] T. Hatsukade, K.P. Kuhl, E.R. Cave, D.N. Abram, T.F. Jaramillo, Insights into the electrocatalytic reduction of CO_2 on metallic silver surfaces, *Physical Chemistry Chemical Physics*, 16 (2014) 13814–13819.
[40] S.B. Liu, H.B. Tao, L. Zeng, Q. Liu, Z.G. Xu, Q.X. Liu, J.L. Luo, Shape-dependent electrocatalytic reduction of CO_2 to CO on triangular silver nanoplates, *Journal of the American Chemical Society*, 139 (2017) 2160–2163.
[41] L. Zeng, J. Shi, H. Chen, C. Lin, Ag nanowires/C as a selective and efficient catalyst for CO_2 electroreduction, *Energies*, 14 (2021) 2840.
[42] Z. Wang, T. Li, Q. Wang, A. Guan, N. Cao, A.M. Al-Enizi, L. Zhang, L. Qian, G. Zheng, Hydrophobically made Ag nanoclusters with enhanced performance for CO_2 aqueous electroreduction, *Journal of Power Sources*, 476 (2020) 228705.
[43] R. Shi, J. Guo, X. Zhang, G.I. Waterhouse, Z. Han, Y. Zhao, L. Shang, C. Zhou, L. Jiang, T. Zhang, Efficient wettability-controlled electroreduction of CO_2 to CO at Au/C interfaces, *Nature Communications*, 11 (2020) 1–10.
[44] B. Qin, Y. Li, H. Fu, H. Wang, S. Chen, Z. Liu, F. Peng, Electrochemical reduction of CO_2 into tunable syngas production by regulating the crystal facets of earth-abundant Zn catalyst, *ACS Applied Materials & Interfaces*, 10 (2018) 20530–20539.
[45] A. Jo, S. Kim, H. Park, H.-Y. Park, J. Hyun Jang, H.S. Park, Enhanced electrochemical conversion of CO_2 to CO at bimetallic Ag-Zn catalysts formed on polypyrrole-coated electrode, *Journal of Catalysis*, 393 (2021) 92–99.
[46] K. Chen, X. Zhang, T. Williams, L. Bourgeois, D.R. MacFarlane, Electrochemical reduction of CO_2 on core-shell Cu/Au nanostructure arrays for syngas production, *Electrochimica Acta*, 239 (2017) 84–89.
[47] S. Sarfraz, A.T. Garcia-Esparza, A. Jedidi, L. Cavallo, K. Takanabe, Cu–Sn bimetallic catalyst for selective aqueous electroreduction of CO_2 to CO, *ACS Catalysis*, 6 (2016) 2842–2851.
[48] W. Guo, J. Bi, Q. Zhu, J. Ma, G. Yang, H. Wu, X. Sun, B. Han, Highly selective CO_2 electroreduction to CO on Cu–Co bimetallic catalysts, *ACS Sustainable Chemistry & Engineering*, 8 (2020) 12561–12567.
[49] P. Moreno-García, N. Schlegel, A. Zanetti, A. Cedeño López, M.d.J. Gálvez-Vázquez, A. Dutta, M. Rahaman, P. Broekmann, Selective electrochemical reduction of CO_2 to CO on Zn-based foams produced by Cu^{2+} and template-assisted electrodeposition, *ACS Applied Materials & Interfaces*, 10 (2018) 31355–31365.
[50] D. Gao, Y. Zhang, Z. Zhou, F. Cai, X. Zhao, W. Huang, Y. Li, J. Zhu, P. Liu, F. Yang, G. Wang, X. Bao, Enhancing CO_2 electroreduction with the metal-oxide interface, *Journal of the American Chemical Society*, 139 (2017) 5652–5655.

[51] Z. Geng, X. Kong, W. Chen, H. Su, Y. Liu, F. Cai, G. Wang, J. Zeng, Oxygen vacancies in ZnO nanosheets enhance CO_2 electrochemical reduction to CO, *Angewandte Chemie*, 130 (2018) 6162–6167.
[52] X. Jiang, H. Wu, S. Chang, R. Si, S. Miao, W. Huang, Y. Li, G. Wang, X. Bao, Boosting CO_2 electroreduction over layered zeolitic imidazolate frameworks decorated with Ag_2O nanoparticles, *Journal of Materials Chemistry A*, 5 (2017) 19371–19377.
[53] K. Ye, T. Liu, Y. Song, Q. Wang, G. Wang, Tailoring the interactions of heterogeneous Ag2S/Ag interface for efficient CO_2 electroreduction, *Applied Catalysis B: Environmental*, 296 (2021) 120342.
[54] T. Möller, W. Ju, A. Bagger, X. Wang, F. Luo, T.N. Thanh, A.S. Varela, J. Rossmeisl, P. Strasser, Efficient CO_2 to CO electrolysis on solid Ni–N–C catalysts at industrial current densities, *Energy & Environmental Science*, 12 (2019) 640–647.
[55] T.N. Huan, N. Ranjbar, G. Rousse, M. Sougrati, A. Zitolo, V. Mougel, F. Jaouen, M. Fontecave, Electrochemical reduction of CO_2 catalyzed by Fe-N-C materials: A structure–selectivity study, *ACS Catalysis*, 7 (2017) 1520–1525.
[56] X. Duan, J. Xu, Z. Wei, J. Ma, S. Guo, S. Wang, H. Liu, S. Dou, Metal-free carbon materials for CO_2 electrochemical reduction, *Advanced Materials*, 29 (2017).
[57] P.P. Sharma, J. Wu, R.M. Yadav, M. Liu, C.J. Wright, C.S. Tiwary, B.I. Yakobson, J. Lou, P.M. Ajayan, X.D. Zhou, Nitrogen-doped carbon nanotube arrays for high-efficiency electrochemical reduction of CO_2: On the understanding of defects, defect density, and selectivity, *Angewandte Chemie*, 127 (2015) 13905–13909.
[58] C. Delacourt, P.L. Ridgway, J. Newman, Mathematical modeling of CO_2 reduction to CO in aqueous electrolytes I. Kinetic study on planar silver and gold electrodes, *Journal of the Electrochemical Society*, 157 (2010) B1902–B1910.
[59] E.L. Clark, M.R. Singh, Y. Kwon, A.T. Bell, Differential electrochemical mass spectrometer cell design for online quantification of products produced during electrochemical reduction of CO_2, *Analytical Chemistry*, 87 (2015) 8013–8020.
[60] E.L. Clark, A.T. Bell, Direct observation of the local reaction environment during the electrochemical reduction of CO_2, *Journal of the American Chemical Society*, 140 (2018) 7012–7020.
[61] M.A. Laguna-Bercero, Recent advances in high temperature electrolysis using solid oxide fuel cells: A review, *Journal of Power Sources*, 203 (2012) 4–16.
[62] Y. Xie, J. Xiao, D. Liu, J. Liu, C. Yang, Electrolysis of carbon dioxide in a solid oxide electrolyzer with silver-gadolinium-doped ceria cathode, *Journal of the Electrochemical Society*, 162 (2015) F397–F402.
[63] H. Lv, Y. Zhou, X. Zhang, Y. Song, Q. Liu, G. Wang, X. Bao, Infiltration of Ce0.8Gd0.2O1.9 nanoparticles on Sr2Fe1.5Mo0.5O6-δ cathode for CO_2 electroreduction in solid oxide electrolysis cell, *Journal of Energy Chemistry*, 35 (2019) 71–78.
[64] X.M. Zhang, Y.F. Song, G.X. Wang, X.H. Bao, Co-electrolysis of CO_2 and H_2O in high-temperature solid oxide electrolysis cells: Recent advance in cathodes, *Journal of Energy Chemistry*, 26 (2017) 839–853.
[65] R. Ferrigno, A.D. Stroock, T.D. Clark, M. Mayer, G.M. Whitesides, Membraneless vanadium redox fuel cell using laminar flow, *Journal of the American Chemical Society*, 124 (2002) 12930–12931.
[66] D.T. Whipple, E.C. Finke, P.J.A. Kenis, Microfluidic reactor for the electrochemical reduction of carbon dioxide: The effect of pH, *Electrochemical and Solid-State Letters*, 13 (2010) D109–D111.
[67] X. Lu, D.Y.C. Leung, H.Z. Wang, J. Xuan, A high performance dual electrolyte microfluidic reactor for the utilization of CO_2, *Applied Energy*, 194 (2017) 549–559.

[68] J.P. Edwards, Y. Xu, C.M. Gabardo, C.-T. Dinh, J. Li, Z. Qi, A. Ozden, E.H. Sargent, D. Sinton, Efficient electrocatalytic conversion of carbon dioxide in a low-resistance pressurized alkaline electrolyzer, *Applied Energy*, 261 (2020).
[69] Y. Kotb, S.-E.K. Fateen, J. Albo, I. Ismail, Modeling of a microfluidic electrochemical cell for the electro-reduction of CO2 to CH3OH, *Journal of The Electrochemical Society*, 164 (2017) E391–E400.
[70] C. Chen, J.F. Khosrowabadi Kotyk, S.W. Sheehan, Progress toward commercial application of electrochemical carbon dioxide reduction, *Chem*, 4 (2018), 2571–2586.
[71] K. Yang, R. Kas, W.A. Smith, T. Burdyny, Role of the carbon-based gas diffusion layer on flooding in a gas diffusion electrode cell for electrochemical CO_2 reduction, *ACS Energy Letters*, 6 (2021) 33–40.
[72] T. Burdyny, W.A. Smith, CO_2 reduction on gas-diffusion electrodes and why catalytic performance must be assessed at commercially-relevant conditions, *Energy & Environmental Science*, 12 (2019) 1442–1453.
[73] Z. Yin, H. Peng, X. Wei, H. Zhou, J. Gong, M. Huai, L. Xiao, G. Wang, J. Lu, L. Zhuang, An alkaline polymer electrolyte CO_2 electrolyzer operated with pure water, *Energy & Environmental Science*, 12 (2019) 2455–2462.
[74] B. Endrődi, E. Kecsenovity, A. Samu, F. Darvas, R.V. Jones, V. Török, A. Danyi, C. Janáky, Multilayer electrolyzer stack converts carbon dioxide to gas products at high pressure with high efficiency, *ACS Energy Letters*, 4 (2019) 1770–1777.
[75] D.M. Weekes, D.A. Salvatore, A. Reyes, A. Huang, C.P. Berlinguette, Electrolytic CO_2 reduction in a flow cell, *Accounts of Chemical Research*, 51 (2018) 910–918.
[76] D.A. Salvatore, D.M. Weekes, J.F. He, K.E. Dettelbach, Y.G.C. Li, T.E. Mallouk, C.P. Berlinguette, Electrolysis of gaseous CO_2 to CO in a flow cell with a bipolar membrane, *ACS Energy Letters*, 3 (2018) 149–154.
[77] C.-T. Dinh, F.P. García de Arquer, D. Sinton, E.H. Sargent, High rate, selective, and stable electroreduction of CO_2 to CO in basic and neutral media, *ACS Energy Letters*, 3 (2018) 2835–2840.
[78] J. Li, G. Chen, Y. Zhu, Z. Liang, A. Pei, C.-L. Wu, H. Wang, H.R. Lee, K. Liu, S. Chu, Y. Cui, Efficient electrocatalytic CO_2 reduction on a three-phase interface, *Nature Catalysis*, 1 (2018) 592–600.
[79] Q. Wan, Q. He, Y. Zhang, L. Zhang, J. Li, J. Hou, X. Zhuang, C. Ke, J. Zhang, Boosting the faradaic efficiency for carbon dioxide to monoxide on a phthalocyanine cobalt based gas diffusion electrode to higher than 99% via microstructure regulation of catalyst layer, *Electrochim. Acta*, 392 (2021) 139023.
[80] E.W. Lees, B.A. Mowbray, D.A. Salvatore, G.L. Simpson, D.J. Dvorak, S. Ren, J. Chau, K.L. Milton, C.P. Berlinguette, Linking gas diffusion electrode composition to CO_2 reduction in a flow cell, *Journal of Materials Chemistry A*, 8 (2020) 19493–19501.
[81] B. Endrodi, G. Bencsik, F. Darvas, R. Jones, K. Rajeshwar, C. Janaky, Continuous-flow electroreduction of carbon dioxide, *Progress in Energy and Combustion Science*, 62 (2017) 133–154.
[82] C. Delacourt, P.L. Ridgway, J.B. Kerr, J. Newman, Design of an electrochemical cell making syngas ($CO.H_2O$) from CO_2 and H_2O reduction at room temperature, *Journal of the Electrochemical Society*, 155 (2008) B42–B49.
[83] L. Ma, S. Fan, D. Zhen, X. Wu, S. Liu, J. Lin, S. Huang, W. Chen, G. He, Electrochemical reduction of CO_2 in proton exchange membrane reactor: The function of buffer layer. *Industrial & Engineering Chemistry Research*, 56 (2017) 10242–10250.
[84] C.M. Gabardo, A. Seifitokaldani, J.P. Edwards, C.-T. Dinh, T. Burdyny, M.G. Kibria, C.P. O'Brien, E.H. Sargent, D. Sinton, Combined high alkalinity and

pressurization enable efficient CO_2 electroreduction to CO, *Energy & Environmental Science*, 11 (2018) 2531–2539.
[85] X.R. Gao, Y. Yu, Q.R. Liang, Y.J. Pang, L.Q. Miao, X.M. Liu, Z.K. Kou, J.Q. He, S.J. Pennycook, S.C. Mu, J. Wang, Surface nitridation of nickel-cobalt alloy nanocactoids raises the performance of water oxidation and splitting, *Appl. Catal. B-Environ*, 270 (2020) 9.
[86] X. Lu, W.-L. Yim, B.H.R. Suryanto, C. Zhao, Electrocatalytic oxygen evolution at surface-oxidized multiwall carbon nanotubes, *Journal of the American Chemical Society*, 137 (2015) 2901–2907.
[87] X. Lu, C. Zhao, Electrodeposition of hierarchically structured three-dimensional nickel-iron electrodes for efficient oxygen evolution at high current densities. *Nature Communications*, 6 (2015) 6616.
[88] Z. Liu, B. Tang, X.C. Gu, H. Liu, L.G. Feng, Selective structure transformation for NiFe/$NiFe_2O_4$ embedded porous nitrogen-doped carbon nanosphere with improved oxygen evolution reaction activity, *Chem. Eng. J.*, 395 (2020) 13.
[89] L. Trotochaud, S.L. Young, J.K. Ranney, S.W. Boettcher, Nickel-iron oxyhydroxide oxygen-evolution electrocatalysts: The role of intentional and incidental iron incorporation. *Journal of the American Chemical Society*, 136 (2014) 6744–6753.
[90] R. Subbaraman, D. Tripkovic, K.C. Chang, D. Strmcnik, A.P. Paulikas, P. Hirunsit, M. Chan, J. Greeley, V. Stamenkovic, N.M. Markovic, Trends in activity for the water electrolyser reactions on 3d m(Ni,Co,Fe,Mn) hydr(oxy)oxide catalysts, *Nature Materials*, 11 (2012) 550–557.
[91] C.T. Dinh, T. Burdyny, M.G. Kibria, A. Seifitokaldani, C.M. Gabardo, F.P.G. de Arquer, A. Kiani, J.P. Edwards, P. De Luna, O.S. Bushuyev, C.Q. Zou, R. Quintero-Bermudez, Y.J. Pang, D. Sinton, E.H. Sargent, CO2 electroreduction to ethylene via hydroxide-mediated copper catalysis at an abrupt interface, *Science*, 360 (2018) 783–787.
[92] J. Resasco, L.D. Chen, E. Clark, C. Tsai, C. Hahn, T.F. Jaramillo, K. Chan, A.T. Bell, Promoter effects of alkali metal cations on the electrochemical reduction of carbon dioxide, *Journal of the American Chemical Society*, 139 (2017) 11277–11287.
[93] K.J.P. Schouten, Y. Kwon, C.J.M. van der Ham, Z. Qin, M.T.M. Koper, A new mechanism for the selectivity to C-1 and C-2 species in the electrochemical reduction of carbon dioxide on copper electrodes, *Chemical Science*, 2 (2011) 1902–1909.
[94] C.I. Shaughnessy, D.J. Sconyers, H.J. Lee, B. Subramaniam, J.D. Blakemore, K.C. Leonard, Insights into pressure tunable reaction rates for electrochemical reduction of co_2 in organic electrolytes, *Green Chemistry*, 22 (2020) 2434–2442.
[95] J.E. Brennecke, B.E. Gurkan, Ionic liquids for CO_2 capture and emission reduction, *Journal of Physical Chemistry Letters*, 1 (2010) 3459–3464.
[96] J.E. Bara, D.E. Camper, D.L. Gin, R.D. Noble, Room-temperature ionic liquids and composite materials: Platform technologies for CO_2 capture, *Accounts of Chemical Research*, 43 (2010) 152–159.
[97] L.Y. Sun, G.K. Ramesha, P.V. Kamat, J.F. Brennecke, Switching the reaction course of electrochemical CO_2 reduction with ionic liquids. *Langmuir*, 30 (2014) 6302–6308.
[98] B.W. Deng, M.X. Gao, R. Yu, X.H. Mao, R. Jiang, D.H. Wang, Critical operating conditions for enhanced energy-efficient molten salt CO_2 capture and electrolytic utilization as durable looping applications. *Applied Energy*, 255 (2019) 10.
[99] G. Kaur, A.P. Kulkarni, S. Giddey, S.P.S. Badwal, Ceramic composite cathodes for CO_2 conversion to co in solid oxide electrolysis cells. *Applied Energy*, 221 (2018) 131–138.

[100] J.P. Edwards, Y. Xu, C.M. Gabardo, C.-T. Dinh, J. Li, Z. Qi, A. Ozden, E.H. Sargent, D. Sinton, Efficient electrocatalytic conversion of carbon dioxide in a low-resistance pressurized alkaline electrolyzer, *Applied Energy*, 261 (2020) 114305.
[101] K. Hara, T. Sakata, Large current density CO_2 reduction under high pressure using gas diffusion electrodes, *Bulletin of the Chemical Society of Japan*, 70 (1997) 571–576.
[102] C.M. Gabardo, A. Seifitokaldani, J.P. Edwards, C.-T. Dinh, T. Burdyny, M.G. Kibria, C.P. O'Brien, E.H. Sargent, D. Sinton, Combined high alkalinity and pressurization enable efficient CO_2 electroreduction to CO. *Energy & Environmental Science*, 11 (2018) 2531–2539.
[103] S. Hernández, M.A. Farkhondehfal, F. Sastre, M. Makkee, G. Saracco, N. Russo, Syngas production from electrochemical reduction of CO_2: Current status and prospective implementation. *Green Chemistry*, 19 (2017) 2326–2346.
[104] A. Löwe, C. Rieg, T. Hierlemann, N. Salas, D. Kopljar, N. Wagner, E. Klemm, Influence of temperature on the performance of gas diffusion electrodes in the CO_2 reduction reaction. *ChemElectroChem*, 6 (2019) 4497–4506.
[105] E.J. Dufek, T.E. Lister, M.E. McIlwain, Bench-scale electrochemical system for generation of CO and syn-gas. *J. Appl. Electrochem.*, 41 (2011) 623–631.
[106] S. Hernández, M.A. Farkhondehfal, F. Sastre, M. Makkee, G. Saracco, N. Russo, Syngas production from electrochemical reduction of CO2: Current status and prospective implementation, *Green Chemistry*, 19 (2017) 2326–2346.
[107] W. Luc, B.H. Ko, S. Kattel, S. Li, D. Su, J.G. Chen, F. Jiao, SO_2-induced selectivity change in CO_2 electroreduction, *Journal of the American Chemical Society*, 141 (2019) 9902–9909.
[108] B.H. Ko, B. Hasa, H. Shin, E. Jeng, S. Overa, W. Chen, F. Jiao, The impact of nitrogen oxides on electrochemical carbon dioxide reduction, *Nature Communications*, 11 (2020) 1–9.
[109] Y.E. Kim, H.S. Yun, S.K. Jeong, Y.I. Yoon, S.C. Nam, Park K.T., Electrocatalytic stability of tin cathode for electroreduction of CO_2 to formate in aqueous solution. *Journal of Nanoscience and Nanotechnology*, 18 (2018) 1266–1269.
[110] X. Peng, S. Karakalos, W. E. Mustain, Preferentially Oriented Ag Nanocrystals with Extremely High Activity and Faradaic Efficiency for CO_2 Electrochemical Reduction to CO, *ACS Applied Materials & Interfaces*, 10 (2018) 1734–1742.
[111] C. Rogers, W.S. Perkins, G. Veber, T.E. Williams, R.R. Cloke, F.R. Fischer, Synergistic enhancement of electrocatalytic CO_2 reduction with gold nanoparticles embedded in functional graphene nanoribbon composite electrodes. *Journal of the American Chemical Society*, 139 (2017) 4052–4061.
[112] Y. Wu, Z. Jiang, X. Lu, Y. Liang, H. Wang, Domino electroreduction of CO_2 to methanol on a molecular catalyst. *Nature*, 575 (2019) 639–642.
[113] U.O. Nwabara, E.R. Cofell, S. Verma, E. Negro, P.J. Kenis, Durable cathodes and electrolyzers for the efficient aqueous electrochemical reduction of CO_2. *ChemSusChem*, 13 (2020) 855–875.
[114] B. Endrődi, A. Samu, E. Kecsenovity, T. Halmágyi, D. Sebők, C. Janáky, Operando cathode activation with alkali metal cations for high current density operation of water-fed zero-gap carbon dioxide electrolysers, *Nature Energy*, 6 (2021) 439–448.

13 Improving the Electrocatalytic Performance by Defect Engineering and External Field Regulation

Jing Hu, Yuanyuan Zhang, and Ping Xu
School of Chemistry and Chemical Engineering,
Harbin Institute of Technology, Harbin,
People's Republic of China

CONTENTS

Recently, researchers have focused on the electrochemical energy conversion devices, which can realize the conversion between electric energy and chemical energy by converting the earth-abundant molecules (e.g., water, carbon dioxide, nitrogen) into high-value products or utilizing the oxidation of fuel molecules

DOI: 10.1201/9781003133971-13

(e.g., methanol, formic acid, and ethanol) to generate electric energy. In these energy conversion systems, electrocatalysts are essential due to the roles in decreasing the reaction activation energy and improving the reaction rate and energy efficiency [1]. Hence, exploiting electrocatalysts with high efficiency, low cost, and high stability is of primary importance for renewable energy systems.

In general, electrocatalytic reactions occur on the electrode-electrolyte interfaces, usually along with the mass transport of gas, liquid molecules, and ions. In these heterogeneous electrocatalytic processes, electrocatalysts play critical roles in promoting the adsorption/desorption of the reactant/intermediate/product and increasing the reaction efficiency [2]. On the basis of the Sabatier principle, the adsorption of reaction species should not be too keen to limit the product desorption or too weak to restrict the reactant activation. Previous works indicated that adsorption energy is strongly related to the electronic structure of surface atoms of electrocatalysts [3,4]. For these reasons, modulating the adsorption energy by tuning surface electronic structures is a practical approach to improve the catalytic activity of electrocatalysts. Recently, researchers have exploited many strategies to prepare highly efficient electrocatalysts, including controlling the crystal facet, alloying, doping heteroatoms, defect engineering, interface engineering, application of the external field, and so on [5–7]. Researchers have also developed a series of theoretical computation methods and related catalytic mechanisms, which can effectively understand the interaction between reactants with catalysts, especially simulating the intermediate adsorption processes [8].

13.1 DEFECTS ENGINEERING OF ELECTROCATALYSTS

Defects that exist widely in nanomaterials can alter the electronic structure and surface property of nanomaterials [9,10]. Recently, studies have demonstrated that the existence of defects in the catalysts is beneficial to the performance improvement [11–16]. For example, Wang and co-workers have summarized the recent exciting findings of defects in catalysts for oxygen reactions and explained the effects of defects on the electrocatalytic performance [17]. In addition to the electrochemical reactions activated by the defects themselves, the defects are also used to provide unique anchor sites for the capture of metal species [18].

Among surface engineering strategies, there is no doubt that defect engineering is of significance for electrocatalysts. Defects exist in almost all of the heterogeneous catalysts and can directly break the periodic crystalline structure of materials. Sequentially, defects modulate the surface electronic structure with localized electron re-distribution. With different defect types, numbers, and locations, materials could be endowed with many different properties in electricity, optics, and chemistry. Thus, defective materials could be widely applied in various fields, including electronics/optoelectronics, catalysis, sensor, energy storage, and conversion devices [19].

The d-band theory has been used to explain the effect of surface energy band change on electrocatalytic performance by Nørskov's group. For the compounds formed by iron, cobalt, nickel, and other elements, surface defects will lead to the change of energy band, and then affect the adsorption of active substances, as well as to achieve the purpose of regulating the electrocatalytic performance. It is well

Idealized *d*- band filling

V_{ad}^2 [Relative to Cu]

0.1 20.8 **Ca** -0.48 4.12	0.2 7.90 **Sc** -0.61 3.43	0.3 4.65 **Ti** -0.46 3.05	0.4 3.15 **V** -0.38 2.82	0.5 2.35 **Cr** -0.01 2.68	0.6 1.94 **Mn** 0.10 2.70	0.7 1.59 **Fe** 0.29 2.66	0.8 1.34 **Co** 0.37 2.62	0.9 1.16 **Ni** 0.47 2.60	1.0 1.0 **Cu** 0.59 2.67	1.0 0.46 **Zn** 2.65
0.1 36.5 **Sr** -0.28 4.49	0.2 17.3 **Y** -0.55 3.76	0.3 10.9 **Zr** -0.44 3.35	0.4 7.73 **Nb** -0.37 3.07	0.5 6.62 **Mo** 0.01 2.99	0.6 4.71 **Tc** 0.24 2.84	0.7 3.87 **Ru** 0.49 2.79	0.8 3.32 **Rh** 0.75 2.81	0.9 2.78 **Pd** 0.81 2.87	1.0 2.26 **Ag** 0.68 3.01	1.0 1.58 **Cd** 3.1
0.1 41.5 **Ba** -0.44 4.65	0.2 17.1 **Lu** -0.55 3.62	0.3 11.9 **Hf** -0.51 3.30	0.4 9.05 **Ta** -0.50 3.07	0.5 7.27 **W** -0.08 2.95	0.6 6.04 **Re** 0.15 2.87	0.7 5.13 **Os** 0.48 2.83	0.8 4.45 **Ir** 0.90 2.84	0.9 3.90 **Pt** 1.08 2.90	1.0 3.35 **Au** 1.08 3.00	1.0 2.64 **Hg** 3.1

Bulk Wigner-Seitz radius, *s* [au]

$d\epsilon_d/d\ ln s$ [au]

FIGURE 13.1 Section of the periodic table with the 3d, 4d, and 5d transition metals and the noble metals. Shown in lower-right corner: bulk Wigner–Seitz radius, s. Lower-left corner: the $d\epsilon_d/d \ln s$. Note how this changes its sign through a row at about half d-band filling as expected from the rigid band model. In the upper-right corner: behavior of the adsorbate s or p-metal d coupling matrix element squared, V_{ad}^2. The V_{ad}^2's generally decrease for increasing nuclear charge within a row and increase down the groups. All above numbers except for the properties of Zn, Cd, and Hg, have been compiled from the above ref. [20]. Finally, upper-left corner: the idealized d-band fillings. These are found to be close to the actual, calculated bulk d-band fillings considering the uncertainties in interpreting these [20].

known from LMTO theory that the d-band centers, ϵ_d, depend sensitively on the Wigner-Seitz radius *s*. In fact, $d\epsilon_d/d \ln s$ have been tabulated for all the elemental metals [20]. Values of *s* and $d\epsilon_d/d \ln s$ for all 3d, 4d, and 5d transition and noble metals are given in Figure 13.1. The concept of a Wigner-Seitz radius containing a neutral atom in a perfect solid can be generalized by the neutral radius, defined to contain a neutral atom for any configuration. For a surface atom, the neutral radius is larger than in the bulk because the contribution to the electron density around a surface atom from the neighbors is smaller than in the bulk. It is seen in Figure 13.1 that $d\epsilon_d/d \ln s$ is positive for all the metals to be considered here Fe-Cu, Ru-Ag, and Ir-Au. Consequently, the d-band centers at the surface are higher in energy than for the bulk. This is in accordance with experience from the self-consistent density functional calculations.

A great deal of defect electrocatalysts, including noble metals, carbon materials, transition metal oxides/hydroxides, and transition metal dichalcogenides, have been prepared and applied in many electrocatalytic reactions, such as oxygen reduction reaction (ORR), oxygen evolution reaction (OER), hydrogen evolution reaction (HER), hydrogen oxidation reaction (HOR), carbon dioxide reduction reaction (CO_2 RR), nitrogen reduction reaction (NRR), and methanol oxidation reaction (MOR). Nevertheless, the diversity and complexity of defects prevented the in-depth understanding of the defect-catalysis. Besides, a specific defect in a typical environment for various materials and reaction systems could play a completely different role. There are still many questions that have not been answered yet. How can defects influence

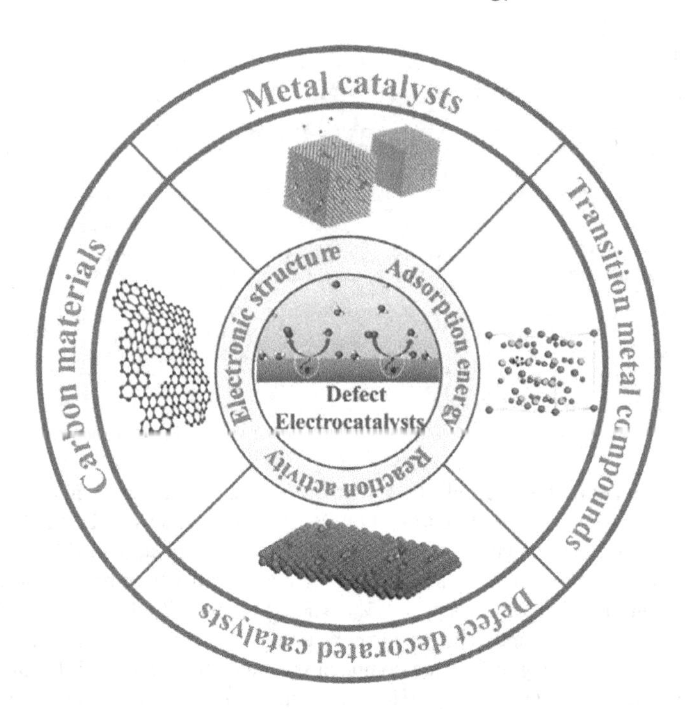

FIGURE 13.2 Schematic illustration of various defect electrocatalysts and important factors including electronic structure, adsorption energy of intermediates, and reaction activity in the design of defect electrocatalysts.

the electrocatalytic activity? How to you rationally utilize the defect to improve the electrocatalytic activity and what principles should be followed when designing the defect electrocatalysts? Even though researchers have reviewed the application of defect engineering in electrocatalytic reactions from different aspects to provide some guidance of design of defect catalysts [16,17,21–26], it is still necessary to systematically study the correlation between defect and catalytic activity when designing defect catalysts. A further understanding of the relationship between defects and electronic structure, adsorption energy, and reaction path in electrocatalysis requires the combination of experimental and theoretical researches simultaneously (Figure 13.2). It is expected that the rational design of defect electrocatalysts should be emphasized more and the defect-catalysis relationship could be closely linked by the electronic structure and adsorption energy. In view of this, several kinds of catalysts have been chosen for further exploration.

13.1.1 Defects in Carbon Materials and Transition Metal Carbides

Carbon materials are kinds of traditional electrode materials with simple preparation and diverse morphology. Moreover, carbon materials also have many intrinsic defects, which can be used to improve the catalytic performance of the electrode.

The conventional defect sites in carbon-based nanomaterials include intrinsic carbon defects (direct processing in the conjugated network without any dopants, such as edges, defects, holes, or topological defects), extrinsic defects (mainly heteroatoms or single metal atoms doping), and combination sites between each other [25,27]. Some typical catalytic centers comprising intrinsic carbon defects are illustrated in Figure 13.3. Defect engineering first refers to constructing the targeted defect species into the synthesized carbon framework effectively by means of various elaborate processing methods, aimed at obtaining defect-induced carbon-based catalysts with high activity and durability. Once obtaining the defective carbon nanomaterials, the specific configuration and defect density need to be clarified through ex-, quasi-, and in-situ evaluation. Thus, it also covers physical characterizations of the defect species, which can be realized by electron microscope images or spectroscopic analysis results. Then, the activity assessment of defective carbon-based electrocatalysts should be conducted by stressing the inherent contribution of defect sites and excluding the effect of other contributors. Most importantly, it is intended to understand the influence of specific defect sites on the charge state and practical functional mechanism of catalysts, accordingly clarifying the activity origins of different defect sites for different electrocatalytic reactions.

Lastly, defect engineering should also involve the scale preparation of defective carbon-based electrocatalysts, in an attempt to promote their industrial application.

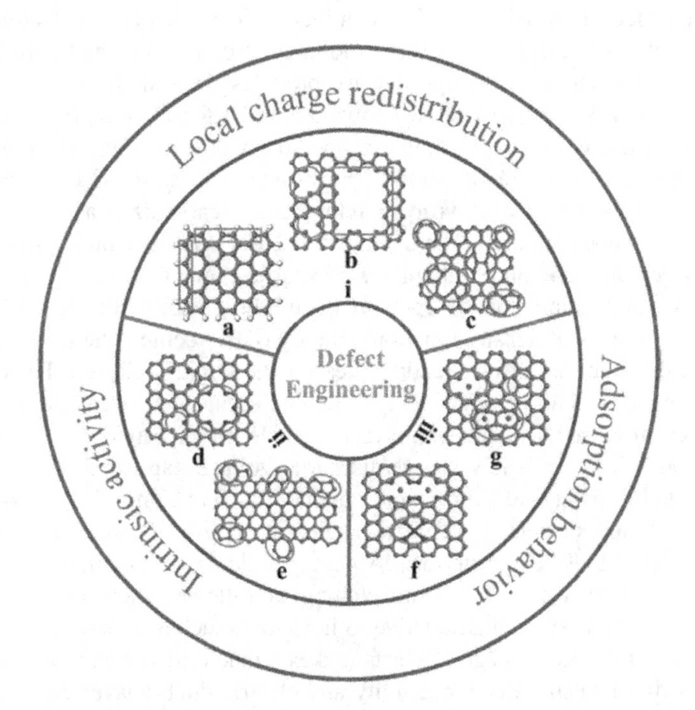

FIGURE 13.3 Schematic illustration of typical intrinsic carbon defect-involved catalytic sites. (*i*) intrinsic carbon defects (a: edge sites, b: defects and holes, c: topological defects); (*ii*) synergetic sites of intrinsic defects and heteroatom dopants (d: N-vacancy synergistic sites, e: heteroatom-topological defect sites); (*iii*) collaborative sites of intrinsic defects and single metal atoms (f: M-N-C sites, g: M-C sites).

Defect engineering is destined to be the central topic of future research about carbon-based catalysts. Wang's group used HOPG (highly oriented pyrolytic graphite) as a model to investigate the defects and the defect-induced charge behavior and correlated the electrocatalytic activities with the surface charge of the active sites. HOPG owns a perfect surface structure [28]. With HOPG as a starting substrate and ideal model, the defect level can be easily controlled through an effective plasma etching strategy. Via simple plasma irradiation, numerous defects are generated, which would induce charge re-distribution on the surface of HOPG. The electrochemical characterizations indicated that the charge enhanced the activity for electrocatalytic reactions (ORR, OER, and HER). The high activity induced by surface charge is confirmed by the DFT calculation. This work provides an efficient way to charge the active sites by edging carbon-based electrocatalysts and proposing a direct relationship between surface charge and the electrocatalytic activity. Lei's group also studied carbon-based catalyst with abundant edge defects [29]. A facile alkali activation method has been proposed to prepare carbon with a large specific surface area and optimized porosity. In addition, subsequent nitrogen-doping leads to high pyridinic-N and graphitic-N contents and abundant edge defects, further enhancing electrochemical activities.

Theoretical modeling via first-principles calculations has been conducted to correlate the electrocatalytic activities with their fundamental chemical structure of N doping and edge defect engineering. The integration of optimized porosity, rich active sites, and optimized charge density provides the sample with a potential difference of 0.92 V and good stability toward both ORR and OER. Jiang et al. [30] reported a spontaneous gas-foaming method to prepare nitrogen-doped ultrathin carbon nanosheets (NCNs) by simply pyrolysing a mixture of citric acid and NH_4Cl. Under the optimized pyrolysis temperature (carbonized at 1000 °C) and mass ratio of precursors (1:1), the synthesized NCN-1000-5 sample possesses an ultrathin sheet structure, an ultrahigh specific surface area (1793 $m^2\ g^{-1}$), and rich edge defects, and exhibits low overpotential and robust stability for the ORR, OER, and HER. By means of density functional theory (DFT) computations, the author's group revealed that the intrinsic active sites for the ORR, OER, and HER are the carbon atoms located at the armchair edge and adjacent to the graphitic N dopants. When practically used as a catalyst in rechargeable Zn-air batteries, a high energy density (806 Wh kg^{-1}), a low charge/discharge voltage gap (0.77 V), and an ultralong cycle life (over 330 hours) were obtained at 10 mA cm^{-2} for NCN-1000-5. The effects of intrinsic pentagon defects on electrocatalysis of carbon nanomaterials had been studied by Mu's group [31]. A pentagon-defect-rich carbon nanomaterial was constructed by means of in-situ etching of fullerene molecules (C_{60}). The electrochemical tests show that, relative to hexagons, such a carbon-based material with abundant intrinsic pentagon defects makes a much greater contribution to the electrocatalytic oxygen reduction activity and electric double layer capacitance. It shows a four-electron-reaction mechanism similar to commercial Pt/C and other transition-metal-based catalysts, and a higher specific capacitance than many reported metal-free carbon-based materials. Density functional theory (DFT) calculations show that the intrinsic pentagons in a carbon matrix possess splendid electrochemical reactivity due to large charge densities and high binding affinity toward oxygen; thus, they can serve as the potential active sites in electrochemical reactions.

Charge transfer is the process with charge relocated permanently between the "donor" and "acceptor", [32] which can occur by doping/lattice regroup in molecular (intramolecular charge transfer) or two different molecules with different work function (intermolecular charge transfer). Electrocatalytic activity of carbon materials can be regulated by both the intramolecular charge transfer (heteroatoms doping, defect engineering, and single metal atom coordination) and intermolecular charge transfer (molecular functionalization and heterojunction) [33]. The charge localization on the electrocatalyst surface can regulate the state of adsorbed molecules or reaction intermediates [34] by changing the adsorption model and reducing the adsorption energy, which can result in tunable catalytic activity. In a carbon-based electrocatalyst, intramolecular charge transfer effects mean the in-plane charge transfer including heteroatom doping, defect/vacancy engineering, and single metal atom coordination.

Since Dai and co-workers employed nitrogen-doped carbon nanotube as ORR electrocatalysts in 2009 [35], as an efficient way, regulating the electronic structure and electrocatalytic activity of carbon materials by heteroatom doping has been widely investigated [36,37]. Through substituting certain carbon atoms by heteroatoms, as shown in Figure 13.4(a), the charge and spin of the neighboring carbon atoms would redistribute due to the different electronegativity between heteroatoms and carbon atoms, and the change of charge and spin of the carbon can regulate its work function and the adsorption energy of reactants or intermediates at specific sites [38]. Nitrogen doping would cause a charge transfer with the neighbor carbon atoms, which can then cause the charge redistribution in the p-conjugated system around the dopants in carbon materials. As shown in Figure 13.4(b), after doping, the carbon atoms adjacent to the electron-accepting nitrogen atoms would be positively charged, and the chemisorption mode of O_2 on the positive carbon atoms would change to a side-on bridge-type (Yeager model) from the end-on adsorption model (Pauling mode) (Figure 13.4(c)). As a result, the adsorption barriers of the doped carbon materials can be significantly reduced [35,39]. In addition, nitrogen doping could also efficiently reduce the bandgap of carbon materials by lifting the highest-occupied molecular orbital (HOMO), which will lengthen (thus weaken) oxygen-oxygen bond to increase the charge transfer between the doped carbon materials and absorbed oxygen molecule (Figure 13.4(d)) [40]. By employing HOPG as the model catalyst with a well-controlled N species doping type (Figure 13.4(e)), the density functional theory (DFT) calculation and experiment have demonstrated that the Lewis basicity carbon atoms near the pyridine nitrogen atoms in the nitrogen-doped carbon are ORR active sites (active sites) [41]. Figure 13.4(f) is the typical schematic for ORR pathway on heteroatom-doped carbon materials; the oxygen molecule is first adsorbed on the carbon site near the N atoms (species B), then reduced by a proton-coupled electron and forms the intermediate OOH species (C), and then the OOH species would be reduced by a direct four-electron mechanism or a sluggish 2+2-electron pathway. In a N-doped carbon, the Lewis basic sites created by pyridine-N and the activated adjacent carbon atoms are the ORR active sites. In general, the charge transfer effect induced

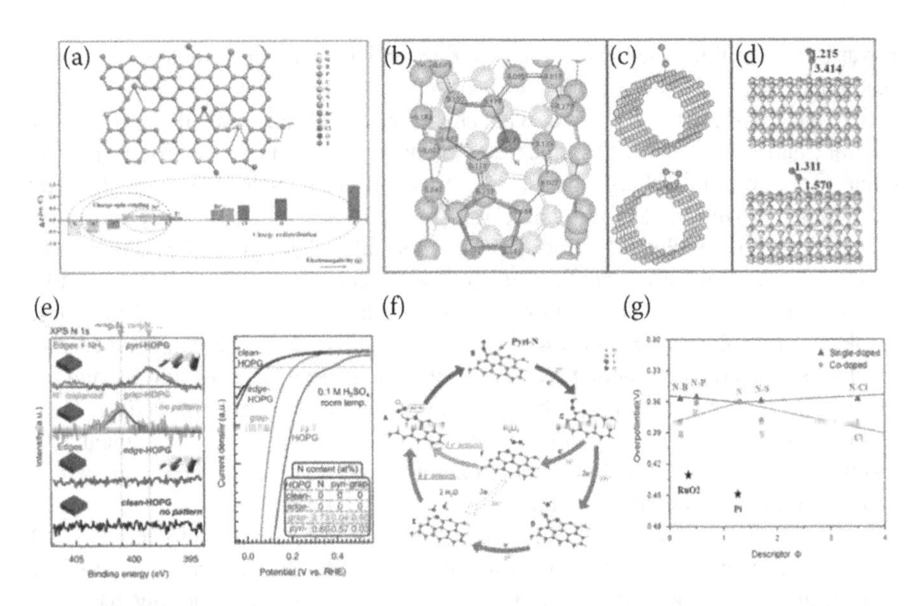

FIGURE 13.4 (a) Schematic illustrates of carbon materials with different heteroatoms doping. Upper graph: the model of carbon network with different heteroatoms doping. Lower graph: the relative electronegative of heteroatoms compared to carbon atoms. Reproduced with permission [26]. Copyright 2018, Wiley-VCH. (b) The charge density redistribution of nitrogen-doped CNTs calculated by theoretical calculation. (c) The change of the oxygen molecule adsorbed modes, the Pauling mode for undoped CNT (top), and the Yeager model for nitrogen-doped CNT. Reproduced with permission [35]. Copyright 2009, AAAS. (d) The geometric configuration and the bond lengths (Å) optimize of O_2 on ((5, 5)-12.9) CNT and N-doped CNT systems. Reproduced with permission [39]. Copyright 2010, American Chemical Society. (e) N1s XPS spectrum of HOPG model catalysts in different state (left). ORR results for model catalysts with different (right). (f) Schematic illustrate of ORR pathway on nitrogen-doped carbon materials. Reproduced with permission [41]. Copyright 2016, AAAS. (g) The ORR/OER activity for single heteroatoms doped, and codoped (N–X) carbon nanomaterials as bifunctional catalysts. Reproduced with permission [42]. Copyright 2016, American Chemical Society.

by nitrogen doping can increase the ORR performance of carbon materials through changing the oxygen chemisorption ways, increasing the active sites, and electrical conductivity.

In addition, a great deal of experimental data has demonstrated that intrinsic defect-rich carbon electrocatalysts sometimes show even better activity than carbon materials doped with many popular heteroatoms (such as F, S, P, B, etc.), clearly indicating the critical role of the intrinsic defects. When studying heteroatom-doped carbon materials, the improved ORR activity has been attributed to charge redistribution after introducing the heteroatom dopants, which can change the O_2 chemisorption mode to effectively weaken the O-O bonding and further facilitate the ORR process [43]. The heteroatom dopants can tune the electron properties of the nearby carbon atoms because their atom sizes and electronegativities, among other properties, differ from those of carbon. Inspired by this understanding, Wang et al.

observed the distribution of the election density of a graphene model and determined the exciting result that the electron distribution on a graphene sheet varies widely [44]. It was clearly seen that the edge carbon is highly charged compared to the basal plane (shown in Figure 13.5(a)). This observation encouraged them to determine whether the edge carbon was more active than the basal plane. To examine their proposal, Wang's group designed a micro-electrochemical testing system (as shown in Figure 13.5(b)) to study the ORR, using highly oriented pyrolytic graphite (HOPG) as the working electrode. An air-saturated droplet with a diameter of approximately 15 μm was deposited on a specified location on the HOPG surface using a micro-injection tool. The standard three electrodes connected the overall system.

Then, the tip could be easily moved to any specified location, such as the edge-rich areas or the basal plane surface. On different locations, the electrochemical signals could be collected, and the data are shown in Figure 13.5(c). It can be clearly seen that the edge carbon is more active than the basal plane and that the

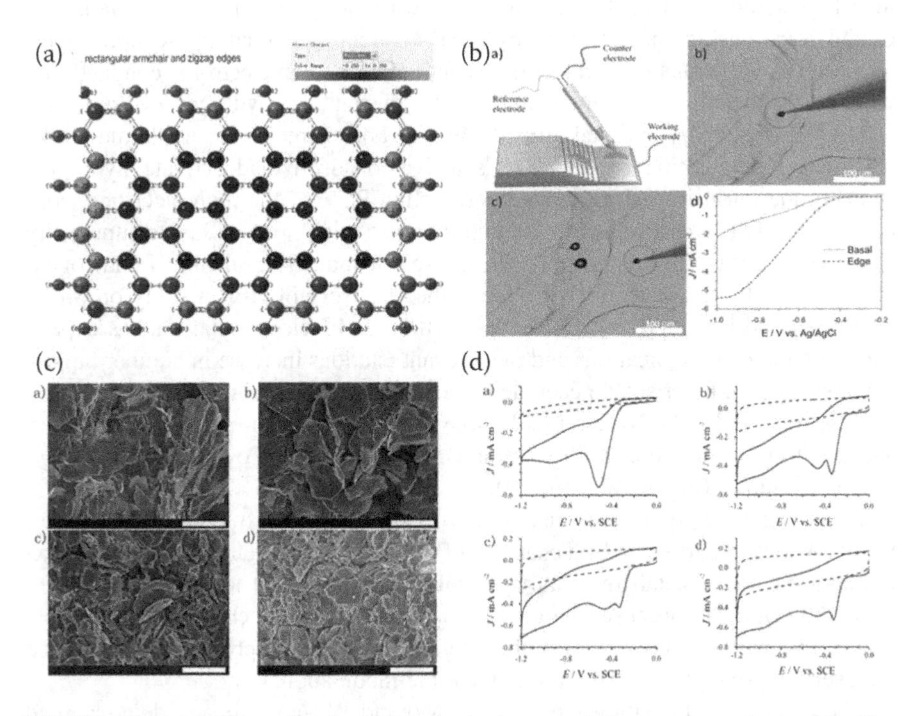

FIGURE 13.5 (a) Charge distribution on the two representative graphene sheets of $C_{72}H_{24}$. b-a) Micro apparatus for the ORR electrochemical experiment, b-b, c) Optical photographs of the HOPG as the working electrode with the air-saturated droplet deposited on the edge and basal plane of the HOPG. (b-d) LSV (linear sweep voltammetry) curves of the ORR tested for a droplet located either on the edge or on the basal plane of the HOPG. (c) SEM images of graphite material that had been ball-milled for different times. (d) CV (cyclic voltammetry) curves of graphite that have been ball-milled for different times. Reproduced with permission [44]. Copyright 2014, Wiley-VCH.

edge carbon can be considered as the intrinsic defects of the carbon for the ORR. This was the first piece of evidence that shows that the edge carbon is more active than the basal plane and demonstrates the critical role of intrinsic defects of carbon; that is, that carbon materials with more defects have better activities than defect-free carbon. This conclusion provides a general principle to design efficient, defect-rich, carbon-based, metal-free electrocatalysts for the ORR. Guided by this concept, commercial graphite materials were subjected to a ball-milling treatment, and more edges/defects were created than in pristine graphite with increasing treatment time because of the decreasing size of the graphite sheets. A series of characterizations confirmed that the defect level of the ball-milled graphite increased greatly after the ball-milling treatment. The electrochemical data clearly showed that the ORR activity improved significantly with the increase in defects, which provides a guideline for increasing the activity of ORR catalysts by edge/defect engineering.

Moreover, remarkable advances have been made regarding the use of transition metals and their carbides, nitrides, chalcogenides, and phosphides, which feature their high activity, outstanding stability, and earth abundance [45–48]. Among the candidates, transition metal carbides (TMCs) are highlighted by several superiorities, e.g., noble-metal-like electronic configuration, high electronic conductivity, wide pH applicability, and more importantly outstanding hydrogen evolution reaction (HER) activity and stability [27,49,50]. For example, the noble-metal-like electronic configuration renders TMCs active to adsorb and activate hydrogen, showing the intrinsic electrocatalysis resembling Pt. And the high electronic conductivity of TMCs resulting from interstitial alloys highlights an innate superiority as electrocatalysts, in comparison with the semi-conductivity of metal chalcogenides, phosphides, and even nitrides. Since the discovery of HER activity on Mo_2C by Vrubel and Hu in 2012 [51], the research on TMC electrocatalysts has experienced a rapid rise. Publications and the relevant citations increase in number rapidly (Figure 13.6(a)). Efforts have been devoted to designing TMC nanostructures (M = Mo, W, V, Ta, Fe, Co, Ni, etc.) with large surface area, well-exposed active sites, and good electronic conductivity, remarkably reducing the overpotential for hydrogen evolution (Figure 13.6(b)) [50].

Meanwhile, the progressive insight into its structure-activity relationship further uncovers the properties and mechanism of TMCs, showing guidelines for the exploration of noble-metal-free catalysts. Following the recent achievements, it appears mandatory to propose a comprehensive, authoritative, critical, and readable review that will emphasize their noble-metal-like electronic property/catalysis and uncover the principles for structural/electronic modulation.

The carbides of transition metals (e.g., Mo and W) have already demonstrated the noble-metal-like catalytic behaviors in petro- and coal-chemical reactions [52], e.g., hydrodesulfurization [53], hydrodenitrogenization [54], Fischer-Tropsch synthesis [55], methanol reforming/decomposition [56,57], etc. Density functional theoretical (DFT) calculation indicated that the hybridization between d-orbitals of M and s- and p-orbitals of C obviously broadened the d-band of M, as exemplified by Mo_2C (Figure 13.7) [58]. The resulting d-band state similar to that of Pt contributed to noble-metal-like catalysis, as well as strong metal-support interactions in carbide-supported catalysts [59–63]. The electronic structures of TMCs are

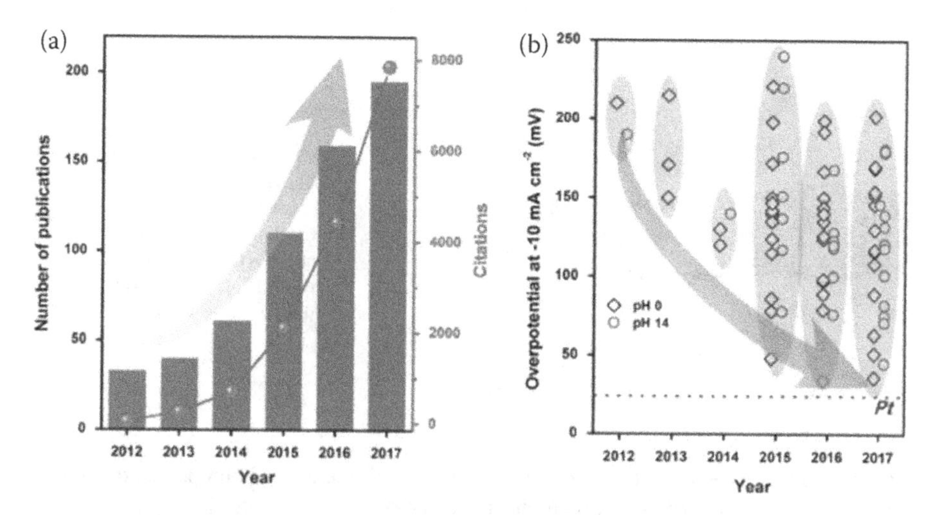

FIGURE 13.6 (a) Numbers of publications and relevant citations during 2012–2017, which are obtained via searching the keywords of "metal carbides electrocatal*" in Web of Science. (b) Evolution of overpotential required to reach a current density of −10 mA cm^{-2} on typical TMC electrocatalysts in 2012–2017.

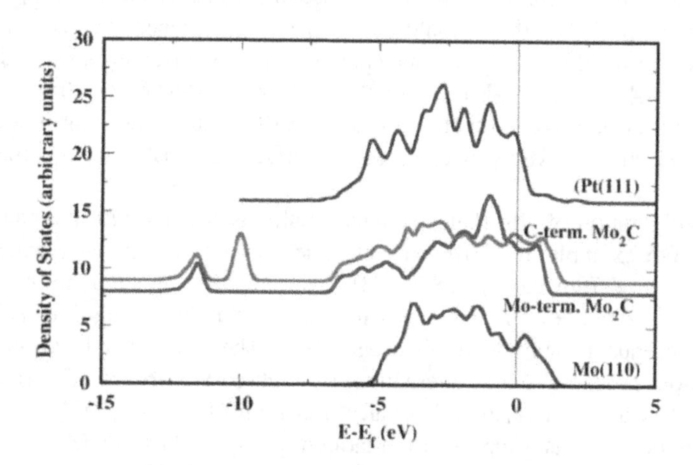

FIGURE 13.7 Comparison of d-band structures of Pt(111), β-Mo_2C(001), and Mo(110). Reproduced with permission [23]. Copyright 2005, Elsevier.

correlated with the interactions between metal and carbon [64], as illustrated in Figure 13.8. First, the surface d-band of metal carbide appears broader than that of pure metal surface, accompanied by a downshifting d-band center (ε_d), because of the hybridization with carbon orbitals [64,65]. Second, the incorporation of carbon increases metal-metal distance [58,66], causing a tensile strain in the metal lattice. Accordingly, the overlap between the d electrons on neighboring metal atoms becomes smaller, and the ε_d moves up in energy [67]. And third, metal-carbon bonds often lead to a charge transfer from metal to carbon due to their different

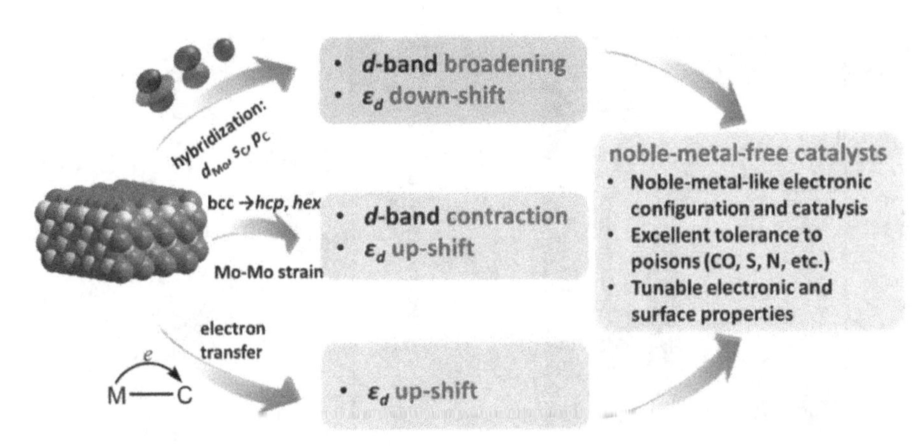

FIGURE 13.8 Schematic illustration of the noble-metal-like catalytic properties of TMCs depending on the interactions between metal and carbon.

electronegativity [65,68]. The decreasing d-band filling will upshift the ε_d. Obviously, the first effect downshifts the ε_d away from the E_F, while the latter two oppositely upshift it toward the E_F. The combination of the above three effects would be responsible for the attractive noble-metal-like electronic properties of metal carbides [64], and they should be taken into comprehensive account to improve the catalysis of TMCs. For instance, the obviously different ε_d was calculated on δ-MoC (−4.79 eV), α-MoC (−3.18 eV), and β-Mo_2C (−3.02 eV) [69]. Accordingly, the low work function on β-Mo_2C(011) suggested that it was more suited for (electro)catalytic processes where surface-adsorbate electron transfer is essential.

The configuration of adsorbates is substantially associated with the band states on TMCs. For example, the HBE on metals decreases nearly linearly when the ε_d downshifts away from the E_F [58,70–73]. The alteration of the ε_d by controlling composition and phase of TMCs is possible to modulate the strength of H-binding and the subsequent catalytic (de)hydrogenation. The noble-metal-like electronic configuration renders TMCs alternative to high-cost Pt-based electrocatalysts [46,49]. It has been established that early transition (Ti, V, Ta, W, Mo, Zr, etc.) TMCs can be used as supports for platinum-group metals (PGMs) in electrocatalysis [49,59,74]. Economically, any replacement of PGMs with TMCs can result in a great reduction in catalyst cost. More importantly, TMCs can promote electronic metal-support interactions to enhance the stability and activity of loading metals [27,75,76]. Indeed, TMCs attract more interest beyond catalyst supports. Their direct contribution to electrocatalysis as active centers will avoid the use of Pt. As verified by experimental and theoretical investigations, the carbides of Mo and W show the ΔG_{H*} similar to that of Pt are highly active in a wide pH range [77]. More importantly, the tunable d-bands in TMCs, relying on composition and crystallography, will enable the optimization on H-binding to boost HER kinetics. For example, there are a dozen phases in molybdenum carbides (Mo_xC), i.e., α-Mo_2C, β-Mo_2C, γ-MoC, η-MoC, δ-MoC, α-MoC_{1-x}, etc., that show quite a different HER performance associated with the varied electronic configuration [78]. In the past years, remarkable advances have been made for emerging TMC electrocatalysts. As a good indicator of the HER activity, η_{10} on TMCs has been evidently reduced from >200 mV in 2012 to ≈30 mV currently. To achieve high-performance TMC electrocatalysts, the first issue is exposing a highly active surface via nano-crystallization. In the model of β-Mo_2C, DFT calculation identified the high activity of Mo-terminated (001) than that of C-terminated surface [58], indicating surface Mo atoms as active sites. However, the M-terminated surface is usually covered by inert carbon layers as control is absent during TMC generation at high temperature. Meanwhile, effective electronic modulation, as another crucial issue in catalyst design, is highly desired to boost the intrinsic activity of TMCs because their binding with H intermediate is so strong that HER kinetics will be severely prohibited [79,80]. So far, at least four approaches can be summed up for TMC catalysts design: (*i*) increasing surface density of active sites via building specific nanostructures, (*ii*) supporting with conducting and high surface-area supports to promote the transportation of electrons and ions, (*iii*) doping by heteroatoms to effectively modulate electronic configuration and thus optimize the physicochemical properties of active sites, and (*iv*) engineering heterointerfaces of TMCs to optimize interfacial electronic configurations or fulfill synergic enhancement in HER kinetics on coupled active components.

13.1.2 Oxygen Defects in Transition Metal Oxides

Oxygen defects are the most popular anion defects in transition-metal oxides because of their low formation energy. With the existence of oxygen defects, the physicochemical properties of oxides may be changed. For example, the surface electronic properties and gap states can be tailored by the concentration of oxygen defects [81–84]. Many researchers have reported that defects in spinel-type oxides and perovskite oxides can greatly improve the OER performance [85,86]. For example, Zheng and co-workers demonstrated a facile method to prepare mesoporous Co_3O_4 nanowires (NWs) with abundant oxygen defects (Figure 13.9(a)) [87], where the pristine mesoporous Co_3O_4 NWs were treated with $NaBH_4$ at room temperature to reduce Co_3O_4 NWs for the formation of oxygen defects. Both X-ray diffraction (XRD) and XPS results suggested that the main body of the NWs remained as the Co_3O_4 structure, and only the surface was reduced to CoO by $NaBH_4$. The current density at a potential of 1.65 V (vs RHE) for the pristine Co_3O_4 NWs, reduced NWs, Pt/C, and IrOx, are 1.8, 13.1, 0.4, and 11.5 mA cm^{-2}, respectively, from the LSV curves (Figure 13.9(b)), which suggested that the reduced Co_3O_4 NW catalyst outperformed the pristine Co_3O_4 by a factor greater than 7 and was even better than IrO_x. The Tafel slopes also showed that the reduced Co_3O_4 had a more facile charge transfer at the solution interface than the pristine Co_3O_4. DFT calculation was used to further investigate the positive role of the oxygen defects in the electrocatalytic performance. The partial charge density was calculated to better understand the oxygen defects state, and the distribution of the two electrons previously associated with the three Co-O bonds is displayed in yellow. As shown in Figure 13.9(d), after losing an oxygen atom, the two electrons that previously occupied the oxygen 2p orbitals become delocalized around the adjacent three Co^{3+} atoms and the O atoms,

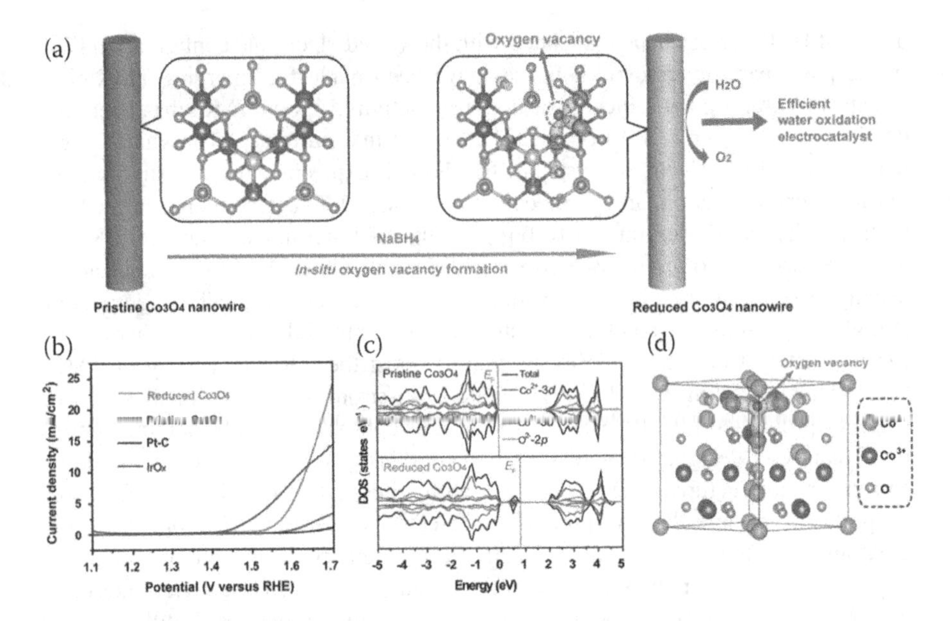

FIGURE 13.9 (a) Schematic of the NaBH4 reduction for in situ creation of oxygen defects in Co_3O_4 NWs for efficient catalysis of OER. (b) LSV currents of the reduced Co_3O_4 NWs (red curve), pristine Co_3O_4 NWs (blue curve), IrO_x (brown curve), and Pt/C (black curve). (c) TDOS and PDOSs of the pristine Co_3O_4 and the reduced Co_3O_4 (with oxygen defects). (d) Partial charge density of the reduced Co_3O_4. The states of VO^{2+} are displayed in yellow. Reproduced with permission [87]. Copyright 2014, Wiley-VCH.

which would increase the degree of electron delocalization from the pristine situation. Both the experimental and theoretical results showed that the oxygen defects formed by reduction contributed to improving the OER activity.

Qiao's group also reported other different metal oxides with abundant oxygen defects [88,89]. Recently, they found that single-crystal cobalt(II) oxide nanorods with oxygen defects have very good OER and ORR activities [11]. The existence of oxygen defects can favorably adjust the electronic structure of CoO, which may assure a fast charge transfer and optimal adsorption energies for intermediates, leading to the desirable ORR or OER performance. In addition to these spinel-type oxides, perovskite oxides have also shown excellent OER performance and have recently been widely studied as representative OER electrocatalysts [85,90,91]. In the structure of the ABO_3-type perovskite, both the A- and the B-sites can be adjusted through doping with different metal cations. By doping with different metals, the electronic, chemical, and physical properties of the perovskite oxides can be adjusted controllably. The existence of oxygen defects in perovskite oxides clearly affects their OER performance. For example, Yang and co-workers reported that $Ca_2Mn_2O_5$ with an oxygen-deficient perovskite electrocatalyst showed enhanced oxygen-evolution activity compared to $CaMnO_3$ [91]. The unit-cell structure and the electron distribution would be changed by removing oxygen atoms from the

$CaMnO_3$. This structure and electron change arising from oxygen defects should help the adsorption of OH^- and further promote the OER activity of the catalyst.

In some cases, when increasing the temperature of the hydrogen treatment to some extent, excessive oxygen defects would undisputedly render the structure unstable and result in lower conductivity. In order to solve this problem, Wu's group reported a combined method by hydrogen treatment together with metal doping to improve the e_g electron-filling degree to 0.81 and demonstrated greatly enhanced OER electrocatalytic activity (Figure 13.10(a)) [92]. Suitable oxygen defects can regulate the electronic state of the perovskite, and the hydrogen treatment can additionally maintain the crystal structure and optimize the material conductivity. Both by doping and with oxygen defects, the e_g electron-filling degree and the conductivity of the catalyst can be regulated. Therefore, hydrogen-treated $Ca_{0.9}Yb_{0.1}MnO_3$ exhibited significantly enhanced OER activity (Figure 13.10(b)) and excellent catalyst stability (Figure 13.10(d)). These oxygen-vacancy strategies representing the philosophy of "less is more" can guide us to the design of novel electrocatalysts with more oxygen defects.

To further improve the activity of metal oxides, it is desirable to combine oxygen defects with other characteristics, which are beneficial to OER performance, such as surface area and conductivity, to greatly enhance the activity. Recently, Wang reported an efficient Co_3O_4-based OER electrocatalyst prepared by a plasma-engraving strategy [93]. Plasma engraving can not only generate oxygen defects on the surface of Co_3O_4 nanosheets to adjust the electronic states, but can also increase the surface area to expose a greater number of active sites (Figure 13.11a). The existence of oxygen defects from plasma engraving can be demonstrated by XRD and XPS data. The SEM images in Figure 13.11(b) show that the pristine Co_3O_4 nanosheets present a continuous and compact surface. Excitingly, after plasma engraving, the sur-face was observed to become loose and porous, and the porous structure consisted of interconnected Co_3O_4 nanoparticles, which increased the surface area and exposed a greater number of active sites. No obvious change of the bulk phase of Co_3O_4 after the plasma treatment was observed from the HRTEM images, indicating that oxygen defects were only present on the sur-face of Co_3O_4. Oxygen defects have been shown to create a greater number of defect states located in the bandgap of Co_3O_4, and the two electrons in the defect states can be easily excited, which would improve the conductivity of Co_3O_4 and its OER performance. The plasma-engraved Co_3O_4 nanosheets, which have high surface areas and high number of oxygen defects, contributed to the enhanced OER performance. Not surprisingly, the plasma-engraved Co_3O_4 nanosheets only needed a potential of 1.53 V vs RHE (reversible hydrogen electrode) to reach a current density of 10 mA cm^{-2}, while the pristine nanosheets required 1.77 V, indicating the high OER activity of the plasma-engraved nanosheets (Figure 13.11(c)). Plasma technology may also be used to treat other OER catalysts, which opens a new method to study the defects of metal compounds.

13.1.3 Defects in Metal-Organic Frameworks (MOFs)

Metal-organic frameworks (MOFs) have recently drawn much attention because of their remarkable advantages, such as enormous surface area, uniform open cavities,

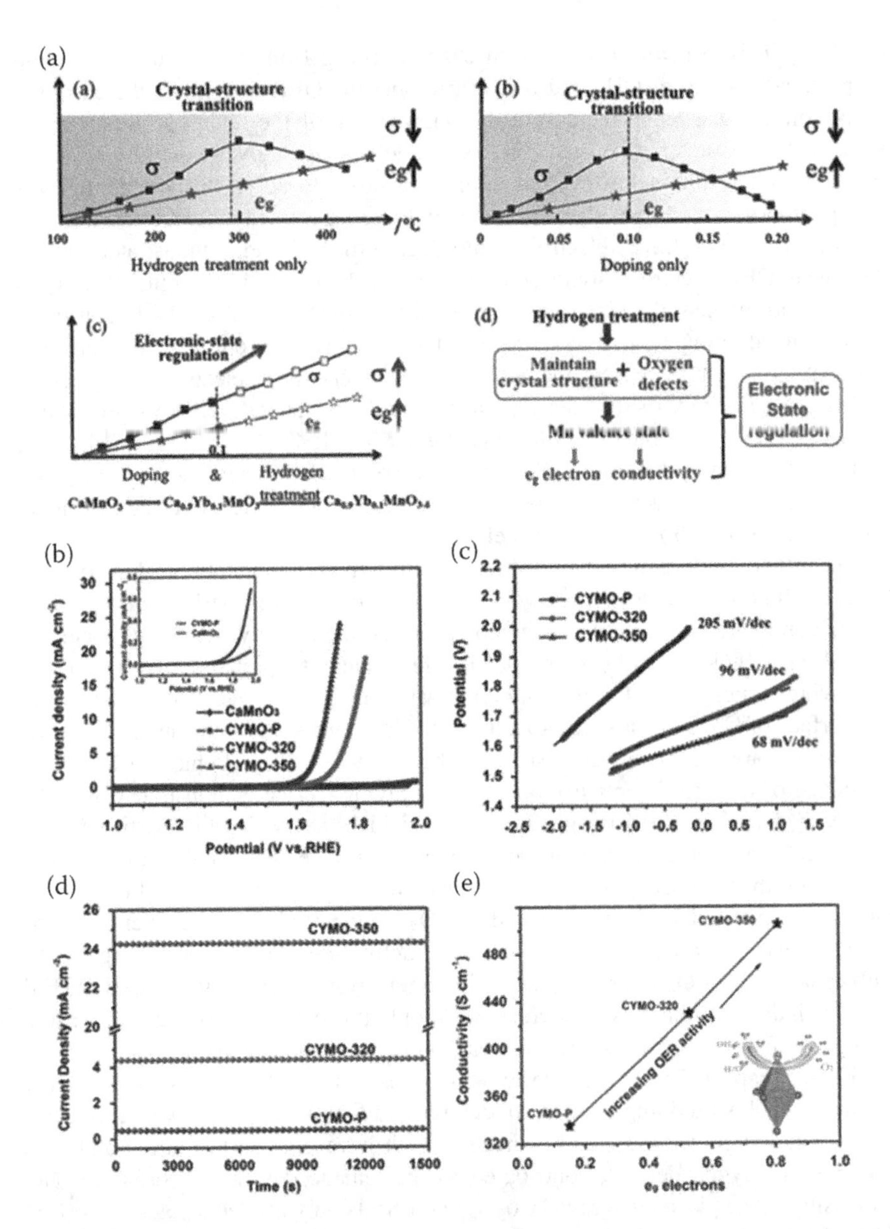

FIGURE 13.10 (a) Schematic illustration of perovskite catalyst optimization: only hydrogen treatment (a), only doping (b), electronic-state regulation by the combination of light doping and hydrogen treatment for ABO_3 perovskite (c). Schematic illustration of how hydrogen treatment influences the electronic state (d). (b) IR-corrected OER polarization curves. (c) Tafel plots for all electrocatalysts. (d) Chronoamperometric responses at an overpotential of 520 mV. (e) The OER activity increased with conductivity and e_g electron-filling status optimized by hydrogen treatment. Inset: Jahn–Teller distortion promoted the formation of oxygen defects, resulting in an optimal Mn e_g filling state and better electrical conductivity. Reproduced with permission [92]. Copyright 2015, Wiley-VCH.

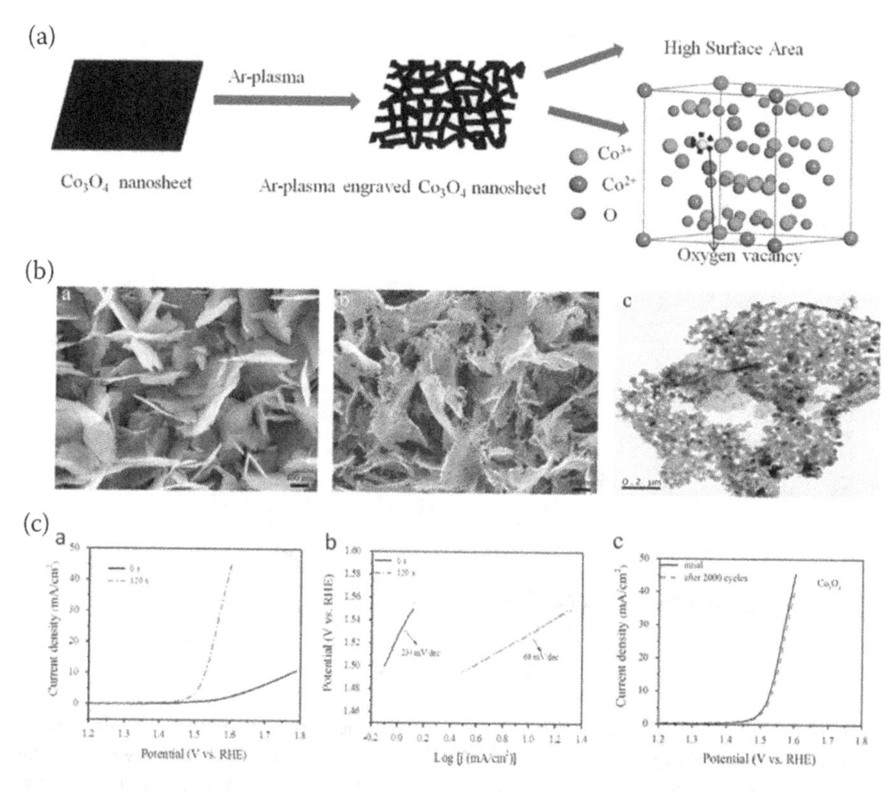

FIGURE 13.11 (a) Illustration of the preparation of the Ar-plasma-engraved Co_3O_4 with oxygen defects and high surface area. (b) SEM images of pristine (a), Ar-plasma-engraved (b) Co_3O_4, and TEM images of Ar-plasma-engraved Co_3O_4 (c). (c-a) The polarization curves of OER on pristine Co_3O_4 (0 seconds) and the plasma engraved Co_3O_4 (120 seconds). (c-b) Tafel plots. (c-c) Stability of the plasma-engraved Co_3O_4 after 2,000 cycles. Reproduced with permission [93]. Copyright 2016, Wiley-VCH.

and exposed metal or organic sites. According to various studies, many different MOF catalysts have shown great potential as electrocatalysts. As discussed previously, oxygen defects would have a great impact on the properties of transition-metal oxides. The defects in MOFs can also provide some exciting results for various applications. Tang's group has developed many excellent studies related to MOFs, and they recently reported novel ultrathin metal–organic framework nanosheets (UMOFNs), which showed high performance for the OER in alkaline solution [94]. They used an easy ultrasonic method to prepare different UMOFNs at room temperature. The UMOFNs were formed by Co^{2+}, Ni^{2+}, or a mixture of Co^{2+} and Ni^{2+} coordinating with benzenedicarboxylic acid (BDC). They demonstrated that the nanosheets were only three coordination structural layers or four metal coordination layers thick by many different characterization methods, such as TEM, atomic force microscopy (AFM), and DFT. These ultrathin nanosheets could create many crystal defects of the unsaturated metal sites on the surfaces, resulting from

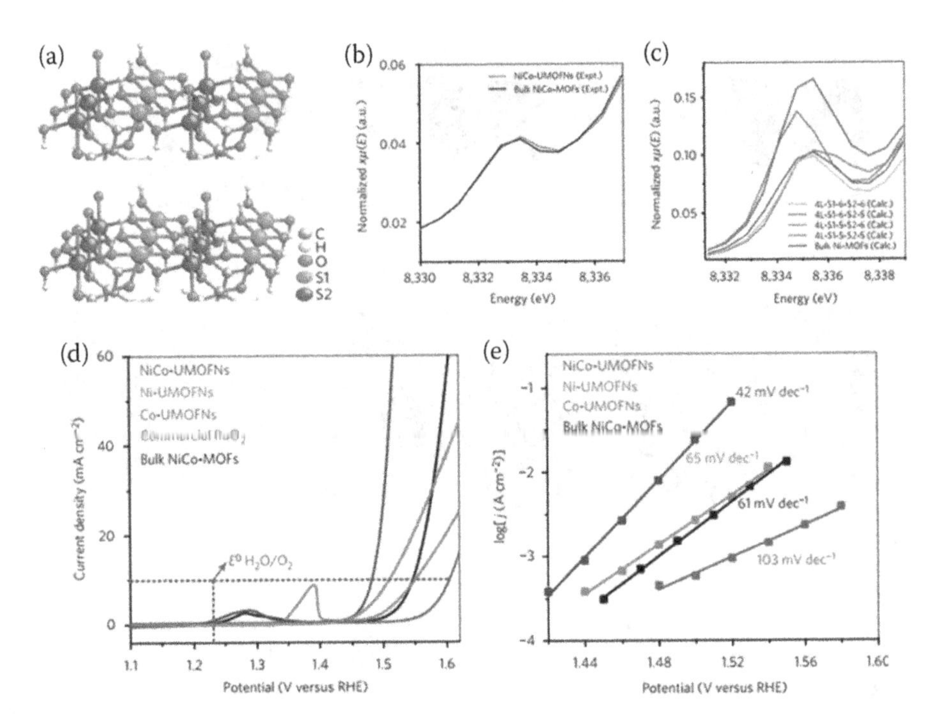

FIGURE 13.12 (a) Fully unsaturated (upper image) and partially saturated (lower image) models for the metals on the surfaces of UMOFNs. The two sets of non-equivalent metal sites are denoted as S1 (green ball) and S2 (purple ball). (b,c) Magnified figures of the experimental and simulated pre-edge peaks, respectively. (d) Polarization curves of NiCo–UMOFNs, Ni–UMOFNs, Co–UMOFNs, RuO_2, and bulk NiCo–MOFs in 1 M KOH. (e) Tafel plots of NiCo–UMOFNs, Ni–UMOFNs, Co–UMOFNs, and bulk NiCo–MOFs in 1 M KOH solution. Reproduced with permission [94]. Copyright 2016, Nature Publishing Group.

the partially terminated BDC coordination bonding with surface metal atoms, as shown in Figure 13.12(a).

To further analyze the true structure of the UMOFNs, the authors used extended X-ray absorption fine structure (EXAFS) to compare UMOFNs with their bulk crystals. However, unexpected results (Figure 13.12(b)) indicated that the ultrathin MOFs and crystal MOFs showed almost identical X-ray absorption near-edge structure (XANES) spectra. Then, the authors used XANES calculations to try to understand this unanticipated result. As shown in Figure 13.12(c), it can be concluded that the 4L-S1–5-S2–6 structural model would represent the surface coordination condition of UMOFNs. With these defects of coordination unsaturated metal sites on the surfaces, the UMOFNs would show higher OER performance than the bulk MOFs. The electrochemical polarization curves of the different samples are shown in Figure 13.12(d), clearly showing that the NiCo-UMOFNs electrode had the best OER performance among the samples, based on its coordination unsaturated metal sites and the coupling effect between Co and Ni. Such coordination unsaturated metal sites are considered to be the active centers [95]. In the near future, these researchers may pay further attention to these types of defects in other MOF materials, which may yield exciting results.

13.1.4 Single-Atom Catalysts

Single-atom catalysts (SACs) with atomically dispersed metal species on solid supports have recently drawn researchers' attention due to the maximum atomic utilization efficiency and high specific/mass activity [96–102]. The unique electronic structure and unsaturated coordination environments of the active centers in SACs have been proven to improve catalytic activity in a variety of reactions [103,104]. However, the preparation of single metal atoms is often accompanied by particle agglomeration due to the extremely high surface energies. In actual practice, the ideal single catalysts are often subject to the high activity, good stability, and high atom density [105,106]. Therefore, it is challenging to fabricate SACs and achieve high atomic dispersion of metal species under realistic synthesis and reaction conditions. SACs always involve the strong interaction between the atomic metal species and the neighboring atoms on supports, which implies that the active sites of SACs can be attributed to the effect of coordination environment between isolated metal atoms and surrounding atoms of the supports [107]. Thus, understanding the interaction between metal atoms and local environment on supports is very important for designing the new generation of SACs with high performance. During the past decades, many researchers have focused on the stabilizing methods of SACs on substrates that are required to possess anchor sites and expose the active sites, thus enhancing the charge transfer [108,109]. For example, Zheng and coworkers have summarized the approaches for stabilizing single metal atoms on various supports [110]. Based on these methods to improve the metal-support interaction and stabilize supported single atomic catalysts to prevent the agglomeration of metal atoms. Therefore, it can be considered an effective method to enhance the electrochemical performance by providing more active sites of SACs and improving their intrinsic reactivity, which are closely related to the intrinsic electronic structure.

The interesting combination of defects and single metal atoms can achieve well stability and high activity of SACs; that is, the defects and SACs can collaboratively promote the catalytic activity [111]. Moreover, the existence of defects can tune the surrounding electronic structure and coordination environment, resulting in the appearance of defects and unsaturated coordination sites, which can serve as anchor sites and disperse single metal atoms on various supports. At present, the defects on carbon materials and metal compounds are found as suitable anchor sites to trap single atomic metal and prevent isolated metal atoms from aggregation [112–114]. Although there are already many nice review articles on SACs in succession, most of these reviews have focused on the preparation, characterization, and application of SACs [115–123], the effect of defects on designing the single atomic catalysts have not been summarized in detail. Therefore, in this article, the authors present a comprehensive discussion about how defects of supports effect on the formation of SACs.

It is a remarkable fact that the strong metal-support interactions play an indispensable role in the SACs [124]. Reducible oxides are always considered as active supports to directly induce the strong metal-support interactions which relate to the presence of defects. Recently, Jones et al. demonstrated that atomically dispersed Pt could be trapped by supports such as ceria. As shown in Figure 13.13(a), Pt nanoparticles on the alumina support in-situ formed PtO_2 species and transferred to various ceria surface facets through physically mixing and aging at 800°C. In these structures, the polyhedral and nanorods ceria showed similar performance, while ceria cubes possessed the lowest reactivity. They indicated that Pt atoms were

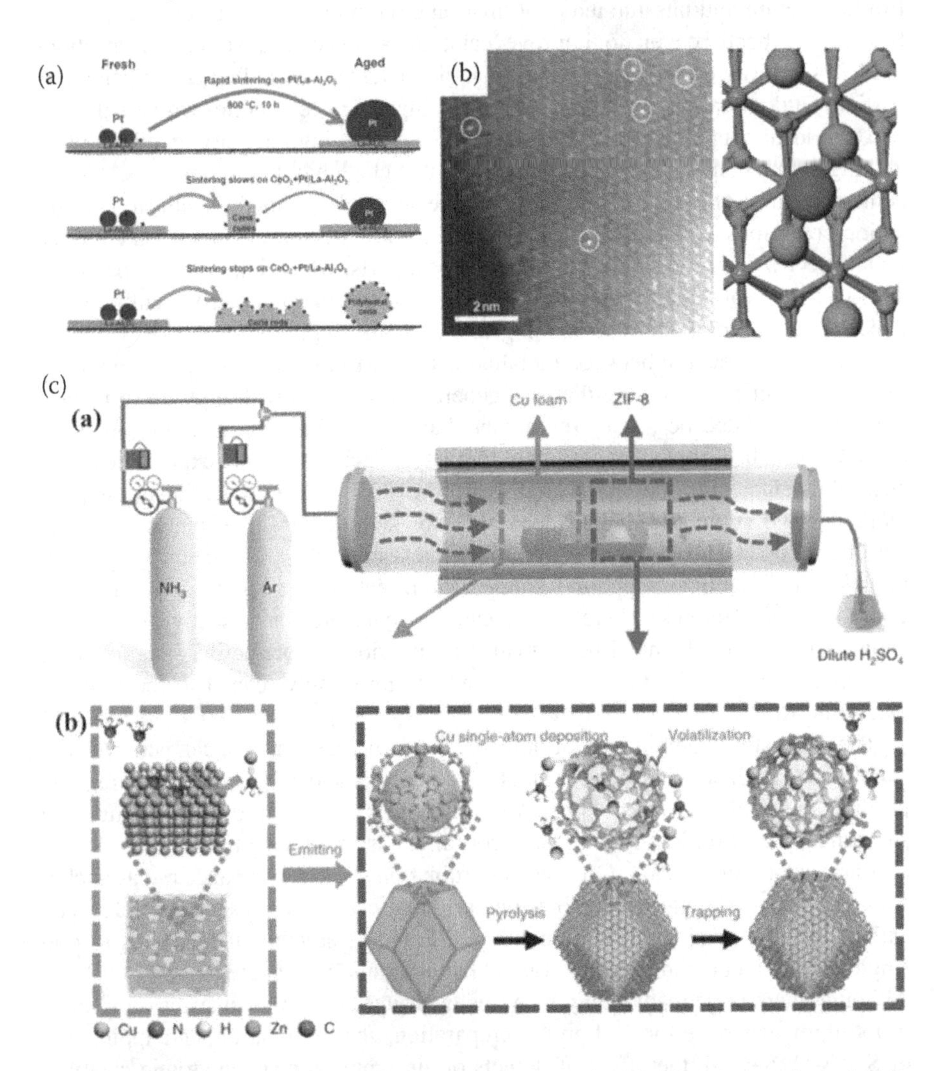

FIGURE 13.13 (a) Illustration of Pt nanoparticle on the alumina support formed PtO_2 species and transferred to ceria surface facets. Reproduced with permission [125]. Copyright 2016. American Association for the Advancement of Science. (b) Atomic structures of Pt_1/FeO_x catalysts. Reproduced with permission [126]. Copyright 2011. Nature Publishing Group. (c) The synthesis of Cu-sas/N-C. Reproduced with permission [132]. Copyright 2018. Nature Publishing Group.

tightly trapped on ceria step-edge sites to create atomically dispersed catalysts [125]. Zhang and co-workers reported the SACs that consist of single Pt atoms dispersedly anchored on defective FeO_x substrate, which has definitely high atomic utilization and exhibits high performance for CO oxidation [126]. The high-angle annular dark field scanning transmission electron microscopy (HAADF-STEM) images showed that the single Fe atoms on the surface were replaced by isolated Pt atoms, thus indicating that the Pt atoms undoubtedly occupied the location of the Fe vacancy (Figure 13.13(b)). Extended X-ray absorption fine structure (EXAFS) spectroscopy investigation revealed that Pt single atoms uniformly anchored on the FeO_x. With the increase of Pt loading, the Pt-O coordination bonding decreased while the coordination bonding of Pt-Pt started to appear and then gradually increased. Both experimental and computation results together show the vital role of the electron interactions between the single Pt atoms and the defective FeO_x, resulting in the excellent catalytic activity. In order to determine the bonding strength, theoretical studies on the geometric configuration and electronic structure of SACs have been performed, which were benefit for revealing the mechanical insight of the reaction kinetics [127]. For example, Jung and co-workers first reported the electronic structure of various single transition metal atoms stabilized on defective graphene with defects by density functional theory (DFT) calculations, revealing that most single metal atoms have strong binding with the vacancy sites of defective graphene [128]. The differential charge density map between Pt@dv-Gr and the defective graphene also demonstrated that the strong metal-support interaction between Pt atoms and the defective substrate can create different behaviors on the transition metal surface. Qiao et al. calculated the binding energies of the Au atom occupied at Fe vacancy sites and the nearest-neighboring oxygen vacancy sites, respectively, and demonstrated that the isolated Au atoms anchored at Fe vacancy sites have higher binding energies. Theoretical studies revealed that single Au atoms anchored on Fe vacancy sites have a high positive charge, and the strong Au-O covalent bonds were found at a FeO_x lattice, leading to the stabilizing of single Au atoms on the support surface [129]. Therefore, the strong metal-support interactions directly relate to the existence of surface defects which were easily created on the reducible oxides. Then the coordination geometric effects have been proven as the enhanced interaction between metal atoms and coordinatively unsaturated sites, which is good for stabilization of single metal atoms. For instance, Cong and co-workers prepared amorphous alumina with coordinatively unsaturated Al^{3+} sites that offered abundant defects to trap single Ru atoms and alter the electronic and geometric properties of metal atoms. The high catalytic performance is considered closely correlated with the strong coordination geometric interactions [130]. It has been demonstrated that unsaturated Al^{3+} sites can be considered as anchor sites to accomplish the atomic dispersion of metal species [131]. Recently, Li and co-workers reported single copper catalysts through directly emitting atoms from bulk metals, and then anchoring on N-doped carbon through the auxiliary of ammonia (NH_3). As shown in Figure 13.13(c), firstly, based on strong Lewis acid-base interactions, the volatile $Cu(NH_3)_x$ species were formed by the coordination between NH_3 and the Cu atoms. Then the $Cu(NH_3)_x$ species were anchored by the defective N-doped carbon support under the NH_3 atmosphere, thus forming the single copper catalysts [132]. Therefore, the surrounding electronic structure and coordination environment can be helpful to stabilize single atomic metals.

13.2 EXTERNAL FIELD REGULATION OF ELECTROCATALYSTS

In recent years, electrochemical technology has made great progress and achieved remarkable results. However, the current research tends to focus on the catalyst material itself, and a complete electrochemical system is composed of multiple parts. In addition to ameliorating the catalyst materials, there are many factors that may affect the overall performance of the electrocatalytic system, such as electrolyte, three-phase interface, electrolytic cell, etc. On the basis of existing catalyst performance, the introduction of some new technologies can significantly improve the overall performance of the electrocatalytic system. This section introduces the application of some new assistive technologies, including electric field, magnetic field, strain, light, electrolytic cell, and electrolyte on electrocatalytic reactions are systematically discussed, focusing on the latest research results.

13.2.1 Electric Field–Enhanced Electron Transport

As for electrocatalytic HER, the activity of catalysts is significantly influenced by their electrical conductivity. Theoretical calculation has proven that molybdenum disulfide (MoS_2) possesses an adequate Gibbs free energy of adsorbed atomic hydrogen, and it has the merit of element abundance; thus, MoS_2 is regarded as a promising alternative of Pt. However, it is hindered by the fewer numbers of thermodynamically unstable Mo edges (act as the active sites) and poor conductivity of the catalyst. These conditions can be ameliorated by compositing with conductive substrates, doping, and phase engineering [133–135]. In addition, electrocatalytic HER is also affected by the contact resistance between electrode and catalyst. As is well known, the basal plane of 2H transition metal dichalcogenides (TMDCs) is relatively inactive for HER. However, its activity can be boosted to be comparable to the 2H phase's metallic edges or 1T phase by improving the electron injection from electrodes to active sites [136]. On the other hand, the transport properties of semiconducting low-dimensional materials exhibit pronounced gate tunability because of their weak electrostatic screening capabilities induced by atomic-scale characters. For example, the on/off ratio of field-effect transistors (FETs) built on 2-D TMDCs have been extensively demonstrated to be beyond 10^6 [137–139].

Based on the above research, researchers have recently revealed that the activity of low-dimensional semiconducting electrocatalysts is sensitive to an external electric field. Wang et al. conducted an in-depth investigation on the influence of a gate electric field on the electrocatalytic HER based on a back-gated 2-D MoS_2 working electrode (Figure 13.14) [140]. Figure 13.14(a) is the schematic diagram of a three-electrode system coupled with the electric field using MoS_2 as a work electrode. The polarization curves of the individual MoS_2 nanosheet-based HER device at different gate voltages are shown in Figure 13.14(b). The catalytic performance is greatly improved, and the Tafel slope decreases with the increase of

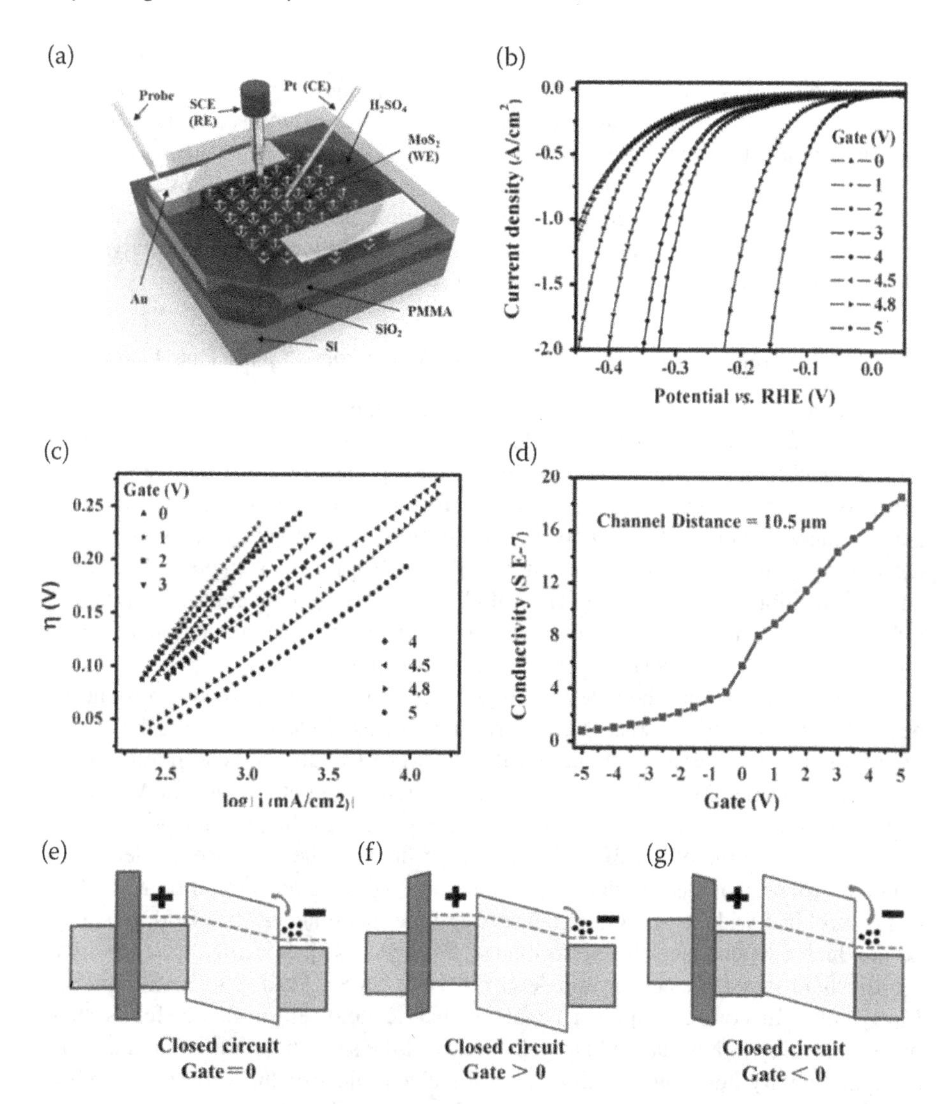

FIGURE 13.14 (a) Schematic diagram of a three-electrode microreactor with a backgated MoS_2 working electrode for HER. (b) Polarization curves of HER upon different backgate voltages and (c) the corresponding Tafel curves. (d) Electrical conductivity of a MoS_2 nanosheet as a function of backgate voltage. (e) The normal state of device under closed circuit without any gate voltage. (f) The energy band of the device with the positive gate voltage bias and (g) with the negative gate voltage bias. Reproduced with permission [140]. Copyright 2017, John Wiley and Sons, Inc.

gate voltage (Figure 13.14(c)). In addition, the channel conductance of individual MoS_2 nanosheet device keeps rising from 0.0073 to 0.1776 S m^{-1} when increasing the gate voltage from –5 to 5 V. It was proposed that the potential mechanism for such a considerable activity improvement should be attributed to a positive gate voltage to reduce energy barrier and then the electrical conductance increase explained by energy band theory. Upon a positive gate voltage, electrons accumulate in the MoS_2 channel through capacitive coupling, resulting in an upshift of Fermi level toward conduction band. Since MoS_2 is an n-type semiconductor, its electrical conductivity is increased and the Au-MoS_2 contact resistance is reduced (Figure 13.14(e–g)). As a result, the charge injection can be tuned over a broad range within a moderate electric field. Furthermore, an increase in electrical conductivity also helps to reduce the iR loss. Therefore, a superior electrocatalysis of MoS_2 for HER is triggered by applying a positive backgate voltage.

13.2.2 Electric Field–Tuned Hydrogen Adsorption Gibbs Free Energy

From the perspective of the enhanced catalytic effect, electrocatalysts accelerate the reaction by facilitating the required electron transfer, as well as the formation and rupture of chemical bonds [141]. Hydrogen adsorption affinity is another crucial metric of electrocatalytic HER. Generally, a moderate interaction is favorable since there is an intricate trade-off between adsorption of hydrogen intermediate and desorption of hydrogen. However, it is still challenging to achieve a hydrogen adsorption Gibbs free energy (ΔG_H) of 0 eV based on conventional strategies [142–144]. It is mainly ascribed to the difficulty in precise regulation of morphology, structure, composition, and defect of nanocatalysts.

Interestingly, the intermediate binding energy of the reaction was also found to be sensitive to electric fields. The electric field–assisted electrocatalysis HER has the advantage of accurately regulating the kinetics of hydrogen adsorption, which has been confirmed by many theoretical calculations. Ling et al.'s use of MoS_2/Au system for a probe, with density functional theory (DFT) the primary principle of research to study the effect of external electric field on the interface of electronic structure, found that electric field easy to atomic hydrogen adsorption on MoS_2/Au can control in a wide range within the scope of the moderate electric field, this may be conducive to enhance the performance of HER [145]. Specifically, a positive electric field induces electron transfer from H to MoS_2, leading to a stronger H-MoS_2 bond. In contrast, upon a negative electric field, electron transfer is suppressed, leading to a weakened H-MoS_2 bond. At the same time, they found a good catalytic activity descriptor by analyzing the electronic structures of H adsorption on free-standing and Au(111) supported MoS_2 edges [146]. An external electric field tunes the position of S p-resonance states, which effectively adjusts the binding strength of H and an optimal Gibbs free energy of H adsorption can be achieved. Notably, under the optimal electric field intensity, an ideal Gibbs free energy can be achieved at the S(111) edge.

Recently, Wu et al. found that the HER performance of pristine and defective MoS_2 were enhanced by applied an electric field from the polarization data (Figure 13.15(a)) [147]. The activity of defective MoS_2 is always superior to that of pristine MoS_2. The overpotential and Tafel slope of both catalysts decrease as gate voltage increases from 0 to 3 V (Figure 13.15(b,c)). Subsequent theoretical simulations revealed that grid-induced excess electron injection can stabilize hydrogen adsorption (Figure 13.2(d,e)). It reduces the potential barrier of the rate-limiting Volmer thermochemical reaction. Excitingly, by introducing excess electrons into

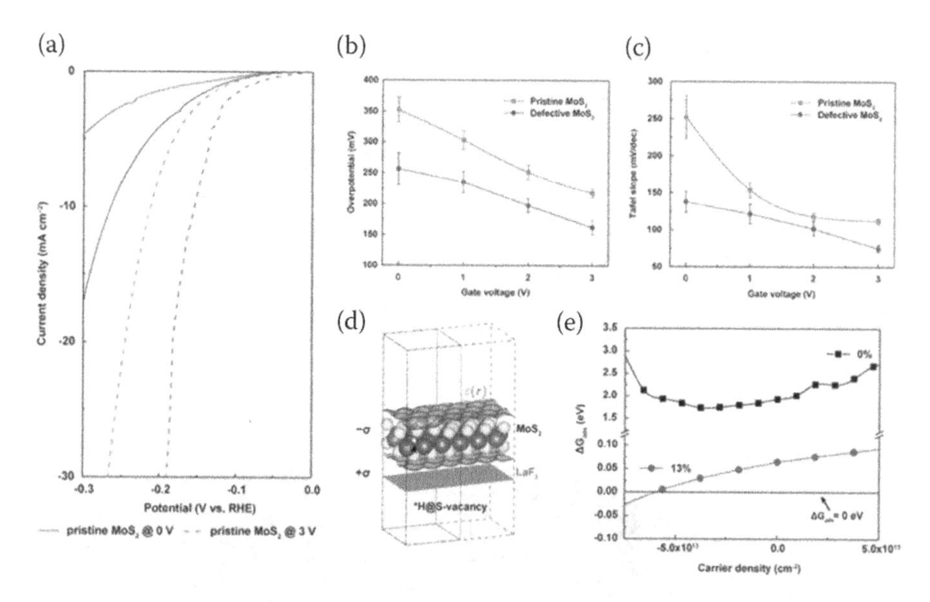

FIGURE 13.15 (a) Polarization curves measured from the pristine (red) and defective (blue) MoS_2 device at 0 V (solid) and 3 V (dash) gate voltage without iR correction. (b) Overpotential and (c) Tafel slope of pristine (red) and defective (blue) MoS_2 nanosheets as a function of gate voltage. (d) Schematic setup of a MoS_2 nanosheet embedded into a dielectric continuum with a counter charge that represents a LaF_3-MoS_2 capacitor. (e) Hydrogen adsorption Gibbs free energy of pristine MoS_2 (black) and defective MoS_2 (13% S vacancies, blue) as a function of carrier density. Reproduced with permission [147]. Copyright 2019, American Chemical Society.

the system under a high hydrogen coverage and high vacancy density, we found ΔG_{ads} approach a value of zero under the range of experimentally estimated carrier densities, revealing general availability of such gate-tuned hydrogen adsorption.

13.2.3 Magnetohydrodynamic (MHD) and Micro-MHD Convection-Promoted Catalysis

MHD and micro-MHD effects are the macroscopic and microscopic convection caused by the interaction between magnetic field and local current density, respectively, which is driven by Lorentz force ($F_L = j_* B$), where j is current density and B is magnetic strength. The form of magnetic convection on the electrode surface is shown in Figure 13.16. F_L reaches a maximum when $B \perp j$, generating magnetic convection parallel to the conventional electrode surface (Figure 13.16(a)). F_L is zero when $B // j$. Besides, the distortion of the current lines will also produce the azimuthal F_L due to the edge effect of the disk electrode and the bulge of insulating bubbles (Figure 13.16(b,c)) [148–150].

The influence of MHD and micro-MHD effects on the electrocatalytic system is mainly targeted at the gas-escaping electrode. For conventional electrodes, the release size and residence time of bubbles depend on the net magnetic force, as

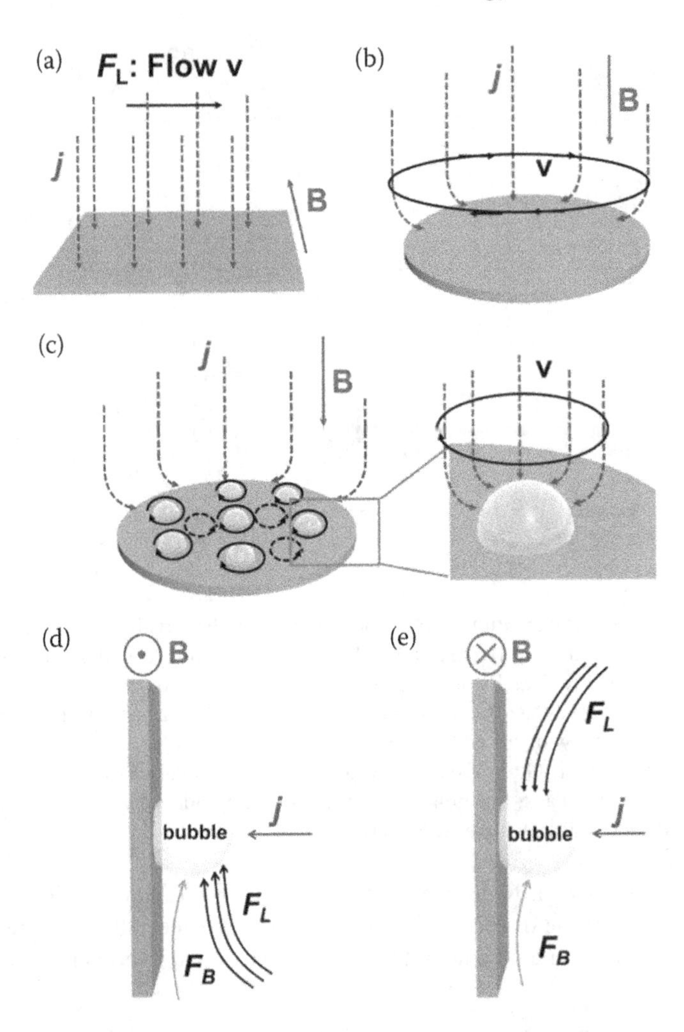

FIGURE 13.16 MHD flow types under homogeneous magnetic field and the schematic diagram of MHD effect for conventional electrodes. (a) MHD flow parallel to the electrode surface occurred when $B \perp j$. (b) The MHD whirlpool around the edge when $B//j$. (c) The micro-MHD generated around the bubbles when $B//j$. The case of F_L and F_B in the same direction (d) and opposite direction (e).

demonstrated in Figures 13.1(e) and 13.16(d). If the direction of F_L and buoyancy force (F_B) are the same ($F_{net} = F_L + F_B$), the upward pumping effect of MHD convection will reduce the average size of bubble detachment, shorten the remaining time on the electrode and increase the rising speed (Figure 13.16(d)), while the opposite phenomenon will occur when the F_L goes downward ($F_{net} = F_L - F_B$) (Figure 13.16(e)) [151–154]. Apparently, the macroscopic MHD effect will bring about enhanced response of the gas-evolving electrodes.

The current controversial point is the micro-MHD. There are experimental data supporting that the micro-MHD effect is conducive to the release of bubbles, but some work has reached the opposite conclusion [155–157]. Follow-up research on this phenomenon found that the promoted or suppressed release of bubbles would depend on the location of the low-pressure region in the bubble. In short, the micro-MHD effect is beneficial to the release of bubbles on the surface of the conventional electrode and the fixation of bubbles on the surface of micro-electrode. Surprisingly, the micro-MHD effect was observed to promote the release of paired bubbles when the nucleation site of bubbles was fixed. The current understanding of the micro-MHD effect is not comprehensive, especially its influence on the behavior of single bubble on the electrode surface and the interaction of multiple bubbles. It is necessary to study the role of micro-MHD effect on the behavior of bubbles by designing different types of electrodes and corresponding numerical simulation analysis.

The influence of MHD and micro-MHD on electrocatalytic reactions is mainly manifested in the following three aspects: (i) reducing the ohmic polarization. Generally, the Bruggeman equation ($\kappa = \kappa_0(1 - f)^{1.5}$ (4), f is the void fraction) is used to calculate the conductivity of the gas bubbles dispersed in the electrolyte solution [158,159]. The MHD disturbance alleviated the supersaturated accumulation of bubbles on the electrode surface, leading to a decrease of the void fraction of the electrolyte and an increase in conductivity, thereby reducing the ohmic voltage drop; (ii) reducing the activation potential. Specifically speaking, due to the adhesion of gaseous products on the electrode surface, the effective active area was inhibited, while the MHD and micro-MHD effect arising from the magnetic field can significantly reduce the coverage of bubbles, avoiding the requirement of extra activation potential; (iii) reducing the concentration polarization. On the basis of the Butler-Volmer equation, the expression of concentration overpotential is $\varepsilon = \frac{RT}{\beta F} \ln \frac{I_d}{I_d - I}$ (5), while the limiting current density (I_d) is related to the diffusion layer (δ) according to Fick's first law: $I_d = nFD_i\frac{c_i^B}{\delta}$ ($c_i^S = 0$) (6). D_i is the ion diffusion coefficient of component i (cm^2/s), c_i^B is the concentration of i in bulk solution (mol cm^{-3}), and c_i^S is the concentration of i on the electrode surface (mol cm^{-3}). The forced flow produced by MHD was inclined to impair the thickness of diffusion layer and enhance mass transfer, thus increasing the I_d and weakening the concentration polarization.

Typically, the work of reducing the overpotential of HER with the MHD effect was done by Hegde and his co-workers (Figure 13.17), where the efficiency of HER was improved by Ni-W alloys by applying a proper magnetic field (0.1–0.4 T) [160]. The CV curves of HER illustrates the onset potential decreased by 20 mV, and the volume of hydrogen obtained by a chronopotentiometry (CP) response within 300 seconds increased by 30.8% compared with that obtained in the absence of applied magnetic field (Figure 13.17(b,c)). The authors ascribed the HER activity enhancement to the MHD effect caused by F_L ($F_L = qvB \sin \theta$) perpendicular to the direction of the magnetic field and a decline in the ohmic drop via H_2 bubble disentanglement. Specifically, F_L acts on the ionic species to form a circular path motion, generating convection flow in the vicinity of the electrode surface, which is proportional to the magnetic field strength to enhance mass transport (Figure 13.17(d)). Posteriorly, the buoyancy of H_2 and MHD force are upward. The

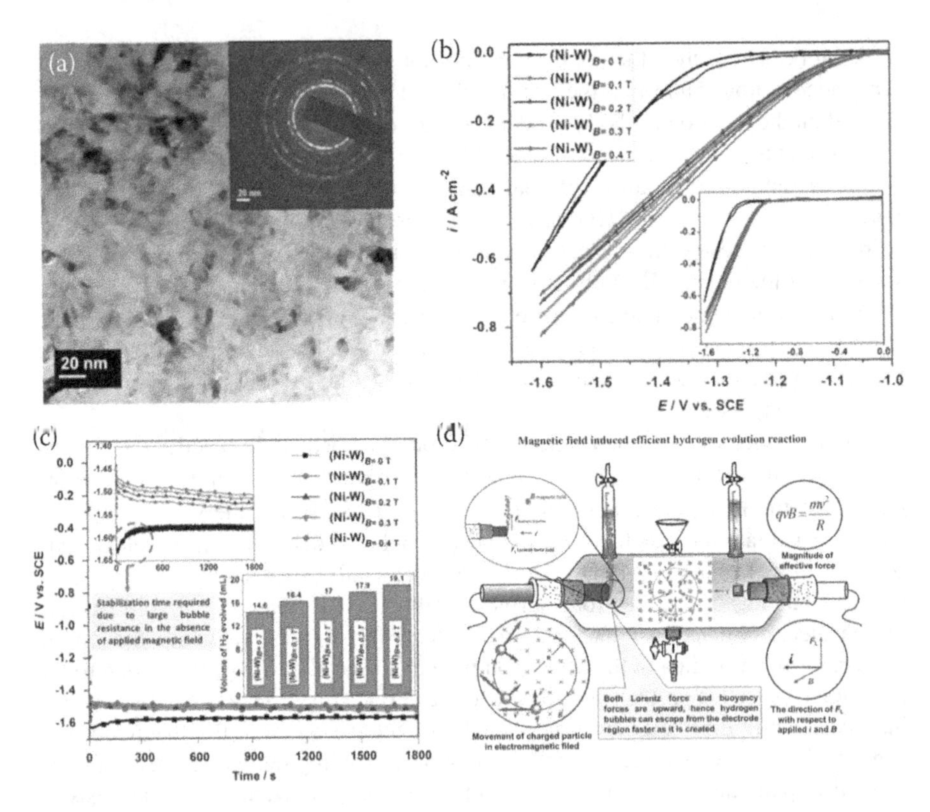

FIGURE 13.17 (a) TEM image of Ni-W alloy and selected area electron diffraction (inset). (b) CV response and (c) chronopotentiometry measurement of Ni-W alloy and evolution analysis of H_2 within 300 seconds under different magnetic field intensities (0.1–0.4 T). (d) Schematic diagram of mechanism of hydrogen desorption accelerated by Lorentz force in the same direction as buoyancy under vertical magnetic field. Reproduced with permission [160]. Copyright 2017, Springer Nature.

rate of H_2 bubbles easily escaping from the electrode surface is accelerated. As a result, the effective active area exposed increases, resulting in the limiting current density is increased and the ohmic drop is decreased; thus, the HER efficiency is improved. This work provided a reasonable mechanism analysis for the magnetic field–enhanced HER and an idea for the design of a new type of electrolytic cell.

13.2.4 Magnetic Hyperthermia-Promoted Catalysis

The magnetothermal effect in electrochemical reactions is derived from the magnetic hyperthermia in cancer therapy, which is generated by the external high-frequency alternating magnetic field (AMF) acting on magnetic nanoparticles (MNPs) with the characteristic of heat being localized [161]. According to Arrhenius equation ($k = Ae^{-E_a/RT}(1)$), temperature is a non-negligible factor in electrocatalytic reactions [162]. Increasing the operating temperature can change the apparent activation energy in the electrocatalytic reactions and reduce the overpotential. Several experimental systems have shown that increase in electrolyte temperature has a positive impact on the electrode kinetics, mass transfer of ions and gaseous species. Furthermore, important parameters for evaluating electrocatalytic reactions (charge transfer resistance, overpotential, Tafel slope, and electrochemical double-layer capacitance) are temperature-dependent under specific conditions [163–165]. Although an increase in temperature promotes the electrode reaction, the conventional heating mode frequently leads to grievous corrosion of the electrolyzer and gas pollution. Inspired by the ability of magnetic hyperthermia to locally heat MNPs, a reasonable prediction is that continuous local heating of magnetic electrodes can be realized under a magnetic field, which can bring fresh vitality to electrocatalytic reactions and provide a possible alternative to traditional heating methods.

Based on the advantages of the magnetothermal effect, the magnetothermal efficiency of materials needs to be discussed. The heating efficiency of MNPs can be described as specific loss power (SLP, heat dissipation per unit mass of magnetic material) is related to two factors. One is the magnitude and frequency of the magnetic field, the SLP value is proportional to the H^2 according to $SLP = \frac{\mu_0 \pi \chi''(f) H^2 f}{\rho \Phi}$ (2), where f and H are the frequency and magnitude of an applied magnetic field, respectively, ρ the particle density, Φ the particle volume fraction, μ_0 the permeability of free space, and χ" the imaginary part of the susceptibility [166]. Another factor is the nano-magnetic parameters of MNPs, such as magnetic anisotropy (K) and saturation magnetization (M_S) [167]. These parameters can be achieved by adjusting the size, shape, and composition of the MNPs. Therefore, finding the critical value of MNPs' heating power in electrocatalysis without damaging the catalytic active sites is drastically important.

Marian Chatenet and his colleagues applied for the first time a high-frequency AMF to in-situ heat the FeC-Ni catalyst by the magnetothermal effect (Figure 13.18) [168]. Galvanostatic measurements demonstrated a very positive impact on the OER behavior immediately when the magnetic field was switched on with different amplitudes. The breakthrough of this work is to realize the local heating mode near the catalyst by using the FeC with high heating-power property, while the temperature of the electrolyte has not changed drastically. It is worth noting that a rigorous numerical method has not been established to calculate the kinetic improvement. Besides, it is uncertain whether the enhancement of catalytic activity is solely due to thermal effect, or such local heating will damage the long-term stability of the catalyst.

13.2.5 Magnetic Field–Engineered Spin States

The concept of spin selectivity effect can be traced back to the pioneering discovery in 1999 that chiral molecules can be used as the filter of electron spin orientation [169]. It allows the preferential spin orientation of electrons to be transmitted through chiral materials, and this specific transport is called chiral-induced spin selectivity (CISS) [170,171]. Notably, the CISS catalysts have been proved to have glorious electrocatalytic activity [172,173]. Relevant experiments in oxygen-

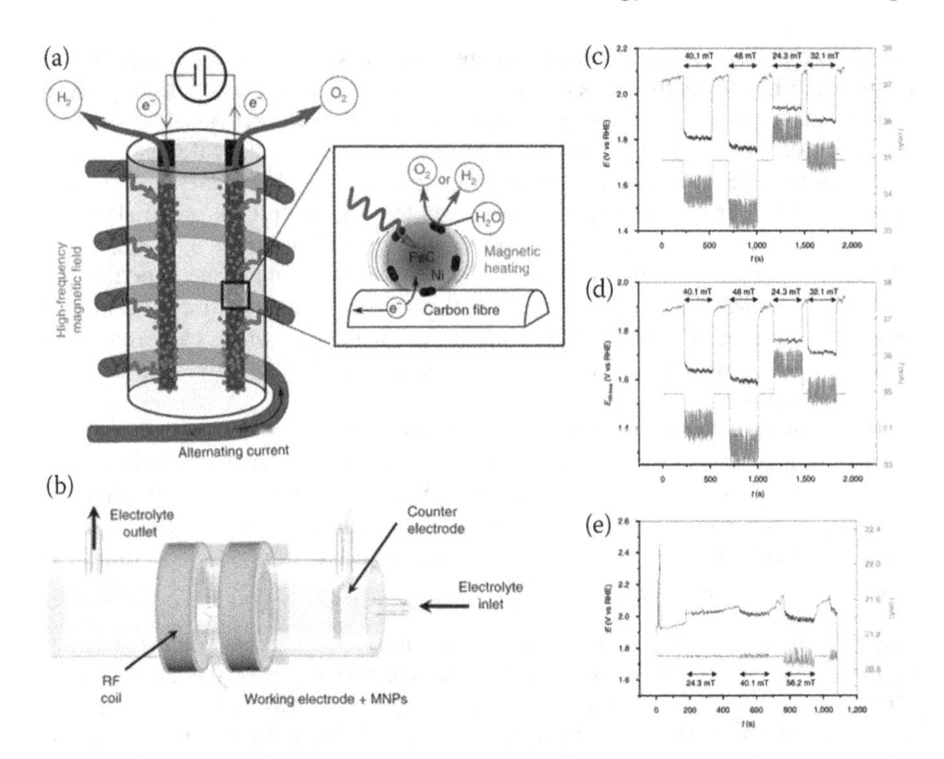

FIGURE 13.18 (a) The FeC-Ni catalyst was heated locally in a high frequency alternating magnetic field (AMF). (b) Schematic diagram of the coupling device of alkaline electrolytic water (AEW) and magnetic field. Galvanostatic measurement for the OER at 35 mA in different field amplitudes with (c) no IR correction and (d) IR correction. (e) Similar measurement for carbon felt at a 21 mA AMF of varying amplitude applied. Reproduced with permission [168]. Copyright 2018, Springer Nature.

involved reactions have demonstrated that the CISS effect can effectively inhibit the formation of H_2O_2 by controlling the intermediates on the catalyst surface to arrange in a spin parallel manner (OH↑×↑OH), and theoretical study of ferromagnetic electrode also shows a spin selectivity effect [174–177]. Similarly, the magnetic field has also been confirmed to have a significant effect on spin selectivity in electrocatalytic reactions, including two aspects, as shown in Figure 13.19. Electrocatalytic reactions usually involve the combination of spin-related free radical pairs, singlet (↑↓) or triplet states (↑↑). Applying magnetic field can cause the interconversion of two spin states, which determines the final product (Figure 13.19(a)). On the other hand, the magnetic field can induce the spin flip of the intermediates adsorbed on the catalyst surface, optimizing the reaction path, and improving the reaction efficiency (Figure 13.19(b)). From the perspective of electrochemistry, the spin effect establishes the internal relationship between catalytic activity and microstructure. Magnetic field–induced spin selectivity (FISS) will adopt the optimal path to improve the electrochemical reaction efficiency in the form of constrained or unconstrained spin state.

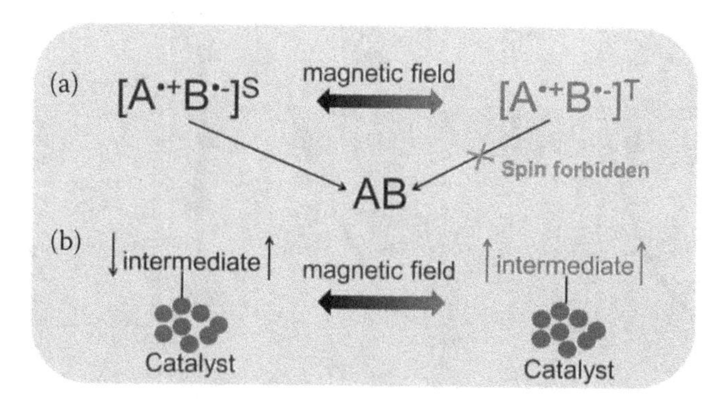

FIGURE 13.19 Effect of magnetic field on spin-controlled catalytic reactions. (a) Magnetic field improves the yield of the final product of electrocatalytic reaction by controlling the spin state of free radical pair. (b) Magnetic field changes the electrocatalytic reaction path by controlling the spin state of the intermediate adsorbed on the catalyst surface.

The representative and systematic work in the field of magnetic effect–induced OER response was completed by Galán-Mascarós and his co-workers, where the FISS effect improved the OER performance on different magnetic catalysts, as illustrated in Figure 13.20 [178]. In this work, the polarization curves of different magnetic materials (IrO_2, NiO, Raney Ni, $Ni_2Cr_2FeO_x$, $NiFe_2O_x$, $FeNi_4O_x$, $ZnFe_2O_x$, $NiZnFe_4O_x$, and $NiZnFeO_x$) with and without a magnetic field confirmed that the enhancement of non-magnetic IrO_2 is negligible, while the rest have more or less positive responses to the magnetic field. The enhancement of magnetic current can be rationalized by the fact that the magnetic field induces O atom adsorption in a triple state (↑O···O↑ and ↑O-O↑) on the catalyst surface, and the speed determination step of OER in alkaline conditions is thermodynamically more inclined to spin parallel. This work has led to a more profound understanding of the mechanism of magnetic field–enhanced electrocatalysis. The reasons for the distinction in the activity of different magnetic catalysts have not been well explained, although a direction for the mechanism research in the field of magnetoelectrochemistry is put forward. There are still many unexplained problems that require more experiments to explore.

13.2.6 Strain Field

One common strategy to enhance the activity of electrocatalysts is to tune their surface electronic structure. Strain engineering, expanded, or compressed arrangements of atoms, are one of the promising routes to manipulate the surface electronic structure of electrocatalysts [179]. However, most studies introduce strain through defect engineering, construction of unmatched heterostructures, and fabrication of core-shell nanostructures [180–182]. These approaches suffer from lack of accuracy for strain construction. In addition, the strain is fixed once catalysts are produced. As a result, it is difficult to achieve active strain tuning of catalysis during operation. Furthermore, applying strain via defects, heterostructures and core-shell

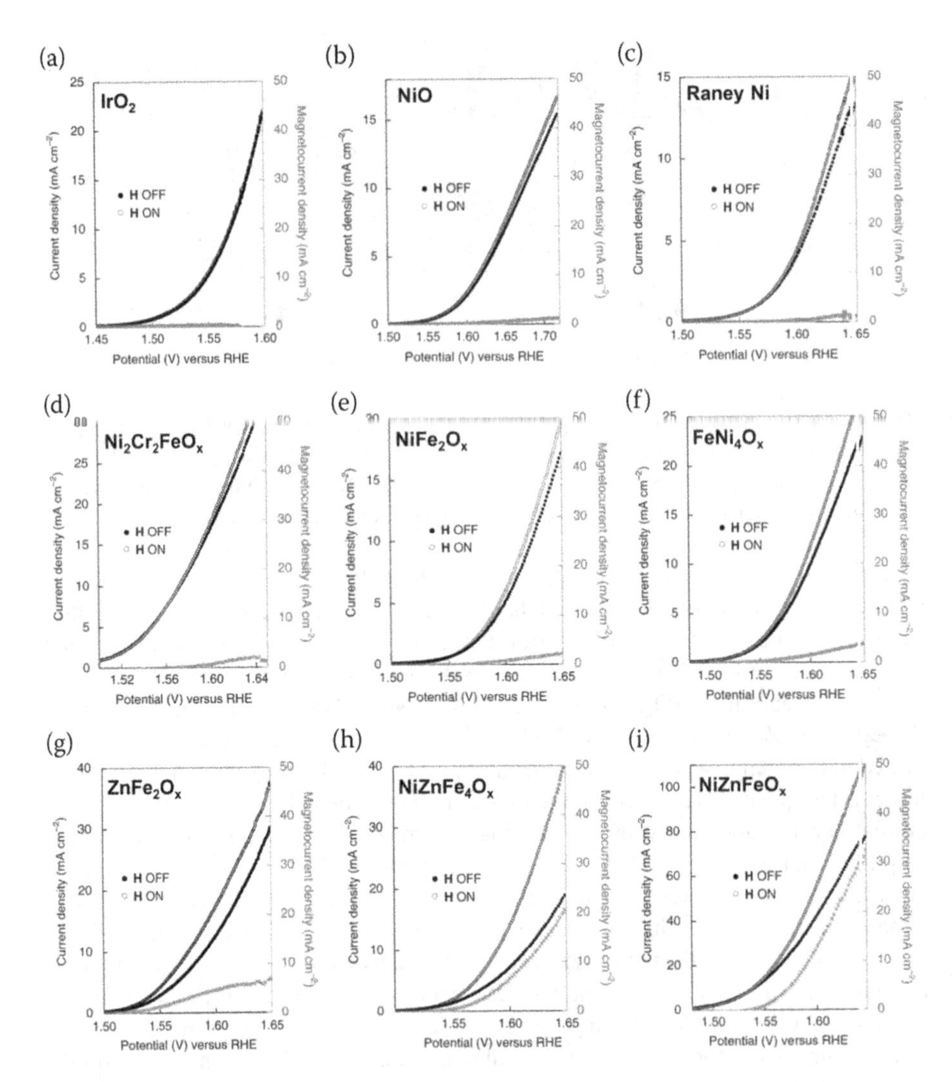

FIGURE 13.20 Polarization curves when the magnetic field is turned on and off and the increasing trend of magnetic current density of (a) IrO_2, (b) NiO, (c) Raney Ni, (d) $Ni_2Cr_2FeO_x$, (e) $NiFe_2O_x$, (f) $FeNi_4O_x$, (g) $ZnFe_2O_x$, (h) $NiZnFe_4O_x$, and (i) $NiZnFeO_x$. Reproduced with permission [178]. Copyright 2019, Springer Nature.

nanostructures are inevitably accompanied with ligand effects and variation of the number of active sites. Therefore, conventional strategies are limited for establishing an explicit correlation between strain and catalytic activity.

In order to solve the previous problem, Lee et al. developed an alternative methodology by introducing strain through external substrates of electrocatalysts. The vacuum-filtered MoS_2 film was then contact-printed onto a 0.2 mm thick Ag-coated PET substrate as a work electrode performed using a three-electrode configuration (Figure 13.21(a,b)) [183]. They are contact-printed on a flexible Ag/polyethylene

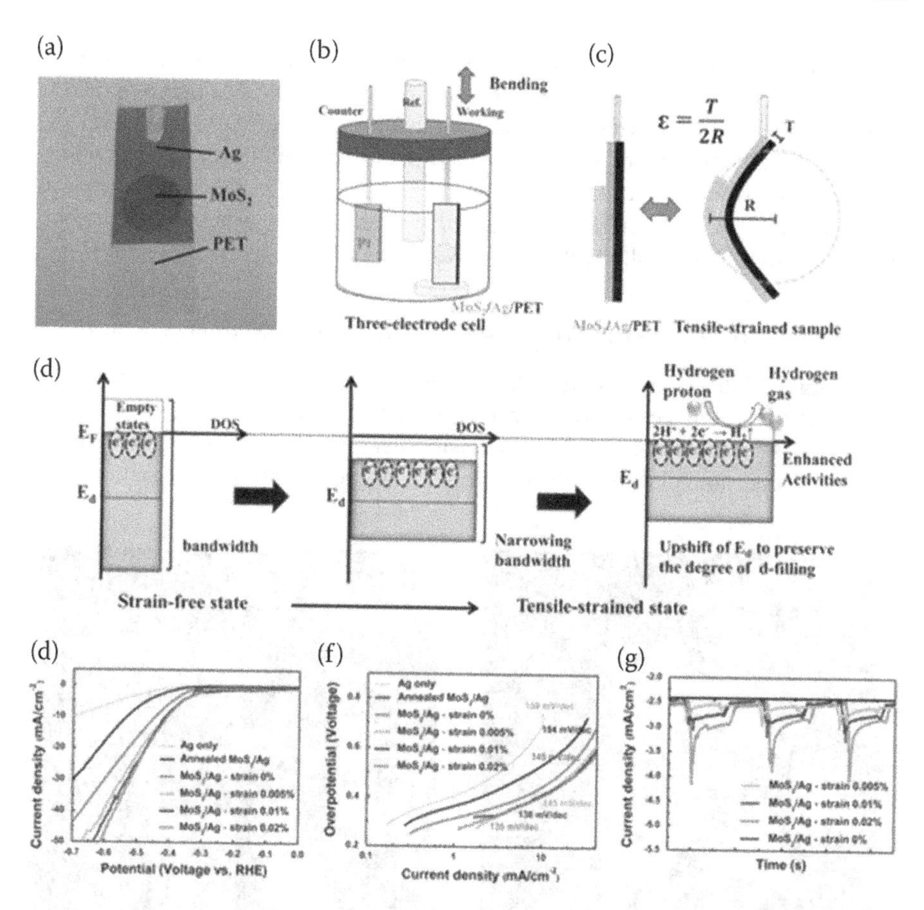

FIGURE 13.21 (a) Fabricated test sample: MoS_2 NS film/Ag electrode/PET substrate. (b) Schematic diagram of the experimental setup for strain-assisted HER based on a three-electrode reactor. (c) Schematic illustration of the mechanism to introduce a tensile strain. (d) Schematic illustrating mechanism through which mechanical strain affects Mo atoms. (e) Tensile-strain-ratio-dependent polarization curves. (f) The corresponding Tafel plots. Chronoamperometric curves recorded in the presence of a periodic strain of various amplitudes. (g) Fluctuation of tensile-strain-ratio-dependent current density over time. Reproduced with permission [183]. Copyright 2014, American Chemical Society.

terephthalate (PET) substrate. Tensile strain (ε) is applied by bending the PET substrate (Figure 13.21(c)). Its value is determined by the thickness of PET substrate (T) and the bending radius (R), which is expressed by $\varepsilon = T/2R \times 100\%$ [184]. Polarization curves in Figure 13.21(e) show that tensile strain increases from 0% to 0.02% and the cathodic current density at an overpotential of –400 mV increases from –5.95 to –9.24 mA cm^{-2}. Correspondingly, the Tafel slope decreased from 145 to 135 mV dec^{-1} (Figure 13.21(f)). The tensile strain-promoted HER is further confirmed by chronopotentiometry measurements, where cathodic current density is boosted once bearing tensile strains (Figure 13.21(g)). Figure 13.8(d) briefly illustrates how tensile strain affects dynamic bands. Mo atoms move far away from each other when MoS_2 NSs acquire mechanical-bending-induced tensile strain. The degree of d-state overlap between neighboring Mo atoms decreases because of the expanded Mo lattice; hence, the bandwidth narrows. The density of d-states near the center of the d-band level (E_d) increases simultaneously and shows upshifts of E_d to maintain the degree of d-filling. The increased density of d-states near the Fermi level (E_F) facilitates the supply of electrons from the electrode to the active-edge sites, thereby increasing the electrochemical activity toward the HER.

13.2.7 Light Field: Hot Electron-Promoted Electrocatalysis Reaction

Under light illumination, excited charge carriers, which are also denoted as hot charge carriers, will be generated in photosensitive materials via light-matter interactions. As discussed in previous sections, energetic charge carriers are liable to overcome the potential barriers of carrier transport and redox reactions. By

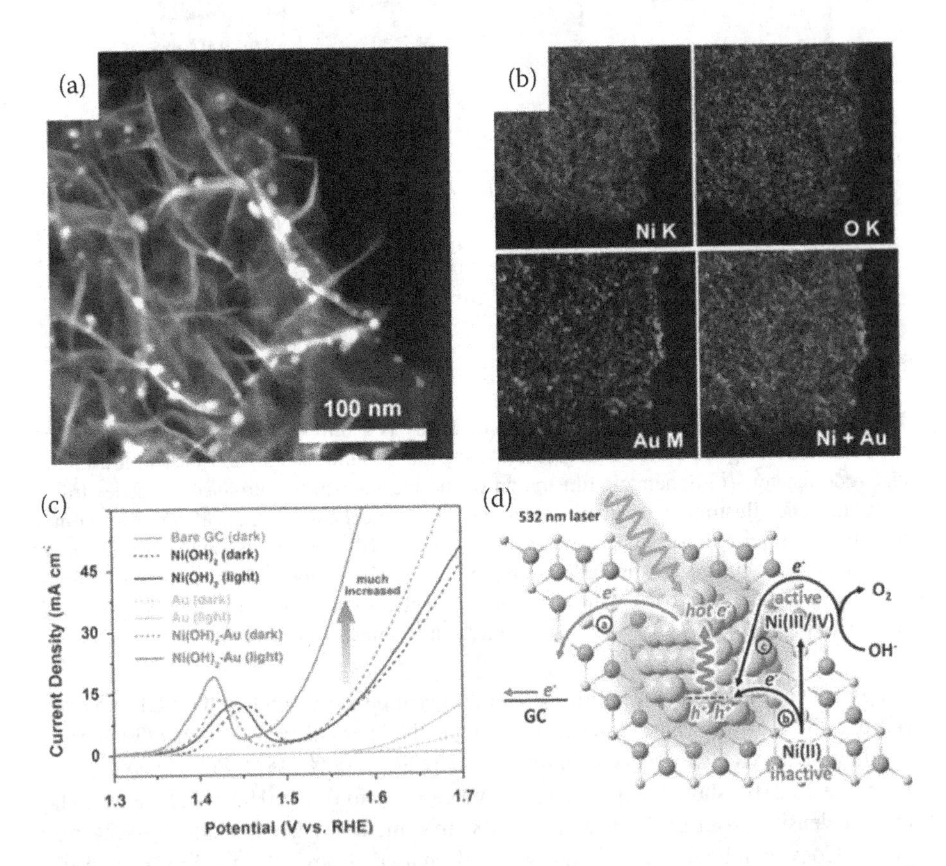

FIGURE 13.22 (a) HAADF-STEM images and (b) corresponding EDS mapping of Ni $(OH)_2$-Au hybrid catalysts. (c) OER polarization curves of $Ni(OH)_2$, Au, and hybrid catalyst with and without light irradiation (532 nm laser). (d) Schematic of plasmon-enhanced OER over $Ni(OH)_2$-Au catalyst. Reproduced with permission [187]. Copyright 2016, American Chemical Society.

separating and directing them to the surface of catalysts, light illumination-induced hot charge carriers can be coupled to various chemical/electrochemical reactions and accelerate the reaction rates [185,186]. In addition, sunlight is an environmentally friendly and inexhaustible resource in nature. Therefore, harvesting optical energy by grafting photosensitive materials onto catalysts represents a promising route to promote electrocatalytic activity.

The pioneer work in plasmon-enhanced OER was accomplished on Au/$Ni(OH)_2$ by Ye and co-workers (Figure 13.22) [187]. Although Au is inert for OER, Au/$Ni(OH)_2$ exhibits higher activity than $Ni(OH)_2$, thanks to the instinct electron transfer from $Ni(OH)_2$ to Au. When light (532 nm laser) is turned on, the overpotential and Tafel slope of Au/$Ni(OH)_2$ for OER are further reduced. Besides, remarkable enhancement in generation of $Ni^{III/IV}$ is confirmed according to the cyclic voltammetry curves upon light irradiation, suggesting that Ni^{II} is oxidized by hot holes generated during the LSPR damping process of Au NPs. The authors further utilized Au/CoO and Au/FeOOH systems to verify the generality of the concept of plasmon-enhanced OER. Chen and co-workers demonstrated the plasmon-enhanced OER on Au-decorated MnO_2 nanosheets and investigated the mechanism of charge transfer with X-ray photoelectron spectroscopy (XPS) and electron paramagnetic resonance (EPR). In addition, Au supported on MOFs and layered double hydroxides (LDH) were also able to better catalyze OER under light irradiation following similar mechanism. In our opinion, in-situ spectroscopic study under illumination is needed for better understanding of the complex mechanism of plasmon-enhanced OER, because the valence state and coordination environment should be observed in situ under light irradiation.

REFERENCES

[1] Z. W. Seh, J. Kibsgaard, C. F. Dickens, I. Chorkendorff, J. K. Nørskov and T. F. Jaramillo, Combining Theory and Experiment in Electrocatalysis: Insights Into Materials Design [J], *Science*, 2017, 355, eaad4998.

[2] V. R. Stamenkovic, D. Strmcnik, P. P. Lopes and N. M. Markovic, Energy and Fuels from Electrochemical Interfaces [J], *Nature Materials*, 2017, 16, 57–69.

[3] A. Ruban, B. Hammer, P. Stoltze, H. L. Skriver and J. K. Nørskov, Surface Electronic Structure and Reactivity of Transition and Noble Metals [J], *Journal of Molecular Catalysis A: Chemical*, 1997, 115, 421–429.

[4] J. R. Kitchin, J. K. Nørskov, M. A. Barteau and J. G. Chen, Role of Strain and Ligand Effects in the Modification of the Electronic and Chemical Properties of Bimetallic Surfaces [J], *Physical Review Letters*, 2004, 93, 156801.

[5] L. D. Burke and J. A. Collins, Role of Surface Defects in the Electrocatalytic Behaviour of Copper in Base [J], *Journal of Applied Electrochemistry*, 1999, 29, 1427–1438.

[6] K. J. J. Mayrhofer, B. B. Blizanac, M. Arenz, V. R. Stamenkovic, P. N. Ross and N. M. Markovic, The Impact of Geometric and Surface Electronic Properties of Pt-Catalysts on the Particle Size Effect in Electrocatalysis [J], *The Journal of Physical Chemistry B*, 2005, 109, 14433–14440.

[7] V. R. Stamenkovic, B. S. Mun, M. Arenz, K. J. J. Mayrhofer, C. A. Lucas, G. Wang, P. N. Ross and N. M. Markovic, Trends in Electrocatalysis on Extended and Nanoscale Pt-bimetallic Alloy Surfaces [J], *Nature Materials*, 2007, 6, 241–247.

[8] A. J. Medford, A. Vojvodic, J. S. Hummelshøj, J. Voss, F. Abild-Pedersen, F. Studt, T. Bligaard, A. Nilsson and J. K. Nørskov, From the Sabatier Principle to a Predictive Theory of Transition-Metal Heterogeneous Catalysis [J], *Journal of Catalysis*, 2015, 328, 36–42.
[9] Z. Fang, B. Bueken, D. E. De Vos and R. A. Fischer, Defect-Engineered Metal-Organic Frameworks [J], *Angewandte Chemie International Edition*, 2015, 54, 7234–7254.
[10] J. Hong, C. Jin, J. Yuan and Z. Zhang, Atomic Defects in Two-Dimensional Materials: From Single-Atom Spectroscopy to Functionalities in Opto-/Electronics, Nanomagnetism, and Catalysis [J], *Advanced Materials*, 2017, 29, 1606434.
[11] T. Ling, D.-Y. Yan, Y. Jiao, H. Wang, Y. Zheng, X. Zheng, J. Mao, X.-W. Du, Z. Hu, M. Jaroniec and S.-Z. Qiao, Engineering Surface Atomic Structure of Single-Crystal Cobalt (II) Oxide Nanorods for Superior Electrocatalysis [J], *Nature Communications*, 2016, 7, 12876.
[12] L. Xu, Q. Jiang, Z. Xiao, X. Li, J. Huo, S. Wang and L. Dai, Plasma-Engraved Co_3O_4 Nanosheets with Oxygen Vacancies and High Surface Area for the Oxygen Evolution Reaction [J], *Angewandte Chemie*, 2016, 128, 5363–5367.
[13] R. Liu, Y. Wang, D. Liu, Y. Zou and S. Wang, Water-Plasma-Enabled Exfoliation of Ultrathin Layered Double Hydroxide Nanosheets with Multivacancies for Water Oxidation [J], *Advanced Materials*, 2017, 29, 1701546.
[14] Z. Xiao, Y. Wang, Y.-C. Huang, Z. Wei, C.-L. Dong, J. Ma, S. Shen, Y. Li and S. Wang, Filling the Oxygen Vacancies in Co_3O_4 with Phosphorus: An Ultra-Efficient Electrocatalyst for Overall Water Splitting [J], *Energy & Environmental Science*, 2017, 10, 2563–2569.
[15] Y. Jia, L. Zhang, A. Du, G. Gao, J. Chen, X. Yan, C. L. Brown and X. Yao, Defect Graphene as a Trifunctional Catalyst for Electrochemical Reactions [J], *Advanced Materials*, 2016, 28, 9532–9538.
[16] D. Yan, H. Li, C. Chen, Y. Zou and S. Wang, Defect Engineering Strategies for Nitrogen Reduction Reactions under Ambient Conditions [J], *Small Methods*, 2019, 3, 1800331.
[17] D. Yan, Y. Li, J. Huo, R. Chen, L. Dai and S. Wang, Defect Chemistry of Nonprecious-Metal Electrocatalysts for Oxygen Reactions [J], *Advanced Materials*, 2017, 29, 1606459.
[18] J. Zhang, X. Wu, W.-C. Cheong, W. Chen, R. Lin, J. Li, L. Zheng, W. Yan, L. Gu, C. Chen, Q. Peng, D. Wang and Y. Li, Cation Vacancy Stabilization of Single-Atomic-Site Pt1/Ni(OH)x Catalyst for Diboration of Alkynes and Alkenes [J], *Nature Communications*, 2018, 9, 1002.
[19] Y. Liu, C. Xiao, Z. Li and Y. Xie, Vacancy Engineering for Tuning Electron and Phonon Structures of Two-Dimensional Materials [J], *Advanced Energy Materials*, 2016, 6, 1600436.
[20] O. K. Anderson, Highlights of Condensed Matter Theory [J], *Internat. School of Phys. Enrico Fermi, Course*, 1985, 89.
[21] C. Tang and Q. Zhang, Nanocarbon for Oxygen Reduction Electrocatalysis: Dopants, Edges, and Defects [J], *Advanced Materials*, 2017, 29, 1604103.
[22] Y. Wang, P. Han, X. Lv, L. Zhang and G. Zheng, Defect and Interface Engineering for Aqueous Electrocatalytic CO_2 Reduction [J], *Joule*, 2018, 2, 2551–2582.
[23] X. Yan, Y. Jia and X. Yao, Defects on Carbons for Electrocatalytic Oxygen Reduction [J], *Chemical Society Reviews*, 2018, 47, 7628–7658.
[24] Y. Jia, J. Chen and X. Yao, Defect Electrocatalytic Mechanism: Concept, Topological Structure and Perspective [J], *Materials Chemistry Frontiers*, 2018, 2, 1250–1268.

[25] Y. Zhang, L. Guo, L. Tao, Y. Lu and S. Wang, Defect-Based Single-Atom Electrocatalysts [J], *Small Methods*, 2019, 3, 1800406.
[26] H. Zhang and R. Lv, Defect Engineering of Two-Dimensional Materials for Efficient Electrocatalysis [J], *Journal of Materiomics*, 2018, 4, 95–107.
[27] Y. C. Kimmel, X. Xu, W. Yu, X. Yang and J. G. Chen, Trends in Electrochemical Stability of Transition Metal Carbides and Their Potential Use As Supports for Low-Cost Electrocatalysts [J], *ACS Catalysis*, 2014, 4, 1558–1562.
[28] L. Tao, M. Qiao, R. Jin, Y. Li, Z. Xiao, Y. Wang, N. Zhang, C. Xie, Q. He, D. Jiang, G. Yu, Y. Li and S. Wang, Bridging the Surface Charge and Catalytic Activity of a Defective Carbon Electrocatalyst [J], *Angewandte Chemie International Edition*, 2019, 58, 1019–1024.
[29] Q. Wang, Y. Lei, Y. Zhu, H. Wang, J. Feng, G. Ma, Y. Wang, Y. Li, B. Nan, Q. Feng, Z. Lu and H. Yu, Edge Defect Engineering of Nitrogen-Doped Carbon for Oxygen Electrocatalysts in Zn–Air Batteries [J], *ACS Applied Materials & Interfaces*, 2018, 10, 29448–29456.
[30] H. Jiang, J. Gu, X. Zheng, M. Liu, X. Qiu, L. Wang, W. Li, Z. Chen, X. Ji and J. Li, Defect-Rich and Ultrathin N Doped Carbon Nanosheets as Advanced Trifunctional Metal-Free Electrocatalysts for the ORR, OER and HER [J], *Energy & Environmental Science*, 2019, 12, 322–333.
[31] J. Zhu, Y. Huang, W. Mei, C. Zhao, C. Zhang, J. Zhang, I. S. Amiinu and S. Mu, Effects of Intrinsic Pentagon Defects on Electrochemical Reactivity of Carbon Nanomaterials [J], *Angewandte Chemie International Edition*, 2019, 58, 3859–3864.
[32] H. J. Wörner, C. A. Arrell, N. Banerji, A. Cannizzo, M. Chergui, A. K. Das, P. Hamm, U. Keller, P. M. Kraus, E. Liberatore, P. Lopez-Tarifa, M. Lucchini, M. Meuwly, C. Milne, J.-E. Moser, U. Rothlisberger, G. Smolentsev, J. Teuscher, J. A. van Bokhoven and O. Wenger, Charge Migration and Charge Transfer in Molecular Systems [J], *Structural Dynamics*, 2017, 4, 061508.
[33] L. Dai, Carbon-Based Catalysts for Metal-Free Electrocatalysis [J], *Current Opinion in Electrochemistry*, 2017, 4, 18–25.
[34] R. Long, K. Mao, X. Ye, W. Yan, Y. Huang, J. Wang, Y. Fu, X. Wang, X. Wu, Y. Xie and Y. Xiong, Surface Facet of Palladium Nanocrystals: A Key Parameter to the Activation of Molecular Oxygen for Organic Catalysis and Cancer Treatment [J], *Journal of the American Chemical Society*, 2013, 135, 3200–3207.
[35] K. Gong, F. Du, Z. Xia, M. Durstock and L. Dai, Nitrogen-Doped Carbon Nanotube Arrays with High Electrocatalytic Activity for Oxygen Reduction [J], *Science*, 2009, 323, 760.
[36] Y. Zheng, Y. Jiao and S. Z. Qiao, Engineering of Carbon-Based Electrocatalysts for Emerging Energy Conversion: From Fundamentality to Functionality [J], *Advanced Materials*, 2015, 27, 5372–5378.
[37] K. Gao, B. Wang, L. Tao, B. V. Cunning, Z. Zhang, S. Wang, R. S. Ruoff and L. Qu, Efficient Metal-Free Electrocatalysts from N-Doped Carbon Nanomaterials: Mono-Doping and Co-Doping [J], *Advanced Materials*, 2019, 31, 1805121.
[38] L. Zhang, C.-Y. Lin, D. Zhang, L. Gong, Y. Zhu, Z. Zhao, Q. Xu, H. Li and Z. Xia, Guiding Principles for Designing Highly Efficient Metal-Free Carbon Catalysts [J], *Advanced Materials*, 2019, 31, 1805252.
[39] X. Hu, Y. Wu, H. Li and Z. Zhang, Adsorption and Activation of O_2 on Nitrogen-Doped Carbon Nanotubes [J], *The Journal of Physical Chemistry C*, 2010, 114, 9603–9607.
[40] S. Wang, L. Zhang, Z. Xia, A. Roy, D. W. Chang, J.-B. Baek and L. Dai, BCN Graphene as Efficient Metal-Free Electrocatalyst for the Oxygen Reduction Reaction [J], *Angewandte Chemie International Edition*, 2012, 51, 4209–4212.

[41] D. Guo, R. Shibuya, C. Akiba, S. Saji, T. Kondo and J. Nakamura, Active Sites of Nitrogen-Doped Carbon Materials for Oxygen Reduction Reaction Clarified Using Model Catalysts [J], *Science*, 2016, 351, 361.

[42] Z. Zhao and Z. Xia, Design Principles for Dual-Element-Doped Carbon Nanomaterials as Efficient Bifunctional Catalysts for Oxygen Reduction and Evolution Reactions [J], *ACS Catalysis*, 2016, 6, 1553–1558.

[43] J. Zhang and L. Dai, Heteroatom-Doped Graphitic Carbon Catalysts for Efficient Electrocatalysis of Oxygen Reduction Reaction [J], *ACS Catalysis*, 2015, 5, 7244–7253.

[44] A. Shen, Y. Zou, Q. Wang, R. A. W. Dryfe, X. Huang, S. Dou, L. Dai and S. Wang, Oxygen Reduction Reaction in a Droplet on Graphite: Direct Evidence that the Edge Is More Active than the Basal Plane [J], *Angewandte Chemie International Edition*, 2014, 53, 10804–10808.

[45] X. Zou and Y. Zhang, Noble Metal-Free Hydrogen Evolution Catalysts for Water Splitting [J], *Chemical Society Reviews*, 2015, 44, 5148–5180.

[46] C. G. Morales-Guio, L.-A. Stern and X. Hu, Nanostructured Hydrotreating Catalysts for Electrochemical Hydrogen Evolution [J], *Chemical Society Reviews*, 2014, 43, 6555–6569.

[47] J. R. McKone, S. C. Marinescu, B. S. Brunschwig, J. R. Winkler and H. B. Gray, Earth-Abundant Hydrogen Evolution Electrocatalysts [J], *Chemical Science*, 2014, 5, 865–878.

[48] M. Zeng and Y. Li, Recent Advances in Heterogeneous Electrocatalysts for the Hydrogen Evolution Reaction [J], *Journal of Materials Chemistry A*, 2015, 3, 14942–14962.

[49] Y. Liu, T. G. Kelly, J. G. Chen and W. E. Mustain, Metal Carbides as Alternative Electrocatalyst Supports [J], *ACS Catalysis*, 2013, 3, 1184–1194.

[50] M. Miao, J. Pan, T. He, Y. Yan, B. Y. Xia and X. Wang, Molybdenum Carbide-Based Electrocatalysts for Hydrogen Evolution Reaction [J], *Chemistry-A European Journal*, 2017, 23, 10947–10961.

[51] H. Vrubel and X. Hu, Molybdenum Boride and Carbide Catalyze Hydrogen Evolution in both Acidic and Basic Solutions [J], *Angewandte Chemie International Edition*, 2012, 51, 12703–12706.

[52] A.-M. Alexander and J. S. J. Hargreaves, Alternative Catalytic Materials: Carbides, Nitrides, Phosphides and Amorphous Boron Alloys [J], *Chemical Society Reviews*, 2010, 39, 4388–4401.

[53] H.-M. Wang, X.-H. Wang, M.-H. Zhang, X.-Y. Du, W. Li and K.-Y. Tao, Synthesis of Bulk and Supported Molybdenum Carbide by a Single-Step Thermal Carburization Method [J], *Chemistry of Materials*, 2007, 19, 1801–1807.

[54] A. Celzard, J. F. Marêché, G. Furdin, V. Fierro, C. Sayag and J. Pielaszek, Preparation and Catalytic Activity of Active Carbon-Supported Mo2C Nanoparticles [J], *Green Chemistry*, 2005, 7, 784–792.

[55] L. Zhong, F. Yu, Y. An, Y. Zhao, Y. Sun, Z. Li, T. Lin, Y. Lin, X. Qi, Y. Dai, L. Gu, J. Hu, S. Jin, Q. Shen and H. Wang, Cobalt Carbide Nanoprisms for Direct Production of Lower Olefins from Syngas [J], *Nature*, 2016, 538, 84–87.

[56] Q. Gao, C. Zhang, S. Wang, W. Shen, Y. Zhang, H. Xu and Y. Tang, Preparation of Supported Mo_2C-based Catalysts from Organic-Inorganic Hybrid Precursor for Hydrogen Production from Methanol Decomposition [J], *Chemical Communications*, 2010, 46, 6494–6496.

[57] L. Lin, W. Zhou, R. Gao, S. Yao, X. Zhang, W. Xu, S. Zheng, Z. Jiang, Q. Yu, Y.-W. Li, C. Shi, X.-D. Wen and D. Ma, Low-Temperature Hydrogen Production from Water and Methanol Using Pt/α-MoC Catalysts [J], *Nature*, 2017, 544, 80–83.

[58] J. R. Kitchin, J. K. Nørskov, M. A. Barteau and J. G. Chen, Trends in the Chemical Properties of Early Transition Metal Carbide Surfaces: A Density Functional Study [J], *Catalysis Today*, 2005, 105, 66–73.
[59] T. G. Kelly and J. G. Chen, Metal Overlayer on Metal Carbide Substrate: Unique Bimetallic Properties for Catalysis and Electrocatalysis [J], *Chemical Society Reviews*, 2012, 41, 8021–8034.
[60] S. Yao, X. Zhang, W. Zhou, R. Gao, W. Xu, Y. Ye, L. Lin, X. Wen, P. Liu, B. Chen, E. Crumlin, J. Guo, Z. Zuo, W. Li, J. Xie, L. Lu, C. J. Kiely, L. Gu, C. Shi, J. A. Rodriguez and D. Ma, Atomic-Layered Au Clusters on α-MoC as Catalysts for the Low-Temperature Water-Gas Shift Reaction [J], *Science*, 2017, 357, 389.
[61] J. A. Rodríguez, L. Feria, T. Jirsak, Y. Takahashi, K. Nakamura and F. Illas, Role of Au-C Interactions on the Catalytic Activity of Au Nanoparticles Supported on TiC(001) toward Molecular Oxygen Dissociation [J], *Journal of the American Chemical Society*, 2010, 132, 3177–3186.
[62] S. He, Z.-J. Shao, Y. Shu, Z. Shi, X.-M. Cao, Q. Gao, P. Hu and Y. Tang, Enhancing Metal-Support Interactions by Molybdenum Carbide: An Efficient Strategy toward the Chemoselective Hydrogenation of α,β-Unsaturated Aldehydes [J], *Chemistry—A European Journal*, 2016, 22, 5698–5704.
[63] Y. Shu, S. He, L. Xie, H. C. Chan, X. Yu, L. Yang and Q. Gao, Ni/Mo_2C Nanowires and Their Carbon-Coated Composites as Efficient Catalysts for Nitroarenes Hydrogenation [J], *Applied Surface Science*, 2017, 396, 339–346.
[64] H. H. Hwu and J. G. Chen, Surface Chemistry of Transition Metal Carbides [J], *Chemical Reviews*, 2005, 105, 185–212.
[65] P. Liu and J. A. Rodriguez, Catalytic Properties of Molybdenum Carbide, Nitride and Phosphide: A Theoretical Study [J], *Catalysis Letters*, 2003, 91, 247–252.
[66] V. Heine, s-d Interaction in Transition Metals [J], *Physical Review*, 1967, 153, 673–682.
[67] M. Mavrikakis, B. Hammer and J. K. Nørskov, Effect of Strain on the Reactivity of Metal Surfaces [J], *Physical Review Letters*, 1998, 81, 2819–2822.
[68] L. Ramqvist, Electronic Structure of Cubic Refractory Carbides [J], *Journal of Applied Physics*, 1971, 42, 2113–2120.
[69] J. R. d. S. Politi, F. Viñes, J. A. Rodriguez and F. Illas, Atomic and Electronic Structure of Molybdenum Carbide Phases: Bulk and Low Miller-Index Surfaces [J], *Physical Chemistry Chemical Physics*, 2013, 15, 12617–12625.
[70] N. A. Khan, M. B. Zellner and J. G. Chen, Cyclohexene As a Chemical Probe of the Low-Temperature Hydrogenation Activity of Pt/Ni(111) Bimetallic Surfaces [J], *Surface Science*, 2004, 556, 87–100.
[71] J. R. Kitchin, J. K. Nørskov, M. A. Barteau and J. G. Chen, Modification of the Surface Electronic and Chemical Properties of Pt(111) by Subsurface 3d Transition Metals [J], *The Journal of Chemical Physics*, 2004, 120, 10240–10246.
[72] J. R. Kitchin, N. A. Khan, M. A. Barteau, J. G. Chen, B. Yakshinskiy and T. E. Madey, Elucidation of the Active Surface and Origin of the Weak Metal–Hydrogen Bond on Ni/Pt(111) Bimetallic Surfaces: A Surface Science and Density Functional Theory Study [J], *Surface Science*, 2003, 544, 295–308.
[73] J. K. Nørskov, T. Bligaard, A. Logadottir, S. Bahn, L. B. Hansen, M. Bollinger, H. Bengaard, B. Hammer, Z. Sljivancanin, M. Mavrikakis, Y. Xu, S. Dahl and C. J. H. Jacobsen, Universality in Heterogeneous Catalysis [J], *Journal of Catalysis*, 2002, 209, 275–278.
[74] D. V. Esposito and J. G. Chen, Monolayer Platinum Supported on Tungsten Carbides as Low-cost Electrocatalysts: Opportunities and Limitations [J], *Energy & Environmental Science*, 2011, 4, 3900–3912.

[75] S. T. Hunt, M. Milina, A. C. Alba-Rubio, C. H. Hendon, J. A. Dumesic and Y. Román-Leshkov, Self-Assembly of Noble Metal Monolayers on Transition Metal Carbide Nanoparticle Catalysts [J], *Science*, 2016, 352, 974.

[76] I. J. Hsu, Y. C. Kimmel, X. Jiang, B. G. Willis and J. G. Chen, Atomic Layer Deposition Synthesis of Platinum-Tungsten Carbide Core–Shell Catalysts for the Hydrogen Evolution Reaction [J], *Chemical Communications*, 2012, 48, 1063–1065.

[77] P. C. K. Vesborg, B. Seger and I. Chorkendorff, Recent Development in Hydrogen Evolution Reaction Catalysts and Their Practical Implementation [J], *The Journal of Physical Chemistry Letters*, 2015, 6, 951–957.

[78] C. Wan, Y. N. Regmi and B. M. Leonard, Multiple Phases of Molybdenum Carbide as Electrocatalysts for the Hydrogen Evolution Reaction [J], *Angewandte Chemie International Edition in English*, 2014, 53, 6407–6410.

[79] S. Wirth, F. Harnisch, M. Weinmann and U. Schröder, Comparative Study of IVB-VIB Transition Metal Compound Electrocatalysts for the Hydrogen Evolution Reaction [J], *Applied Catalysis B: Environmental*, 2012, 126, 225–230.

[80] R. Michalsky, Y.-J. Zhang and A. A. Peterson, Trends in the Hydrogen Evolution Activity of Metal Carbide Catalysts [J], *ACS Catalysis*, 2014, 4, 1274–1278.

[81] M. Guan, C. Xiao, J. Zhang, S. Fan, R. An, Q. Cheng, J. Xie, M. Zhou, B. Ye and Y. Xie, Vacancy Associates Promoting Solar-Driven Photocatalytic Activity of Ultrathin Bismuth Oxychloride Nanosheets [J], *Journal of the American Chemical Society*, 2013, 135, 10411–10417.

[82] R. Schaub, E. Wahlström, A. Rønnau, E. Lægsgaard, I. Stensgaard and F. Besenbacher, Oxygen-Mediated Diffusion of Oxygen Vacancies on the TiO2(110) Surface [J], *Science*, 2003, 299, 377.

[83] S. Polarz, J. Strunk, V. Ischenko, M. W. E. van den Berg, O. Hinrichsen, M. Muhler and M. Driess, On the Role of Oxygen Defects in the Catalytic Performance of Zinc Oxide [J], *Angewandte Chemie International Edition*, 2006, 45, 2965–2969.

[84] H. Li, J. Shang, H. Zhu, Z. Yang, Z. Ai and L. Zhang, Oxygen Vacancy Structure Associated Photocatalytic Water Oxidation of BiOCl [J], *ACS Catalysis*, 2016, 6, 8276–8285.

[85] J. Suntivich, K. J. May, H. A. Gasteiger, J. B. Goodenough and Y. Shao-Horn, A Perovskite Oxide Optimized for Oxygen Evolution Catalysis from Molecular Orbital Principles [J], *ChemInform*, 2012, 43.

[86] Jian Bao, Xiaodong Zhang, Bo Fan, Jiajia Zhang, Min Zhou, Wenlong Yang, Xin Hu, Hui Wang, Bicai Pan and Y. Xie, Ultrathin Spinel-Structured Nanosheets Rich in Oxygen Deficiencies for Enhanced Electrocatalytic Water Oxidation [J], *Angewandte Chemie*, 2015, 127, 7507–7512.

[87] Y. Wang, T. Zhou, K. Jiang, P. Da, Z. Peng, J. Tang, B. Kong, W.-B. Cai, Z. Yang and G. Zheng, Reduced Mesoporous Co_3O_4 Nanowires as Efficient Water Oxidation Electrocatalysts and Supercapacitor Electrodes [J], *Advanced Energy Materials*, 2014, 4, 1400696.

[88] R. Gao, Z. Li, X. Zhang, J. Zhang, Z. Hu and X. Liu, Carbon-Dotted Defective CoO with Oxygen Vacancies: A Synergetic Design of Bifunctional Cathode Catalyst for Li–O_2 Batteries [J], *ACS Catalysis*, 2016, 6, 400–406.

[89] T. Y. Ma, Y. Zheng, S. Dai, M. Jaroniec and S. Z. Qiao, Mesoporous $MnCo_2O_4$ with Abundant Oxygen Vacancy Defects as High-Performance Oxygen Reduction Catalysts [J], *Journal of Materials Chemistry A*, 2014, 2, 8676–8682.

[90] Y. Zhu, W. Zhou, Z. G. Chen, Y. Chen, C. Su, M. O. Tade and Z. Shao, $SrNb_{0.1}Co_{0.7}Fe_{0.2}O_{3-d}$ Perovskite As a Next-Generation Electrocatalyst for

Oxygen Evolution in Alkaline Solution [J], *Angewandte Chemie International Edition in English*, 2015, 127, 3969–3973.
[91] J. Kim, X. Yin, K.-C. Tsao, S. Fang and H. Yang, $Ca_2Mn_2O_5$ as Oxygen-Deficient Perovskite Electrocatalyst for Oxygen Evolution Reaction [J], *Journal of the American Chemical Society*, 2014, 136, 14646–14649.
[92] Y. Guo, Y. Tong, P. Chen, K. Xu, J. Zhao, Y. Lin, W. Chu, Z. Peng, C. Wu and Y. Xie, Engineering the Electronic State of a Perovskite Electrocatalyst for Synergistically Enhanced Oxygen Evolution Reaction [J], *ChemInform*, 2015, 46.
[93] L. Xu, Q. Jiang, Z. Xiao, X. Li, J. Huo, S. Wang and L. Dai, Plasma-Engraved Co_3O_4 Nanosheets with Oxygen Vacancies and High Surface Area for the Oxygen Evolution Reaction [J], *Angewandte Chemie International Edition*, 2016, 55, 5277–5281.
[94] S. Zhao, Y. Wang, J. Dong, C.-T. He, H. Yin, P. An, K. Zhao, X. Zhang, C. Gao, L. Zhang, J. Lv, J. Wang, J. Zhang, A. M. Khattak, N. A. Khan, Z. Wei, J. Zhang, S. Liu, H. Zhao and Z. Tang, Ultrathin Metal-Organic Framework Nanosheets for Electrocatalytic Oxygen Evolution [J], *Nature Energy*, 2016, 1, 16184.
[95] P. F. Liu, S. Yang, B. Zhang and H. G. Yang, Defect-Rich Ultrathin Cobalt–Iron Layered Double Hydroxide for Electrochemical Overall Water Splitting [J], *ACS Applied Materials & Interfaces*, 2016, 8, 34474–34481.
[96] H. Fei, J. Dong, Y. Feng, C. S. Allen, C. Wan, B. Volosskiy, M. Li, Z. Zhao, Y. Wang, H. Sun, P. An, W. Chen, Z. Guo, C. Lee, D. Chen, I. Shakir, M. Liu, T. Hu, Y. Li, A. I. Kirkland, X. Duan and Y. Huang, General Synthesis and Definitive Structural Identification of MN_4C_4 Single-Atom Catalysts with Tunable Electrocatalytic Activities [J], *Nature Catalysis*, 2018, 1, 63–72.
[97] C. Zhao, X. Dai, T. Yao, W. Chen, X. Wang, J. Wang, J. Yang, S. Wei, Y. Wu and Y. Li, Ionic Exchange of Metal–Organic Frameworks to Access Single Nickel Sites for Efficient Electroreduction of CO_2 [J], *Journal of the American Chemical Society*, 2017, 139, 8078–8081.
[98] J. Zhang, J. Liu, L. Xi, Y. Yu, N. Chen, S. Sun, W. Wang, K. M. Lange and B. Zhang, Single-Atom Au/NiFe Layered Double Hydroxide Electrocatalyst: Probing the Origin of Activity for Oxygen Evolution Reaction [J], *Journal of the American Chemical Society*, 2018, 140, 3876–3879.
[99] P. Yin, T. Yao, Y. Wu, L. Zheng, Y. Lin, W. Liu, H. Ju, J. Zhu, X. Hong and Z. Deng, Single Cobalt Atoms with Precise N-Coordination as Superior Oxygen Reduction Reaction Catalysts [J], *Angewandte Chemie*, 2016, 128(36), 10958–10963.
[100] P. N. Duchesne, Z. Y. Li, C. P. Deming, V. Fung, X. Zhao, J. Yuan, T. Regier, A. Aldalbahi, Z. Almarhoon, S. Chen, D.-e. Jiang, N. Zheng and P. Zhang, Golden Single-Atomic-Site Platinum Electrocatalysts [J], *Nature Materials*, 2018, 17, 1033–1039.
[101] G. Sun, Z.-J. Zhao, R. Mu, S. Zha, L. Li, S. Chen, K. Zang, J. Luo, Z. Li, S. C. Purdy, A. J. Kropf, J. T. Miller, L. Zeng and J. Gong, Breaking the Scaling Relationship via Thermally Stable Pt/Cu Single Atom Alloys for Catalytic Dehydrogenation [J], *Nature Communications*, 2018, 9, 4454.
[102] H. Zhang, L. Yu, T. Chen, W. Zhou and X. W. Lou, Surface Modulation of Hierarchical MoS_2 Nanosheets by Ni Single Atoms for Enhanced Electrocatalytic Hydrogen Evolution [J], *Advanced Functional Materials*, 2018, 28, 1807086.
[103] P. Liu, Y. Zhao, R. Qin, S. Mo, G. Chen, L. Gu, D. M. Chevrier, P. Zhang, Q. Guo, D. Zang, B. Wu, G. Fu and N. Zheng, Photochemical Route for Synthesizing Atomically Dispersed Palladium Catalysts [J], *Science*, 2016, 352, 797.

[104] Y. Xue, B. Huang, Y. Yi, Y. Guo, Z. Zuo, Y. Li, Z. Jia, H. Liu and Y. Li, Anchoring Zero Valence Single Atoms of Nickel and Iron on Graphdiyne for Hydrogen Evolution [J], *Nature Communications*, 2018, 9, 1460.

[105] L. Liu and A. Corma, Metal Catalysts for Heterogeneous Catalysis: From Single Atoms to Nanoclusters and Nanoparticles [J], *Chemical Reviews*, 2018, 118, 4981–5079.

[106] X.-F. Yang, A. Wang, B. Qiao, J. Li, J. Liu and T. Zhang, Single-Atom Catalysts: A New Frontier in Heterogeneous Catalysis [J], *Accounts of Chemical Research*, 2013, 46, 1740–1748.

[107] X.-P. Yin, H.-J. Wang, S.-F. Tang, X.-L. Lu, M. Shu, R. Si and T.-B. Lu, Engineering the Coordination Environment of Single-Atom Platinum Anchored on Graphdiyne for Optimizing Electrocatalytic Hydrogen Evolution [J], *Angewandte Chemie International Edition*, 2018, 57, 9382–9386.

[108] A. Wang, J. Li and T. Zhang, Heterogeneous Single-Atom Catalysis [J], *Nature Reviews Chemistry*, 2018, 2, 65–81.

[109] E. J. Peterson, A. T. DeLaRiva, S. Lin, R. S. Johnson, H. Guo, J. T. Miller, J. Hun Kwak, C. H. F. Peden, B. Kiefer, L. F. Allard, F. H. Ribeiro and A. K. Datye, Low-Temperature Carbon Monoxide Oxidation Catalysed by Regenerable Atomically Dispersed Palladium on Alumina [J], *Nature Communications*, 2014, 5, 4885.

[110] R. Qin, P. Liu, G. Fu and N. Zheng, Strategies for Stabilizing Atomically Dispersed Metal Catalysts [J], *Small Methods*, 2018, 2, 1700286.

[111] L. Zhang, J. M. T. A. Fischer, Y. Jia, X. Yan, W. Xu, X. Wang, J. Chen, D. Yang, H. Liu, L. Zhuang, M. Hankel, D. J. Searles, K. Huang, S. Feng, C. L. Brown and X. Yao, Coordination of Atomic Co-Pt Coupling Species at Carbon Defects as Active Sites for Oxygen Reduction Reaction [J], *Journal of the American Chemical Society*, 2018, 140, 10757–10763.

[112] G. Liu, A. W. Robertson, M. M.-J. Li, W. C. H. Kuo, M. T. Darby, M. H. Muhieddine, Y.-C. Lin, K. Suenaga, M. Stamatakis, J. H. Warner and S. C. E. Tsang, MoS_2 Monolayer Catalyst Doped with Isolated Co Atoms for the Hydrodeoxygenation Reaction [J], *Nature Chemistry*, 2017, 9, 810–816.

[113] S. Yang, J. Kim, Y. J. Tak, A. Soon and H Lee, Single-Atom Catalyst of Platinum Supported on Titanium Nitride for Selective Electrochemical Reactions [J], *Angewandte Chemie*, 2016, 55, 2058–2062. 10.1002/ange.201509241.

[114] L. Tao, C.-Y. Lin, S. Dou, S. Feng, D. Chen, D. Liu, J. Huo, Z. Xia and S. Wang, Creating Coordinatively Unsaturated Metal Sites in Metal-Organic-Frameworks as Efficient Electrocatalysts for the Oxygen Evolution Reaction: Insights Into the Active Centers [J], *Nano Energy*, 2017, 41, 417–425.

[115] C. Zhu, S. Fu, Q. Shi, D. Du and Y. Lin, Single-Atom Electrocatalysts [J], *Angewandte Chemie International Edition*, 2017, 56, 13944–13960.

[116] Z. Li, D. Wang, Y. Wu and Y. Li, Recent Advances in the Precise Control of Isolated Single-site Catalysts by Chemical Methods [J], *National Science Review*, 2018, 5, 673–689.

[117] Y. Peng, B. Lu and S. Chen, Single Atom Catalysts: Carbon-Supported Single Atom Catalysts for Electrochemical Energy Conversion and Storage (Adv. Mater. 48/2018) [J], *Advanced Materials*, 2018, 30, 1870370.

[118] Y. Chen, S. Ji, C. Chen, Q. Peng, D. Wang and Y. Li, Single-Atom Catalysts: Synthetic Strategies and Electrochemical Applications [J], *Joule*, 2018, 2, 1242–1264.

[119] H. Zhang, G. Liu, L. Shi and J. Ye, Single-Atom Catalysts: Emerging Multifunctional Materials in Heterogeneous Catalysis [J], *Advanced Energy Materials*, 2018, 8, 1701343.

[120] X. Sun, Y. Chen, K. Liu, Y. Ding, M. Zeng and L. Fu, Atomic Scale Catalysts: Atomic Scale Materials for Emerging Robust Catalysis [J], *Small Methods*, 2018, 2, 1800049.

[121] S. Mitchell, E. Vorobyeva and J. Pérez-Ramírez, The Multifaceted Reactivity of Single-Atom Heterogeneous Catalysts [J], *Angewandte Chemie International Edition*, 2018, 57, 15316–15329.

[122] C. Zhu, Q. Shi, S. Feng, D. Du and Y. Lin, Single-Atom Catalysts for Electrochemical Water Splitting [J], *ACS Energy Letters*, 2018, 3, 1713–1721.

[123] H. Yan, C. Su, J. He and W. Chen, Single-atom Catalysts and Their Applications in Organic Chemistry [J], *Journal of Materials Chemistry A*, 2018, 6, 8793–8814.

[124] P. Hu, Z. Huang, Z. Amghouz, M. Makkee, F. Xu, F. Kapteijn, A. Dikhtiarenko, Y. Chen, X. Gu and X. Tang, Electronic Metal-Support Interactions in Single-Atom Catalysts [J], *Angewandte Chemie*, 2013, 53, 3418–3421. 10.1002/ange.2013 09248.

[125] J. Jones, H. Xiong, A. T. DeLaRiva, E. J. Peterson, H. Pham, S. R. Challa, G. Qi, S. Oh, M. H. Wiebenga, X. I. Pereira Hernández, Y. Wang and A. K. Datye, Thermally Stable Single-Atom Platinum-On-Ceria Catalysts via Atom Trapping [J], *Science*, 2016, 353, 150.

[126] B. Qiao, A. Wang, X. Yang, L. F. Allard, Z. Jiang, Y. Cui, J. Liu, J. Li and T. Zhang, Single-Atom Catalysis of CO Oxidation using Pt_1/FeO_x [J], *Nature Chemistry*, 2011, 3, 634–641.

[127] J. Liang, Q. Yu, X. Yang, T. Zhang and J. Li, A Systematic Theoretical Study on FeOx-Supported Single-Atom Catalysts: M1/FeOx for CO Oxidation [J], *Nano Research*, 2018, 11, 1599–1611.

[128] S. Back, J. Lim, N.-Y. Kim, Y.-H. Kim and Y. Jung, Single-Atom Catalysts for CO_2 Electroreduction with Significant Activity and Selectivity Improvements [J], *Chemical Science*, 2017, 8, 1090–1096.

[129] B. Qiao, J.-X. Liang, A. Wang, C.-Q. Xu, J. Li, T. Zhang and J. J. Liu, Ultrastable Single-Atom Gold Catalysts with Strong Covalent Metal-Support Interaction (CMSI) [J], *Nano Research*, 2015, 8, 2913–2924.

[130] N. Tang, Y. Cong, Q. Shang, C. Wu, G. Xu and X. Wang, Coordinatively Unsaturated Al^{3+} Sites Anchored Subnanometric Ruthenium Catalyst for Hydrogenation of Aromatics [J], *ACS Catalysis*, 2017, 7, 5987–5991.

[131] J. H. Kwak, J. Hu, D. Mei, C.-W. Yi, D. H. Kim, C. H. F. Peden, L. F. Allard and J. Szanyi, Coordinatively Unsaturated Al^{3+} Centers as Binding Sites for Active Catalyst Phases of Platinum on γ-Al_2O_3 [J], *Science*, 2009, 325, 1670.

[132] Y. Qu, Z. Li, W. Chen, Y. Lin, T. Yuan, Z. Yang, C. Zhao, J. Wang, C. Zhao, X. Wang, F. Zhou, Z. Zhuang, Y. Wu and Y. Li, Direct Transformation of Bulk Copper into Copper Single Sites Via Emitting and Trapping of Atoms [J], *Nature Catalysis*, 2018, 1, 781–786.

[133] X. Ren, Q. Ma, H. Fan, L. Pang, Y. Zhang, Y. Yao, X. Ren and S. Liu, A Se-doped MoS_2 nanosheet for improved hydrogen evolution reaction [J], *Chemical Communications*, 2015, 51, 15997–16000.

[134] M. A. Lukowski, A. S. Daniel, F. Meng, A. Forticaux, L. Li and S. Jin, Enhanced Hydrogen Evolution Catalysis from Chemically Exfoliated Metallic MoS2 Nanosheets [J], *Journal of the American Chemical Society*, 2013, 135, 10274–10277.

[135] A. Mammoottil Abraham, G. Bharath, A. Hai and F. Banat, Preparation of MoS2/ graphene nanostructures and their supercapacitor and hydrogen evolution reaction (HAB - Herein, a simple and facile two-step sonication method was developed for the preparation of nanosheets-like MoS2 wrapped graphene nanohybrids for energy storage and hydrogen evolution reaction (HER) applications. TEM and HR-TEM

images revealed that the 120 nm sized lateral diameter of sheet-like MoS2 was completely immobilized on the surfaces of graphene via the probe sonication method. The resultant nanohybrid exhibited bifunctional activities of supercapacitor and HER. With this tailored bifunctional nanoarchitecture, the nanohybrid based electrode showed an improved specific capacitance of 350 F g−1 for a current load of 1 A g−1 with a cyclic efficiency of 85 % in 6 M KOH electrolyte solution. Additionally, the developed MG11 nanohybrid was tested as an electrocatalyst for HER. The latter exhibited a low onset potential of ~125 mV with a Tafel slope of 41 mV/decade. Thus, MG11 nanohybrid presents an excellent prospect for a low-cost electrode for a supercapacitor and an electrocatalyst for hydrogen evolution reaction. ER) performances [J], *Journal of Physics D: Applied Physics*, 2019, 53, 065501.

[136] D. Voiry, R. Fullon, J. Yang, C. de Carvalho Castro e Silva, R. Kappera, I. Bozkurt, D. Kaplan, M. J. Lagos, P. E. Batson, G. Gupta, Aditya D. Mohite, L. Dong, D. Er, V. B. Shenoy, T. Asefa and M. Chhowalla, The role of electronic coupling between substrate and 2D MoS2 nanosheets in electrocatalytic production of hydrogen [J], *Nature Materials*, 2016, 15, 1003–1009.

[137] B. Radisavljevic, A. Radenovic, J. Brivio, V. Giacometti and A. Kis, Single-layer MoS2 transistors [J], *Nature Nanotechnology*, 2011, 6, 147–150.

[138] S. Larentis, B. Fallahazad and E. Tutuc, Field-effect transistors and intrinsic mobility in ultra-thin MoSe2 layers [J], *Applied Physics Letters*, 2012, 101, 223104.

[139] T. Georgiou, R. Jalil, B. D. Belle, L. Britnell, R. V. Gorbachev, S. V. Morozov, Y.-J. Kim, A. Gholinia, S. J. Haigh, O. Makarovsky, L. Eaves, L. A. Ponomarenko, A. K. Geim, K. S. Novoselov and A. Mishchenko, Vertical field-effect transistor based on graphene–WS2 heterostructures for flexible and transparent electronics [J], *Nature Nanotechnology*, 2013, 8, 100–103.

[140] J. Wang, M. Yan, K. Zhao, X. Liao, P. Wang, X. Pan, W. Yang and L. Mai, Field Effect Enhanced Hydrogen Evolution Reaction of MoS2 Nanosheets [J], *Advanced Materials*, 2017, 29, 1604464.

[141] H. N. Nong, L. J. Falling, A. Bergmann, M. Klingenhof, H. P. Tran, C. Spöri, R. Mom, J. Timoshenko, G. Zichittella, A. Knop-Gericke, S. Piccinin, J. Pérez-Ramírez, B. R. Cuenya, R. Schlögl, P. Strasser, D. Teschner and T. E. Jones, Key role of chemistry versus bias in electrocatalytic oxygen evolution [J], *Nature*, 2020, 587, 408–413.

[142] J. Zhu, Z.-C. Wang, H. Dai, Q. Wang, R. Yang, H. Yu, M. Liao, J. Zhang, W. Chen, Z. Wei, N. Li, L. Du, D. Shi, W. Wang, L. Zhang, Y. Jiang and G. Zhang, Boundary activated hydrogen evolution reaction on monolayer MoS2 [J], *Nature Communications*, 2019, 10, 1348.

[143] S.-Y. Lu, S. Li, M. Jin, J. Gao and Y. Zhang, Greatly boosting electrochemical hydrogen evolution reaction over Ni3S2 nanosheets rationally decorated by Ni3Sn2S2 quantum dots [J], *Applied Catalysis B: Environmental*, 2020, 267, 118675.

[144] C. Liu, T. Gong, J. Zhang, X. Zheng, J. Mao, H. Liu, Y. Li and Q. Hao, Engineering Ni2P-NiSe2 heterostructure interface for highly efficient alkaline hydrogen evolution [J], *Applied Catalysis B: Environmental*, 2020, 262, 118245.

[145] F. Ling, T. Zhou, X. Liu, W. Kang, W. Zeng, Y. Zhang, L. Fang, Y. Lu and M. Zhou, Electric field tuned MoS2/metal interface for hydrogen evolution catalyst from first-principles investigations [J], *Nanotechnology*, 2017, 29, 03LT01.

[146] F. Ling, X. Liu, H. Jing, Y. Chen, W. Zeng, Y. Zhang, W. Kang, J. Liu, L. Fang and M. Zhou, Optimizing edges and defects of supported MoS2 catalysts for hydrogen evolution via an external electric field [J], *Physical Chemistry Chemical Physics*, 2018, 20, 26083–26090.

[147] Y. Wu, S. Ringe, C.-L. Wu, W. Chen, A. Yang, H. Chen, M. Tang, G. Zhou, H. Y. Hwang, K. Chan and Y. Cui, Two-Dimensional MoS2 Catalysis Transistor by Solid-State Ion Gating Manipulation and Adjustment (SIGMA) [J], *Nano Letters*, 2019, 19, 7293–7300.
[148] S. Mohan, G. Saravanan and A. Bund, Role of magnetic forces in pulse electrochemical deposition of NinanoAl2O3 composites [J], *Electrochimica Acta*, 2012, 64, 94–99.
[149] L. M. A. Monzon and J. M. D. Coey, Magnetic fields in electrochemistry: The Lorentz force. A mini-review [J], *Electrochemistry Communications*, 2014, 42, 38–41.
[150] K. Scott, Process intensification: An electrochemical perspective [J], *Renewable and Sustainable Energy Reviews*, 2018, 81, 1406–1426.
[151] H. Matsushima, T. Iida and Y. Fukunaka, Observation of bubble layer formed on hydrogen and oxygen gas-evolving electrode in a magnetic field [J], *Journal of Solid State Electrochemistry*, 2012, 16, 617–623.
[152] M. F. Kaya, N. Demir, M. S. Albawabiji and M. Taş, Investigation of alkaline water electrolysis performance for different cost effective electrodes under magnetic field [J], *International Journal of Hydrogen Energy*, 2017, 42, 17583–17592.
[153] M.-Y. Lin, L.-W. Hourng and J.-S. Hsu, The effects of magnetic field on the hydrogen production by multielectrode water electrolysis [J], *Energy Sources, Part A: Recovery, Utilization, and Environmental Effects*, 2017, 39, 352–357.
[154] M.-Y. Lin, L.-W. Hourng and C.-H. Wu, The effectiveness of a magnetic field in increasing hydrogen production by water electrolysis [J], *Energy Sources, Part A: Recovery, Utilization, and Environmental Effects*, 2017, 39, 140–147.
[155] D. Fernández, M. Martine, A. Meagher, M. E. Möbius and J. M. D. Coey, Stabilizing effect of a magnetic field on a gas bubble produced at a microelectrode [J], *Electrochemistry Communications*, 2012, 18, 28–32.
[156] J. A. Koza, S. Mühlenhoff, P. Żabiński, P. A. Nikrityuk, K. Eckert, M. Uhlemann, A. Gebert, T. Weier, L. Schultz and S. Odenbach, Hydrogen evolution under the influence of a magnetic field [J], *Electrochimica Acta*, 2011, 56, 2665–2675.
[157] J. A. Koza, M. Uhlemann, A. Gebert and L. Schultz, Desorption of hydrogen from the electrode surface under influence of an external magnetic field [J], *Electrochemistry Communications*, 2008, 10, 1330–1333.
[158] R. E. D. L. Rue and C. W. Tobias, On the Conductivity of Dispersions [J], *Journal of the Electrochemical Society*, 1959, 106, 827.
[159] H. Vogt, A hydrodynamic model for the ohmic interelectrode resistance of cells with vertical gas evolving electrodes [J], *Electrochimica Acta*, 1981, 26, 1311–1317.
[160] L. Elias and A. Chitharanjan Hegde, Effect of Magnetic Field on HER of Water Electrolysis on Ni–W Alloy [J], *Electrocatalysis*, 2017, 8, 375–382.
[161] B. Seo and S. H. Joo, A magnetic boost [J], *Nature Energy*, 2018, 3, 451–452.
[162] K. Onda, T. Kyakuno, K. Hattori and K. Ito, Prediction of production power for high-pressure hydrogen by high-pressure water electrolysis [J], *Journal of Power Sources*, 2004, 132, 64–70.
[163] S. Piontek, C. Andronescu, A. Zaichenko, B. Konkena, K. Junge Puring, B. Marler, H. Antoni, I. Sinev, M. Muhler, D. Mollenhauer, B. Roldan Cuenya, W. Schuhmann and U.-P. Apfel, Influence of the Fe:Ni Ratio and Reaction Temperature on the Efficiency of (FexNi1–x)9S8 Electrocatalysts Applied in the Hydrogen Evolution Reaction [J], *ACS Catalysis*, 2018, 8, 987–996.
[164] T. P. Heins, N. Harms, L.-S. Schramm and U. Schröder, Development of a new Electrochemical Impedance Spectroscopy Approach for Monitoring the Solid Electrolyte Interphase Formation [J], *Energy Technology*, 2016, 4, 1509–1513.

[165] S. I. Fletcher, F. B. Sillars, R. C. Carter, A. J. Cruden, M. Mirzaeian, N. E. Hudson, J. A. Parkinson and P. J. Hall, The effects of temperature on the performance of electrochemical double layer capacitors [J], *Journal of Power Sources*, 2010, 195, 7484–7488.

[166] R. E. Rosensweig, Heating magnetic fluid with alternating magnetic field [J], *Journal of Magnetism and Magnetic Materials*, 2002, 252, 370–374.

[167] A. E. Deatsch and B. A. Evans, Heating efficiency in magnetic nanoparticle hyperthermia [J], *Journal of Magnetism and Magnetic Materials*, 2014, 354, 163–172.

[168] C. Niether, S. Faure, A. Bordet, J. Deseure, M. Chatenet, J. Carrey, B. Chaudret and A. Rouet, Improved water electrolysis using magnetic heating of FeC–Ni core–shell nanoparticles [J], *Nature Energy*, 2018, 3, 476–483.

[169] K. Ray, S. P. Ananthavel, D. H. Waldeck and R. Naaman, Asymmetric Scattering of Polarized Electrons by Organized Organic Films of Chiral Molecules [J], *Science*, 1999, 283, 814.

[170] P. C. Mondal, W. Mtangi and C. Fontanesi, Chiro-Spintronics: Spin-Dependent Electrochemistry and Water Splitting Using Chiral Molecular Films [J], *Small Methods*, 2018, 2, 1700313.

[171] P. C. Mondal, N. Kantor-Uriel, S. P. Mathew, F. Tassinari, C. Fontanesi and R. Naaman, Chiral Conductive Polymers as Spin Filters [J], *Advanced Materials*, 2015, 27, 1924–1927.

[172] W. Mtangi, V. Kiran, C. Fontanesi and R. Naaman, Role of the Electron Spin Polarization in Water Splitting [J], *The Journal of Physical Chemistry Letters*, 2015, 6, 4916–4922.

[173] Y. Jiao, R. Sharpe, T. Lim, J. W. H. Niemantsverdriet and J. Gracia, Photosystem II Acts as a Spin-Controlled Electron Gate during Oxygen Formation and Evolution [J], *Journal of the American Chemical Society*, 2017, 139, 16604–16608.

[174] W. Zhang, K. Banerjee-Ghosh, F. Tassinari and R. Naaman, Enhanced Electrochemical Water Splitting with Chiral Molecule-Coated Fe3O4 Nanoparticles [J], *ACS Energy Letters*, 2018, 3, 2308–2313.

[175] W. Mtangi, F. Tassinari, K. Vankayala, A. Vargas Jentzsch, B. Adelizzi, A. R. A. Palmans, C. Fontanesi, E. W. Meijer and R. Naaman, Control of Electrons' Spin Eliminates Hydrogen Peroxide Formation During Water Splitting [J], *Journal of the American Chemical Society*, 2017, 139, 2794–2798.

[176] K. B. Ghosh, W. Zhang, F. Tassinari, Y. Mastai, O. Lidor-Shalev, R. Naaman, P. Möllers, D. Nürenberg, H. Zacharias, J. Wei, E. Wierzbinski and D. H. Waldeck, Controlling Chemical Selectivity in Electrocatalysis with Chiral CuO-Coated Electrodes [J], *Journal of Physical Chemistry C*, 2019, 123, 3024–3031.

[177] S. Bhattacharjee and S.-C. Lee, Controlling Oxygen-Based Electrochemical Reactions through Spin Orientation [J], *Journal of Physical Chemistry C*, 2018, 122, 894–901.

[178] F. A. Garcés-Pineda, M. Blasco-Ahicart, D. Nieto-Castro, N. López and J. R. Galán-Mascarós, Direct magnetic enhancement of electrocatalytic water oxidation in alkaline media [J], *Nature Energy*, 2019, 4, 519–525.

[179] T. Ling, D.-Y. Yan, H. Wang, Y. Jiao, Z. Hu, Y. Zheng, L. Zheng, J. Mao, H. Liu and X.-W. Du, Activating cobalt (II) oxide nanorods for efficient electrocatalysis by strain engineering [J], *Nature Communications*, 2017, 8, 1–7.

[180] J. R. Petrie, H. Jeen, S. C. Barron, T. L. Meyer and H. N. Lee, Enhancing perovskite electrocatalysis through strain tuning of the oxygen deficiency [J], *Journal of the American Chemical Society*, 2016, 138, 7252–7255.

[181] D. Zhou, S. Wang, Y. Jia, X. Xiong, H. Yang, S. Liu, J. Tang, J. Zhang, D. Liu and L. Zheng, NiFe hydroxide lattice tensile strain: enhancement of adsorption of

oxygenated intermediates for efficient water oxidation catalysis [J], *Angewandte Chemie International Edition*, 2019, 58, 736–740.

[182] J. R. Petrie, V. R. Cooper, J. W. Freeland, T. L. Meyer, Z. Zhang, D. A. Lutterman and H. N. Lee, Enhanced bifunctional oxygen catalysis in strained LaNiO3 perovskites [J], *Journal of the American Chemical Society*, 2016, 138, 2488–2491.

[183] J. H. Lee, W. S. Jang, S. W. Han and H. K. Baik, Efficient Hydrogen Evolution by Mechanically Strained MoS2 Nanosheets [J], *Langmuir*, 2014, 30, 9866–9873.

[184] N. Petrone, I. Meric, J. Hone and K. L. Shepard, Graphene field-effect transistors with gigahertz-frequency power gain on flexible substrates [J], *Nano Letters*, 2013, 13, 121–125.

[185] W.-Q. Chen, M.-C. Chung, J. A. A. Valinton, D. P. Penaloza, S.-H. Chuang and C.-H. Chen, Heterojunctions of silver–iron oxide on graphene for laser-coupled oxygen reduction reactions [J], *Chemical Communications*, 2018, 54, 7900–7903.

[186] X.-W. Liao, S.-S. Wang, G.-Y. Xu and C. Wang, Enhanced Electrocatalysis via Boosted Separation of Hot Charge Carriers of Plasmonic Gold Nanoparticles Deposited on Reduced Graphene Oxide [J], *ChemElectroChem*, 2019, 6, 1419–1426.

[187] G. Liu, P. Li, G. Zhao, X. Wang, J. Kong, H. Liu, H. Zhang, K. Chang, X. Meng, T. Kako and J. Ye, Promoting Active Species Generation by Plasmon-Induced Hot-Electron Excitation for Efficient Electrocatalytic Oxygen Evolution [J], *Journal of the American Chemical Society*, 2016, 138, 9128–9136.

Index